Springer Series in Brain Dynamics 1

Erol Başar (Ed.)

Dynamics of Sensory and Cognitive Processing by the Brain

Integrative Aspects of Neural Networks, Electroencephalography, Event-Related Potentials, Contingent Negative Variation, Magnetoencephalography, and Clinical Applications

Based on a Conference in West Berlin in August 1985

With Editorial Assistance
by Theodore Melnechuk

With 182 Figures, Some in Color

Springer-Verlag
Berlin Heidelberg NewYork
London Paris Tokyo

Professor Dr. Erol Başar
Institute of Physiology
Medical University Lübeck
Ratzeburger Allee 160
D-2400 Lübeck 1, FRG

ISBN-13:978-3-642-71533-4 e-ISBN-13:978-3-642-71531-0
DOI: 10.1007/978-3-642-71531-0

Library of Congress Cataloging-in-Publication Data. Dynamics of sensory and cognitive processing by the brain: integrative aspects of neural networks, electroencephalography, event-related potentials, contingent negative variation, magnetoencephalography, and clinical applications/edited by Erol Başar; with editorial assistance by Theodore Melnechuk. – (Springer series in brain dynamics; 1) "Based on a conference in West Berlin in August 1985." Includes bibliographies and index.
ISBN-13:978-3-642-71533-4
1. Electroencephalography – Congresses. 2. Evoked potentials (Electrophysiology) – Congresses. 3. Brain – Magnetic fields – Congresses. 4. Neural circuitry – Congresses. 5. Neurophysiology – Congresses. I. Başar, Erol. II. Melnechuk, Theodore. III. Series. [DNLM: 1. Brain physiology – congresses. 2. Electroencephalography – congresses. 3. Electrophysiology – congresses. 4. Evoked Potentials – congresses. 5. Magnetics – congresses. WL 102 D997 1985] QP376.5.D96 1988 612'.8255 – dc19

Softcover reprint of the hardcover 1st edition 1988

2125/3130-543210

Preface

In neurophysiology, the emphasis has been on single-unit studies for a quarter century, since the sensory work by Lettwin and coworkers and by Hubel and Wiesel, the central work by Mountcastle, the motor work by the late Evarts, and so on.

In recent years, however, field potentials – and a more global approach generally – have been receiving renewed and increasing attention. This is a result of new findings made possible by technical and conceptual advances and by the confirmation and augmentation of earlier findings that were widely ignored for being controversial or inexplicable.

To survey the state of this active field, a conference was held in West Berlin in August 1985 that attempted to cover all of the new approaches to the study of brain function. The approaches and emphases were very varied: basic and applied, electric and magnetic, EEG and EP/ERP, connectionistic and field, global and local fields, surface and multielectrode, low frequencies and high frequencies, linear and nonlinear. The conference comprised sessions of invited lectures, a panel session of seven speakers on "How brains may work," and a concluding survey of relevant methodologies. The conference showed that the combination of concepts, methods, and results could open up new important vistas in brain research.

Included here are the proceedings of the conference, updated and revised by the authors. Several attendees who did not present papers at the conference later accepted my invitation to write chapters for the book.

This book presents the broadest possible picture of current Western research on this subject and includes contributions by most of the recognized scientists in this field. Therefore, it should be of interest to neuroscientists, neurologists, and all basic and clinical investigators concerned with new techniques of monitoring and analyzing the brain's electromagnetic activity, including the application of the new understanding of nonlinear dynamics to brain function.

The conference was held in West Berlin's modern International Conference Center. I am grateful for the support from the Senate of West Berlin that made the conference possible.

I also thank Theodore Melnechuk for assisting me in planning and co-chairing the session on "How brains may work" and for editing its transcript, as well as for his general editorial assistance with the entire book. For assistance in planning and administering the conference, I thank my secretary Miss Kristine Kay, Dr. Joachim Röschke, and my laboratory staff members.

The organizer wishes to thank the following institutions for providing financial support: the Senate of Berlin, Berlin; Bayer AG, Leverkusen; Data-Analysis Computer Systeme GmbH, Ottobrunn; Drägerwerk AG, Lübeck; Gesellschaft der Freunde und Förderer der Medizinischen Universität zu Lübeck, Lübeck; Ernst Leitz KG, Hamburg; Science Trading GmbH, Frankfurt/M.; S.H.E. GmbH, Aachen; and Sternkopf GmbH, Lübeck.

EROL BAŞAR

Contents

V. Magnetic Fields of the Brain

VI. Clinical Applications

VII. Workshop: How Brains May Work

Organizers: E. Başar and T. Melnechuk

Addresses of Participants

ADEY, W. R., Pettis Memorial Veterans Administration Hospital, 11201 Benton Street, Loma Linda, CA 92357, USA

BABLOYANTZ, A., Faculté des Sciences, Université Libre de Bruxelles, Campus Plaine CP 231, Boulevard du Triomphe, B-1050 Bruxelles, Belgium

BAK, C. K., Physics Laboratory I, Technical University of Denmark, DK-2800 Lyngby, Denmark

BAŞAR, E., Institut für Physiologie, Medizinische Universität Lübeck, Ratzeburger Allee 160, D-2400 Lübeck 1, FRG

BULLOCK, T. H., Department of Neurosciences, A-001, University of California, San Diego, La Jolla, CA 92093, USA

ECKHORN, R., Biophysik, Phillips-Universität, Renthof 7, D-3550 Marburg, FRG

FREEMAN, W. J., Department of Physiology-Anatomy, University of California, Berkeley, CA 94720, USA

GALAMBOS, R., Children's Hospital Research Center, 8001 Frost Street, San Diego, CA 92123, USA

GEVINS, A. S., EEG Systems Laboratory, 1855 Folsom Street, San Francisco, CA 94103, USA

HARI, R., Low Temperature Laboratory, Helsinki University of Technology, SF-02150 Espoo, Finland

HEINZE, H. J., Neurologische Klinik mit Klinischer Neurophysiologie, Medizinische Hochschule Hannover, Konstanty-Gutschow-Straße 8, D-3000 Hannover 61, FRG

HOKE, M., Division of Experimental Audiology, Ear, Nose, and Throat Clinic, University of Münster, D-4400 Münster, FRG

JOHN, E. R., Brain Research Laboratories, New York University Medical Center, 550 First Avenue, New York, NY 10016, USA

KYDD, R. R., Department of Psychiatry and Behavioural Science, University of Auckland School of Medicine, Auckland, New Zealand

Künkel, H., Neurologische Klinik mit Klinischer Neurophysiologie, Medizinische Hochschule Hannover, Konstanty-Gutschow-Straße 8, D-3000 Hannover 61, FRG

Lebech, J., Physics Laboratory I, Technical University of Denmark, DK-2800 Lyngby, Denmark

Lees, G.J., Department of Psychiatry and Behavioural Science, University of Auckland School of Medicine, Auckland, New Zealand

Leinfellner, W., Institut für Volkswirtschaftslehre und Volkswirtschaftspolitik, Technische Universität Wien, Argentinierstraße 8/175, A-1040 Wien, Austria

Makeig, S., Children's Hospital Research Center, 8001 Frost Street, San Diego, CA 92123, USA

Maurer, K., Universitäts-Nervenklinik, D-8700 Würzburg, FRG

Melnechuk, T., Department of Neurosciences, School of Medicine, University of California, San Diego, La Jolla, CA 92093, USA

Mitzdorf, U., Institut für Medizinische Psychologie der Universität München, Goethestraße 31, D-8000 München 2, FRG

Nunez, P.L., Department of Biomedical Engineering, Tulane University, New Orleans, LA 70118, USA

Papakostopoulos, D., Burden Neurological Institute, Bristol, England

Petsche, H., Institut für Neurophysiologie der Universität Wien, Währinger Straße 17, A-1090 Wien, Austria

Picton, T.W., University of Ottawa, Division of Neurology, Ottawa General Hospital, 501 Smyth Road, Ottawa, Ontario K1H 8L6, Canada

Pockberger, H., Institut für Neurophysiologie der Universität Wien, Währinger Straße 17, A-1090 Wien, Austria

Rappelsberger, P., Institut für Neurophysiologie der Universität Wien, Währinger Straße 17, A-1090 Wien, Austria

Reitboeck, H.J., Biophysik, Phillips-Universität, Renthof 7, D-3550 Marburg, FRG

Röschke, J., Institut für Physiologie, Medizinische Hochschule Lübeck, Ratzeburger Allee 160, D-2400 Lübeck 1, FRG

Saermark, K., Physics Laboratory I, Technical University of Denmark, DK-2800 Lyngby, Denmark

Scholz, M., Neurologische Klinik mit Klinischer Neurophysiologie, Medizinische Hochschule Hannover, Konstanty-Gutschow-Straße 8, D-3000 Hannover 61, FRG

Speckmann, E.-J., Physiologisches Institut der Universität, Robert-Koch-Straße 27a, D-4400 Münster, FRG

STAMPFER, H. G., The University of Western Australia, Department of Psychiatry and Behavioural Science, Queen Elizabeth II Medical Centre, Nedlands, Western Australia 6009, Australia

WALDEN, J., Physiologisches Institut der Universität, Robert-Koch-Straße 27a, D-4400 Münster, FRG

WERNER, G., Department of Psychiatry, University of Pittsburgh School of Medicine, Pittsburgh, PA 15261, USA

WRIGHT, J. J., Department of Psychiatry and Behavioural Science, University of Auckland School of Medicine, Auckland, New Zealand

I. Compound Potentials of the Brain: From the Invertebrate Ganglion to the Human Brain

Compound Potentials of the Brain, Ongoing and Evoked: Perspectives from Comparative Neurology

T. H. BULLOCK

1 Introduction

From the point of view of general biology, it is not surprising that one can record a compound fluctuating field potential from the brain. Our expectations come from several directions:

1. Fluctuating membrane potentials and various kinds of episodic potentials of action or oscillation are of general occurrence among nerve cells as well as other kinds of cells, for example, the cells of the blastula (Burr and Bullock 1941), the skin, gland, gut, muscle, and blood vessels.
2. At least six different kinds of active potentials are known in neurons and parts of neurons, including synaptic potentials with various properties, hyperpolarizations with long duration and decreased conductance, plateau potentials, pacemaker potentials, spikes, and negative and positive afterpotentials. The power spectrum of all these processes extends from dc to several kHz.
3. Lamination or other geometric biases can be expected to influence the summing of these cellular sources in special situations.
4. The null hypothesis of the independence of generators predicts a certain level of coincidence, depending on the duration of the cellular event and the definition of simultaneity.
5. Therefore large numbers of generators, small and large, are operative, in all orientations, some rhythmically but many episodically and generating broad-band signals, mostly spreading decrementally. The composite will therefore have a lot of spatial microstructure and less and less structure at macro levels. What cannot be predicted is the amplitude, frequency, and spatial and temporal structure.

The last-named features will depend on the severe *anisotropy* of typical nervous tissue, with a labyrinth of low-resistance spaces between high-resistance cells, many high-resistance clefts and high-capacitance membranes. In addition, they depend on the *heterogeneity of kinds* of nerve cells and the inhomogeneity of their processes. A wide spectrum of velocity of spread of excitation will influence the composite field potentials. The degree and extent of *synchrony* of activity among contiguous or non-contiguous somata, or among dendritic arbors or axonal collaterals and terminals, will be an important determinant. The variable participation of each of several classes of neuroglial cells, and the possible contributions from ependyma, pia mater, and blood vessels, besides sources not yet recognized, cannot be neglected. Even in a well-ordered, relatively simple nervous tissue, with quite regular neurons, input impulses will arrive stochastically, output impulses will arise in some staggered tempo,

Springer Series in Brain Dynamics 1
Edited by Erol Başar

and synaptic currents and other subthreshold events of various kinds will come and go without averaging to a smooth vector sum in the external current path, if measured in a manner that does not alter the field structure.

The purpose of this essay is to add to the five starting points above some propositions about the fundamental nature of ongoing and evoked compound field potentials of the brain, designed to be heuristic by increasing the sources of new data, both technically and in respect to variety of species. Starting from first principles, let us re-examine the phenomena we study every day from the general physiological perspective and look at the manifestations of ongoing and evoked activity in brains of lower vertebrates and invertebrates. After all, it was in these taxa in which most of the stages of evolution of this most evolved of all systems took place.

The first thesis I would like to propose, therefore, from general biological considerations, is that we should not think of ongoing activity in the brain as simply the sum of synaptic potentials, as some authors imply and others explicitly state. Nor should we think of the EEG as simply the sum of synaptic and spike potentials, or the result of diffuse synchronization or of a driving rhythm, unless good evidence for such a limitation is adduced. Instead, we should start with the assumption that all the sources I have listed, plus others not yet recognized, are contributing. That means very large numbers of *microscopic dipoles*, most of them ac or fluctuating sources (in addition to being dc sources), and a basic or ground state of near independence of these sources. Any degree of synchronization above the level of coincidence requires proof of its reality, as well as special explanation. This position seems to me the most tenable as well as suggestive of new approaches.

2 Hierarchy of Complexity; Bottom-up Approach to Information Richness

As a long-time observer in this field, I am impressed by the divergence of frames of reference or perspectives among workers. The first level of divergence is in what we expect or aim for; what represents an explanation or insight.

The EEG seems to mean quite different things to different workers, each confident because his/her view is based on hard data. Some appear virtually to identify the EEG with alpha or with 8–12 Hz activity and practically ignore the rest of the power spectrum. Some appear to consider the human EEG as the basic phenomenon and ignore species differences. Some appear to consider the scalp-recorded activity as the phenomenon of interest and think only on occasion about the variety of forms of cellular activity in different cortical laminae and subcortical nuclei, or about the local inhomogeneities glossed over by electrodes that must look through the scalp and skull. Some think of the EEG as essentially similar to travelling waves at the surface of a pond. Few authors are concerned with the *microstructure of the field potentials* in the brain on a scale below 5 mm, or the temporal nonstationarity in the domain below about 2 s.

Of course there are notable exceptions and I appreciate many insightful authors such as Petsche, Lopes da Silva, Creutzfeldt, Adey, Elul, Freeman, and others, and the pioneers, like Adrian, Bremer, Gerard, and a goodly list of others. In this introductory paper I am unable to reference most authors who should be cited in a full-

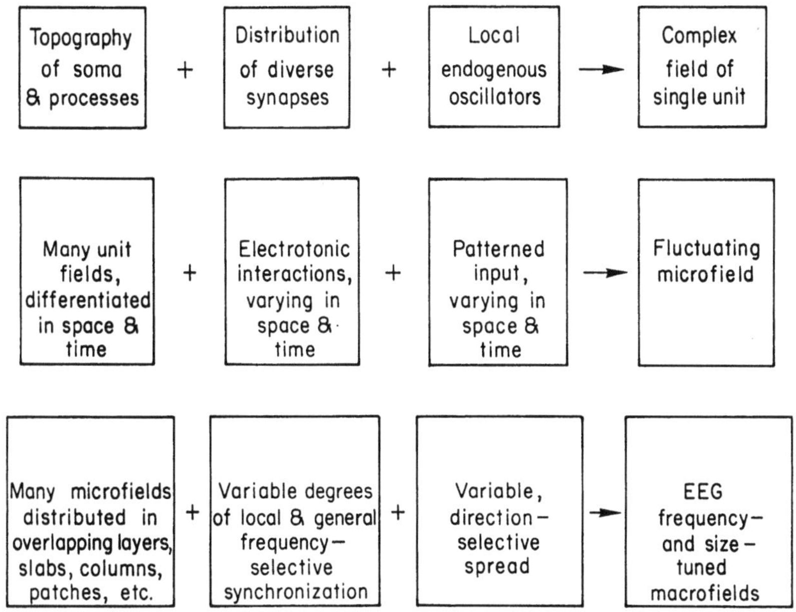

Fig. 1. Scheme of some of the supposed generators of the constituent fields that sum to the observed electrical activity of the brain. The sum of each line is a constituent of the next

dress history, or to give the data for most of the assertions I make. My purpose is to convey a research agenda by pointing out some opportunities and needs based on perceptions of the shortcomings of our present picture.

Let me start with a graded series of abstractions of some underlying factors that must contribute to the EEG (Fig. 1). These are not novel but fully compatible with classical notions (Adrian and Matthews 1934; Creutzfeldt and Houchin 1974; Lopes da Silva and van Rotterdam 1982).

Note that *synchronization* can increase the power at certain frequencies but its consequences are not simple. We are speaking of events with a significant duration, so that the general case is not one of perfect coincidence but some degree of overlap; moreover, the effect depends on the dipole orientations. Both of these factors are influenced by the low-velocity spread of even subthreshold events, a spread which reorients the field as it turns corners and extends into branches.

Synchrony is probably much less than 100% in the usual range of normal states. It may be only slightly above the *chance level of the null hypothesis.* In view of the difficulties of knowing the number of potentially independent generators active at any time, we may say it is actually impossible to decide how much agreement among units is the chance level. *Cooperativity* as distinct from independence is not adequately measured by synchrony in the usual sense of overlap in time; delayed dependence may be common. Synchrony may take several cycles to become established.

I am taking an approach diametrically opposite to the classical effort to dissect the components of the EEG, assess their importance, and attempt to state the main contributors – an approach that has been called *top-down,* starting as it does with the observed EEG. Instead, I am trying to recognize all the potential contributors and the geometric factors that may determine their summation – a kind of *bottom-up approach*. The reason is, I believe, that we must assume all the factors I have mentioned – plus, no doubt, others I have omitted – are relevant somewhere. I think we have been significantly limited in our thinking and hence in research by identifying the phenomenon with the output of our particular measurement techniques, recalling the blind men describing the elephant.

I want to lift up the possibility that the ongoing compound field potentials in the brain, in their fine structure in space and time, are much more *information-rich* than we have generally assumed or acted upon. The usual assumption has been that the neural representation of behavior and perception is only to be sought in the temporospatial pattern of unit activity – usually considered to be simply spike activity. Instead, my understanding of the anatomy and its ever increasing order and decreasing chaos leads me to propose that also at the levels of assemblages or masses of smaller or larger dimensions, the composite activity, especially prespike activity, will have significant local differentiation and is not entirely blurred. Columns or slabs, laminae, patches, and glomeruli are some of the reminders of order, even though they may consist of heterogeneous arrays of neurons.

3 Needs and Opportunities for New Approaches to Tap Information Richness

In discussing what kind of information richness there could be that is not already well sampled by the large battery of sophisticated techniques now in use in EEG analysis, I will speak of only two domains, but each is a major challenge, technically and conceptually. A prime aim of this paper is to suggest a research agenda worthy of major effort and support.

The first domain of opportunity and need is *multichannel, wide-band recording* with controlled electrode spacing in the *millimeter range* or less. The need is to extend the elegant pioneering work such as that of Petsche and his colleagues to the three-dimensional array instead of the one-dimensional row of electrodes. The *hardware* challenge is fabrication, placement, control, and verification of electrode positions with such a density of electrodes, without damaging too much tissue. Microelectrodes displace a very small volume at their tips but when recording at any depth, the usual tapering form results in a significant displacement along the shank or length if the electrodes are only a millimeter or less apart. New electrode design is urgently needed for three- and even for two-dimensional analysis. The *software* challenge is to deal with enough data to reveal pattern or meaningful distribution over a sufficient volume of tissue for a sufficient period of time and to reduce it in useful ways and display it so as to optimize the use of the human faculties.

Several laboratories are making strides in one or another of these respects (Kruger and Bach 1981; Kuperstein and Eichenbaum 1985; Petsche et al. 1984; Pickard 1979; Praetorius et al. 1977; Prohaska et al. 1979), yet an adequate and practical solution of the several severe demands still requires some major development. Most effort is currently directed at unit spikes, discarding the low frequencies. This is partly because the information overload on the investigator is serious, even limit-

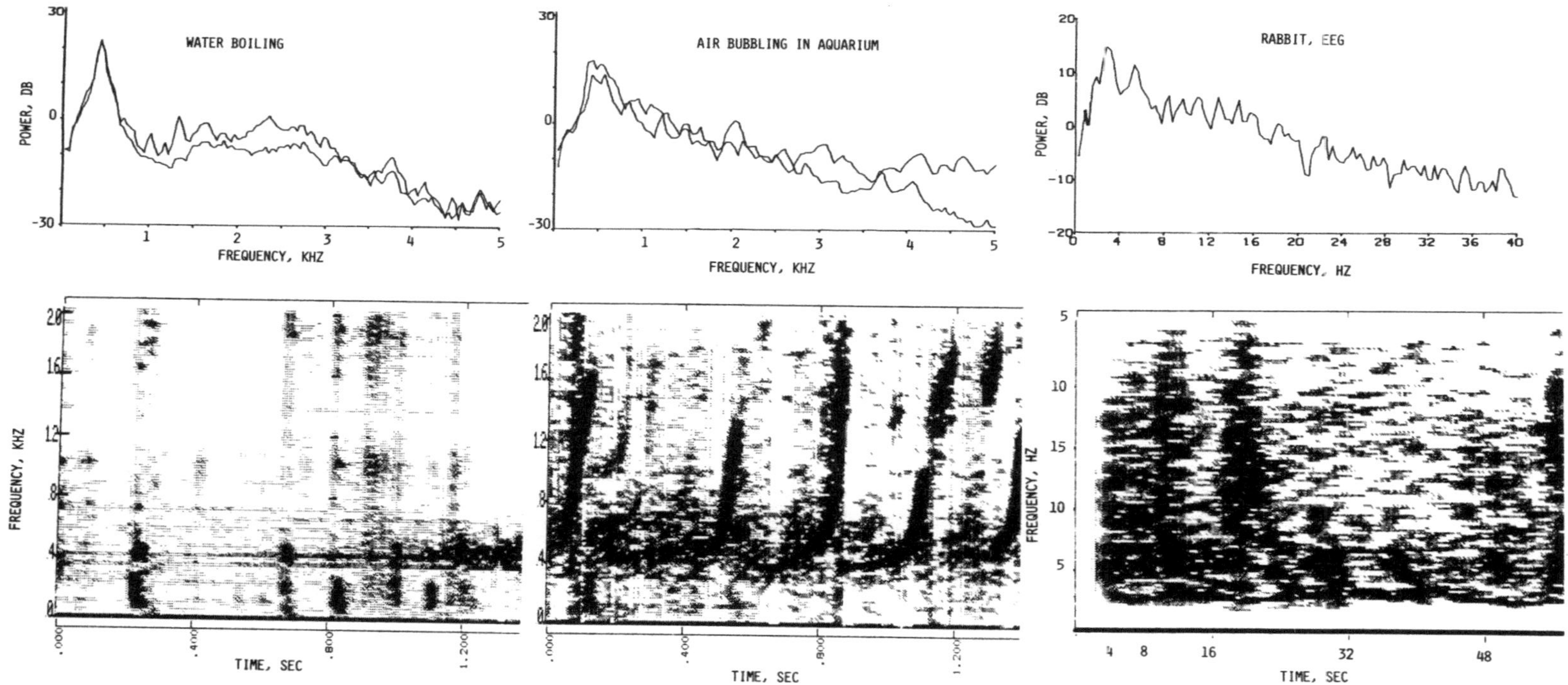

Fig. 2. Spectrograms in two forms, for sounds and EEG. Pairs of averaged power spectra of the familiar, wide-band sounds, categorized in less than 1 s by the human auditory system, show little evidence that consistent, essential fine structure of the power spectrum is the basis of recognition. Indeed, severe distortion of the power spectrum by filters would not interfere with recognition. The sound spectrograms, preserving the time domain structure, presumably contain the necessary and sufficient clues for categorization, even though successive samples do not look superficially alike. The suggestion is strong that hitherto unnoticed differences between the EEG of different species, parts, and states of the brain may inhere in the temporal structure on a fractional second scale

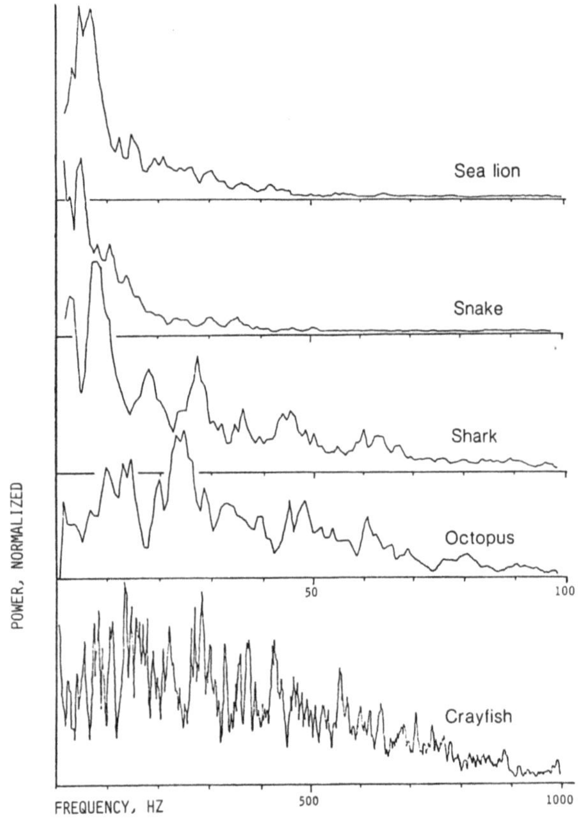

Fig. 3. Power spectra of the EEG of several species. The vertebrate ongoing activity was recorded from the cerebrum (sea lion, *Zalophus* and rattlesnake, *Crotalus*) or tectum (shark, *Carcharhinus*) in the quiet, unanesthetized, awake state with bipolar electrodes. The octopus activity was from the vertical lobe in a similar state. Crayfish *(Cambarus)* activity was from the circumesophageal connectives near the brain. Note the difference in scale on the *abscissas*. (From Bullock 1983)

ing ourselves to the presently familiar forms of wide-band analysis, such as power spectra, cross spectra, current source density, and coherence.

The second domain of opportunity and need is even more formidable, namely *imaginative extension of the forms of analysis*. Even for the single-channel record, my strong feeling is that we might be missing the main characteristics of the information flow and this is vastly more true for the multichannel case. It often helps to think of commonplace analogies. It is not only the skilled garage mechanic but also you and I who can distinguish and even identify in less than 1 s many wide-band, noisy sounds, such as the rustling of a newspaper, rubbing of hands, soft and heavy breathing, the breeze in the pine trees, the flowing brook (Fig. 2), and a virtual infinity of others. These examples are simple, nearly continuous signals. Think what we do with speech, even whispered or hoarse-voiced speech without prominent carrier frequencies, even in the presence of high levels of disturbing background, even when that background is quite structured. Think of the difficulty and the limited success achieved in computer analysis of speech; I mean the recognition of words or signals of meaning, so as to control a typewriter, for example. At the least, this analogy suggest that we should not treat the ongoing brain activity as a stationary process even for periods as short as 1 s and we should advance beyond arbitrary averaging of successive segments (Fig. 3).

Another analogy comes to mind: the discovery of whale songs. Think of the meters and meters of sound spectrograms – noisy, gray smudges – that Payne and McVay (1971) pored over before they noticed a pattern repeating every 12 min! The human eyes – and I would add the ears working together with the eyes – are better than any computer program at recognizing unpredictable patterns.

Likewise for the more adequate treatment of *evoked potentials,* we should advance from the assumption that they arise from a background of irrelevant noise to be averaged out and instead look for the causes of the trial-to-trial fluctuations in form as well as amplitude. This is a difficult demand to carry out and is easy to ignore. My purpose is mainly to remind ourselves that it is so easy to average that we are likely to be bemused into forgetting the reality of highly variable potentials following adequate stimuli. Efforts in the past to discover categories of evoked potentials to the same repeated stimulus should be extended. There is reason to hope that we can bring the variance under control, at least somewhat, as McDonald (1964) did many years ago when he greatly increased the amplitude of the average evoked potential by a simple contingency of giving the stimulus after a 3 s interval plus an additional interval until the EEG was of low amplitude for 80 ms.

4 Justification of the Assumed Information Richness

There are many reasons why I think there might be information rather than simply stochastic machine noise in the details of the temporal and spatial structures between points 1 mm and less apart. The case seems to me overwhelming that such fine structure is present and meaningful, at least as an epiphenomenal sign of what is happening, and perhaps also in part as causal signals for neurons.

The first reason is that fine structure is there. We know from findings such as those of Petsche et al. (1984) that coherence can be very different in tenths of a millimeter across the cortex, and we know from their work and that of others like Mitzdorf (1985) that current sources and sinks can be smaller than a cortical lamina and can change in milliseconds. These are averaged findings, which, no doubt, have suffered some broadening.

A second reason is that fine structure must be there, to judge from our accumulated understanding of how nervous tissue works. We are well past the days when we could think simply in terms of circuits whose connectivity in terms of nerve impulse distribution explained their function. We realize that not only local circuits but also the three-dimensional geometry of arbors and ramifications, the spacing of the hundreds or thousands of synapses in the electron microscopist's usage, which act together to make one synapse in the physiologist's sense, the whole congeries of subthreshold potentials that integrate direct and modulatory influences and control transmitter release without or between spikes are all real, and extremely differentiated among bits of nervous tissue here and there. We realize today that neurons are much more different than we used to think, and the numbers of essentially indistinguishable neurons much smaller; one estimate (Bullock 1980) says that equivalence sets of more than 200 neurons are rare.

The resulting compound field potentials are bound to be a mixture, both stochastic and determinate or predictable; both apparently random and clearly consistent or patterned. Information nearly as rich as perceptions and behavior must inhere not only at the level of neuronal units but also higher, at assemblage levels. It cannot drop to complete disorder, even at the level of 100000 neurons – a number that occupies an area a bit over 1 mm^3 in our cortex, and a volume that is highly coherent in field potential activity with the adjacent 50 or 100 mm^3.

A third reason for believing the EEG is information-rich on the millimeter and millisecond scales is that the fine structure of the ongoing background, from the little that we know of it, is so much like that of the evoked activity, which, we can assume, represents meaningful signals.

A fourth reason is that I have to believe, until shown otherwise, that there are meaningful differences in the ongoing transactions of areas 17, 18, 19, 7, 5, 3, to mention a few on top, and of different patches in the corpus striatum and different laminae and x and y coordinates within them in the lateral geniculate nucleus (LGN), as examples. That means I do not subscribe to the view that the cortex is made of *perfectly equal* modules. Even if there are somewhat similar modules, they must be receiving differently patterned input, even in the quiet, resting subject.

A fifth and equally important reason comes from considering the comparison of species instead of comparing parts of the brain. At least when we compare the alert human species and any other, particularly if that other is a fish or a frog, the vastly more sophisticated operations commonly present in the human species, judging by its cognitive transactions, compel me to consider the possibility that at least some blurred reflection of that difference may be discernible in the compound field potentials, if we look in the right way.

The main message of this essay is to assert that we need to think up new ways of looking at the field potentials, both the stream of waves on the single channel, and the patterns among many closely spaced channels; both the ongoing and the evoked. Of course this will not suffice, even when we solve the enormous problems outlined above. We will only have some clues as to how the brain works – just as single unit studies only give us some clues. We need all the windows we can peer through, and more.

5 Comparison Across Species, Classes, and Phyla

The surprising result of comparing major groups of vertebrates is that, so far, no difference can be said to be consistent between the microelectrode, intracerebral electrogram of elasmobranchs, teleosts, amphibians, reptiles, birds, and mammals. No consistent differences have been established among the classes and no consistent correlations are known with brain size, neuron number or density, lamination, or cortical development. Species peculiarities, such as the prominence of alpha activity in many though not all humans and a few other taxa, are few or little known. In general, the differences between parts of the brain and between individuals and especially between states of the same subject are far greater than the candidate differences between species or classes. To be sure, one difference between classes is that no slow-wave sleep is found in fishes, amphibians, and perhaps even reptiles (Allison 1972). Apparently all vertebrates, from the mesencephalon forwards, have smooth, slow brain waves, mainly below 40 Hz, the energy maximally < 15 Hz and falling off above that maximum more or less steeply, sometimes > 10 db per octave – whether large mammal or tiny fish, with or without cortex, whether in highly ordered centers with regular arrays of cells and processes or in centers not notable for patterned structure (Enger 1957; Hobson 1967; Klemm 1969; Laming 1980, 1981, 1982, 1983; Laming and Savage 1981; Schade and Weiler 1959; Segura and de Juan 1966).

In sharp contrast to the EEG of any vertebrate, most of the *higher invertebrates* (annelids, arthropods, and gastropods) have ongoing CNS activity dominated by spikey, fast, high voltages, whereas low frequencies, though present in the power spectrum, are of such low voltage as to be almost unnoticeable, apart from components of broad spikes and afterpotentials. The power spectra show relatively high activity above 100 Hz. I suspect this striking difference between vertebrates and invertebrates is not due to the size or number of cells but is a consequence of some important underlying difference in the cooperativity of somata, dendrites, and neuropile.

Interestingly, the cephalopods such as octopus have ongoing brain activity, in respect to the prominence of low frequencies, more like that of vertebrates than of other invertebrates (Bullock 1984). They have a peculiarity that has long delayed analysis, namely an ability to switch from an electrically active state to a state of virtual electrical silence.

It seems likely that the *vertebrate-invertebrate contrast* could tell us something about the nature of compound field potentials (Bullock 1945) and that it must have other dimensions besides the power spectrum. The comparison has so far been made almost exclusively on the power spectrum. There may well be a significantly lower root mean square voltage in our samples of elasmobranchs, fish, and frogs, but a careful systematic study of this simple parameter has not been done. It is my contention that our bases for comparison have so far been as inadequate as though we were comparing human languages among different cultures merely by microphone records of voltage against time, power spectra, and amplitudes. Surely a battery of descriptors is needed to find EEG correlates of the well-known and striking differences between, for example, fish and mammals in histological differentiation.

A few suggestions for other dimensions not yet adequately surveyed are the following. These are very inadequate suggestions in the face of the contentions I have put forward and they are not new ideas, yet they are measurements which are still little used as ways of comparing states, parts, or species.

One is *coherence plotted against distance* between recording microelectrodes on or in the cortex (Fig. 4). It is no surprise, but it has not been systematically quantified before, that at each frequency coherence declines with distance in the mammalian cortex (Bullock 1983; Bullock et al. 1983, 1984). It is not a simple space constant or monotonic decline and cannot be anticipated on either theoretical or experiential grounds. Our sample of data is still limited. Coherence on the surface of the resting rabbit cortex is significantly less than 1.0 already with separations of 0.5 mm in some samples. Within about 4–6 mm in typical cases, coherence in the frequency band 1–5 Hz falls to about 0.5, at 5–10 Hz to approximately 0.4, at 10–20 Hz to approximately 0.35, at 20–40 Hz to approximately 0.3. While the variance is large, this trend is quite consistent. The decline with log distance is more sigmoid than linear. Another trend is less consistent – a decline in coherence with frequency at a given

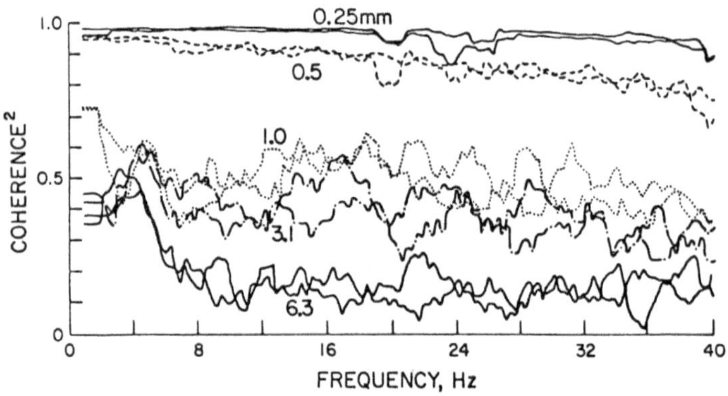

1.0
0.5
0
COHERENCE^2
0.25mm
0.5
1.0
3.1
6.3
0
8
16
24
32
40
FREQUENCY, Hz

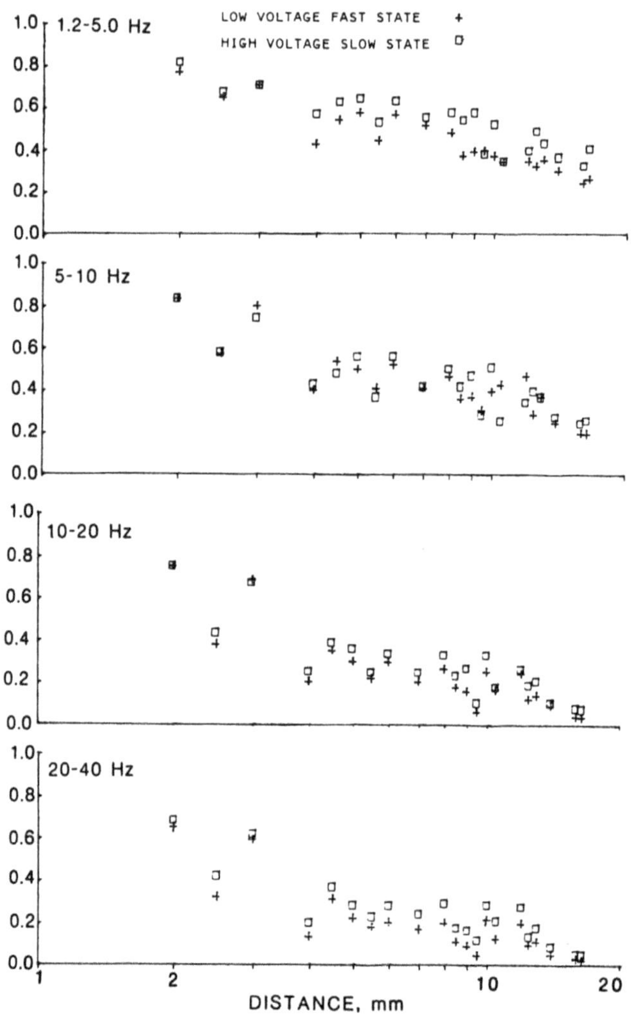

1.2-5.0 Hz
LOW VOLTAGE FAST STATE +
HIGH VOLTAGE SLOW STATE □
5-10 Hz
10-20 Hz
20-40 Hz
1.0
0.8
0.6
0.4
0.2
0.0
1
2
10
20
DISTANCE, mm

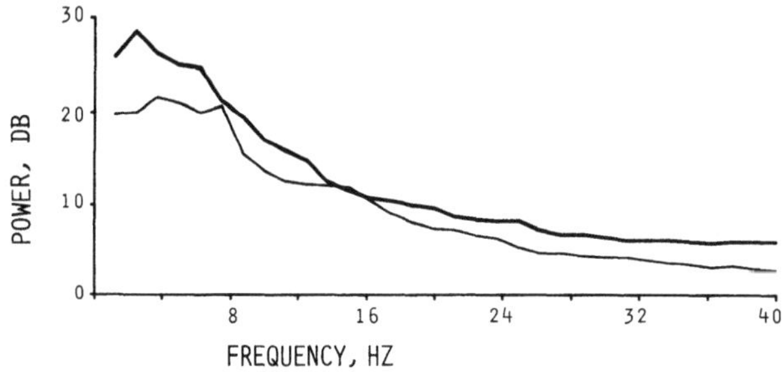

Fig. 5. Power spectra for the "low-voltage fast" *(thin line)* and "high-voltage slow" *(thick line)* sets, averaging from 18 electrodes for at least 2 min each (same experiment as Fig. 3). Clearly, the principal difference between the two states is in the power spectra and much less in the synchrony (Fig. 4)

distance; this often shows plateaus and steep slopes. At 16 mm and sometimes even at 8 mm in the rabbit, coherence is usually indistinguishable from the null hypothesis of independence; put the other way, a small, significant coherence can often be detected for low frequencies at 8 mm and sometimes at 16 mm.

Coherence may decrease with distance much *more gradually in the vertebrates* than in most higher invertebrates, even at low frequencies. We are accumulating experience with this measure but it is still too early to generalize.

Coherence plotted against distance between recording sites represents a *measure of synchrony*, possibly the best one available; the falling curve reflects the volume of tissue above a given level of coherence. As such it does not always agree with the eyeball expectation (Danilova 1975) (Fig. 5). During a high-voltage slow-wave state that might be early sleep in the rabbit, the coherence in each band is only very slightly higher than in a low-voltage, fast, alert state. The so-called synchronized record may reflect mainly a change in the power spectrum in favor of low frequencies (Bullock et al. 1984). In midseizure brought on in rats by metrazol, coherence is not especially high for small distances nor low for larger distances, but bunched at approximately 0.7 in the band of maximal energy, 12–16 Hz, in spite of frequent broad spikes synchronous in all channels (Bullock et al. 1983).

Coherence with distance also contributes to a potentially revealing picture of the *fine structure of the electrogram in space*, in the 0.5–5 mm range. This is especially needed in the tangential plane of the cortex and within noncortical structures such as

Fig. 4. Coherence as a function of frequency *(top)* and distance *(bottom)*, for pairs of electrodes at different separations; alert rabbit, dorsal cortex. *Top,* pairs of averages of 15 epochs of 3.2 s each show that there are consistent differences, according to the separation, but only in the general level, not the fine structure. Coherence tends to fall with distance and, less regularly, with frequency. *Bottom,* using the fall-off with distance as a measure of synchrony, samples are compared from a high-voltage, slow "synchronized" state and a low-voltage, fast, "desynchronized" state in another rabbit implanted with 16 electrodes on the pial surface. The points are grand averages of samples representing the coherence between given pairs for 39 consecutive 3.2 s epochs, repeated up to three times approximately 20 min apart; for some abscissa values only two pairs, for others up to 48 pairs shared the same separation. The differences between the "low-voltage fast" set (+) and the "high-voltage slow" set (□), though small, are highly significant at 1.2–5, 10–20, and 20–40 Hz, but not statistically significant at 5–10 Hz

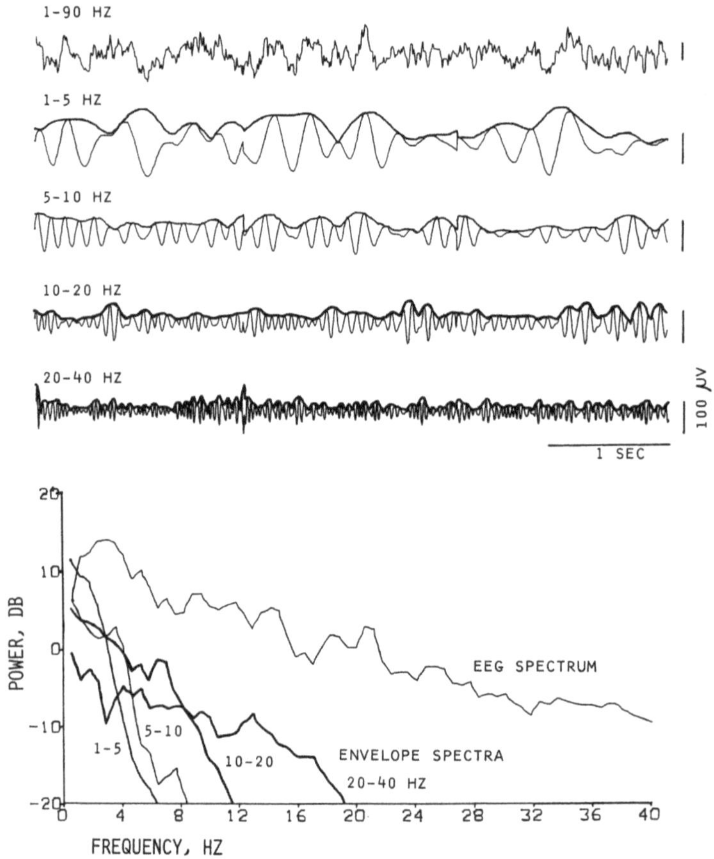

Fig. 6. Other EEG descriptors. *Top,* EEG of turtle hippocampus in the resting state, recorded via wide-band (1–90 Hz) and four narrow-band filters, with the envelopes computed as the magnitude of the "analytic signal" obtained from the data sequence by suppressing the negative frequencies and doubling. *Bottom,* power spectra of the wide-band EEG and the four envelopes. It is typical of nearly all EEG records that narrow-band amplitude waxes and wanes continually and is not constant even for a few seconds

the diencephalon, the basal ganglia, and the nonmammalian pallium. It is more difficult to look for *temporal* changes in synchrony in the range of seconds by the method of coherence decline with distance, since we need a good deal of averaging to get reproducible coherence estimates. Nevertheless, it may be a parameter that is fluctuating importantly.

The quantitative estimation of synchrony might be a useful part of a battery of descriptors for comparing major brain regions and states of the animal and possibly also between classes of vertebrates.

Another group of descriptors with which we are accumulating experience begins with the separation of the wide-band EEG into several, typically four, digitally band-pass-filtered components (Başar 1980), typically approximately one octave wide. We then compute the amplitude envelopes of these components and make comparisons

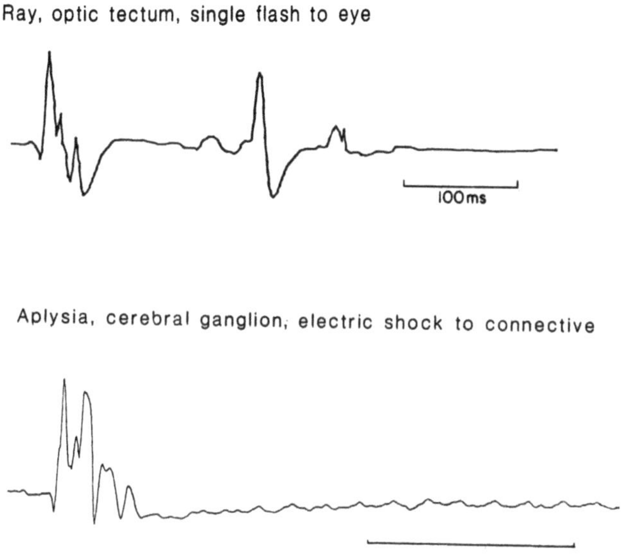

Fig. 7. Evoked potentials of lower forms. *Top,* an elasmobranch, showing that late waves are not necessarily broad and can arise in brain stem levels. Actually, these main peaks arise in the retina with nearly the same latency as observed in the tectum. *Bottom,* a gastropod mollusc, showing that in spite of a predominantly spikey ongoing background activity, the neuropile of an invertebrate ganglion can respond to a synchronized nerve shock with a slow evoked potential, presumably due to several factors including conduction velocity dispersion, wide soma potentials, long afterpotentials and imperfectly synchronized postsynaptic potentials. *100 ms* applies to both time scales

of these (Fig. 6). The power spectra of the envelopes show major modulations, at 1 Hz and even higher, in the amplitudes of all components. Correlations between the envelopes of the different bands measure the congruence of waxing and waning between the frequency components.

Evoked potentials likewise show as much or more variety within a given species of mammal, when we compare recording loci from cord to medulla to cortex, as they do between fishes and mammals. It is not necessary to illustrate this familiar fact, but I show an example to remind you that subcortical flash – evoked potentials can have brief but long latency components in fish (Fig. 7) and another example to show commonality among wide-ranging species in the early click-evoked auditory brain stem responses (Fig. 8). At the moment we are more impressed by the similarities in evoked potentials among all species than any ostensible differences. Perhaps nonhuman species will lack an event-related potential to a bad joke or incongruous sentence, but fish have both early, fast evoked potentials and late, slow evoked potentials. This is not the place to enter into issues such as whether the evoked potential is properly understood as a stabilization or other transform of the ongoing activity or is partly or largely new activity of cells that were previously relatively quiet. Invertebrates (insects, crustaceans, gastropods, cephalopods) also show both fast and slow evoked activity. What characterizes each major group is not the presence versus absence but the details of the level of the brain, the modality, and the form of the stimulus associated with each evoked potential shape.

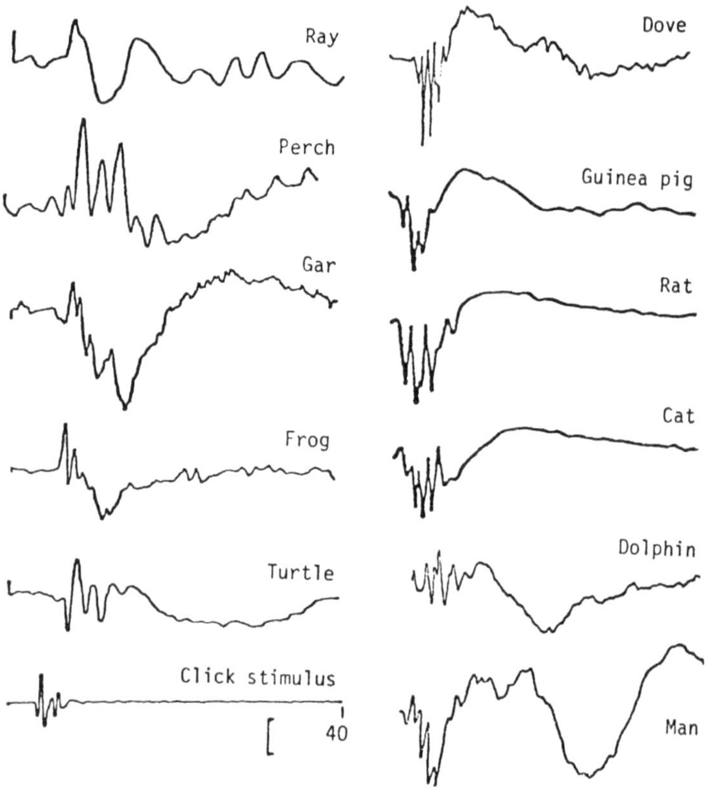

Fig. 8. Auditory brain stem responses (ABR) of several species, recorded outside the brain by averaging from 64 to 2000 responses to airborne clicks at low repetition rate (< 10/per second). A microphone record of the click used for several of the species is shown at lower left but the arrival time is not precisely the same for all species. As long as the main initial wave is above 1 kHz, the exact composition of the click is unimportant in determining the ABR form; for the ray the click has to be < 400 Hz and the response wave form is sensitive to its composition. Recording electrodes in the *left column* and in the dove in the *right column* were just intracranial via fine midline holes through the cranium above the posterior cerebellum and above the rostral end of the cerebrum; rostral electrode negative = upwards deflection. In the mammals, electrodes were near the vertex and the mastoid extracranially; vertex negative = upwards deflection. All records are 40 ms long. Voltage scale mark is approximately 2 μV for ray, perch, rat guinea pig, cat, and dolphin, 5 μV for man. Amplifier filters: 10–3000 Hz except dolphin; 1–5000 Hz. (Modified from Bullock 1983)

6 Summary

Despite more than 50 years of research, we still have little idea of how much synchrony there is in the EEG, or of how it varies among states of the brain, parts of the brain, and major groups of animals, or of how any cooperativity is distributed among the diverse cellular and subcellular generators in a given volume. Among the opportunities and needs, I emphasize *research at the semimicro level* comprising the structural domain of approximately 0.5–5 mm and the time domain of approximately 0.2–

2 s. There is probably an information-rich world in these dimensions; therefore we urgently need many-channelled recording with wide-band amplifiers and some novel analytical algorithms in order to tap this information, at least for descriptors that will reveal differences between states, parts, and species.

Our present methods are as limited as would be those of an anthropologist trying to distinguish commonalities from differences in the vocalizations of football crowds in Britain and bullfight crowds in Spain by using microphones above the crowds and analyses that treat epochs like 1 s as stationary. The analogy is chosen to include fluctuating tendencies for groups of individuals, scattered or all together, to synchronize their voices. Even with the advantage of synchrony, as distinct from the cocktail party, understanding depends on learning the rudiments of the two languages; discerning significant differences depends on hearing many samples and learning the invariants. A parallel effort in the comparative study of EEG and evoked potential, i.e., high temporal and spatial resolution, using human pattern recognizers in both visual and auditory realms on many samples might well turn up species, state- and stage-specific features, or at least those features characteristic of higher taxa (orders, classes) and major segments of the forebrain.

References

Adrian ED, Matthews BHC (1934) The interpretation of potential waves in the cortex. J Physiol (Lond) 81:440–471

Allison T (1972) Comparative and evolutionary aspects of sleep. In: Chase MH (ed) The sleeping brain. Brain Information Service, Brain Research Institute, UCLA, Los Angeles, pp 1–57

Başar E (1980) EEG-brain dynamics, relation between EEG and brain evoked potentials. Elsevier, Amsterdam

Bullock TH (1945) Problems in the comparative study of brain waves. Yale J Biol Med 17:657–679

Bullock TH (1980) Reassessment of neural connectivity and its specification. In: Pinksker HM, Willis WD (eds) Information processing in the nervous system. Raven, New York, pp 199–220

Bullock TH (1983) Electrical signs of activity in assemblies of neurons: compound field potentials as objects of study in their own right. Acta Morphol Hung 31:39–62

Bullock TH (1984) Ongoing compound field potentials from octopus brain are labile and vertebrate-like. Electroencephalogr Clin Neurophysiol 57:473–483

Bullock TH, Lange GD, McClune MC (1983) Spatial structure of cortical EEG: synchrony of small populations can be measured by coherence as function of distance. Neurosci Abstr 9:1194

Bullock TH, Lange GD, McClune MC (1984) A measure of synchrony in the cortical EEG: the slow wave drowsy state is slightly more synchronized horizontally than the low voltage fast state. Neurosci Abstr 10:1143

Burr HS, Bullock TH (1941) Steady state potential differences in the early development of Amblystoma. Yale J Biol Med 14:51–57

Creutzfeldt O, Houchin J (1974) Neuronal basis of EEG waves. In: Remond A (ed) Handbook of electroencephalography and clinical neurophysiology, vol 2, part C. Elsevier, Amsterdam, pp 5–55

Danilova NN (1975) Neuronal mechanisms of synchronization and desynchronization of electrical activity of the brain. In: Sokolov EN, Vinogradova OS (eds) Neuronal mechanisms of the orienting reflex. Erlbaum, Hillsdale; Wiley, New York, pp 178–199

Enger PS (1957) The electroencephalogram of the codfish. Acta Physiol Scand 39:55–72

Hobson JA (1967) Respiration and EEG synchronisation in the frog. Nature 213:988–989

Klemm WR (1969) Animal electroencephalography. Academic, New York

Kruger J, Bach M (1981) Simultaneous recording with 30 microelectrodes in monkey visual cortex. Exp Brain Res 41:191–194

Kuperstein M, Eichenbaum H (1985) Unit activity, evoked potentials and slow waves in the rat hippocampus and olfactory bulb recorded with a 24-channel microelectrode. Neuroscience 15: 703–712

Laming PR (1980) Electroencephalographic studies on arousal in the goldfish *(Carassius auratus).* J Comp Physiol Psychol 94:238–254

Laming PR (1981) The physiological basis of alert behaviour in fish. In: Laming PR (ed) Brain mechanisms of behaviour in lower vertebrates. Cambridge University Press, Cambridge, pp 203–224

Laming PR (1982) Electroencephalographic correlates of behavior in the anurans, *Bufo regularis* and *Rana temporaria.* Behav Neural Biol 34:296–306

Laming PR (1983) Relationships between the responses of visual units, EEGs and slow potential shifts in the optic tectum of the toad. In: Ewert J-P, Capranica RR, Ingle DJ (eds) Advances in vertebrate neuroethology. Plenum, New York, pp 595–602 (NATO ASI Series A: Life sciences, vol 56)

Laming PR, Savage GE (1981) Seasonal differences in brain activity and responsiveness shown by the goldfish *(Carassius auratus).* Behav Neural Biol 32:386–389

Lopes da Silva F, van Rotterdam A (1982) Biophysical aspects of EEG and MEG generation. In: Niedermeyer E, Lopes da Silva F (eds) Electroencephalography: basic principles, clinical applications and related fields. Urban and Schwarzenberg, Baltimore, pp 15–26

McDonald M (1964) A system for stabilizing evoked potentials obtained in the brain stem of the cat. Med Electron Biol Eng 2:417–423

Mitzdorf U (1985) Visually and electrically evoked field potentials and current source densities in the cat visual cortex. In: Morocutti C, Rizzo PA (eds) Evoked potentials. Neurophysiological and clinical aspects. Elsevier, Amsterdam, pp 273–279

Payne RS, McVay S (1971) Songs of humpback whales. Science 173:587–597

Petsche H, Pockberger H, Rappelsberger P (1984) On the search for the sources of the electroencephalogram. Neuroscience 11:1–27

Pickard RS (1979) Printed circuit microelectrodes. Trends Neurosci 2:259–261

Praetorius HM, Bodenstein G, Creutzfeldt OD (1977) Adaptive segmentation of EEG records: a new approach to automatic EEG analysis. Electroencephalogr Clin Neurophysiol 42:84–94

Prohaska O, Pacha F, Pfundner P, Petsche H (1979) A 16-fold semi-microelectrode for intracortical recordings of field potentials. Electroencephalogr Clin Neurophysiol 47:629–631

Schadé JP, Weiler PJ (1959) EEG patterns in goldfish. J Exp Biol 36:435–452

Segura ET, de Juan A (1966) Electroencephalographic studies in toads. Electroencephalogr Clin Neurophysiol 21:373–380

Nonlinear Neural Dynamics in Olfaction as a Model for Cognition

W. J. FREEMAN

1 Introduction

The forebrain of primitive vertebrates is so heavily devoted to olfaction that for half a century investigators were misled into considering the function of the hippocampus as being exclusively olfactory. For example, the anterior third of the forebrain of the tiger salamander forms the bulb, the medial third is hippocampus, and the lateral third comprises the piriform and striato-amygdaloid complex (Herrick 1948). According to Herrick, a transitional zone in the mantel receives thalamic axons that convey input to the forebrain from all other sensory systems. He proposed that with the expansion and increasing dominance of these other systems, the brain expanded by adding new parts while preserving the topology of connections of those parts already existing. This view has survived to the present with modifications; it is as if, seeing that olfaction was a success, other systems moved in and co-opted the machinery of the forebrain. Olfaction remains the simplest among the sensory systems. For this reason, if for no other, the study of sensation and cognition might well begin with the sense of smell. But there are three other good reasons: the parallels that exist between olfaction and other senses in their psychophysics, in the dynamics of the masses of neurons comprising them, and in the types of neural activity that they generate.

2 Psychophysics

The olfactory system resembles other sensory systems in consisting of a surface array of receptors of multiple kinds that project in parallel to arrays of central neurons. Some examples of stimuli that are comparable to an odorant are the sight of a constellation such as Orion in the winter sky, the feeling of putting on a coat that is not the correct size, and the sound of a tone that allows immediate identification of the instrument being played – a piano, oboe, etc. These operations are rapid, spatial, and global, and they depend on past experience. The information is expressed by spatial relationships among activated and equally importantly nonactivated receptors, without reference to simple geometric forms. The time frames are longer than that of an action potential but shorter than a heart beat; according to Efron (1970), on the order of 0.1 s.

All of these systems are legendary for their sensitivity and at the same time for their stability and broad dynamic range, qualities that engineers often find to be anti-

Springer Series in Brain Dynamics 1
Edited by Erol Başar

thetical. Olfactory sensitivity lies in part in the regenerative molecular feedback mechanisms of single cells (Lancet et al. 1985) such that a single odorant molecule may trigger a train of action potentials in a receptor cell. However, sensitivity is also provided, especially in macrosmatic animals, by the immense numbers of receptors. In the cat, for example, there are in the order of 10^8 receptors on each side, a numerosity enhances the likelihood of capture of molecules in turbulent air passed over the turbinate bones. Herein lies a major difficulty in understanding olfaction, which Lashley (1950) identified in vision as the problem of stimulus equivalence. Supposing that there might be in the order of 10–100 types of receptor, then there must be 1–10 million of each type. If an odorant can be identified repeatedly at concentrations ranging over 3–5 orders of magnitude, and if the lowest concentration involves stimulation of 10–100 receptors, how is an invariant constructed by the brain for an odorant over multiple trials, when the spatial pattern of excited receptors is never twice identical? The same type of problem obviously occurs in visual recognition of faces or signatures and auditory recognition of voices or words.

Sensitivity in olfaction is enhanced by experience under reinforcement and is disenhanced without it. Most of us have a limited repertoire of about 16 odorants under absolute discrimination, but the number can be increased without limit by sustained practice (Cain and Engen 1977). We can recognize some odors that were once important to us at intervals over many years in a flash flood of vivid associative memories that impel us to action. These are basic properties that olfaction shares with all other senses, far transcending in importance the decomposition of stimuli into lines, planes, and spectral peaks.

3 Neural Dynamics: Nonlinearity

I propose here that all of these properties inhere in the bulb in a single, comprehensive, nonlinear operation. The main bulbar constituents are large numbers of densely interconnected excitatory neurons (the mitral and tufted cells) and inhibitory neurons (the granule cells). The receptor input (Fig. 1) is spatially coarse-grained into segments corresponding to glomeruli, which form the bulbar equivalent of cortical columns with a mean segment width of about 0.25 mm. There are about 2000 glomeruli in each bulb of the rabbit. The several types of periglomerular interneurons in the outer layers of the bulb perform various janitorial tasks of input dynamic range compression, automatic volume control, spatial contrast enhancement, clipping, holding, and dc offset or bias regulation, among others (Freeman 1975). The negative feedback relation between the mitral and granule cells (Rall and Shepherd 1968) establishes a neural oscillator that receives its input through each glomerulus. These oscillators are coupled by mutually excitatory axosomatic synapses broadly over the bulb (Nicoll 1971) and by mutually inhibitory interactions through cellular mechanisms not yet clearly identified. Their output under coupling is at a frequency in the gamma range of 35–90 Hz, determined in the main by the passive membrane time constants (about 5 ms) and by the gains in the three types of feedback loop. Because of the widespread coupling, the EEGs from all parts of the

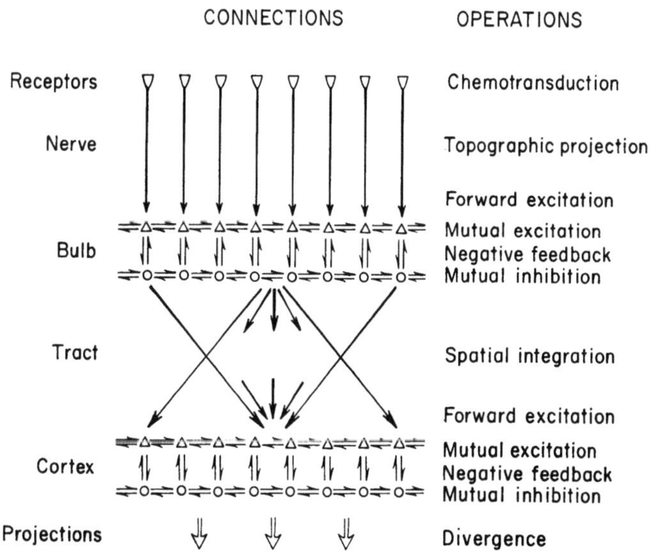

Fig. 1. A flowchart of activity in the olfactory system. Each layer is organized into a sheet of neurons. The state variables are defined for the axonal and dendritic modes in the two surface dimensions. They are discretized at intervals corresponding to the spatial coarse-graining by the glomeruli. Interactions occur laterally in each layer. The primary olfactory nerve provides for topographic projection of the input, whereas the lateral olfactory tract provides for spatial integration of the output. (From Freeman 1983)

main bulb at all times have a common waveform and everywhere a common instantaneous frequency (Freeman 1986).

These oscillators are inherently nonlinear. The nonlinearity stems from the voltage-dependent nonlinearity modeled for the action potential of nerve membrane by the Hodgkin-Huxley equations (Freeman 1979a). In the neural ensemble, it emerges as a sigmoidal function (Fig. 2) that relates pulse density (pulses per second per unit volume of the ensemble) to the density of excitatory dendritic current at the trigger zones. The curve is asymptotic to zero pulse density with inhibitory postsynaptic potential (IPSP) current and to a maximum for the ensemble with excitatory postsynaptic potential (EPSP) current. Two processes combine to give this shape. One is the exponential increase in tendency to fire with increasing depolarization (the sodium permeability or m-factor in the Hodgkin-Huxley equations). The second is the collection of metabolic, restorative, accommodative, and hyperpolarizing processes that establish the upper limit on firing rate, both on the long-term firing of single neurons and, by the ergodic hypothesis, on the entire ensemble over the short term. The nonlinearity is static, as distinct from the time-varying linear relationship that holds between membrane potential and firing rate for regularly firing single neurons. This is because neurons spend 99% of their lifespan below threshold, and because the firing pattern of each neuron closely resembles a Poisson process unrelated to those of its neighbors.

The nonlinear function is determined experimentally by calculating the pulse probability of mitral cell firing conditional on the EEG amplitude. The calculation is repeated for each EEG sample at 1 ms digitized intervals forward and backward in

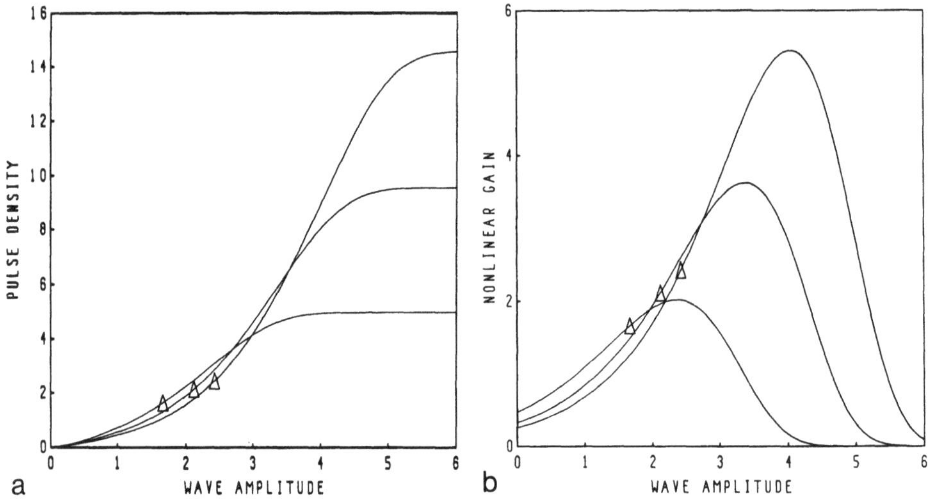

Fig. 2. (a) Three examples of a curve fitted to statistical data showing conversion of dendritic current density to axonal pulse density. (b) Derivatives of the three curves that give the nonlinear gain. *Triangles,* resting or equilibrium values. With increasing current amplitude there is a coupled increase in pulse density and in gain. (From Freeman 1979a)

time ± 25 ms, in order to allow for the time lags in the neural oscillator. The procedure also serves to demonstrate that the firing probability of each mitral cell oscillates at the common EEG frequency, and that the modulation amplitude in firing rate co-varies with the peak-to-peak amplitude of EEG oscillation. Mitral cell firing is statistically closedly related to the EEG at all times and at each point of the bulb.

The nonlinear function for each bulbar ensemble is under centrifugal control. The shape of the sigmoid curve is retained, but the steepness is subject to increase, along with an increase both in mean and maximal firing rates. The derivative of the function represents the nonlinear gain of each local ensemble. The maximal gain is always displaced to the excitatory side. In animals under increased arousal or motivation, the gain is increased and the displacement to the excitatory side is extended, along with the increase in mean firing rate. The centrifugal input is most likely the cholinergic projection to the outer layers of the bulb. On the peripheral side, any receptor input excites the bulb and thereby raises its mean firing rate and its instantaneous gain. The curve is fixed but the operating point changes. Owing to the surge of receptor input with each inhalation, the bulb tends to undergo a recurrent increase and decrease in gain with the respiratory cycle.

Because of the bilateral saturation, the sigmoid curve is the most important mechanism providing for the stability of the bulbar mechanism (Freeman 1979b). The same curve also provides for its remarkable sensitivity, in the main because of the mutually excitatory feedback loop. Excitation of one subset excites another which re-excites the first, now in a more sensitive state, so that a regenerative increase in activity can occur. However, the negative feedback gain is also increased, so that instead of runaway excitation, a burst of oscillation appears. It begins during inhalation and ends during exhalation, and it is seen only in aroused, motivated animals (except occasionally in light stages of anesthesia, and then in an abnormal frequency range).

4 Neural Dynamics: Spatial Properties

Studies of the spatial patterns of these bursts manifested in the EEG have been made in rabbits with arrays of 64 electrodes chronically implanted over the lateral surface of the bulb. The EEG shows no dependence on novel odorants presented to naive animals, other than nonspecific changes associated with orienting responses. The spatial patterns of amplitude and phase modulation of the burst frequency vary within narrow limits about stereotypic mean patterns that are as characteristic for each individual as a handwritten signature. Under classical conditioning to respond differentially to two ordors (Viana di Prisco and Freeman 1985), one reinforced [conditioned stimulus (CS) +] and the other not (CS −), two new spatial patterns of amplitude emerge (Fig. 3), one for each CS. They are present only when the one or the other CS is present (Freeman 1986). For this demonstration, the EEG must be filtered with digital filters designed to conform to the spatial and temporal passbands

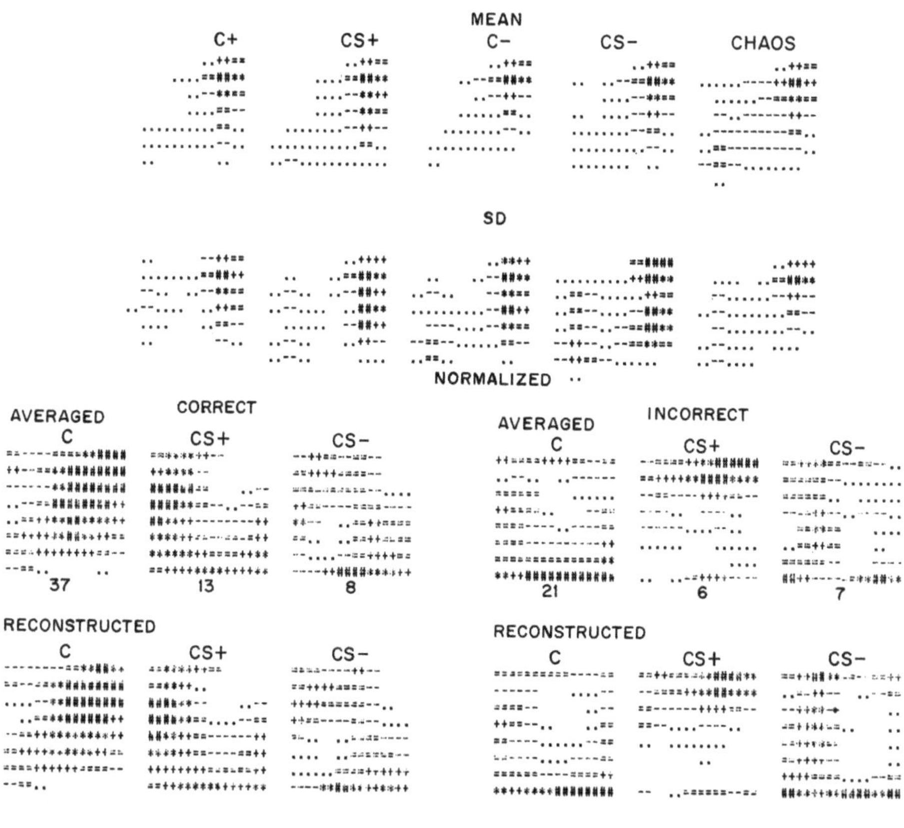

Fig. 3. Density plots (seven levels in descending order of amplitude #* + = − .) of EEG activity. *Upper frames,* means and SDs of amplitudes (*Chaos* refers to the disorderly bursts not subject to classification in respect to odors). *Lower frame,* amplitudes normalized by channel and by group, with those correctly classified on the *left* and those incorrectly classified by discriminant analysis on the *right. Bottom row,* patterns reconstructed from factor scores and loadings that were used for classification. (From Freeman 1986)

of the granule cell contribution to the EEG (Freeman 1986). The resultant patterns serve together with discriminant analysis to classify correctly, on average, 82% of EEG bursts sampled during control and test odor periods (Freeman 1986; Freeman and Viana di Prisco 1986). These patterns cover the entire array and, by inference from surface EEG phase gradients (Freeman 1986) and depth recording (Bressler 1984), the entire main bulb. The information density over the bulb is spatially uniform to within ± 5% (SD) of its mean, as measured by its value for correct classification of bursts.

The results show that insofar as the EEG is concerned, the bulb has the capability of responding selectively to odorants, but only in aroused animals that are trained to detect and respond to the test odors. This is in striking contrast to the results from unit studies in anesthetized or immobilized animals, which show selective responding of single neurons to some odors and not others, irrespective of training (e.g., Moulton 1976). Studies of metabolic activity with 2-deoxyglucose show that different patterns of radiographic density in the glomerular layer result from presentation of different odors (e.g., Lancet et al. 1982). These studies still lack proper controls for individual variation. The method allows only one odor for each animal; the EEG method shows foci of high amplitude activity that are similar to the high-density metabolic foci in size, shape, and location, but the degree of variation in EEG pattern between individuals exceeds that between odorants for each individual. Still, it is reasonable to conclude that input to the bulb from receptors establishes local regions of activity specific to an odor, and the output of the bulb is a global pattern involving all bulbar neurons, provided that the animal has been trained. Otherwise the global bulbar response is not spatially or temporally coherent or reproducible.

This transformation of local input to global output that incorporates past experience is the key to bulbar function. It is best understood by description in terms of nonlinear dynamics (Garfinkel 1983). A set of distributed, coupled, nonlinear oscillators has an infinite number of ways of performing, but within certain conditions of input and interaction strengths it tends to enter a definable state of activity and stay there until perturbed or modified. If under repeated perturbation it tends always to return to the same state, the system dynamics is said to have, or be governed by, an attractor. Attractors fall into three classes. The simplest is that of equilibrium; this occurs in the bulb only under deep anesthesia or in death. Periodic oscillation characterizes the limit cycle attractor; this appears in the EEG during bursts with inhalation. The most complex type is called the strange or chaotic attractor; its manifestation is nonperiodic activity that may appear to be random, of the sort that characterizes the resting EEG in nonmotivated animals and also the low-level EEG activity during exhalation.

Switching from one attractor to another is called a "state change" or "bifurcation". Its occurrence requires a parametric change in the system. Bulbar input provides for this by virtue of the nonlinear gain increase with receptor input during inhalation. The state change is from low-amplitude chaos to a high-amplitude spatially coherent limit cycle, and then back again. Order emerges from chaos and collapses with each cycle of respiration. There may be indefinitely many attractors of each type. Each is characterized by a set of parameter values and by a basin defined by a domain of input. The evidence suggests that a limit cycle attractor may form for each odorant that an animal is trained to respond to.

I believe that a limit cycle attractor is formed in the following way. On each inhalation of an odorant, the subset of the receptors that is sensitive to the odorant coactivates a subset of mitral cells. These are interconnected by excitatory axosomatic synapses that are bidirectional (Willey 1973). In accordance with Hebb's rule (Hebb 1949; Viana di Prisco 1984), these synapses are strengthened under coexcitation, provided that a reinforcing stimulus is paired with the odorant. Reinforcement activates neurons in the locus coeruleus, thus releasing into the bulb (and elsewhere) norepinephrine that enables the synaptic change (Gray et al. 1984). With repeated inhalations in the same and sequential trials, the odorant is delivered by turbulent flow in the nose to an ever-changing fraction of the subset of sensitive receptors, which leads progressively to the ultimate inclusion of all those mitral cells to which they project into a nerve cell assembly. These strengthened, mutually excitatory connections give the property to the assembly that, if any fraction of the sensitive receptors receives the odorant, their input to the bulb excites the entire assembly in a stereotypic manner (Freeman 1979c).

At once this constitutes figure completion, generalization over equivalent stimuli, and sensitization specific to a repeatedly reinforced class of stimulus. Computer simulations (Freeman 1979b) have shown that an increase of 40% on average in synaptic strength may increase the sensitivity of the bulb to a particular odorant by as much as 40000 times above the basal or naive level, because of the combination of mutual excitation and the nonlinear gain. After the completion of training, the subset of receptors activated during the training defines the basin of the attractor, and the nerve cell assembly of mitral cells determines the spatial structure of the limit cycle oscillation, which extends well beyond the assembly to involve the entire bulb. In principle, we can show how one odorant molecule can shape the activity of several hundred thousand second-order neurons.

I conceive the bulb as carrying a repertoire of learned limit cycle attractors, one for each odorant previously reinforced. Each is distinguished by its input basin with respect to receptors and by the spatial amplitude modulation pattern of its output. Random access is facilitated by the chaotic basal state, which keeps the bulb far from equilibrium and ready to move rapidly to any region of optimal convergence. The steadfast spatial pattern of bursts in the control state, in which no reinforced odor is given, indicates that an attractor exists for the background odor complex as well, and that bulbar output then signals the status quo. If a novel odor is given, the result is suppression of orderly burst activity and the appearance of broad-spectrum, spatially irregular, and nonreproducible bursts. Commonly, the highest peak of their multiply peaked spectra is at a frequency about half that of the sharply tuned frequency of the orderly bursts. I call these bursts "disorderly" or "chaotic". The prepyriform cortex to which the bulb projects responds to input as a tuned oscillator with spectral resonances around 18–24 Hz and 40–70 Hz (Freeman 1975). This suggests that the lower transmission frequency of the chaotic bursts can signal the failure of the bulbar mechanism to converge to a limit cycle attractor, and that repeated failures can lead to either of two outcomes: habituation if there is no reinforcement which updates sensitivity to a new status quo, or formation of a new limit cycle attractor under reinforcement. In other words, the bulbar mechanism provides a novelty detector without requiring an exhaustive search through information stored in the bulb.

5 Neural State Variables and Observables

Although the bulb has numerous specialized features not found elsewhere in the brain, these are not responsible for its main properties. At base it consists of a sheet of interconnected excitatory and inhibitory neurons with parallel input and output. This is an elementary description of neocortex as well. The static nonlinearity is a generalizable property of axonal membrane to be expected for every large ensemble in the cerebral cortex. The time and space constants are common to many, if not most, cerebral neurons. Hence the same basic dynamics can be expected to exist in all parts of the cerebral cortex.

I infer that odorant information is conveyed to the bulb by action potentials on particular receptor axons and that excitation is established and integrated among local subsets of mitral cells having apical dendrites within a limited number of glomeruli that correspond to neocortical columns. Following bifurcation, the entire bulb, comprising roughly 1 cm^2 of cortical tissue, goes to a limit cycle attractor in the basin selected by the input. The output is global; the information is conveyed by action potentials on mitral axons, but it is in the form of a macroscopic pulse density function that is continuous in time and the two surface dimensions. The information is imposed as spatial amplitude modulation (in the surface dimensions as distinct from the time envelope) of the limit cycle carrier oscillation that is common to the entire bulb. Each event lasts in the order of 75–100 ms and repeats at the respiratory rate of 1–7 Hz. At the macroscopic level, each event can again be discretized into the surface grain of the glomeruli and the time frame of the burst; that is, olfaction can be treated as a sampled data system analogous to a digital graphic display.

The intrinsic state variables of a model for this system must correspond to the active states of pools of like neurons, which Sherrington identified as their central excitatory states (CES). For this reason, the proposed view might be described as neo-Sherringtonian. These activities are conveyed in local concentrations of action potentials, transmitter substances, and dendritic currents. They are manifested to observers in the forms of unit activity and electromagnetic field potentials. In all instances, the measurements of these observables must be properly filtered, averaged, and otherwise transformed in order to bring them into conformance with the CES, and they must be assigned to the proper elements in the model; for example, in the bulb, the EEG should be assigned to the granule cells and unit activity at the appropriate depth should be assigned to mitral cells.

The parallels to other sensory systems are straightforward. Information is conveyed by action potentials on thalamocortical axons and is established in local regions corresponding to columns, with different kinds of information being established at the microscopic level in each of the multiple cortical subareas comprising a sensory projection area. The neurons onto which the afferent activity is projected consist of excitatory and inhibitory neurons that are known to be densely interconnected by negative feedback and mutual excitation and can be inferred to have mutual inhibition as well. The crucial step for integration in perception may be the bifurcation of the interactive neural mass from a low-level chaotic attractor to a learned limit cycle attractor, such that the output of an extensive area of cortex at the macroscopic level might convey information on the whole in the spatial modulation of the amplitude of the limit cycle frequency. Again, input is local, output is global, and in analogy to the hologram, all parts of the output reflect all parts of the input.

Investigation of this hypothesis is likewise straightforward. The requisite carrier and gating frequencies respectively in the high beta and gamma ranges and in the theta and alpha ranges have been observed in most areas of neocortex. In visual cortex, during alpha suppression, the sequence of bifurcations requisite for trains of bursts might be provided by saccades. The steps that are needed to test the hypothesis are (1) the detailed spectral characterization of these activities, including use of complex demodulation over extended time series of the EEG; (2) the identification of the sources and sinks of the electric currents underlying these spectral peaks; (3) the assignment of these activities as states of variables of identified types of neurons in the cortex (4) measurement of the open loop time and space constants under deep anesthesia (Freeman 1975); (5) establishment of the spectral and spatial domains of neocortex over which commonality of wave form holds, such that chaotic or limit cycle attractors can be sought; and (6) behavioral analysis to determine the dimensions of the activity that relate to the stimulus and response variables selected for testing. Some progress has already been made in relating information content of visual and auditory stimuli to the waveforms of event-related potentials from neocortex. According to the present hypothesis, these correlations are adventitious and secondary, because the information relating to content is to be sought in the spatial dimensions, while the time courses of events are expected to reflect primarily the neural operations being performed on that information (Freeman and Schneider 1982).

None of these six steps is trivial; each may require several years to be brought to fruition. The outcome will be exceedingly important, because these kinds of information are essential to devise, evaluate, and improve macroscopic models of the distributed nonlinear dynamics of the forebrain.

In conclusion, the esence of cognition lies in forming and testing expectations based on past experience. In science it takes the form: if I do X, I expect A or B or the unexpected. Each outcome has predictable consequences. In rabbit olfaction it takes the form: if inhalation, then either status quo (the background), odor A, odor B, or an unexpected odor. Each inhalation is the action of a pattern generator or limit cycle attractor in the brain stem respiratory nuclei; each neural response is mediated by limit cycle attractors in the bulbs. I postulate that licking and sniffing are likewise mediated by limit cycle attractors in motor systems, whose basins receive the output of the bulbs. Basically this is a simple model of simple conditioned reflexes, but it tells us what to look for and how to look for it, as we try to understand how the brain synthesizes a percept from diverse sensory detail in the literal twinkling of an eye or the wriggle of a nose.

Summary

Neurons in cerebral cortex interact synaptically by mutual excitation, mutual inhibition, and negative feedback. Typically the negative feedback connections are locally dense, leading to the formation of local oscillators corresponding to columns. They are interconnected by mutually excitatory connections over large cortical areas. An appropriate model of cortex is a sheet of distributed coupled oscillators; observation is performed with arrays of surface EEG electrodes.

The dynamics of such systems are shaped by tendencies under perturbation to converge to stable states that are identified with attractors of three kinds. An *equilibrium* attractor is manifested in cortex by a steady state under deep anesthesia; a *limit-cycle* attractor is manifested by regular oscillation, and a *strange* attractor is manifested by chaos that appears to be random activity. Transition (bifurcation) from one attractor to another is imposed by a parametric change of the model or cortex.

The EEG of the olfactory bulb at rest appears chaotic. Inhalation excites the bulb, causing a parametric increase in negative feedback gain; the bulb bifurcates to a limit-cycle state. In the control condition of breathing air, the EEG spatial pattern is stereotypic for the background odor complex. With training to discriminate odors, a new spatial pattern appears with each odor, manifesting a learned limit-cycle attractor. These patterns appear to cover the entire bulb; input is local and output is global. The integration of a stimulus with past experience takes less than 0.1 s. Other sensory systems have similar properties; therefore bulbar dynamics may provide a useful model to explore preattentive processing in vision and other cognitive operations in the neocortex.

Acknowledgement. This work was supported by a grant MH06686 from the National Institute of Mental Health.

References

Bressler SL (1984) Spatial organization of EEGs from olfactory bulb and cortex. Electroencephalogr Clin Neurophysiol 57:270–276

Cain WS, Engen T (1977) Olfactory adaptation and the scaling of olfactory intensity. In: Pfaffman C (ed) Olfaction and taste III, Rockefeller University press, New York, pp 127–141

Efron R (1970) The minimum duration of a perception. Neuropsychologia 8:57–63

Freeman WJ (1975) Mass action in the nervous system. Academic, New York

Freeman WJ (1979a) Nonlinear gain mediating cortical stimulus-response relations. Biol Cybern 33:237–247

Freeman WJ (1979b) Nonlinear dynamics of paleocortex manifested in the olfactory EEG. Biol Cybern 35:21–37

Freeman WJ (1979c) EEG analysis gives model of neuronal template-matching mechanism for sensory search with olfactory bulb. Biol Cybern 35:221–234

Freeman WJ (1983) Dynamics of image formation by nerve cell assemblies. In: Başar E, Flohr H, Haken H, Mandell AJ (eds) Synergetics of the brain. Springer, Berlin Heidelberg New York, pp 102–121

Freeman WJ (1986) Analytic techniques used in the search for the physiological basis of the EEG. In: Gevins A, Remond A (eds) Methods of analysis of brain electrical and magnetic signals. Elsevier, Amsterdam (Handbook of encephalography and clinical neurophysiology, vol 3A/2)

Freeman WJ, Schneider WS (1982) Changes in spatial patterns of rabbit olfactory EEG with conditioning to odors. Psychophysiology 19:44–56

Freeman WJ, Viana di Prisco G (1986) EEG spatial pattern differences with discriminated odors manifest chaotic and limit cycle attractors in olfactory bulb of rabbits. Proceedings, conference on brain theory, Trieste 1984. Springer, Berlin Heidelberg New York Tokyo

Garfinkel A (1983) A mathematics for physiology. Am J Physiol 245:R455–R466

Gray CM, Freeman WJ, Skinner JE (1984) Associative changes in the spatial amplitude patterns of rabbit of olfactory EEG are norepinephrine dependent. Neurosci Abstr 10:121
Hebb DO (1949) The organization of behavior. Wiley, New York
Herrick CJ (1948) The brain of the tiger salamander. University of Chicago Press, Chicago
Lancet D, Greer CA, Kauer JS, Shepherd GM (1982) Mapping of odor-related neuronal activity in the olfactory bulb by high-resolution 2-deoxyglucose autoradiogrpahy. Proc Natl Acad Sci USA 79:670–674
Lancet D, Heldman J, Chen Z, Pace U (1985) Odorant-sensitive adenylate cyclase in olfactory cilia. Am Chem Soc Abstr 7
Lashley KS (1950) In search of the engram. Symp Soc Exp Biol 4:454–482
Moulton DG (1976) Spatial patterning of response to odors in the peripheral olfactory system. Physiol Rev 56:578–593
Nicoll RA (1971) Recurrent excitation of secondary olfactory neurons: a possible mechanism for signal amplification. Science 171:824–825
Rall W, Shepherd GM (1968) Theoretical reconstruction of field potentials and dendrodendritic synaptic interactions in olfactory bulb. J Neurophysiol 31:884–915
Viana di Prisco G (1984) Hebb synaptic plasticity. Prog Neurobiol 22:89–102
Viana di Prisco G, Freeman WJ (1985) Odor-related bulbar EEG spatial pattern analysis during appetitive conditioning in rabbits. Behav Neurosci 99:964–978
Willey TJ (1973) The ultrastructure of the cat olfactory bulb. J Comp Neurol 152:211–232

EEG – Dynamics and Evoked Potentials in Sensory and Cognitive Processing by the Brain

E. Başar

1 Preliminary Remarks

One of the main concerns of brain research is to measure the brain's electrical activity and, in this way, to try to detect the coding of behaviorally relevant information in the CNS. It is usually assumed that there is no uniform code for behaviorally relevant information in the neuronal networks that constitute the CNS. There are also no standard methods for clearly describing the functional and behavioral components of the brain's electrical activity. Analyses of the EEG, of evoked potentials (EPs), and of endogeneous potentials (P300 family) are among the most fundamental research tools for understanding the sensory and cognitive information processing in the brain. Since Berger's discovery of the EEG and Adrian's measuring of cortical field potentials, these powerful techniques have been adequately described in several outstanding books (Berger 1938; Freeman 1975; Niedermeyer and Lopes da Silva 1982). The ensemble of reports in this volume shows the broad extent of applications of the EEG, of sensory EPs and event-related potentials (ERPs), and of contingent negative variation (CNV) to the understanding of CNS information processing and of behavior.

The main goal of this report is to elucidate associations between the EEG, EPs, and the cognitive components of ERPs, and to explore the extent to which the combined analysis of these tools might augment our knowledge about sensory and cognitive information processing in the brain. Another important goal of this study is to show that *invariant modes* may exist that could possibly facilitate the exchange of information between various brain structures.

In our earlier publications, we analyzed single epochs of EEG-EP sets in the frequency domain and concluded that ERPs and EPs of the brain have their source in generators that are at least partially in common with those that also generate the EEG. We further assumed that sensory excitation of EEG-generating networks gives rise to enhanced and time-locked EEG fragments that constitute the major components of sensory potentials and ERPs of the brain (Başar 1980). In addition to the core methodology that we previously applied for understanding associations and interactions between the EEG and EPs, we present in this study a chain of paradigms – a battery of new tools – in order to understand EEG dynamics better.

The strategy for interpreting the field potentials of the brain should, in turn, help us to understand various strategies of the brain itself. One of the most fundamental questions that arises is the following: has the brain, independently of its various special functions, some global strategies with the help of which the internal communication and coordination between various neuronal networks is optimized?

Springer Series in Brain Dynamics 1
Edited by Erol Başar

At this point it is most pertinent to quote Fessard (1961), who tried to emphasize the role of neuronal networks in the brain:

> The brain, even when studied from the restricted point of view of sensory communications, must not be considered simply as a juxtaposition of private lines, leading to a mosaic of independent cortical territories, one for each sense modality, with internal subdivisions corresponding to topical differentiations. The track of a single-unit message is doomed to be rapidly lost when one tries to follow it through a neuronal field endowed with network properties, within which the elementary message readily interacts with many others. Unfortunately, we still lack principles that would help us describe and master such operations in which heterosensory communications are involved. These principles may gradually emerge in the future from an extensive use of multiple microelectrode recordings, together with a systematic treatment of data by modern electronic computers, so that pattern-to-pattern transformation matrices can be established and possibly generalized. For the time being, it seems that we should do better to try to clear up such principles as seem to govern *the most general transformations* – or *transfer functions* – of multiunit homogeneous messages during their progression through neuronal networks.

Although new techniques using a combination of multiple-unit electrodes and powerful computers are already emerging for exploration of the cerebral cortex (see, for example, Eckhorn and Reitboeck in this volume), it is still difficult to describe the most general transfer functions of sensory communications in the brain on the basis of single-unit recordings. Therefore, the aim of the investigation presented in this study has been to describe the general transfer function in the brain by analyzing global features of the brain, an analysis in which the field potentials are explored by using an ensemble of multiple neuroelectrodes in various substructures of the brain. Since it has been shown that the EEG is not a simple noise (see Babloyantz, and Röschke and Başar in this volume), the use of EEG and EPs to describe the general transfer functions is quite legitimate.

The question, "what are the neuronal correlates of the EEG and of EPs?" has been treated by several authors (see, for example, Creutzfeld et al. 1966, 1969; Verzeano 1973; Ramos et al. 1976; Freeman 1975, and for reviews see Başar 1980, 1983 a, b). Every model that tries to describe the EEG and field potentials offers a new window on the problem, as discussed elegantly by Bullock (this volume).

In our earlier work, we repeatedly argued that in order to understand sensory EPs, one has to analyze the EEG immediately prior to the sensory stimulation. Therefore we always analyze EPs along with the portion of the EEG that immediately precedes the sensory stimulation. Accordingly, an analysis in the frequency domain has been shown to be adequate for comparing EEG and EPs.

In order to search for any common characteristics that might be contained in sensory EPs, it is adequate to measure EPs to various sensory stimuli, such as photic, auditory, and somatosensory stimuli. Furthermore, EPs as well as the EEG from human brains and brains of cat, other vertebrates, and also invertebrates, can all be compared by searching for common brain strategies, as Bullock (1984) has emphasized for years.

2 Methods

2.1 Mathematical Methods

2.1.1 Combined Analysis Procedure. Frequency Domain Comparison of EEG and EP

Our methodology for the frequency domain analyses of spontaneous activity and EPs can be briefly described as follows:

1. A sample of the spontaneous activity of the studies brain structure just prior to the stimulus is digitized and stored in the core memory of the computer (for the experimental setup see Başar et al. 1975a).
2. A stimulus signal is applied to the subject.
3. The single EP following the stimulation is also digitized in the prememory. The EEG just prior to stimulation and the resulting EP are stored together as a combined record on the digital magnetic tape controlled by the computer (the so-called EEG-EPogram).
4. The first three steps explained above are repeated about 100 times, depending on the nature of the experiment.
5. The power spectrum density function is obtained from the epoch of the spontaneous activity (EEG) recorded prior to the stimulation. A method proposed by Bingham et al. (1967) is used to estimate power spectra. This method consists of the following steps: (a) the mean of the measured values is substracted prior to analysis; (b) a data window of the following form is applied to each datum over a period t; (c) windowed data is Fourier-transformed using a fast Fourier transform (FFT) algorithm and a Fourier periodogram of this modified time series is obtained; (d) a smoothed power spectrum is obtained by filtering the periodogram itself to improve the statistical stability of the power estimate.
6. The single EP of the same epoch is transformed to the frequency domain with the Fourier transform in order to obtain the instantaneous frequency characteristic (which describes the response to a single stimulus), $G(j\omega)$, of the studied brain structure:

   $$G(j\omega) = \int_0^\infty \frac{d\{c(t)\}}{\mathrm{d}t} \exp(-j\omega t)\,\mathrm{d}t$$

   $c(t)$ is the step response of the system, here, the sensory EP or ERP. Details of this method, which we called the transient response – frequency characteristic method (TRFC), are given in references (Başar 1980; Başar et al. 1975a).

2.1.2 Time Domain Comparison of the EEG and EPs by Means of Digital Filtering

The methodology for comparing the brain's spontaneous activity and EPs can be briefly described as follows:

1. A sample of the spontaneous activity of the studied brain structure just prior to stimulus is recorded and stored in the disk memory of the computer.
2. A stimulation signal is applied to the experimental animal (or human subject). This signal may be a light flash or an acoustical stimulation; for example, an auditory step function in the form of a tone burst of 2000 Hz and 80 dB.
3. The single evoked response following the stimulation is also stored in the disk memory. (The EEG just prior to the stimulation and the resulting EP are stored together as a combined record.)
4. The operations explained in the three steps above are repeated about 100 times. (The number of trials depends on the nature of the experiment and the behavior of the subject or the experimental animal.)
5. The EPs stored in the disc memory of the computer are averaged using a selective averaging method described previously (Başar 1980; Başar et al. 1975a; Ungan and Başar 1976).
6. The selectively averaged EP (SAEP) is transformed to the frequency domain with the Fourier transform in order to obtain the amplitude frequency characteristic $[G(j\omega)]$ of the studied brain structure (see also step 6 under Combined Analysis Procedure).
7. The frequency band limits of the amplitude maxima in $G(j\omega)$ are determined and digital pass band filters are determined according to these band limits.
8. The stored and selected epochs of EEP-EP sets (EEG-EPogram) are filtered with the properly chosen filters described in step 7.
9. The voltage of the root mean square (RMS) values of maximal amplitudes existing in the filtered EPs, and the so-called enhancement factor for the given EEG-EPogram, are evaluated.

Definition of the "Enhancement Factor X". In a given experimental record of EEP-EP, the enhancement factor (X) is the ratio of the maximal time-locked response amplitude to the $(2 \cdot \sqrt{2}$ RMS) value of the spontaneous activity just prior to the stimulus, with both signals (spontaneous and evoked activities) being filtered within the same band limits (Fig. 1).

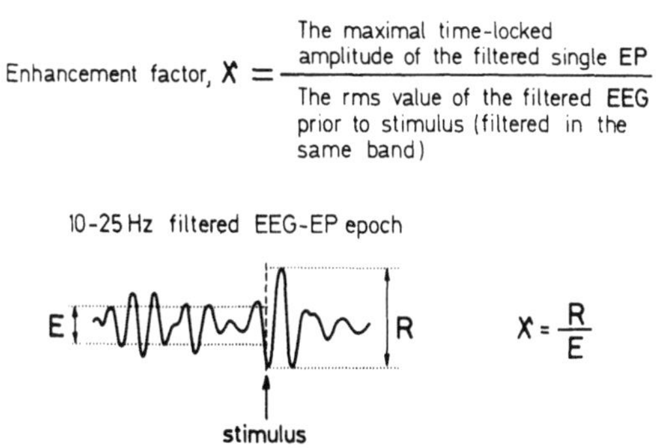

Fig. 1. Definition of the enhancement factor X on a sample component of EEG-EPogram. As shown, the peak-to-peak value of the evoked response is compared with the peak-to-peak value of a sinusoidal signal having the same RMS value as the spontaneous EEG preceding the stimulus onset

2.2 Experimental Method

The measurement of brain potentials to be described in the following sections of this chapter was performed using chronically implanted cats. The recording electrodes were located in various nuclei of the auditory pathway, such as those in the acoustical cortex (gyrus ectosylvian anterior – GEA); medial geniculate nucleus (MG); inferior colliculus (IC); and those in such centers as the mesencephalic reticular formation (RF) and hippocampus (HI), both of which are indirectly related to the auditory pathway. The electrodes were placed in these nuclei according to the stereotaxic atlas by Snider and Niemer (1964) with the following coordinates: MG (Fr.A: 3.5, L: 9, H: 1.5), IC (Fr.P: 2.5, L: 5, H: 3.5), mesencephalic RF (Fr.A: 3, L: 4, H: −1), dorsal HI (Fr.A: 3.5, L: 6.2, H: 8.8), and the GEA. The stainless steel electrodes were 0.2 mm in diameter. The derivations were against a common reference, which consisted of three stainless steel screws in different regions of the skull. A David Kopf 1404 Instrument was used for stereotaxic surgery. During the experiments the cats were moving freely, resting, or sleeping in an echo-free and soundproof room.

3 Application of Strategies

3.1 Experiments with Chronically Implanted and Freely Moving Cats

Figure 2 shows simultaneously recorded typical EPs of five functionally important structures of the cat brain. The stimuli consisted of auditory signals in the form of tone bursts of 2000 Hz and 80 dB. During the experimental session, the cats could move freely in a soundproof and echo-free room. The cats were observed by means of a video camera. The movement artefacts, if any, could be eliminated by using selective averaging. The averaged EPs of Fig. 2 have been transformed to the frequency domain by means of the Fourier transform in order to obtain amplitude frequency characteristics which are presented in Fig. 3 (The Fourier transform and the TRFC method are explained under Mathematical Methods).

At first glance, one sees a dominant maximum (or selectivity) in the frequency range of 10 Hz (the alpha frequency range). Other important selectivities are in the 3–7 Hz range (in most of the structures), 40 Hz (in the cortex and HI), 70 Hz (in the IC) and 60 Hz (in the RF).

Definition of "Selectivity". We define "selectivity" as the ability of brain networks to facilitate (or activate) electrical transmission within determined frequency channels when stimulation signals are applied to the brain. In our earlier studies, the selectivities of 4 Hz, 10 Hz, and 40 Hz have been called „common selectivities", because the selectivities in these frequency channels are common to various brain structures (Başar 1980).

Figure 4 shows mean value curves for about 20 experiments with nine cats. The selectivities in the frequency range between 3 and 8 Hz, which were usually obtained in all of the amplitude characteristics, are greatly reduced or have disappeared, whereas selectivities around the 10–15 Hz range are highly dominant in spite of the mean value evaluation. We call these maxima, which almost always exist in the mean amplitude frequency characteristics, the consistent selectivities. In other words, the

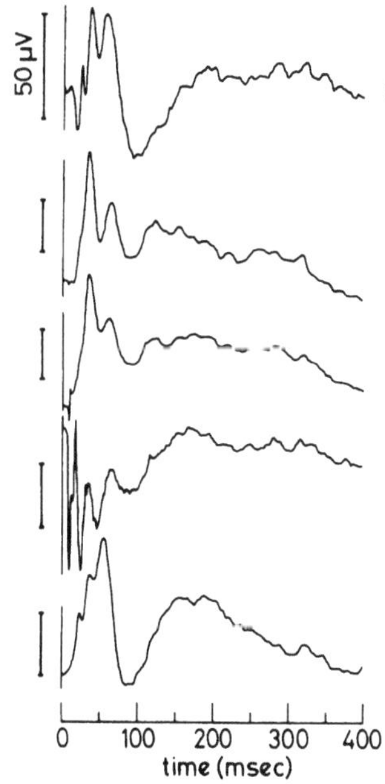

Fig. 2

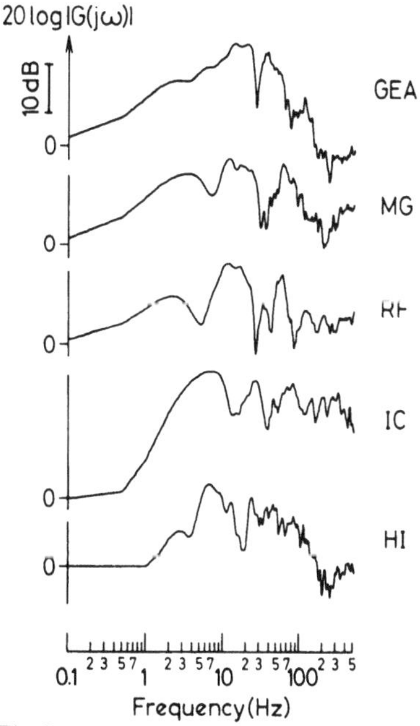

Fig. 3

Fig. 2. Two typical sets of simultaneously recorded and selectively averaged EPs in different brain nuclei of chronically implanted cats, elicited during the waking stage by auditory stimuli in the form of a step function. Direct computer plots. Negativity upwards. (From Başar et al. 1979a)

Fig. 3. Simultaneously obtained amplitude characteristics of different brain nuclei of the cat determined by Fourier transform of averaged EPs elicited by 2000-Hz and 80 dB acoustical stimuli. The *abscissa* shows the input frequency in a logarithmic scale, the *ordinate* shows the potential amplitude, $|G(j\omega)|$, in decibels. (From Başar et al. 1979a)

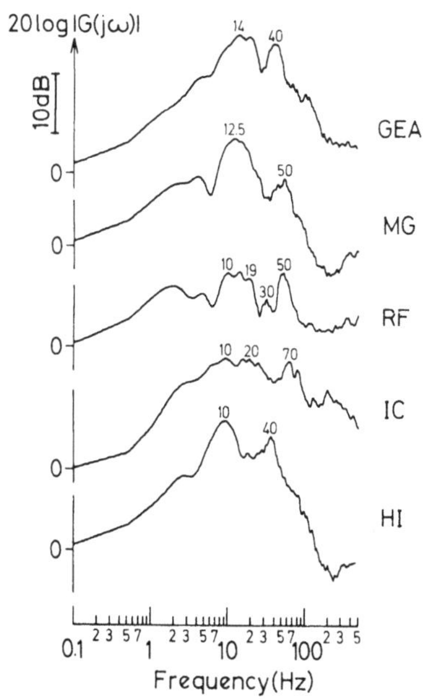

Fig. 4. Mean value curves of simultaneous amplitude characteristics of different brain nuclei obtained from 19 experiments on nine cats given auditory stimuli while awake. Direct computer plots. The *abscissa* shows the input frequency in a logarithmic scale, the *ordinate* the mean value potential amplitude, $|G(j\omega)|$, in decibels. The curves are normalized in such a way that the amplitude at 0 Hz is equal to 1 (or $20 \log 1 = 0$)

probability for maximal signal transfer through the consistent selective frequency channels is high in experiments during the waking state.

However, when one studies amplitude characteristics evaluated from single trial EPs, one can often see differentiation and/or competition between 10 and 20 Hz in a great number of single curves. The frequency characteristics of single EPs have been published elsewhere (Başar et al. 1979a, b; Başar 1980).

3.2 Comparison of EEG and EPs in the Frequency Domain

Independently of the modality of stimulation and of the brain structures studied, we measured almost invariant selectivities in 4 Hz, 10 Hz, and 40 Hz frequency positions in channels where the EEG or changes of the spontaneous activity were also usually recorded. This fact led us to compare ongoing and evoked activities of various brain structures step by step during changes of spontaneous activity. In order to study this relationship, we used the combined analysis procedure in the frequency domain as already described under Combined Analysis Procedure.

In Fig. 5, the prestimulus EEG and poststimulus EP epochs are compared in the frequency domain. Figure 5 presents a typical set of power spectral density functions of the prestimulus EEG immediately prior to stimulation and poststimulus EP epoch following an acoustical stimulation of the cat RF. The power spectra of the prestimulus EEG and the power spectra of the EP have a similar shape at first glance, with peaks in the same frequency channels. However, the strength of the EP power

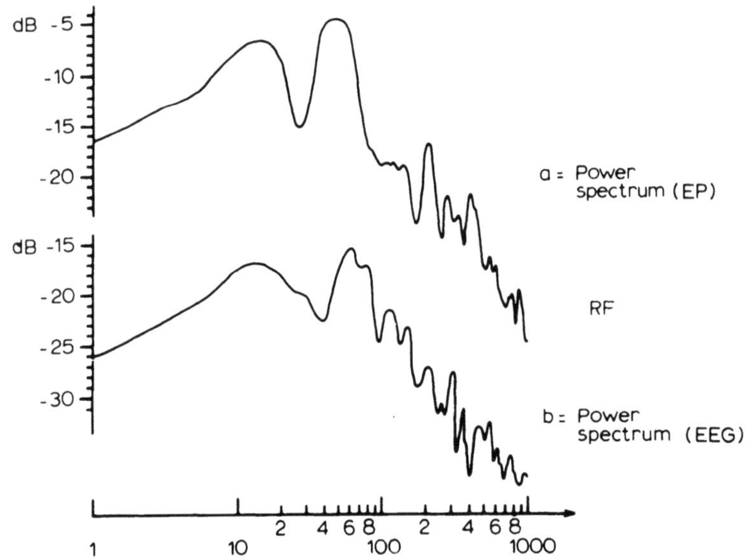

Fig. 5. Two typical sets of power spectral density functions of the prestimulus EEG and poststimulus EP epochs of the reticular formation *(RF)*. The *abscissa* gives the frequency in logarithmic scale, the *ordinate* the power spectral density in decibels. The *curves* are drawn with their absolute magnitudes in order to allow direct comparison between the prestimulus and poststimulus spectral peaks. (From Başar et al. 1979a)

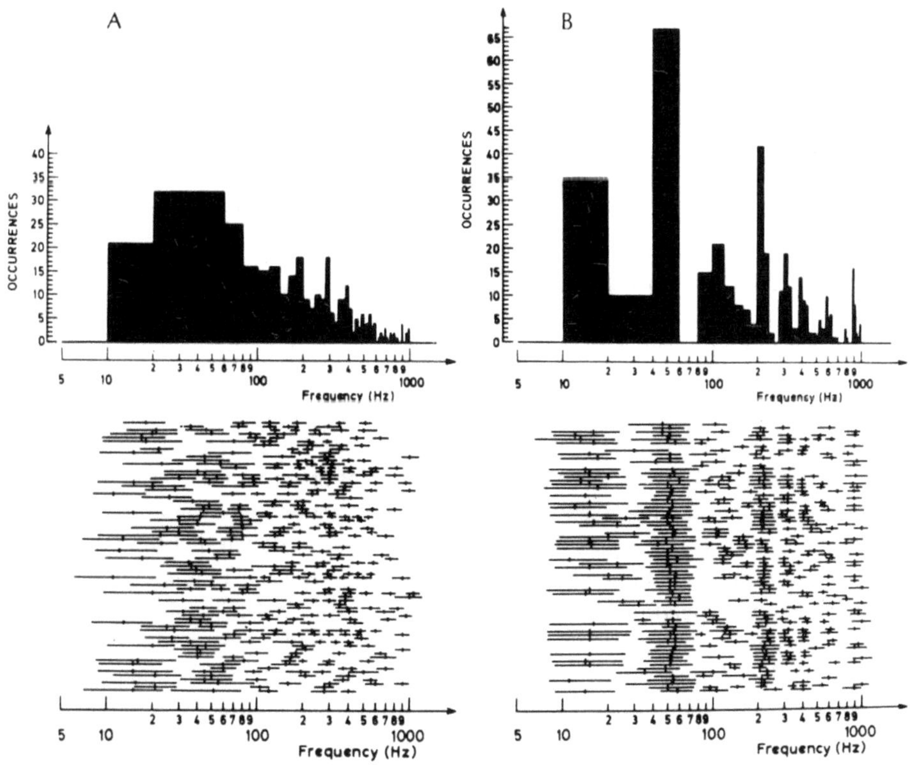

Fig. 6. Histograms showing frequency distribution as revealed by (A) prestimulus power spectra, (B) poststimulus instantaneous frequency characteristics of the RF. At the *bottom* of the figure, each spectral peak (or amplitude maximum) is represented by an approximate band width (horizontal line segment), and a center frequency is indicated. The histograms shown at the *top* of the figure were derived by plotting the number of center frequencies falling into each of a set of 20-Hz slots versus frequency. The number of power density–instantaneous frequency characteristic pairs used to obtain the histograms was 71

spectrum is much higher in comparison to the EEG (the highest peaks are at −7 dB in EPs and at −17 dB for the EEG). Moreover, there are some other differences, which can be studied by analyzing the frequency distribution of the power spectra.

Figure 6 illustrates a comparative analysis of power spectra of ongoing and evoked activities. Since the power spectra of the prestimulus EEGs and those of the single epochs of EPs have randomly varying spectral peaks (and/or amplitude maxima), a reasonably obvious classification scheme is to plot the respective number of events that fall into each of a set of frequency ranges (slots). The bottom of Fig. 6A illustrates the spectral peaks seen in the power spectra, whereas the bottom of Fig. 6B depicts the amplitude maxima seen in the power spectra of the EPs of the RF. Each spectral peaks is represented by an appropriate band width (horizontal line segment) and a center frequency is indicated. The number of center frequencies falling into each of a set of 20 Hz slots around 50 Hz, 70 Hz, 90 Hz, etc., were determined. The resulting histograms, shown at the top of Fig. 6A, B, were presented by plotting the number of center frequencies in each frequency channel.

The histogram for the RF, shown in Fig. 6, was the result of 71 single epochs of EEG and evoked spectra. The analysis of single evoked spectral peaking frequencies demonstrated a *marked alignment* in comparison to the frequency positions of the ongoing activity. In other words, the frequency centers of the EPs are focused to narrow bands in comparison to the frequency distribution of the ongoing activity. The analysis of frequency distribution in the 10–20 Hz activity of the EEG spectra (Fig. 6A) reveals an almost random distribution between 10 and 20 Hz. About 50 slots are distributed over heterogeneous frequencies. In the EP spectral distribution, most of the peaks are centered around 13 Hz. The same phenomenon is also observed for 55 Hz and for higher frequencies up to 1000 Hz. The histogram of the EP power spectra (Fig. 6) shows quantitatively that the substantial frequency stabilization pushes the randomly occurring EEG spectral peaks to sharp frequency channels. Some frequency channels are easily determined in Fig. 6B, where the EP power spectra had to occur most frequently in narrower bands. These are 10–20 Hz, 40–60 Hz, 80–140 Hz, 200–240 Hz, 280–340 Hz, 380–440 Hz, 580–640 Hz, and 880–920 Hz.

Further exact comparisons of evoked and ongoing spectra of the EEG in various structures of the cat brain have been made by Gönder und Başar (1978) and explained in details by Başar (1980). These studies have shown that the *increase* in the power of spectra obtained from EPs can reach values 10 or 15 times greater than the power in the spectra of the ongoing activity. In the studies mentioned, it could be further shown that spontaneous oscillations with smaller magnitudes can be frequency stabilized more efficiently by stimulation signals than can spontaneous oscillations with larger magnitudes. This behavior was then compared with the behavior of a model consisting of an ensemble of coupled oscillators. A population of coupled oscillators that are already in a state of synchrony depicting large amplitudes as the system average cannot be brought to an increased excited stage, whereas if such a system is not in synchrony (desynchronized spontaneous activity), the system response is largely enhanced. The comparison with coupled oscillators is explained in detail in several publications (Başar et al. 1979a; Başar 1980, 1983a, b).

Although the comparison of evoked spectra and EEG spectra allows the drawing of important conclusions about resonance and frequency stabilization effects in the brain, it should be pointed out that neither the power spectra nor the frequency characteristics supply any information about time and phase of single EEG-EP epochs.

Furthermore, whether or not the shape of a transient wave packet in a given EEG frequency range preceding a sensory stimulation may influence the EP cannot be determined by comparing power spectra.

Moreover, transient oscillatory waveforms of various frequencies often have comparable amplitudes in the time domain. In power spectra, the power of an oscillation depends not only on the maximal amplitude of such an oscillation, but also on its duration. Since the duration of low-frequency oscillation is much longer than that of high-frequency oscillation, the low-frequency components in power spectra usually have more weight and mask the high-frequency activity.

For these reasons, a combined analysis of the EEG and EPs in the time domain had to be considered for further understanding of the important relation between the EEG and EPs. The justification for the use of the EEG-EPogram method has been discussed in several longer reviews or monographs (Başar 1976, 1980; Başar et al. 1979a, b).

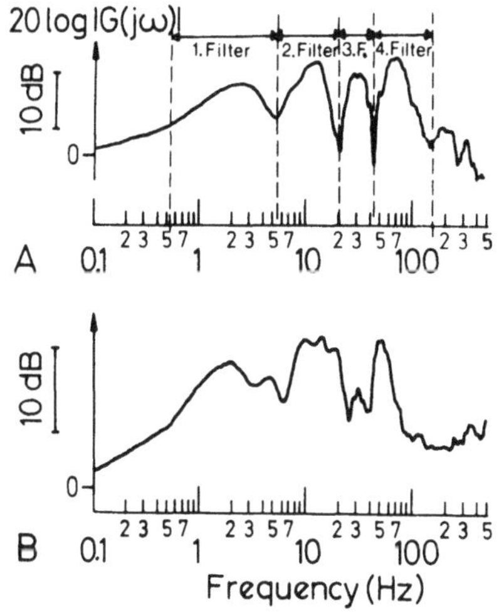

Fig. 7. Determination of band limits of four pass-band filters from the amplitude frequency characteristics of the cat reticular formation (A). These band limits specify the selectivity channels, which are the consistent amplitude maxima, as revealed in the mean value frequency characteristics curve (B)

Figure 7 illustrates the amplitude frequency characteristics of the *reticular formation* of the cat brain, obtained again upon acoustical stimulation of a freely moving cat with tone bursts of 2000 Hz. Since the amplitude characteristics during this experimental session depicted marked selectivities (or resonances) around 10 Hz and 40 Hz, as well as smaller peakings in other frequency channels (for example at 3 Hz or 20 Hz), we have chosen these examples first for the sake of simplicity.

We consider ten randomly chosen EEG-EP epochs filtered with band-pass filters of 8–13 Hz, illustrated in Fig. 8A. This frequency band has been chosen according to the selctivity channels in the amplitude frequency characteristics of Fig. 7A. The 10 Hz activity prior to stimulation is often well synchronized and depicts regular, almost sinusoidal wave packets with large amplitudes up to 70 µV. The stimulus elicited responses are often time locked (sweep nos. 1, 5, 8, 9, 10, and 11); however, the time-locked patterns do not have shapes different from the elementary waveforms depicted in the spontaneous activity. "Spontaneous patterns synchronized without stimulation" observed markedly in the prestimulus activities of sweep nos. 11, 12, 13, and 18, are observed in all relevant frequency regions of the EEG (see the 40 Hz activity in Fig. 8B; several other examples are in Başar 1980). The differences between the spontaneous EEG and evoked patterns are:

1. The synchronized 10-Hz pattern prior to stimulation occurs randomly, whereas the 10-Hz wave packet is triggered regularly upon external sensory stimulation, provided that there is not ample 10-Hz activity immediately prior to stimulation. In cases where ample ongoing 10-Hz activity precedes the stimulation, there is, as a rule, no time-locked enhancement upon stimulation (sweep nos. 9, 12, and 18).

2. The evoked 10-Hz response usually has a large magnitude in comparison to the magnitudes of the spontaneous 10-Hz wave packets. One of the most important points in this kind of analysis is the fact that in the spontaneous activity (although it has a random distribution), patterns that are comparable to the evoked patterns can often be detected. These spontaneous patterns, when they occur, have *comparable* magnitudes, the same frequency, and the same shape as the evoked pat-

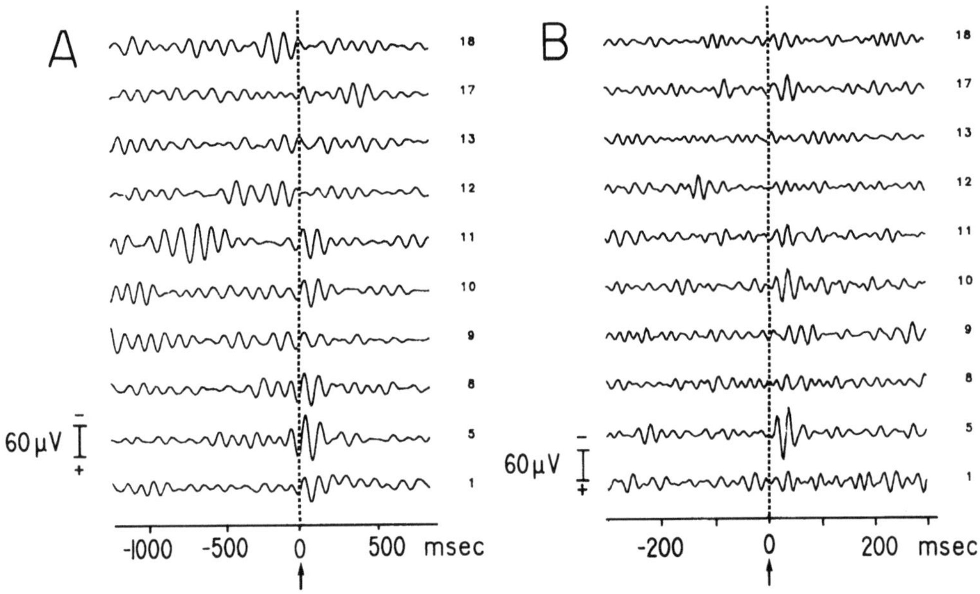

Fig. 8. (A) Filtered EEG-EP epochs in the 8–13 Hz frequency range. (B) Filtered EEG-EP epochs in the 30–60 Hz frequency range

Table 1. Enhancement factor X

Sweep no.	Frequency channels		
	3–8 Hz	8–13 Hz	30–60 Hz
1	2.8	4.3	1.5
2	2.4	3.1	1.4
3	2.0	1.8	1.6
4	1.1	1.8	0.9
5	1.5	1.7	0.7
6	0.5	1.2	1.2
7	1.9	1.6	1.0
8	2.4	1.4	0.7
9	2.0	1.0	1.3
10	1.1	2.2	1.4
11	2.4	1.3	0.8
12	1.8	1.5	1.2
13	1.4	1.4	1.6
14	2.3	2.2	1.2
15	1.6	1.7	1.6
16	1.2	2.1	2.6
17	2.3	1.6	1.4
18	2.1	2.0	1.7
19	1.5	0.8	1.5
20	2.0	1.2	1.2

terns. As Table 1 shows, the evoked wave packets are usually enhanced (in other words, the enhancement factor is a measure of the magnification of wave packets).

Figure 8B illustrates the filtered EEG-EP epochs in the 40-Hz frequency range (the filter limit is 30–60 Hz). The 40-Hz spontaneous and evoked wave packets show the same behavior as do the 10-Hz waveforms. The responses usually show time-locked, large-amplitude wave packets, provided that there is no synchronized 40-Hz activity prior to stimulation. In the spontaneous activity, however, randomly occurring 40-Hz waves are often recorded.

4 Internal EPs and Excitability of the Brain. Comparison with Dissipative Structures

As we have seen in the previous examples in Fig. 8, in the spontaneous activity of the brain some patterns can often be detected that are comparable to the EP patterns of the same frequency. The expression "internal evoked potential" is used for the spontaneous patterns, which occur randomly without any external stimulation (originating probably from hidden internal sources) and which have the same frequency and the same shape with comparable amplitudes as the evoked patterns. Based on various experiments using the methodology of EEG-EPograms, we have tentatively formulated the following working hypothesis, which we call "the excitability rule" (Başar 1980):

> Various brain structures depict spontaneous rhythmic activity in a wide frequency range between 1 Hz and 1000 Hz. Without application of external sensory stimulation, the spontaneous activity of a given brain structure can often show frequency-stable and high-magnitude electrical activity. If regular spontaneous oscillations can be detected in the electrical activity of a defined brain structure during a determined period, it is to be expected that upon external sensory stimulation, this structure will have a response susceptibility in the same frequency channel (for example, 40-Hz activity of the hippocampus and cortex, 250–300 Hz activity of the cerebellum, 10-Hz activity of all brain structures, and 10-Hz activity of human scalp responses). The response susceptibility of a brain structure depends mostly on its susceptibility to its own intrinsic rhythmic activity. Frequency stabilization, time-locking, and amplification of the spontaneous activity upon stimulation contribute greatly to the genesis of large potential changes which are called evoked potentials.

By considering the ensemble of results of frequency characteristics and the combined evaluation of power spectra and of the filtered single trials of ongoing and evoked activities, several rules or principles can be derived. These rules and principles constitute, on the one hand, a dynamical framework for the understanding of field potentials, and on the other hand, they elicit new ideas for describing the dynamic structure of EPs.

Nicolis and Prigogine (1977) described a unified formulation for self-organization phenomena in complex systems, i.e., systems involving a large number of interacting subunits. Such systems can present, under certain conditions, a marked coherent behavior extending well beyond the scale of the individual subunit. Biological order, the generation of coherent light by a laser, the emergence of spatial or temporal patterns of activity in chemical kinetics and in fluid dynamics, or finally the functioning of an animal ecological system or even of a human society, provide some striking illustrations of the coherent occurrence of such self-organization phenomena (Haken 1977).

Although our experiments were not systematically analyzed in terms of the concept of dissipative structures, here are some examples in which the schemata of Nicolis and Prigogine (1977) and of Katchalsky et al. (1974) can be partly adapted to our empirical findings.

Goldbeter (1980) analyzed the behavior of two biological systems for which experimental evidence exists for excitable and/or oscillatory behavior. The first system is that of glycolytic oscillations; the second is the adenosine 3′,5′-cyclic monophosphate (cAMP) signaling system, which controls periodic aggregation in the cellular slime mold *Dictyostelium discoideum.* Goldbeter and Caplan (1976) stated that sustained oscillations and excitability are closely associated in chemical systems. In other words, if a biological enzyme system demonstrates occasional or sustained oscillations, then this system is also excitable in the frequencies of the sustained oscillations. Later, Goldbeter and Segal (1980) described the ability to relay signals as an example of what in more general contexts is called excitability, i.e., the ability of a system to amplify a small perturbation in a pulsatory manner. Support for their contention that in slime molds a single mechanisms underlies both excitability and oscillating ability is the fact that both phenomena occur under closely related conditions in chemical systems such as the Belousov-Zhabotinsky reaction and in models for the nerve membrane and for an autocatalytic pH-controlled enzyme reaction. The statement of Goldbeter presents an excellent analogy to the findings or statements concerning the excitability of various brain structures in various frequency ranges.

As stated above, a brain structure is excitable if it shows sustained oscillations in a given frequency channel. This is one of the striking examples of the usefulness of the study of dissipative structures in understanding the brain's excitability.

These are some of the basic rules and principles which have been derived from learning the results of several experiments with field potentials of cat and human brains. In this report, only a concise description has been given by giving a few examples.

5 Some Applications of Basic Principles to EEG-EP Dynamics

The excitability rule and the relationship between the EEG and EPs described in the previous sections can be used in a variety of applications in the evaluation, analysis, and interpretation of sensory EPs and cognitive components of ERPs.

5.1 The Use of the Enhancement Factor in Order to Augment the Signal/Noise Ratio of the Sensory EP

An analysis of the single EEG-EP epochs of Fig. 8 filtered in the 8–13 Hz frequency range showed that if a large amplitude oscillatory waveform precedes the stimulation, there usually exists no time-locked response oscillation. Normally, the largest responses are detected in the stages where low-amplitude activity is observed as stimulation has been applied.

In our earlier observations on human EPs, using EEG-EP epochs in various frequency ranges, we also observed that in slow-frequency ranges, the phase angle of the prestimulus EEG waves might be likely to make an important contribution to the

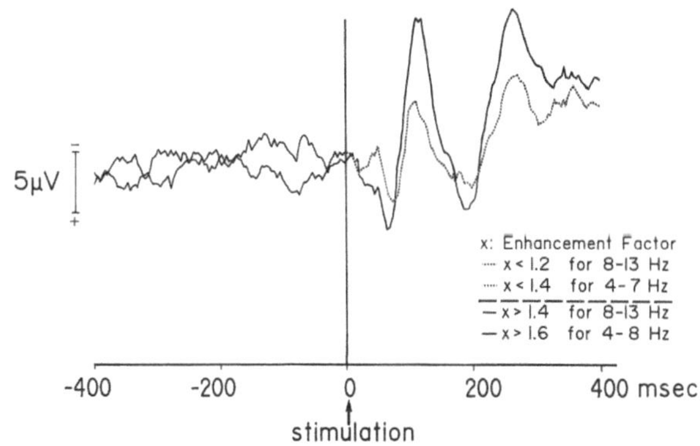

Fig. 9. The AEP from human vertex evaluated by using two different criteria. *Dashed curve*, EP evaluated by using the ensemble of single trials of EEG-EPs with small enhancement factors ($X < 1.2$ for 8–13 Hz and $X < 1.4$ for 4–8 Hz). *Solid curve*, the EP evaluated by using the ensemble of single trials of EEG-EPs with large enhancement factors ($X > 1.4$ for 8–13 Hz and $X > 1.6$ for 4–8 Hz)

amplitude of the EP signal in a given frequency range (Başar 1980). According to these observations, it should be possible to find an ensemble of optimal EEG-EP sets of high responses which could give rise to an averaged potential with large amplitudes. We now cite, for example, an experiment during which a healthy subject received a stimulus of 2000-Hz tones of 1-s duration. An analysis of single EEG-EP epochs showed that for the 8–13 Hz (alpha) frequency range, the enhancement factor X_α had a mean value of 1.3, and that for the 4–7 Hz (theta) frequency range, X_θ had a mean value of 1.5. Accordingly, we chose two ensembles of EEG-EP epochs. The first ensemble contained epochs with enhancement factors $X_\alpha > 1.4$ for 8–13 Hz and $X_\theta > 1.6$ for 4–8 Hz. The second ensemble contained epochs with enhancement factors $X_\alpha < 1.2$ for 8–13 Hz and $X_\theta < 1.4$ for 4–8 Hz.

The dashed curve of Fig. 9 shows the EP evaluated by using the ensemble with small enhancement factors, whereas the solid curve shows the EP evaluated by using the ensemble with large enhancement factors. The wave complex containing waves at positions P100–200 showed much larger magnitudes by more than twofold in the solid curve in comparison to the dashed curve.

This type of analysis can be very useful for experiments during which only a small number of sweeps can be obtained, such as experiments with babies, with psychiatric patients, or with freely moving animals, or experiments on stimulation threshold, as for hearing or pain. By choosing sweeps with high enhancement factors, a relevant signal/noise ratio can be obtained with only a very small number of sweeps. A word of caution should be added here, to say that this type of analysis can be useful for detection of EPs in noisy experiments where ongoing activity or experimental conditions suppress the EP activity and where the aim of the experiments is the detection of EPs or some defined transienst EP components. A more *causal* and *physiologically* relevant evaluation of selected EEG-EP epochs will be described below.

5.2 Comparison of the Human Auditory EP by Selecting EEG Epochs with Various Levels of EEG Activity

Since the enhancement factor is a function of the prestimulus EEG, the next step in analyzing the influence of the EEG has been to choose selection criteria by considering various levels of the EEG activity just preceding the stimulus. The example in

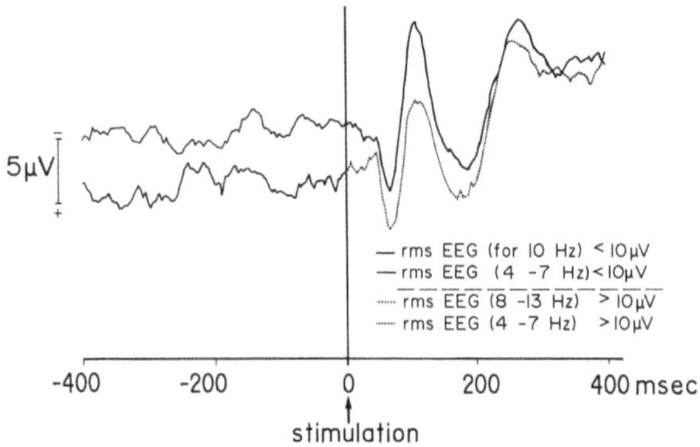

Fig. 10. The AEP from human vertex evaluated by using two different criteria. *Dashed curve,* the selectively averaged EP evaluated by using only EEG-EP sets with RMS values of prestimulus EEG greater than 10 μV for filtered EEGs in the 8–13 Hz and 4–7 Hz frequency ranges. *Solid curve,* the selectively averaged EP evaluated by using only EEG-EP sets with RMS values of prestimulus EEG less than 10 μV for filtered EEGs in the 8–13 Hz and 4–7 Hz frequency ranges

Fig. 10 illustrates two EPs evaluated during the same experimental session at which the auditory EPs (AEPs) of Fig. 9 were recorded. In this analysis, we again selected two different classes of EEG-EP sets. The solid curve was obtained by averaging EEG-EP subsets in which the RMS value of the prestimulus EEG was smaller than 10 µV for the frequency ranges of 8–13 Hz and 4–7 Hz. In other words, for this subset of EEG-EP epochs, the prestimulus EEG should not exceed 10 µV after filtering either with the band-pass filter of 8–13 Hz or with that of 4–7 Hz (two different filters applied!). For the analysis of the 8–13 Hz component, a time period of 250 ms prior to auditory stimulation was used. For the analysis of the 4–7 Hz frequency component, a duration of 400 ms prior to stimulation was evaluated.

In the second stage of analysis, a subset of EEG-EP epochs that have shown RMS EEG values larger than 10 µV was averaged (higher EEG activity just prior to stimulation). Both subsets contained an equal number of EEG-EP epochs. Although the two EPs in Fig. 10 have almost similar shapes, the amplitudes of various deflections are different. The dashed curve, the EP with a high-amplitude prestimulus EEG, is about 40% smaller than the solid curve containing sweeps with a small prestimulus EEG. In other words, there is an inverse relationship between the magnitude of the prestimulus EEG and the amplitudes of the EPs. This example is important for the analyzer of EPs who is not yet accustomed to performing an analysis in the frequency or time domain using concepts of alpha or theta responses. This example shows that the conventional EP is a function of the EEG, which should be studied prior to performing a signal averaging of EPs. We have shown here the influence of theta and alpha components in the EEG. The same analysis has been extended to the 40-Hz range, depicting the same inverse relationship between 40-Hz prestimulus EEG and middle-latency responses of the EP. This analysis will be published elsewhere.

5.3 Superposition Principle of Various EP Components

An analysis of single EEG-EP epochs made during a session of recording sensory EPs showed that the enhancement factors in the frequency channels of 4–8 Hz, 8–13 Hz, 15–25 Hz, and 30–60 Hz do not increase or decrease together. There are very often periods in which the theta enhancement is large while alpha enhancement is low, and vice versa. Table 1 shows enhancement factors in three frequency ranges after the evaluation of single EEG-EP trials with light stimulation. By analyzing such an ensemble of EEG-EP sets, one is usually able to find subensembles showing similar enhancement factors. Our question is now whether it is possible to find in a given ensemble of EEG-EPograms two (or more) subsets that present two types of EPs. More precisely, is it possible to isolate an *alpha-EP* or a *theta-EP* without using filtering analysis, which theoretically (or mathematically) predicts the existence of damped oscillatory waveforms, such as the responses in Fig. 8 present in the 10-Hz and 40-Hz frequency ranges? We chose two subsets of EEG-EP epochs. The first subset contained selected epochs with *high alpha response* and *low theta response* (large enhancement factor $X_\alpha > 1.8$ for the 8–13 Hz range, $X_\theta > 1$ for the 3–8 Hz range. On the contrary, the second subset contained selected epochs with low alpha response and with high theta response ($X_\alpha < 1$ for the 8–13 Hz range, $X_\theta > 1.6$ for the 3–8 Hz range).

The amplitude characteristic of the EP containing unselected sweeps (all the sweeps) showed a maximum of response or selectivity in a broad range between 5 and 14 Hz. Alpha and theta frequency responses were not distinctly separated (Fig. 11A). The amplitude characteristic of *subset 1,* with high alpha response ($X > 1.8$), showed a peak around 10 Hz and a minimum around 5 Hz (Fig. 11B), whereas the amplitude frequency characteristics of the *second subset,* with high theta enhancement, showed a selectivity between 2 and 8 Hz, but without the peaking at 10 Hz (Fig. 11C). These are two completely different frequency characteristics. The averaged transient EPs from the two different subsets are also completely different. The EP (alpha EP) of Fig. 12A can be fitted with the curve filtered in the 8–13 Hz frequency range, whereas the EP (theta EP) in Fig. 12B fits well with the curve filtered in the 3–8 Hz range.

In other words, the AEP evaluated by using selected sweeps with high alpha response and low theta response showed the damped oscillatory waveform predicted by using the 8–13 Hz digital filter. The AEP, which is evaluated by averaging all of the sweeps [without using any selection criterion (Fig. 12A)], has a more complicated waveform. A detailed analysis would show that the existence of a small number of sweeps in the nonselected AEP is caused by the fact that the averaging of the alpha and theta oscillatory waveforms leads to smoothing effects that result from cancellation or superposition of the elementary waveforms. The elementary alpha and theta responses are usually already superimposed in single EPs. In a long recording session, it is usually possible to register some sweeps with only one or two elementary waveforms that are shaped mostly by a unique frequency component. In most of the cases, a larger number of elementary response wave packets are superimposed.

The implications of this *superposition principle* are extremely important for the analysis of EPs, because the possibility that two elementary responses may obscure

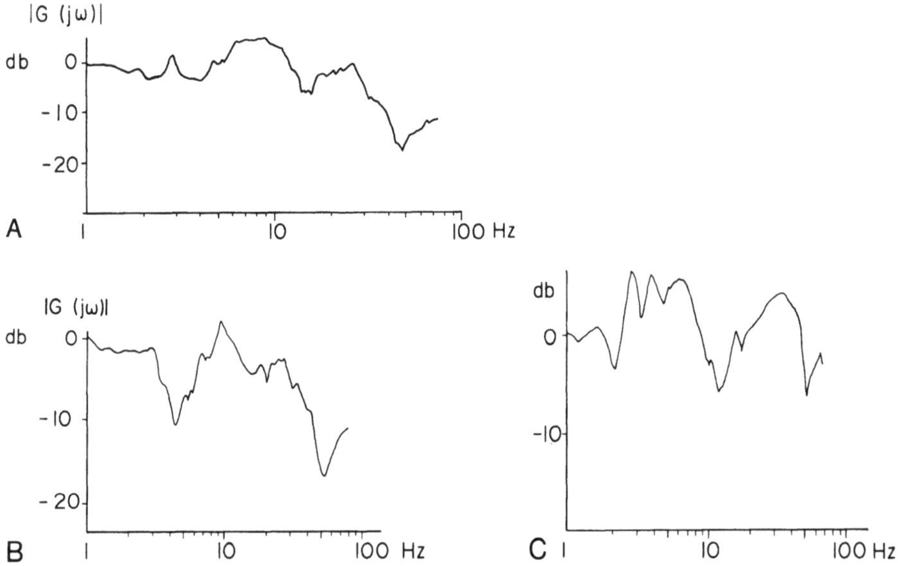

Fig. 11. (A) Amplitude frequency characteristic obtained by application of light stimulation. The subject was stimulated with light stimulation in the form of a step function which has a duration of 1 s. This curve is obtained by means of Fourier transform of the vertex AEP (average of 100 unselected sweeps). (B) The amplitude frequency characteristic of the same subject obtained from the selectively averaged AEP by using *subset 1* of single EPs depicting high alpha response (see text). (C) The amplitude frequency characteristic of the same subject obtained from the selectively averaged AEP by using *subset 2* of single EPs depicting low alpha response (see text)

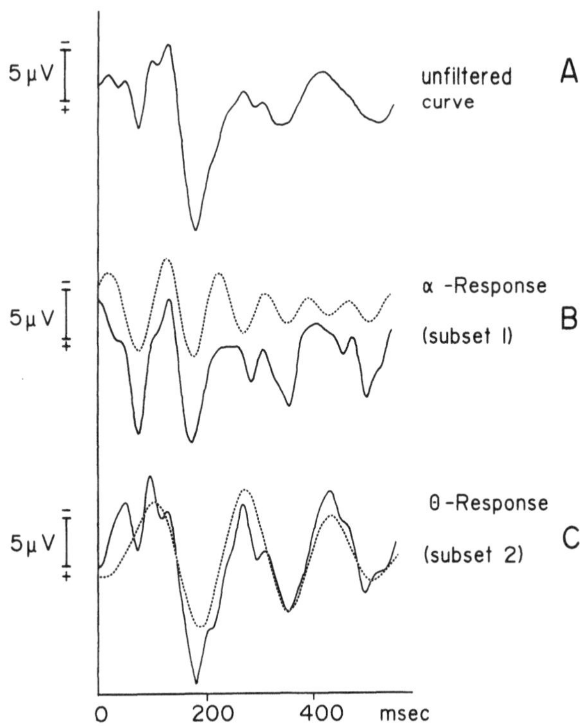

Fig. 12. EPs corresponding to the experiment in Fig. 11

each other is important not only in the analysis of EPs, but also in the analysis of the ongoing EEG! The illustration of the filtered spontaneous activity in Fig. 8 demonstrates that the waning and waxing of 10-Hz activity and of 40-Hz activity are not necessarily interdependent. These observations raise the important question of whether some quiet EEG stages could be caused by superposition of various activities occurring at the same time but depicted as waning and waxing behavior in slightly different frequencies. A superposition principle for the EEG is also pointed out by other authors (Wright and Kydd in this volume; Weiss 1986).

The analysis described in this section was carried out by using the visual human EP, depicting two masked (or interwoven) components, namely alpha and theta responses. In our earlier detailed work on the analysis of the cat hippocampus and other subcortical structures, influences of several components could be recognized. Higher frequency components, for example, in the frequency range of 30–80 Hz (the 40-Hz frequency) have an influence on middle latency components of ERPs (Başar et al. 1987a). One important consequence of the present analysis is the fact that it is possible to select EEG states that would give rise at least to two or three types of EPs during the same recording session. This fact demonstrates that the prestimulus EEG is of major importance in experiments to measure and interpret ERPs.

6 EEG-EP Dynamics and Cognitive Processes

In the foregoing section, we showed that the prestimulus EEG is one of the most important parameters that influence the amplitude, the time behavior and, in general, the shape of the EP. Furthermore, the examples given in the previous sections pointed out possible kinds of analysis and gave hints about the structure of the EP, which is, as a rule, related to the EEG. On the other hand, EEG changes during various tasks and especially during cognitive processes have been described by several scientists since the pioneering work of Berger (see, for example, Creutzfeldt 1983; Giannitrapani 1985). In order to treat the association between the EEG and the cognitive components of ERPs, several steps have been taken by our research group. The general conclusion drawn from the findings of our laboratories is as follows: when subjects perform tasks in relation to regular, frequently presented target stimuli, the regular pattern of stimulation induces more stable and predictable pre- and poststimulus activity. The onset of these changes can vary within and between subjects, as well as between the different frequencies of target stimuli used in the experiments. Furthermore, informing subjects of the stimulus pattern appears to augment the regularity and synchronization of their EEGs, as if such knowledge is somehow able to influence a subject's EEG rhythmicity (Başar et al. 1984; Stampfer and Başar 1985; Başar and Stampfer 1985).

In our studies of P300 endogenous potentials, we modified the conventional auditory "oddball" paradigm to demonstrate the development and variation of prestimulus "preparation changes" in the EEG. The subjects were stimulated with repetitive frequent oddball tones. During the experiments, the subjects learned by themselves the applied sequence of the target and nontarget tones. We measured subjects with all of the various degrees of synchronized EEG patterns during various

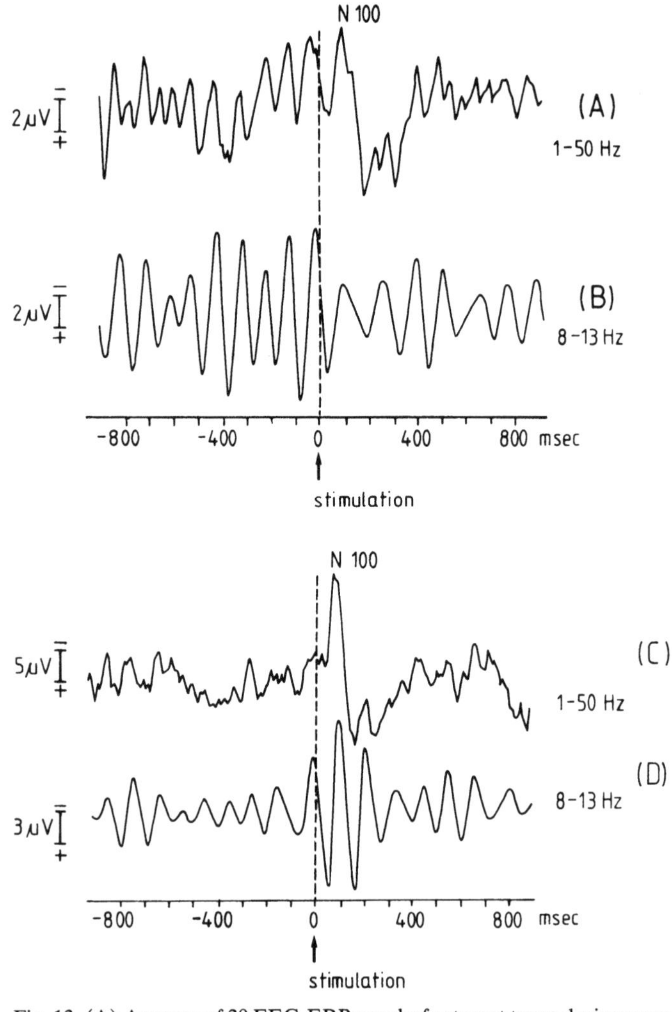

Fig. 13. (A) Average of 20 EEG-ERP epochs for target tones during an experiment in which the subject showed abundant alpha activity. (B) Average of EEG-ERP in A filtered with a band-pass filter of 8–13 Hz. (C) Average of 20 EEG-ERP epochs of the same subject during a period of low-amplitude prestimulus alpha activity. (D) Averaged EEG-ERP in C filtered with a band-pass filter of 8–13 Hz. (Modified from Başar and Stampfer 1985)

stages of the experiments. These changes were recorded in various frequency channels, mostly in the 1–4 Hz, 4–8 Hz, and 8–13 Hz frequency ranges. Figure 13A shows an average of 20 EEG-EP epochs for target tones during the experiment with a subject having abundant alpha activity. During this measurement, a striking rhythmicity of 10 Hz was observed in the prestimulus segment. Figure 13B is a filtered version of Fig. 13A (band pass 8–13 Hz). An obvious "blocking" effect can be seen following the point of stimulation. When each of the 20 single sweeps is filtered in the same band width and compared to the filtered average, it becomes apparent that the prestimulus rhythmicity is not a passive phenomenon; there is evidence of time-locking to the expected stimulus (a kind of *forward time-locking*). We evaluated the mean alpha magnitude of the 20 single sweeps. It amounted to 16 μV. If the prestimulus activity had been random, the averaged amplitude of the 20 sweeps should have been reduced by a factor of $\sqrt{20}$, or about 4.5[1].

We have repeatedly observed an inverse relationship between the amplitude of prestimulus alpha activity and the amplitude of the poststimulus N100 response. Figure 13C shows an average of EEG-EP epochs with low-amplitude prestimulus alpha activity (the average RMS amplitude of alpha activity amounted to 7 μV). Figure 13D is the filtered version of the respective average above in the frequency range of 8–13 Hz. It can be seen that the low-amplitude prestimulus alpha activity is associated with high-amplitude poststimulus enhancement (Fig. 13D), whereas the reverse applies in Fig. 13B. When poststimulus alpha blocking occurs, as in Fig. 13B, there is usually evidence of an "after discharge" around 400 ms poststimulus.

The above results have been interpreted as evidence of an active functional relationship between prestimulus alpha activity and poststimulus alpha enhancement, which bears a highly significant relationship to the N100 peak of averaged data. The change in amplitude of the N100 peak is highly marked, as the comparison of the unfiltered averages in Fig. 13A, C demonstrates.

Here, it should be also emphasized that the analysis above is pertinent for studies in cognitive neurophysiology, in which the N100 peak, as well as changes in the N100 peak (N100 tuning), are analyzed in order to discuss several behavioral reactions. The lesson we might draw from the interaction of the N100 peak with the prestimulus alpha activity is that the behavior change observed with the N100 peak, or the so-called N100 tuning, should be considered along with the alpha behavior of the subject, and that efforts might be undertaken to illuminate the controversies about N100. The discrepancies described by several investigators with regard to N100 tuning could have been understood by examining at the beginning whether or not their subjects were of the alpha type.

7 Brain Event-Related Rhythms Preceding Cognitive Tasks

Our preliminary results, showing that during cognitive tasks, prestimulus EEG tends to attain a phase-ordered pattern prior to repetitively expected stimulation, led us to extend our experiments with a new emphasis. Tones of 2000 Hz and 80 dB and of 800-ms duration were applied at regular intervals of 2600 ms. Every third or fourth tone was omitted. The subjects were asked to predict and *mark mentally* the time of occurrence of the omitted signal. One second of the EEG prior to omitted stimuli was also recorded along with the ERP.

The experiments were carried out with ten healthy volunteer subjects. When subjects learned and successfully followed the rhythmicity contained in the paradigms, they were usually able to increase their attention. Then rhythmic prestimulus EEG patterns could be observed. Most of the subjects reported that at the beginning of an experimental session with repetitive signals, they had difficulty in predicting the time of occurrence of stimulus omission. However, in the second part of the experiment,

[1] From systems theory it is known that during the averaging process time-unlocked random signals are reduced in amplitude by a factor $\sqrt{N}$, N being the number of averaged sweeps.

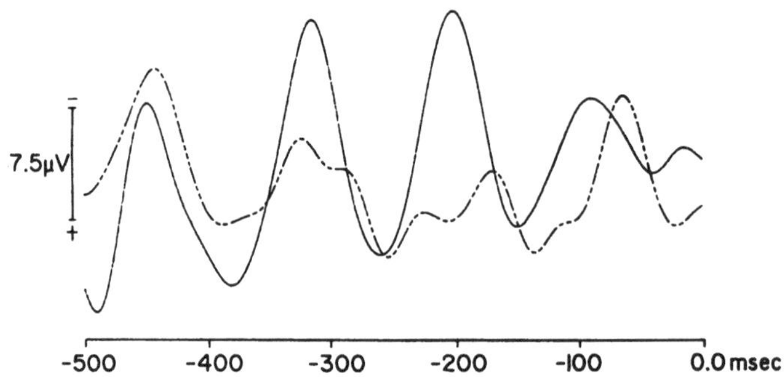

Fig. 14. Averages of the first *(broken line)* and last *(solid line)* ten prestimulus EEG epochs of the experiment, filtered in the 1–25 Hz frequency band. (Başar et al. 1987b)

they were usually able to predict the time of the omitted signal. Accordingly, in our signal analysis, we applied a selective averaging, by grouping the first ten prestimulus sweeps at the beginning of the experiment and the last ten sweeps towards the end of the experiment. Figure 14 illustrates comparatively the averages of the first and the last ten prestimulus EEG epochs (digitally filtered between 1 and 25 Hz), which were measured at the vertex of a subject who reported that at the beginning of the experiment he felt insecure and had diffuse attention. However, he was able to concentrate better on the tasks near the end of the experimental session and could perform his tasks much better. The average of the ten sweeps at the end of the experiment depicted a regular rhythmic behavior with large amplitudes. The first ten sweeps tend also to the same rhythmicity; however, the average is not very regular, the amplitude of the oscillation being low (compare Fig. 14).

Rhythms similar in principle to those illustrated in Fig. 14 have been observed in all of the subjects. However, the alignment and phase reordering were not the same in all the subjects. The exact time of regularity and phase reordering showed fluctuations in a time period 500 ms prior to the event.

Figure 15 shows the results of an experiment with another subject. In this case, the superposition of the last nine sweeps of the experiment is illustrated. Single sweeps have here been digitally filtered in the frequency range between 7 and 13 Hz. The regularity of the single EEG rhythms is in this case such that it is easy to observe that the shape of the EEG prior to stimulation reached a template pattern.

It is widely recognized that the endogenous ERP components are related to cognitive processing of stimulus information or to the organization of behavior, rather than being evoked by the presentation of the stimulus. The event-related rhythms that have been presented in this study reflect the effort performed by the brain in the expectation and prediction of an event. They are purely endogenous, since they are emitted in relation to a mental task, and not following a physical event or prior to a physical motion. Our results differ highly from reports describing EEG changes in frequency and amplitude during cognitive tasks, since our experiments indicate that the EEG of a subject can reach patterns with a constant template during a constant mental task. This pattern has a defined *phase-reordering* and *alignment* during the execution of the mental task, which start approximately 500 ms prior to the event. Accordingly, we want to emphasize that the EEG may play a highly active defined

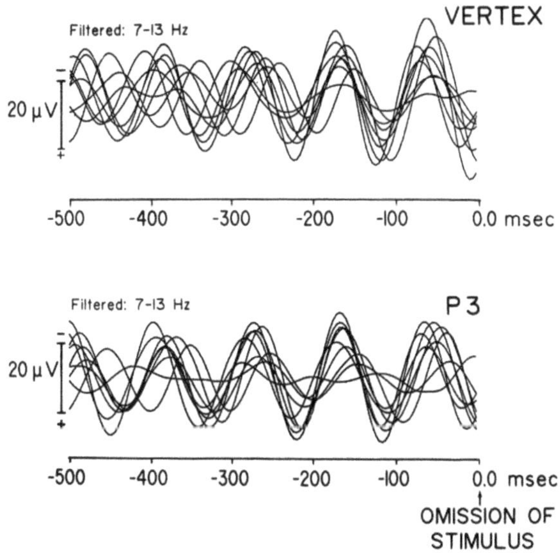

Fig. 15. The last nine sweeps from vertex (Cz) and parietal leads filtered in the 7–13 Hz frequency band. (Başar et al. 1987b)

role in processes of cognition, especially those involved in the generation of perceptions and short-term memory.

Freeman (1975; Freeman and Skarda 1985) has shown that the EEG of the olfactory bulb and cortex in awake, motivated rabbits and cats shows a characteristic temporal pattern consisting of bursts of 40–80 Hz oscillations, superimposed on a surface-negative baseline potential shift synchronized to each inspiration. He has since interpreted this finding as follows.

> The neural activity which is induced by an odour during a period of learning provides the specification for a neural template of strength connections between the neurons made active by that odour. Subsequently when the animal is placed in the appropriate setting, the template may be activated in order to serve as a selective filter for search and detection of the expected odour. (Freeman 1979)
>
> A key step in information processing depends on an orderly transition of cortical activity from a quasi-equilibrium to a limit-cycle (synchronized oscillation state and back again). In this interpretation, the synchronized 40-Hz EEG activity, or the 40 Hz limit-cycle activity, serves as an operator on sensory input, to abstract and generalise aspects of the input into pre-established categories, thereby creating information for further central processing. (Freeman 1983)

In other words, the limit cycle may represent an image of the input that the animal expects to identify or is searching for. The results of the present study and that of Başar et al. (1984) support this interpretation and the possibility of its generalization to other sensory modalities as well to as a wider range of EEG frequencies. We have presented evidence that expectancy and selective attention, associated with regular, frequent target stimuli, result in highly synchronized EEG activity. This regular "limit cycle" activity occurs in various frequency ranges between 1 and 40 Hz. In this study we have focused our analysis on the 1–13 Hz range. Accordingly, we conclude tentatively that the 1–4 Hz, 4–7 Hz, and 8–13 Hz activities serve as "operators" in the selective filtering of expected target stimuli. We suggest that Freeman's concept can be generalized to various sensory systems and to EEG frequencies such as delta, theta, and alpha. Furthermore, experiments in this study have shown that a regular pattern of stimulation can induce a "preferred" phase angle, which appears to facilitate an optimal brain response to the sensory input. When subjects were not informed that target stimuli would be presented regularly and alternately, their EEG acitivty appears to have developed an "operator state" spontaneously. At the beginning of the experiment, the prestimulus EEG activity was usually poorly synchronized. The duration of alpha and theta enhancement was prolonged and delta activity showed a delayed latency shift. These P300-like changes were seen in both target and nontarget responses. Once subjects "discovered" the regular pattern of stimulation, there was a progressive reduction in the delta delay as well as in the duration of theta and alpha oscillation. The ERP response began to look almost like a sensory EP (Stampfer and Başar 1985). The important conclusion drawn from these findings is that EEG operators appear to modulate the response characteristics as a function of learning. Başar et al. (1984) used the expression "operative stage" for degrees of brain synchronization in delta, theta, and alpha activity during brain information processing associated with sensory and cognitive inputs. In contrast to the statements made by Freeman (1983), Başar (1980), and Başar et al. (1984) suggested that EEG activity also contains response fragments to internal sensory and cognitive inputs (more details and examples concerning the EEG template pattern are analyzed by Başar et al. 1987b).

8 Concluding Remarks

The experiments and results presented in this study have several implications and application possibilities. Some aspects of the concepts and methods presented have been explained in detail earlier. In this short conclusion we will emphasize only some of the generalizable results and new avenues.

The experiments described in the previous sections and our earlier results (Başar 1983) have shown that during experiments with cognitive tasks it is possible to measure reproducible EEG patterns. In other words, the use of modern computer techniques makes it possible to find quasi-"deterministic EEG" patterns related to defined brain functions. The specific brain function consists in this case of the generation of reproducible EEG patterns related to learning and short-term memory. We will call this type of memory a "dynamic memory." Engrams of this dynamic memory are manifested in the form of oscillatory waveforms or reproducible templates (see also the Epilogue to this volume).

The use of the expression "deterministic EEG" finds its legitimacy in several results of Babloyantz, Röschke and Başar, and Freeman (this volume). Babloyantz has shown that the human sleep EEG contains a strange attractor with a dimension of approximately 5 [for a definition of a "strange attractor," see Babloyantz (this volume), Röschke and Başar (this volume), and Başar 1983]. Röschke and Başar have described the dimensionality of strange attractors in various structures of the brain as between 4 and 5.

We have emphasized for a long time that the EEG should not be considered as a simple noise to be eliminated in experiments on sensory EPs and ERPs. The EEG is most probably a memory-related and cognition-related operator that seems to govern the most important common transfer functions of the brain. The question of Fessard (1961) that we emphasized in the preliminary remarks of this paper is partly answered by our description of frequency characteristics. Theta, alpha, beta, and 40-Hz frequency channels appear to be fundamental transmission channels in which heterogeneous messages during sensory and cognitive processes occur. Accordingly, we propose that the most general transfer functions of the brain's sensory and cognitive communication are partly reflected in the general resonance phenomena that occur in distributed neural networks throughout the brain. It is also important to emphasize that sensory and cognitive tasks seem to use the same frequency channels, but with different weights (compare also Stampfer and Başar in this volume). Even by analyzing the endogenous potentials of ERPs (P300 wave, N100 tuning), it has been shown that the cognitive changes in the brain's response activity reflect a transition to coherent stages of the already existing resonance transmission lines. There are no new frequencies. But the changes do occur as tuning of the existing resonance properties (Stampfer and Başar 1985; Başar et al. 1984). Various cognitive tasks or sensory communications use a specific combination of various resonant modes. The resulting ERPs reflect the (resonant) responses in a given quasi-*invariant modes of resonances.* The immense variability of single EPs should in this case reflect the most important general transfer in the brain to cognitive and sensory input. In preliminary results, Başar and Bullock indicated that even invertebrate ganglia (Aplysia) show transfer functions similar to those of mammalian brain (Başar and Bullock 1986). Even in a small number of resonance channels, a responding brain possesses enormous flexibility because of the possible combinations of various channels. The formulation of such a working hypothesis is ambitious and demands a still larger number of experiments (see also the remarks of Başar in the chapter on "How Brains May Work," this volume). However, the results of the present analysis support the possibility of such a generalization.

We also want to point out that studies of single EEG and EP trials may open new avenues of research with a great capacity for inspiring new interpretations of how the brain processes sensory and cognitive information. The averaged EPs indicate only a global image of the brain. On the contrary, analyses of the dynamics of EEG fragments of short duration and of single trial EPs can contribute to augmenting our understanding of cognitive tasks.

References

Başar E (1976) Biophysical and physiological systems analysis. Addison-Wesley, Reading

Başar E (1980) EEG-Brain dynamics. Relation between EEG and brain evoked potentials. Elsevier/North-Holland, Amsterdam

Başar E (1983a) Synergetics of neuronal populations. A survey on experiments. In: Başar E, Flohr H, Haken H, Mandell AJ (eds) Synergetics of the brain. Springer, Berlin Heidelberg New York, pp 183–200

Başar E (1983b) Toward a physical approach to integrative physiology. I. Brain dynamics and physical causality. Am J Physiol 245(4): R510–R533

Başar E, Bullock TH (1986) A comparative analysis of compound potentials in the cat brain and in the Aplysia cerebral ganglion. Pflugers Arch 70:406, R13

Başar E, Stampfer HG (1985) Important associations among EEG-dynamics, event-related potentials, short-term memory and learning. Int J Neurosci 26:161–180
Başar E, Gönder A, Özesmi C, Ungan P (1975a) Dynamics of brain rhythmic and evoked potentials. I. Some computational methods for the analysis of electrical signals from the brain. Biol Cybern 20:137–143
Başar E, Gönder A, Özesmi C, Ungan P (1975b) Dynamics of brain rhythmic and evoked potentials. II. Studies in the auditory pathway, reticular formation, and hippocampus during the waking stage. Biol Cybern 20:145–160
Başar E, Demir N, Gönder A, Ungan P (1979a) Combined dynamics of EEG and evoked potentials. I. Studies of simultaneously recorded EEG-EPograms in the auditory pathway, reticular formation and hippocampus of the cat brain during the waking stage. Biol Cybern 34:1–19
Başar E, Durusan R, Gönder A, Ungan P (1979b) Combined dynamics of EEG and evoked potentials. II. Studies of simultaneously recorded EEG-EPograms in the auditory pathway, reticular formation and hippocampus of the cat brain during sleep. Biol Cybern 34:21–30
Başar E, Başar-Eroglu C, Rosen B, Schütt A (1984) A new approach to endogenous event-related potentials in man: relation between EEG and P300-Wave. Int J Neurosci 24:1–21
Başar E, Başar-Eroglu C, Greitschus F, Rosen B (1987a) Int J Neurosci 33:103–117
Başar E, Başar-Eroglu C, Röschke J, Schutt A (1987b) EEG Journal
Berger H (1929) Über das Elektroencephalogramm des Menschen. Arch Psychiatr Nervenkr 87: 527–570
Berger H (1938) Das Elektroencephalogramm des Menschen. Nova Acta Leopoldina 6 (38)
Bingham C, Godfrey MD, Tukey JW (1967) Modern techniques of power spectrum estimation. IEEE Trans AU 15:56–66
Bullock TH (1984) Comparative neuroscience holds promise for quiet revolutions. Science 225:473–478
Bullock TH (1987) Compound potentials of the brain, ongoing and evoked: perspectives from comparative neurology. In: Başar E (ed) Dynamics of sensory and cognitive processing of the brain. Springer, Berlin Heidelberg New York Tokyo
Creutzfeld OD (1983) Cortex Cerebri. Springer, Berlin Heidelberg New York
Creutzfeld OD, Watanabe S, Lux HD (1966) Relations between EEG-phenomena and potentials of single cortical cells. I. Evoked responses after thalamic and epicortical stimulation. Electroenceph Clin Neurophysiol 20:1–18
Creutzfeld OD, Rosina A, Ito M, Probst W (1969) Visual evoked response of single cells and of EEG in primary visual area of the cat. J Neurophysiol 32:127–139
Fessard A (1961) The role of neuronal networks in sensory communications within the brain. In: Rosenblith WA (ed) Sensory communication. MIT Press, Cambridge
Freeman WJ (1975) Mass action in the nervous system. Academic, New York
Freeman WJ (1979) Nonlinear gain mediating cortical stimulus-response relations. Biol Cybern 33: 237–247
Freeman WJ (1983) Dynamics of image formation by nerve cell assemblies. In: Başar E, Flohr H, Haken H, Mandell AJ (eds) Synergetics of the brain. Springer, Berlin Heidelberg New York, pp 102–121
Freeman WJ, Skarda CA (1985) Spatial EEG patterns, non-linear dynamics and perception: the neo-Sherringtonian view. Brain Res Rev 10:147–175
Giannitrapani D (1985) The electrophysiology of intellectual functions. Karger, Basel
Goldbeter A (1980) Models for oscillations and excitability in biochemical systems. In: Segel LA (ed) Mathematical models in molecular and cellular biology. Cambridge University Press, Cambridge, pp 250–291
Goldbeter A, Segel A (1980) Control of developmental transitions in the cyclic AMP signalling system of *Dictyostelium discoideum.* Differentiation 17:127–135
Haken H (1977) Synergetics, an introduction. Springer, Berlin Heidelberg New York
Nicolis G, Prigogine I (1977) Self-organization in nonequilibrium systems: from dissipative structures to order through fluctuations. Wiley and Sons, New York
Niedermeyer E, Lopes da Silva F (1982) Electroencephalography. Basic principles, clinical applications and related fields. Urban and Schwarzenberg, Baltimore
Ramos A, Schwartz E, John ER (1976) Evoked potential-unit relationship in behaving cats. Brain Res Bull 1:69–75

Stampfer HG, Başar E (1985) Does frequency analysis lead to better understanding of human event related potentials. Int J Neurosci 26:181–196

Ungan P, Başar E (1976) Comparison of Wiener filtering and selective averaging of evoked potentials. Electroencephalogr Clin Neurophysiol 40:516–520

Verzeano M (1973) The study of neuronal networks in the mammalian brain. In: Thompson RF, Patterson MM (eds) Bioelectric recording techniques. Part A: Cellular processes and brain potentials. Academic, New York

Weiss V (1986) Memory as a macroscopic ordered state by entrainment and resonance in energy pathways. In: Weiss V, et al. (eds) Psychogenetik der Intelligenz. Modernes Lernen Borgmann, Dortmund

Wright JJ, Kydd RR, Lees GJ (1987) Gross electrocortical activity as a linear wave phenomenon, with variable temporal damping. In: Başar E (ed) Dynamics of sensory and cognitive processing of the brain. Springer, Berlin Heidelberg New York Tokyo

Electrophysiological and Radiographic Evidence for the Mediation of Memory by an Anatomically Distributed System

E. R. JOHN

1 Historical Introduction

A guiding principle in studies of brain-behavior relationships in general, and the mediation of memory in particular, has been that knowledge of the mediating anatomical structure often provides invaluable insight into how a function is performed. Recently, my colleagues and I have been using a double-labelled 2-deoxyglucose radioautographic technique to study the anatomical distribution of the metabolic correlates of memory activation in split-brain cats. Before presenting the results of these current studies, I will summarize the salient findings of our 35 years of prior research on this problem, so that our present metabolic studies can be integrated into the perspective of the preceding neurophysiological evidence.

1.1 Radioautography

Our current studies with labelled deoxyglucose constitute a return to my initial approach to the study of memory. Figure 1 is a radioautograph of a mid-sagittal section of a rabbit brain, showing the distribution of radioactive phosphorus (^{32}P) in a food-deprived animal after a 4-h feeding period following injection of the radioactive tracer. The high concentration of radioactivity in the hypothalamus reflects the uptake of (^{32}P)-labelled glucose-6-phosphate by neurons in that region, activated during the ingestion of food by the hungry rabbit. Interpretation of such radioautographs was restricted by several obstacles; in particular, the rapid spread of the labelled phosphate into many of the metabolic pathways of the Krebs cycle raised serious doubt that the observed diffuse distribution of the radioactivity corresponded well to

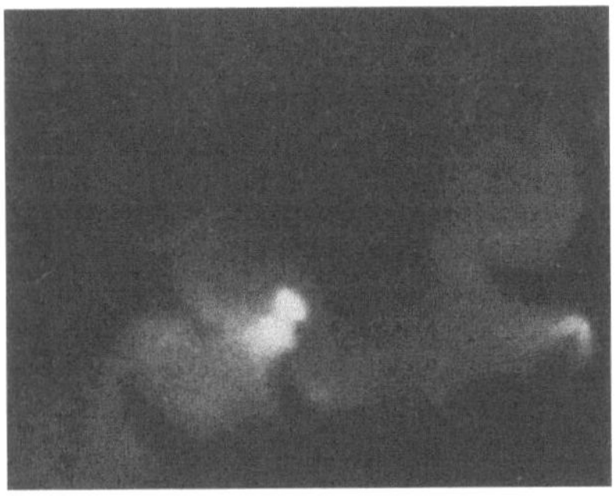

Fig. 1. Enlargement of a radioautograph of rabbit brain section. (From John 1954)

Springer Series in Brain Dynamics 1
Edited by Erol Başar

the locus of the cells where uptake initially occurred. Inability to find support to pursue this approach further compelled the development of other strategies.

1.2 Early Electrophysiological Evidence of Distributed Memory

We devised an electrophysiological analogue to radiotracer technique (John and Killam 1959). Visual or auditory stimuli at specific repetition rates, called *tracer stimuli,* were used as discriminanda in conditioning experiments in cats with multiple electrodes chronically implanted in a wide sample of brain structures. The spontaneous electrical activity, or EEG, recorded from these brain structures in the unanesthetized behaving animal was examined for rhythmic activity at the repetition rate or frequency of the tracer stimuli. Such frequency-specific activity was identified as *labelled rhythms* and was interpreted as reflecting the involvement of any brain region where labelled rhythms appeared in processing information about the tracer stimuli.

When tracer stimuli were first presented to naive animals at the beginning of conditioning, labelled rhythms were observed in a limited set of regions, largely corresponding to the sensory-specific structures mediating the modality of the stimulus (see. Fig. 2). As the conditioned response became established, labelled rhythms appeared in many additional brain structures. Such observations suggested that during learning, an anatomically extensive system was organized involving many nonspecific as well as sensory-specific regions and mediating subsequent performance of the learned behavior.

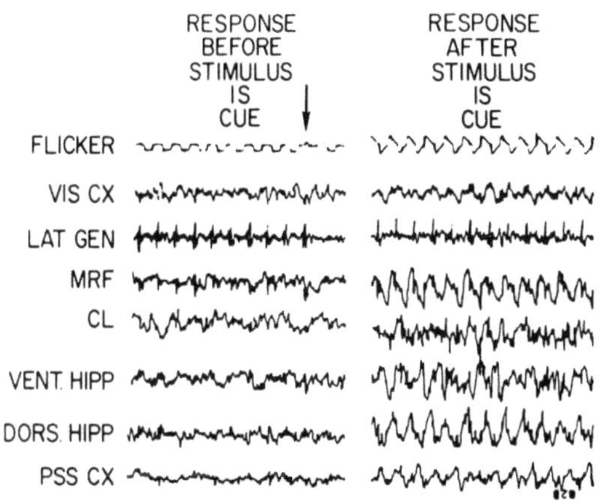

Fig. 2. *RESPONSE BEFORE STIMULUS IS CUE* shows the effect of photic (6-Hz flicker) on a cat after it had learned that milk could be obtained whenever a lever was pressed. The flicker had no signal value at this stage. Little "labelled" activity was elicited by the flicker except in the lateral geniculate *(LAT GEN)*. The response disappeared in the lateral geniculate because of the internal inhibition which occurred as the cat pressed the lever and waited for milk. *RESPONSE AFTER STIMULUS IS CUE* shows records during presentation of flicker (7 Hz) after the cat had learned a frequency discrimination. The cat was responding correctly by not pressing the lever down. There is marked enhancement of labelled responses at the stimulus frequency. *FLICKER*, 6-Hz photic stimulation; *VIS CX*, visual cortex; *MRF*, mesencephalic reticular formation; *CL*, nucleus centralis lateralis; *VENT HIPP*, ventral hippocampus; *DORS HIPP*, dorsal hippocampus; *PSS CX*, posterior suprasylvian cortex). (Data from John 1967).

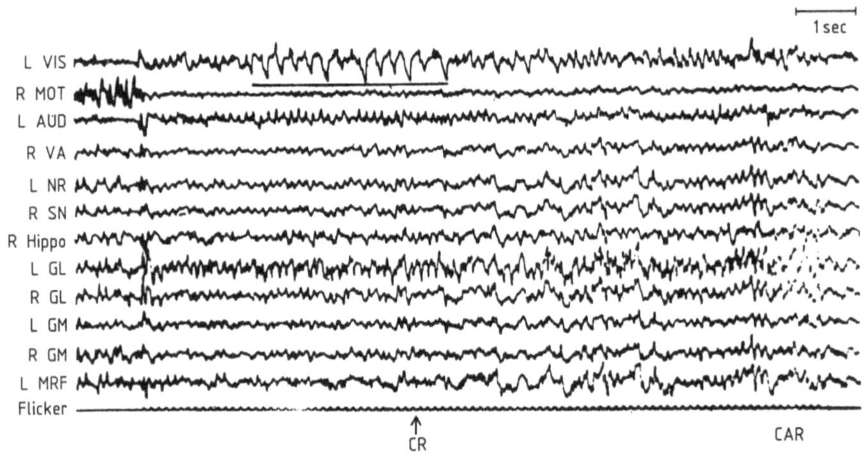

Fig. 3. Differentially trained cat responding to a 7.7-Hz flicker conditioned stimulus with an error followed by self-correction. Note the appearance of waves *(underlined)* in the visual cortex *(L VIS)* approximating the frequency of the absent cue (3.1-Hz flicker) just prior to the erroneous performance *(CR* and *arrow),* and the disappearance of those waves as the animal self-corrects by performing appropriately to the cue *(CAR)*. Bipolar derivations. (Data from John et al. 1975)

It rapidly became apparent that the pervasive labelled rhythms that appeared during learning had two distinct origins. A part of such activity clearly reflected the frequency of the external physical event, and was *exogenous*. Another part, however, seemed to reflect the activation of a memory rather than external reality, and was *endogenous*. One manifestation of endogenous activity was often seen in differentially trained animals during behavioral errors, when a discriminative response appropriate to an absent stimulus was performed after presentation of a different discriminanda. An example of such an *error of commission* (John and Killam 1960) is presented in Fig. 3. This cat was trained to perform a conditioned avoidance response (CAR) upon presentation of a 7.7-Hz flicker and a conditioned approach response (CR) to a 3.1-Hz flicker. The figure shows a marked 3.1-Hz rhythmic activity in the visual cortex while the cat performs an erroneous CR during presentation of the 7.7-Hz cue for a CAR. When the flicker cue remained on after the error, the labelled rhythm changed to the appropriate 7.7-Hz frequency and the correct behavioral response ensued.

Abundant literature reviewed elsewhere (John 1967) shows that an anatomically extensive *representational system* is established during experience. Representational systems may be activated by diverse internal as well as external cues and release a mode of electrophysiological activity, often coupled with behavioral acts, suggestive of the activation of a memory. Evidence of such released patterns can be seen in errors of commission, errors of omission, differential generalization, and so-called assimilation of the rhythm accompanied by behavioral performance.

1.3 Similarity of Widely Distributed Evoked Potentials

With the advent of average-response computers, it became possible to visualize these processes with better resolution. Figure 4 shows the evolution of visual evoked potential (EP) waveshapes elicited by tracer stimuli in multiple anatomical regions, throughout several stages in the elaboration of differentiated conditioned responses in a cat. Note that all of the waveshapes here shown come from *bipolar* derivations. The first column of waveshapes, recorded at the onset of conditioning, shows a diversity of responses with marked differences among regions. The most marked EPs are in the visual cortex and lateral geniculate body. When a simple CAR begins to be performed (column 2), rather similar waveshapes appear in many brain regions, nonsensory as well as sensory-specific. With establishment of a differentiated CAR, the power of the EP is markedly increased and the waveshapes in different regions become extremely similar. With substantial overtraining, these waveshapes undergo

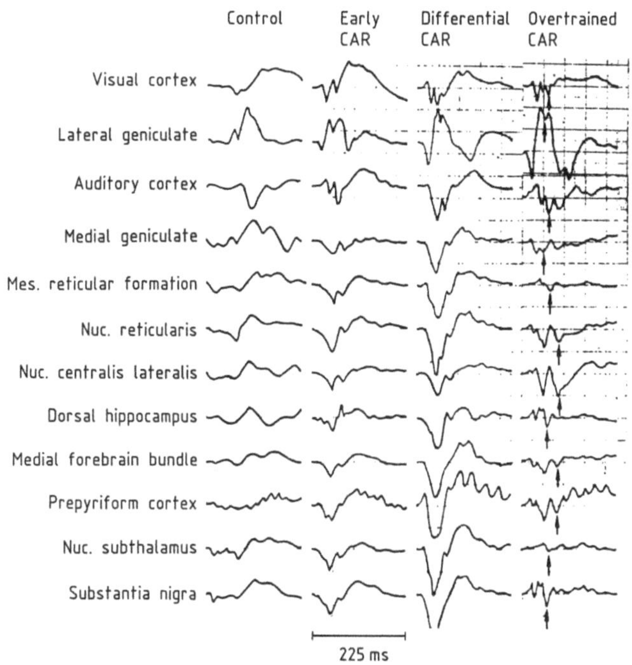

Fig. 4. Evolution of visual evoked response. *CONTROL*, average responses evoked in different brain regions of a naive cat by presentation of a novel flicker stimulus. Several regions show little or no response, and different regions display differing types of response. *EARLY CR*, responses to the same stimulus shortly after elaboration of a simple conditioned approach response (CAR, a lever press to obtain food). A definite response with similar features can now be discerned in most regions. *DIFFERENTIAL CAR*, responses to the same flicker conditioned stimulus shortly after establishment of a differential avoidance response to flicker at a second frequency. As usual, discrimination training has greatly enhanced the response amplitude, and the similarity between responses in different structures has become more marked. *OVERTRAINED CAR*, after many months of overtraining on the differentiation task, the waveshapes undergo further changes. The *arrows* point to a component usually absent or markedly smaller in behavioral trials on which this animal failed to perform the CAR. Monopolar derivations. (Data from John et al. 1975)

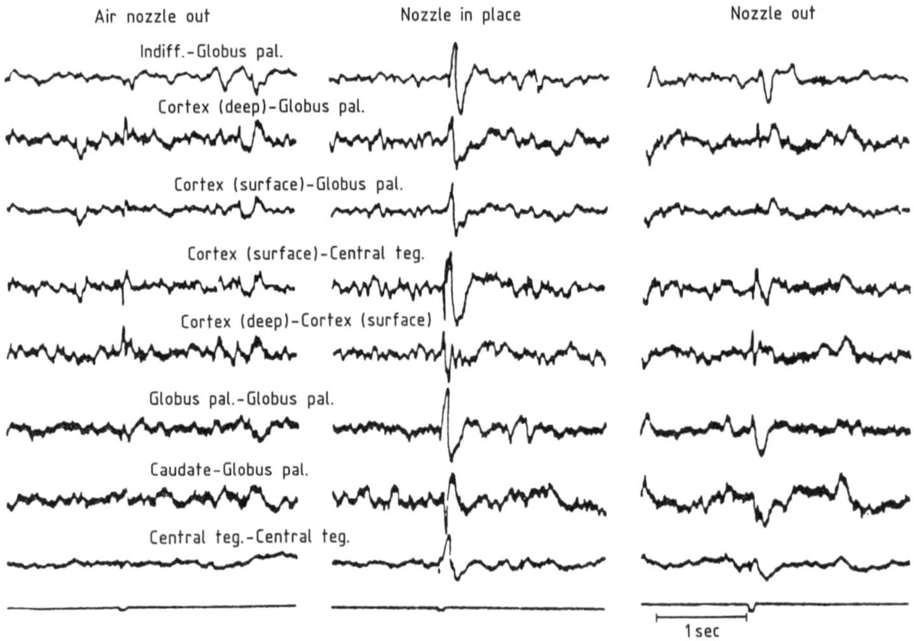

Fig. 5. Recordings from surface and deep bipolar derivations in a monkey that had repeatedly experienced an air puff delivered to the cornea through a nozzle whenever a click occurred. The *left* set of recordings shows the EPs elicited by the click after conditioning, when the air nozzle was removed from the test chamber. The click presentation, indicated by the deflection on the *bottom line*, causes a small primary component to appear which is localized to the auditory cortex. The *middle* set of recordings shows the EPs elicited by the same click when the air nozzle was in place, although no air puff was delivered. The EP now has a very similar waveshape and latency in all structures and contains a very prominent late component. The *right* set of records shows the reappearance of smaller, more localized EPs with different waveshapes when the nozzle is removed. (Data from Galambos and Sheatz 1962)

further changes. A prominent late component appears in many regions. This component, which we called the *readout* component, was usually absent if presentation of the conditioned stimuli failed to elicit behavioral performance (John 1972). Figure 5 illustrates a very similar phenomenon observed independently by Galambos. Such observations show that the similarity of EP waveshapes in different regions is dynamic in origin, and does not reflect static new connections.

Figure 6 illustrates EP waveshapes and single-unit activity recorded from the same microelectrode in the auditory cortex of a curarized animal. Before conditioning, the EP displays only a primary component, accompanied by an increased rate of discharge in a neuron relatively close to the microelectrode, which produces the larger amplitude spike discharge. Smaller spikes are also visible in the record, produced by a slightly more distant neuron which seems unresponsive to the click. After conditioning, the latency of the primary EP component decreases and a secondary EP component appears. The neuron producing the large spike discharge continues to display increased activity during the primary, but the neuron slightly more distant (producing the smaller spike) now displays a brisk increase in activity during the

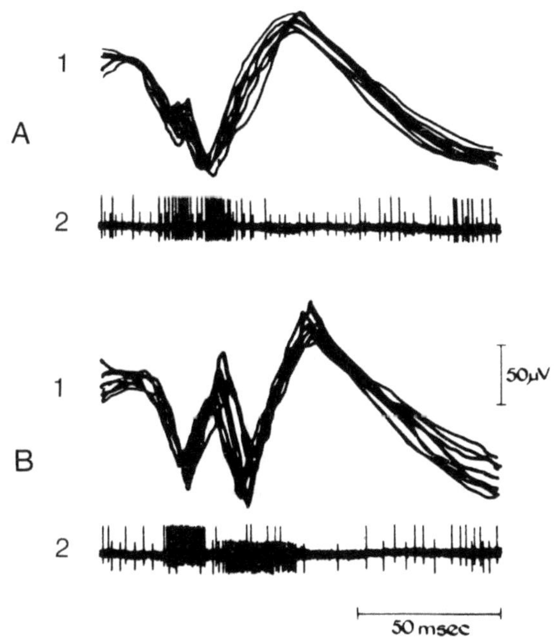

Fig. 6A, B. EP waveshapes and single-unit activity recorded from the same microelectrode in the auditory cortex of a curarized animal. The *upper* pair of tracings shows superimposed single EPs (A1) and unit discharges (A2) elicited by repeated click stimuli prior to conditioning. The *lower* pair of tracings shows superimposed single EPs (B1) and unit discharges (B2) elicited by repeated click stimuli after systematic pairing of the click with unavoidable shock. (From F. Morrell 1970, unpublished).

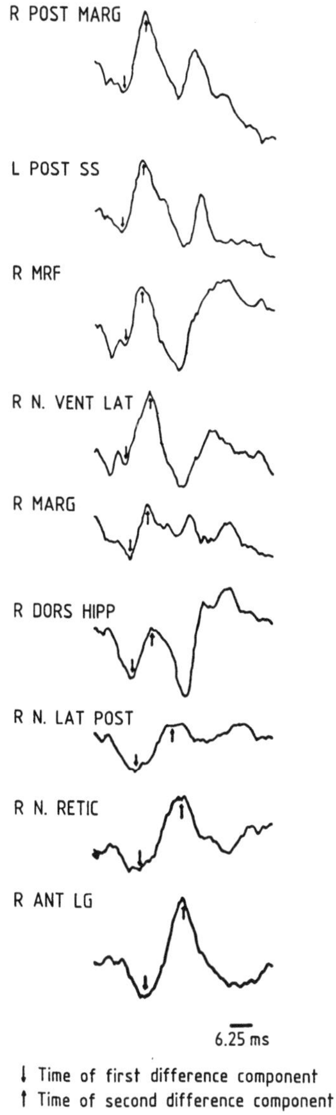

Fig. 7 ▶

Fig. 7. "Difference" waveshapes constructed by subtraction of averaged responses evoked by 7.7-Hz test stimulus during trials resulting in no behavioral performance from averaged responses evoked by the same stimulus when generalization occurred. Each of the original averages was based on 200 evoked potentials providing a sample from five behavioral trials. Analysis epoch was 62.5 ms. The onset and maximum of the difference wave has been marked by *two arrows* on each waveshape. The structures have been arranged from *top* to *bottom* in rank order with respect to latency of the difference wave. Note that the latency and shape of the initial component of the difference wave is extremely similar in the first four structures, and then appears progressively later in the remaining regions. *POST MARG*, posterior marginal gyrus; *POST SS*, posterior suprasylvian gyrus; *MRF*, mesencephalic reticular formation; *N. VENT LAT*, nucleus ventralis lateralis; *MARG*, marginal gyrus; *DORS HIPP*, dorsal hippocampus; *N. LAT POST*, nucleus lateralis posterior; *N. RETIC*, nucleus reticularis; *ANT LG*, anterior lateral geniculate; *R*, right side; *L*, left side. (Data from John 1967)

secondary EP component (F. Morrell 1970, unpublished). Such observations suggest that different cells may mediate exogenous and endogenous activity, as will be shown later.

By algebraic subtraction of EP waveshapes recorded during stimulus presentations resulting in no behavioral response (NR) from EPs recorded during trials resulting in conditioned response (CR), it was possible to obtain a computed estimate of the endogenous components of the EP in various regions. Figure 7 illustrates a set of such difference waveshapes constructed from CR (exogenous plus endogenous) and NR (exogenous only) EP waveshapes recorded from many regions at the same time (John 1967). Inspection of these estimates of the endogenous process shows a remarkable similarity in waveshape and simultaneity of time course in the top four derivations, recorded from bipolar electrodes located in regions quite remote from one another (marginal gyrus, suprasylvian gyrus, mesencephalic reticular formation, nucleus ventralis lateralis). Other regions showed a later appearance of this process, possibly reflecting centrifugal propagation of the process from an initial set of structures to other regions in the system. We proposed that this released process reflected *neural readout from memory*.

1.4 Evidence for Resonance Between Neuronal Ensembles

More precise examination of difference waveshapes recorded simultaneously from bipolar electrodes with small tip separations (< 1 mm) in different anatomical regions separated by considerable distances and characterized by very different morphology disclosed a remarkable synchronization of these endogenous processes. Figure 8 illustrates difference waveshapes photographed from the screen of an average response computer with a resolution of 1.25 ms per bin. Inspection of these data shows that the endogenous process was released in the visual cortex, lateral geniculate body, and mesencephalic reticular formation within less than the 1.25-ms time interval represented by each bin. Since synaptic traverse time is about 1 ms, and axonal transmission between these distant regions is known to require several milliseconds at least, it is difficult to explain such observations in conventional terms. These data suggest that after sufficient afferent information reaches different neuronal ensembles, they enter a coupled *common mode of oscillation* as if some kind of *resonance* had taken place between the separate ensembles. The extreme synchronization of that resonance is easier to reconcile with the action of an electromagnetic field than with synaptic transactions.

It was possible to demonstrate that endogenous activity could produce a released facsimile of the usual response to an absent event (Ruchkin and John 1966; John et al. 1969). Presentation of a neutral test stimulus midway in repetition rate between two differentiated tracer stimuli results in performance of the behavioral response, sometimes appropriate to one and sometimes to the other of the familiar but absent cues. Figure 9 shows that when the neutral visual test stimulus V_3 resulted in performance of the conditioned approach response, the waveshape V_3CR resembled the waveshape V_1CR elicited during correct response to the visual approach cue V_1 with a correlation coefficient of 0.59. When V_3 resulted in performance of the conditioned avoidance response, the waveshape V_3CAR resembled the waveshape V_2CAR elicited during correct response to the visual avoidance cue V_2 with a correlation coeffi-

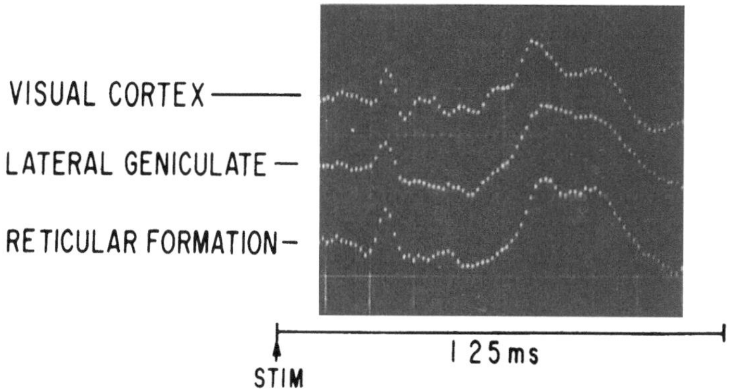

Fig. 8. Difference waveshapes obtained by subtracting average responses computed during three trials resulting in no performance (NR) from average responses computed during five trials resulting in correct performance of the conditioned avoidance response (CAR). All recordings were bipolar, and 75 evoked potentials were used in each of the constituent averages. Note the correspondence in latency and waveshape of the difference process in these various regions. (Data from John 1967)

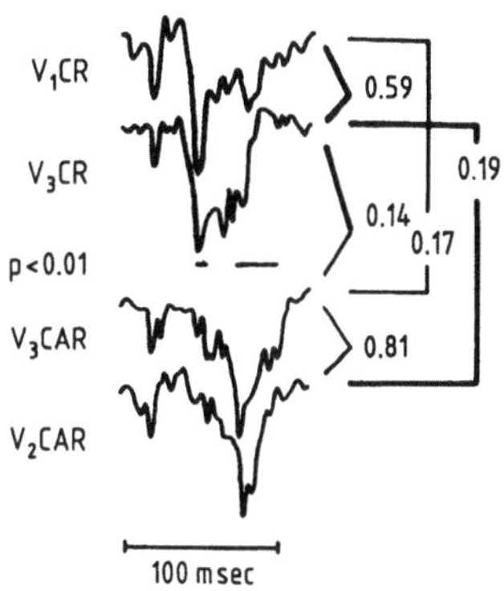

Fig. 9. Response waveshapes with averages based upon sequences of evoked potentials selected by the experimenter from the last 4 s of multiple behavioral trials. Average sample size, 15. See text for explanation. (Data from John et al. 1969)

cient of 0.81. The two waveshapes V_3CR and V_3CAR elicited by the same physical event activating two different memories were drastically different, with a correlation coefficient of 0.14.

Similar tests of *differential generalization* were conducted in many animals, using auditory as well as visual cues, and a wide variety of differentiated instrumental responses involving approach-approach, approach-avoidance and avoidance-avoidance discriminations. Under all of these conditions, the same release of facsimile waveshapes was observed. Cross-correlation of single EPs elicited by neutral stimuli against templates obtained by averaging EPs to differentiated stimuli could accomplish accurate prediction of responses in differential generalization tests of this sort (John 1972; John et al. 1973). Such observations suggested that the anatomically distributed representational system established during learning acquired the property of entering a *specific mode of oscillation* characteristic of its previous behavior under the conditions of the learning situation, if a sufficient subset of those conditions were reproduced. These specific modes of oscillation were released, but not caused by the actual physical stimulus.

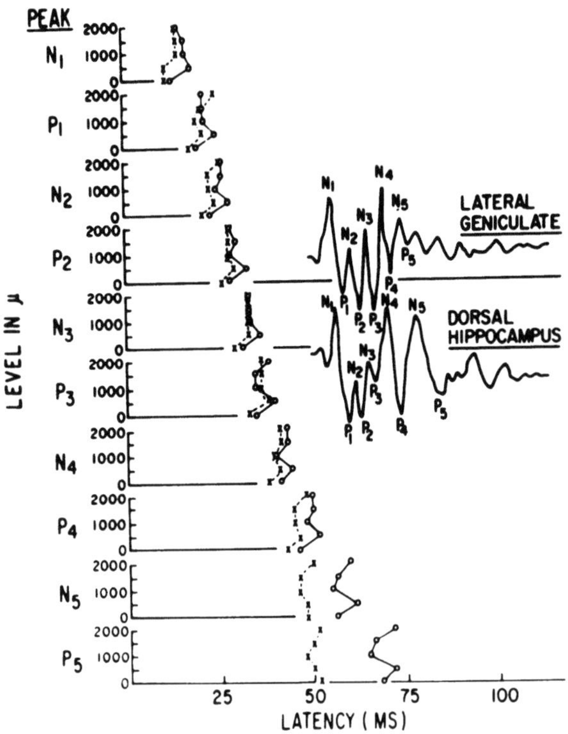

Fig. 10. Analysis showing similar latencies of acoustic evoked response (AER) components recorded from the lateral geniculate *(crosses)* and the dorsal hippocampus *(circles)*. Latency of component is plotted along the *abscissa* versus depth of penetration along the *ordinate*. Successively later components are depicted by *graphs* from *top* *(N_1)* to *bottom* *(P_5)*. Each point is based on an average response to 500 stimulus presentation in multiple behavioral trials. (Data from John and Morgades 1969b)

The anatomical distribution of coupled modes of oscillation between remote anatomical regions characterized by grossly different structure was studied further using multiple chronically implanted moving microelectrodes. Figure 10 illustrates data obtained from two such electrodes roving through the left lateral geniculate body and the right dorsal hippocampus of a trained, performing cat . A total of 500 EPs recorded during behavioral trials resulting in correct discriminated performance was averaged at each electrode position. The typical EPs obtained from the two anatomical regions are illustrated on the right. Crosses represent latencies of lateral geniculate components and circles those of the dorsal hippocampus. Up to component P_4, regions in these two structures display latency differences of less than 1 ms. Subensembles at different levels within the same structure reveal greater latency differences than exist between some ensembles in two different regions. These data suggest that the spatial structure of the hypothesized resonance underlying common modes of oscillation is locally inhomogenous in some domains.

Simultaneous monopolar recording of EPs and multiple units were obtained from these multiple moving microelectrodes (John and Morgades 1969b). Figure 11 illustrates data simultaneously recorded from the lateral geniculate (LG) and dorsal hippocampus of a trained cat under various circumstances. In each of the four examples, the solid line represents the EP waveshape and the stippled region represents the poststimulus histogram of a group of several neurons. Example A shows similar EP waveshapes in the two brain regions, with poststimulus histograms displaying peaks and troughs that indicate that voltage fluctuations in the EP modulate the *probability* of firing in the local neural populations. In trials resulting in correct per-

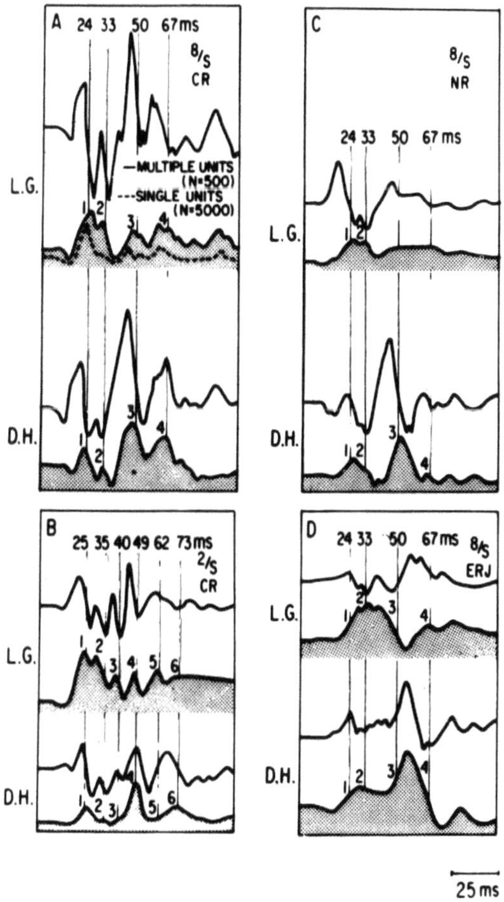

Fig. 11. (A) AERs *(solid curves)* and PSHs *(shaded areas)* simultaneously recorded from microelectrodes in the lateral geniculate body on the left side *(L.G.)* and the dorsal hippocampus on the right side *(D.H.)* during correct performance *(CR)* to the 8 Hz stimulus by cat 2. *Numbered vertical lines* indicate components considered to correspond with respect to relative latency. These and all other responses illustrated in this figure computed from 500 stimulus presentations, except for the PSH derived from a single unit in LG, shown as a *dotted line* ($N = 5000$). Note the correspondence between the curve describing the probability of firing of this single neuron observed over a long period of time and the PSH for the neural ensemble observed for one-tenth that time. (B) AERs and PSHs simultaneously recorded from LG and DH during correct performance *(CR)* to the differential 2-Hz conditioned stimulus. (C) AERs and PSHs simultaneously recorded from LG and DH during presentations of the 8-Hz conditioned stimulus which resulted in no behavioral performance *(NR)*. (D) AERs and PSHs simultaneously recorded from LG and DH during presentation of a novel stimulus illuminated by the 8-Hz flicker *(ERJ)*. (Data from John and Morgades 1969b)

formance to an 8-Hz visual cue for an approach response (food) requiring that the left lever on a work panel be pressed, the two regions show close correspondence both in EP morphology and neural firing probability. Example B shows a different temporal pattern but still good correspondence between the two regions during trials resulting in correct performance of a right lever press to a 2-Hz visual cue for a food reward. Example C shows that the similarity vanishes when presentation of the learned cue fails to elicit appropriate behavioral performance. Example D further illustrates the dynamic nature of a representational system, showing the gross alterations in temporal patterning and dissociation of the two anatomical regions when the cat looks at the experimenter's face illuminated by the 8-Hz flicker which has previously been the cue for an approach response.

1.5 Interim Conclusions Regarding Mechanisms

These data permit several conclusions:

1. The temporal patterns of deviation from random firing in local neural ensembles is modulated by the voltage of the local field potential. EPs are not epiphenoma,

nor do EP similarities between different regions reflect volume conduction from distant dipole generators. Rather, they reflect the nonrandom coherence in firing patterns of local neural populations.

2. The correspondence between large samples of the firing patterns of well-isolated single units (dotted line in Example A, $n = 5000$) and smaller samples of the firing patterns of multiple units (stippled area in Example A, $n = 500$) suggests a property of *ergodicity;* that is, the nonrandom behavior of a single neural element across a long period of information processing converges to the nonrandom behavior of a neural ensemble across a short period of processing the same information. This implies that the brain computes reliable information quickly by some sort of spatial averaging across large numbers of elements each of which is unreliable in the short term.
3. The correspondence in temporal pattern and close synchronization between EPs and ensemble firing observed in trained animals processing learned information is of dynamic origin and arises from some field property of the representational system, rather than from the static function of new synaptic connections established during learning.

1.6 Further Evidence for the Distribution of Meaning

The synchronization between neural ensembles was studied by measuring the amplitude and latency of peaks in multiple unit poststimulus histograms as the microelectrodes were moved across extensive anatomical trajectories over the course of many

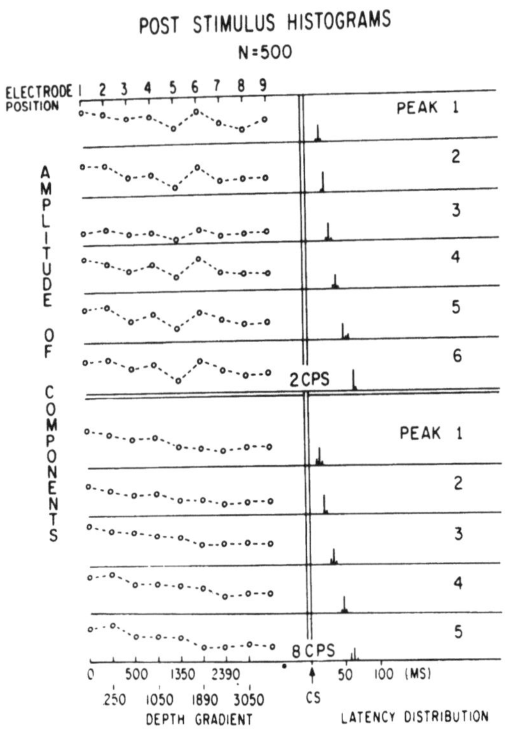

Fig. 12. Graphs on the *left side* illustrate the amplitude gradients of PSH peaks computed for a 2-Hz positive and 8-Hz negative flicker cue. Successively later components are depicted from *top* to *bottom*. In each graph, amplitude of the component is plotted as vertical displacement, while depth of electrode penetration is plotted as horizontal displacement. Each *graph* on the *right* shows the latency distribution of the PSH peak whose amplitude gradients are found in the graph at the same level on the *left side* of the figure. Component latency is plotted along the *abscissa,* while number of positions showing a peak at that latency is represented as the *ordinate.* (Data from John and Morgades 1969b)

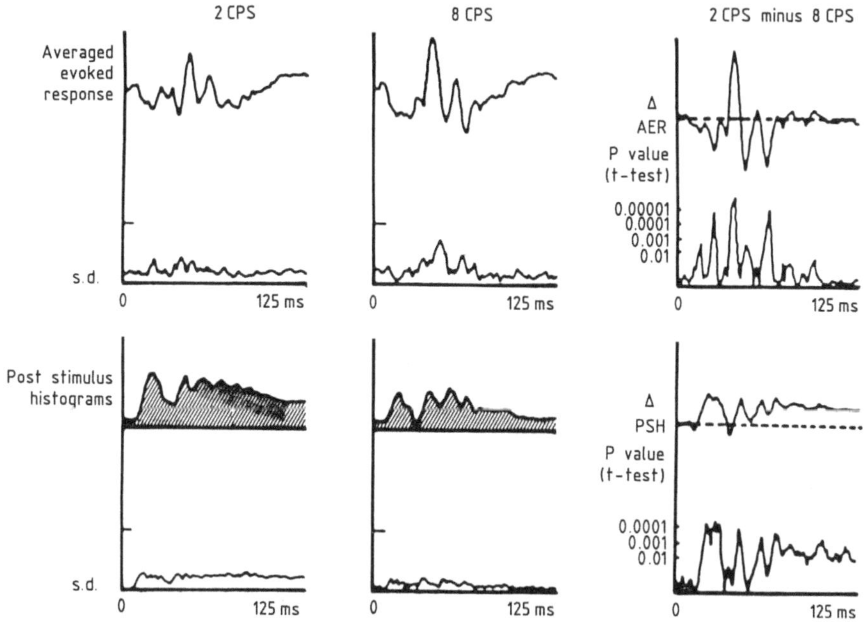

Fig. 13. *Upper left,* the *top curve* in this graph shows the *average* of the AERs elicited by the 2-Hz conditioned stimulus (CS) across all electrode positions in the mapped regions, while the *lower curve* shows the standard deviation of the group of AERs. *Lower left,* the *top curve* shows the *average* of the PSHs elicited by the 2-Hz CS across the same electrode positions and the lower curve shows the standard deviation. *Upper middle,* the *average* AER to the 8-Hz CS and the corresponding standard deviation. *Lower middle,* the *average* PSH to the 8-Hz CP and its standard deviation. *Upper right,* the *top curve* shows the difference waveshape resulting from the subtraction of the average AER to the 8-Hz CS from the average AER to the 2-Hz CS. The *lower curve* shows the *p* value, as computed by the *t* test, for each point of the difference wave; *lower right,* the *top curve* shows the difference waveshape resulting from the subtraction of the average PSH to the 8-Hz CS from the average PSH to the 2-Hz CS. The lower curve shows the *p* value for each point of the difference. (Data from John and Morgades 1969b)

months of study in each animal (John and Morgades 1969b). The chronically implanted electrodes were left in each position for several weeks in order to study each local ensemble adequately. Figure 12 illustrates the remarkable similarity in the temporal probability of nonrandom firing patterns observed across a trajectory of more than 3000 μ in the lateral geniculate body of a differentially trained cat. The amplitude of each peak of the poststimulus histogram at successive electrode positions is shown in the graphs on the left and the latency of the corresponding peak is shown on the right. These data establish the impressive synchronization of nonrandomness in the firing patterns of these extensive neural populations. Comparison of the upper and lower halves of the figure, representing the responses to the two discriminated stimuli, demonstrates that each stimulus elicits its own characteristic firing pattern across this anatomical domain.

Data from such trajectories were used to evaluate the likelihood that different conditioned responses were mediated by dedicated local circuits mediated by a restricted set of neurons. For the two discriminated conditioned stimuli, the grand average EP and poststimulus histogram and the corresponding standard deviations were computed across the full range of anatomical sites explored in the trajectory of microelectrode positions. These data were used to compute the *t* test between the two firing patterns across the whole anatomical domain, shown on the right side of Fig. 13 (John and Morgades 1969b). These measurements established that the variance between the firing patterns elicited by the two conditioned stimuli within any neural ensemble was *greater* than the variance of firing pattern induced by presentation of each of the stimuli across the full set of neural populations in the trajectory. These results support the proposition that the information about the *meaning* of each of the stimuli is distributed throughout the responsive neural elements in each anatomical region rather than by the specific responses of particular cells in a selective circuit which is different for each learned event.

1.7 Multipotentiality of Different Brain Regions

Using algebraic operations performed upon EPs obtained in correct and incorrect responses to differentiated conditioned stimuli, recorded from 34 electrodes placed in diverse anatomical loci in each of 22 trained cats, it was possible to compute the variance in the EP waveshape contributed by exogenous and endogenous processes in many different brain regions (Bartlett and John 1973). Figure 14 presents the results of these computations, plotting the log value of the variance due to endogenous components versus the variance due to exogenous components for different anatomical systems. The results show that endogenous components are present in all anatomical regions, in an amount which is logarithmically proportional to the amount of exogenous components. These results indicate that memory processes are diffusely represented in all brain regions, and that the participation of any neural population in the representational system for a specific memory is logarithmically proportional to the participation of that population in responses elicited by the external stimuli during the learning experience. This relationship strongly suggests some biochemical linkage between the discharges of neurons caused by afferent input and the modification of the responding cells in some fashion that times them so that they can be readily recruited into a resonating common mode of oscillation characteristic of the representational system for that experience.

The data represented by solid circles in Fig. 14 were derived from studies using visual conditioned stimuli, while open circles represent data derived from experiments using auditory stimuli. It is noteworthy that the slope of the best-fit line for stimuli in these two sensory modalities is the same. However, structures in the visual system lie above those in the auditory system for visual cues, and structures in the auditory system lie above those in the visual system for auditory cues. This suggests that so-called sensory-specific regions have a higher signal-to-noise ratio for events in the corresponding sensory modality, but such information is also represented in nonspecific regions with a lower signal-to-noise ratio. Different brain regions are therefore not *equipotential,* as suggested by Lashley, but *multipotential,* with different signal-to-noise ratios for different events. All regions, however, participate to some extent in the representation of all classes of information, at least after learning experiences establish a representational system.

Later studies, not illustrated here, established the existence of two kinds of

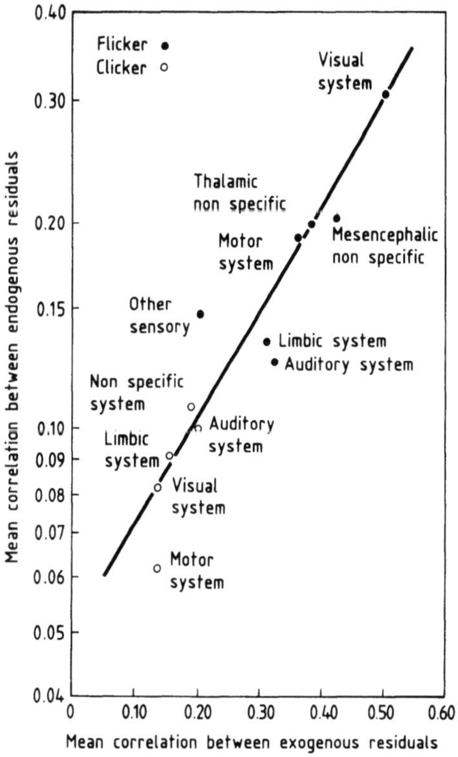

Fig. 14. Plot of mean correlation coefficients between exogenous residuals vs. endogenous residuals for different neural systems and for different cue modalities. *Closed circles,* flicker frequencies as stimuli. Auditory system: $N = 305$; auditory cortex (16 cats); medial geniculate (16): brachium colliculi inferioris (1). Limbic system: $N = 303$; hippocampus (16); dentate (5); cingulate (5); septum (5); prepyriform (6); medial forebrain bundle (6); mammilary bodies (5); hypothalamus (7). Mesencephalic nonspecific: $N = 158$; reticular formation (18); central gray (1); central tegmental tract (1). Motor system: $N = 146$; motor cortex (4); substantia nigra (10); nucleus ruber (4); nucleus ventralis anterior (9); subthalamus (5). Other sensory: $N = 54$; sensorimotor cortex (4); nucleus lateralis posterior (1); nucleus ventralis postero lateralis (5); nucleus ventralis postero medialis (1). Thalamic nonspecific: $N = 139$; nucleus centralis lateralis (13); nucleus reticularis (6), nucleus reuniens (1); medialis dorsalis (5); pulvinar (1). Visual system: $N = 394$; visual cortex (18); lateral geniculate (18); brachium colliculi superioris (2). *Open circles,* click frequencies as stimuli. Auditory system: $N = 48$; auditory cortex (5); medial geniculate (5). Limbic system: $N = 69$; hippocampus (5); dentate (3); cingulate (3); septum (3) prepyriform (2); medial forebrain bundle (3); mammilary bodies (3); hypothalamus (2). Motor system: $N = 37$; motor cortex (1); substantia nigra (4); nucleus ruber (1); nucleus ventralis anterior (5); subthalamus (2). Nonspecific system: $N = 50$; mesencephalic reticular formation (6); central gray (1); central tegmental tract (1); nucleus centralis lateralis (3); nucleus reticularis (3). Visual system: $N = 55$; visual cortes (6); lateral geniculate (6); brachium colliculi superioris (1). N denotes the number of independent measurements within the designated system. Date from monopolar and bipolar derivations were combined. Replications varied across cats and structures (Data from Bartlett et al. 1975)

neurons in each of many regions studied (Ramos et al. 1976). We found neurons whose firing patterns were determined by the physical stimulus, which we called "stable cells," and neurons whose firing patterns were predictive of the subsequent behavioral response independent of the physical stimulus, which we calles "plastic

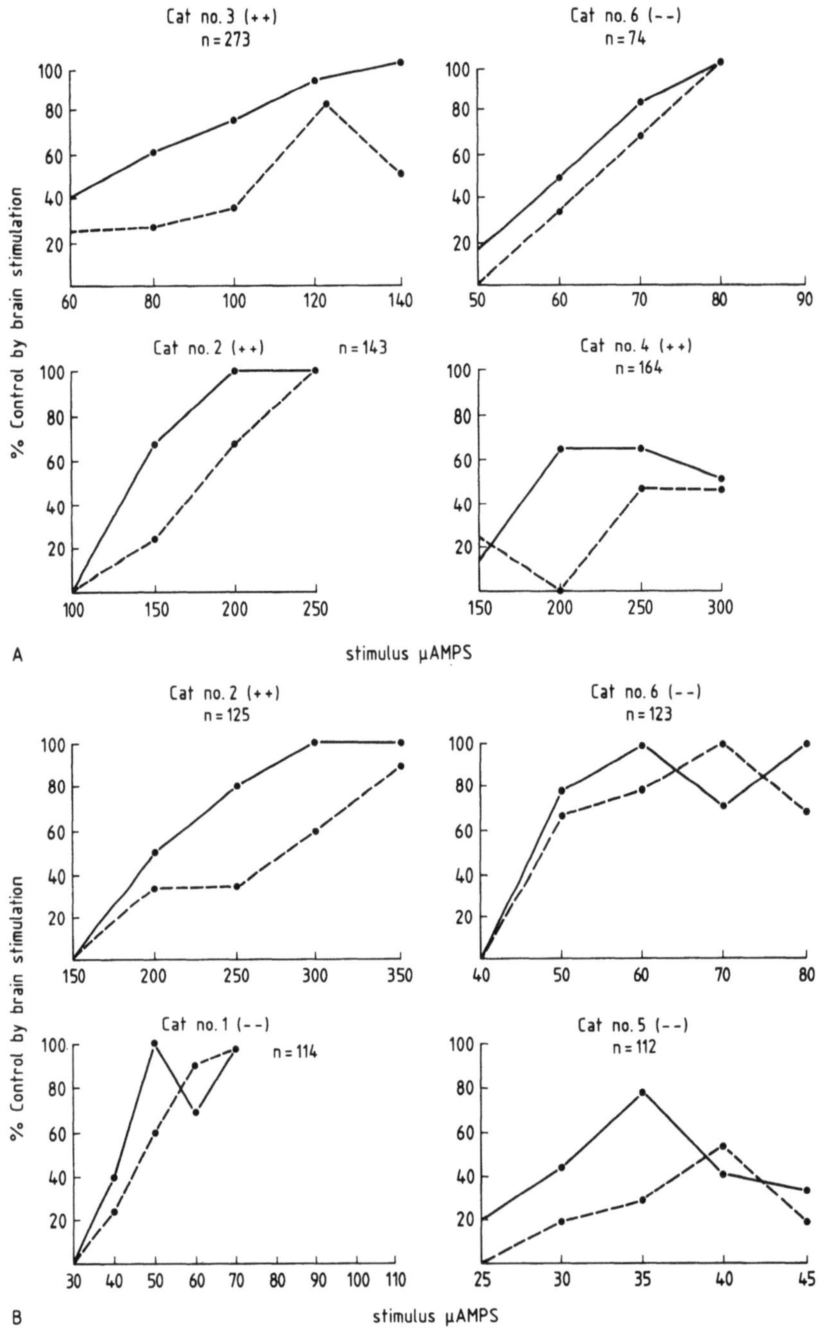

Fig. 15A, B. Contradiction of auditory (A) and visual (B) signals by brain stimuli. Each *graph* in the figure shows the effectiveness with which stimulation of the mesencephalic reticular formation at either of two frequencies (RF_1 and RF_2) contradicted simultaneously presented visual stimuli (V_2 and V_1, *left*) or auditory stimuli (A_2 and A_1, *right*), plotted as a function of increasing current intensity. For cats 1, 3, and 6, frequency 1 was 4 per second and frequency 2 was 2 per second. For cats 2, 4, and 5, frequency 1 was 5 per second and frequency 2 was 1.8 per second. *Solid lines* show the outcomes when peripheral stimulation at the higher frequency (V_1 or A_1) was pitted against RF

cells." Stable cells, which we believe mediate exogenous activity, and plastic cells, which we believe mediate endogenous activity, were found closely intermingled in different anatomical regions. About 30% of the cells examined in these studies were in the plastic category.

1.8 Probabalistic Model of Learning

The various experiments reviewed above suggested that learning established a representational system involving most if not all regions of the brain, with each region participating in the memory to an extent proportional to activation during the learning experience, and with neurons in all regions engaged by the representational system in a probabilistic manner. Population of neurons participated in a common mode of oscillation characterized by nonrandom temporal patterns of firing. Individual neuronal discharges were of importance only insofar as they contributed to the population statistics, rather than because they represented coded events in labelled lines corresponding to specific connections constituting a circuit for a specific memory.

This model, derived by inference from experimental observations, was subjected to direct test by electrical stimulation of the brain (Kleinman and John 1975). Trains of 200-ms bipolar rectangular pulses at a 1-KH_3 repetition rate, modulated by the repetition rate of previously established visual and auditory tracer stimuli and counterbalanced for total current, were introduced via electrodes implanted into diverse brain regions of cats previously trained to perform differentiated approach (+) or avoidance (−) responses to the tracer stimuli. Direct electrical stimuli of the mesencephalic reticular formation with these frequency-modulated pulse trains *immediately* resulted in accurate performance of differentiated conditioned responses appropriate to the modulation frequency; that is, appropriate for the visual or auditory tracer stimulus whose repetion rate corresponded to the frequency modulation of the elctrical pulse train. Figure 15A illustrates the effectiveness of direct electrical stimulation of the mesencephalic reticular formation in determining the outcome of *conflict trials,* in which simultaneous auditory conditioned stimuli were contradicted by brain stimulation at the repetition rate previously established for the alternative behavioral cue. These studies were counterbalanced with respect to central and peripheral stimulus repetition rates. The graphs indicate that, as the current delivered by the pulse train to the reticular formation increased, the percentage of control of the differentiated behavior by the direct electrical input also increased, to 100% in three of the four cats.

Figure 15B illustrates similar results obtained when visual conditioned stimuli were contradicted by direct electrical stimulation of the reticular formation. These

---

stimulation at the lower frequency (RF_2), while the *dotted lines* show the outcomes when the higher-frequency stimulus was delivered to the RF. Cats 1, 5, and 6 were trained to perform avoidance-avoidance discrimination (− −), while cats 2, 3, and 4 were trained to perform approach-approach discrimination (+ +). *N* refers to the total number of conflict trials carried out in each cat, accumulated in three sessions for cats 2, 5, and 6 and four sessions for cat 1 (visual RF conflict), and in three sessions for cat 2, four for cat 6, five for cat 4, and seven for cat 3 (auditory RF conflict). (Data from Kleinman and John 1975)

data support the proposition that learned information is represented by the nonrandom firing patterns of anatomically extensive neural populations, independently of which neurons in the ensemble actually fire at any moment. It is extremely unlikely that these electrical currents, delivered arbitrarily through gross electrodes and causing the discharge of cells around the electrode tip proportional to the gradient of current flow, successfully activated neurons in selective discrete circuits representing specific memories. The increase in control of behavior as a function of the amount of current delivered, which corresponds to the percentage of cells in the population driven by the brain stimulation, further supports the contention that information is represented by the statistical nonrandomness of temporal firing patterns in neural populations rather than by the firing of discrete cells in labelled circuits mediating a specific memory.

This proposition was tested by a further experiment using direct electrical stimulation of the brain. Cats were trained to perform one conditioned response to a 2-Hz visual or auditory tracer stimulus and a different behavioral response for the same reward to a 4-Hz visual or auditory stimulus. After substantial overtraining, electrical pulse trains modulated at 2-Hz were delivered to two brain regions at a time. In one condition, the two regions were stimulated *in phase,* so that two bursts of pulses per second were received by the whole brain. In another condition, the two regions were stimulated *250 ms out of phase,* so that four bursts of pulses were received by the whole brain. When input to the two regions was in phase, the cats performed the behavioral response appropriate to a 2-Hz sensory cue. However, when the input to the two regions was out of phase, the cats performed the behavior response appropriate to a 4-Hz sensory cue, *although each region was receiving a 2-Hz stimulus.* Figure 16 presents the EPs recorded from the visual cortex in trials at thresholds for this phenomenon when the 2-Hz stimuli were delivered 250 ms out of phase to the mesencephalic reticular formation and nucleus medialis dorsalis. The EPs from the visual cortex show that when the behavioral response was appropriate to a 4-Hz stimulus, the visual cortex registered four events per second, while two events per second were registered when spatial summation failed to occur. These data show that the integration of temporal firing patterns across different neural ensembles can be utilized as information by the brain, even though no neural ensemble is directly stimulated by the integrated pattern. The probability that the two out-of-phase 2-Hz electrical pulse trains selectively activated neurons in pathways mediating response to a 4-Hz visual or auditory conditioned stimulus is vanishingly small. This experiment, in my opinion, completed a process which has become universally accepted as essential for the contention that a natural phenomenon is understood:

1. The phenomenon must be *observed.*
2. The observations must be *interpreted.* "Interpretation" means that a hypothetical model must be *constructed.*
3. The phenomenon must be *controlled.* "Control" means that some prediction of phenomena independent of the initial observations must be supported by experimental manipulations of the observed system.

In the present case, a variety of observations of neural electrical activity in trained animals led to the proposition that learning established an anatomically distributed representational system, in which neural ensembles in different brain regions represented previous experience by a shared temporal pattern of nonrandom coherence

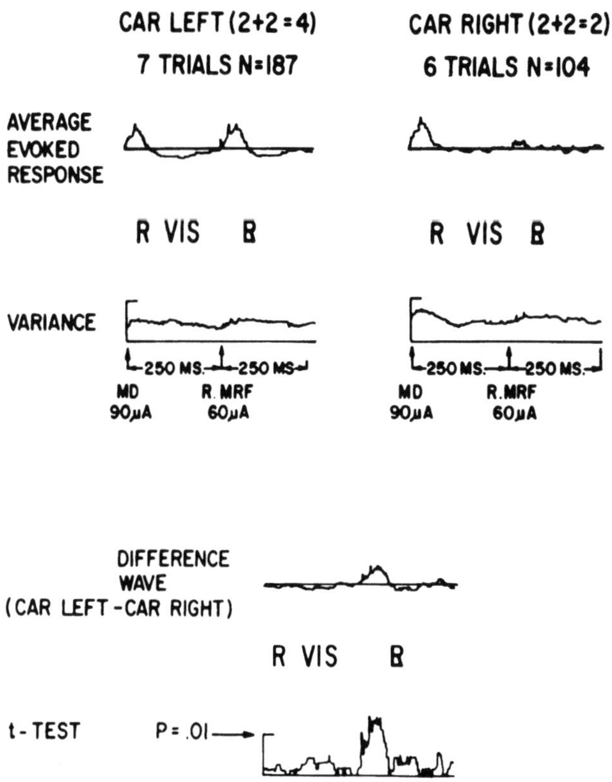

Fig. 16. Averaged evoked responses *(top)* recorded from the visual cortex to two-per-second out-of-phase stimulation delivered to the medialis dorsalis and mesencephalic reticular formation. Evoked response waveshapes *(left* and *right)* are different depending on behavioral outcome. Difference wave and *t* tests between evoked respones on *top, left,* and *right at bottom.* (Data from unpublished observations by Kleinman and John 1975)

in firing, independently of which individual neurons contributed to the statistical behavior of the ensemble. This probabilistic model was directly tested by electrical stimulation of various brain regions, which elicited performance of differentiated conditioned responses in a manner compatible with the model of statistical encoding, and incompatible with a connectionistic model.

2 Metabolic Mapping of Neurons Involved in Memory

After the summary given above of our prior electrophysiological studies of memory, we are now ready to present the results of our current radiotracer experiments. These experiments take advantage of three technical strategies.

Metabolic Trapping of Labelled Glucose. In recent years, the obstacle to early radioautographic studies (John 1951), namely the rapid movement of labelled glucose-6-phosphate into the metabolic pathways of the Krebs cycle, has been overcome. The solution involves the use of 2-deoxyglucose (2-DG) which is transported from the blood into brain cells and phosphorylated like the natural substrate, glucose. Deoxyglucose-6-phosphate, however, cannot be further metabolized because of its altered structure and is thus metabolically trapped inside the cell, providing a tracer for localized glucose uptake (Kennedy et al. 1975; Sokoloff et al. 1977). Since this strategy was devised, numerous studies have utilized it for mapping various functional neural pathways by radioautography. Such studies have often demonstrated striking changes in regional glucose utilization as a result of gross changes induced in the functional state of various brain areas.

These gross changes have been demonstrated by comparing mean values of glucose utilization for selected regions between groups of control and experimental animals. In order for such studies to succeed, the regions selected for comparison must be suspected to mediate the functions under study, the functional changes caused by the experimental variable must be relatively large, and the variability in metabolism small for each brain region of interest within each of the subjects in the control and experimental groups. These conditions are often difficult to satisfy, especially for subtle functions mediated by brain regions not yet identified.

Sequential Double Labelling. Potentially, with appropriate quantification of the radioautographic images, 2-DG mapping can be used to measure more subtle influences on regional neural activity. One limitation on the precision which can be achieved in such studies is the inherent variability of regional neurochemistry, permeability of the blood-brain barrier, and brain blood flow, diffusion, and transport across cell membranes in different brain areas, as well as the variability in metabolic rate of the same brain area between different individual animals. Shortly after the 2-DG metabolic trapping strategy was devised, a sequential double-label DG procedure was described (Agranoff and Altenan 1977; Altenan und Agranoff 1978). This method takes advantage of the fact that 2-DG can be labelled with a variety of radioactive tracers without changing its molecular structure. It is possible to select two such tracers, which differ sufficiently with respect to half-life or energy of the emitted radiation, so that conditions can be devised to make radioautographs which reveal the distribution of only one or of both tracers. By appropriate algebraic manipulation of serial radioautographs, taking advantage of those conditions, the distribution of 2-DG labelled with each of the two tracers can be separately determined. The distribution of 2-DG labelled with one tracer can now be used to obtain a baseline metabolic control for each brain region, while the distribution of 2-DG labelled with the second tracer can be used to study relative increases or decreases in local glucose metabolism under some experimental condition of interest.

The sequential double-label strategy offers an elegant potential solution to the problem of inherent regional variability in glucose metabolism, and to the intraregional differences between animals. However, this solution is valid only if it is possible to ensure that the functional state of all brain regions in the control and experimental conditions to be compared will be different only because of experimental influences and not because of unspecific fluctuations in metabolic rate within each region. Since no such guarantee can be given, the precision potentially available via the double-label strategy requires an estimate of test–retest variability in regional glucose metabolism under unchanged conditions. This requirement is difficult to satisfy,

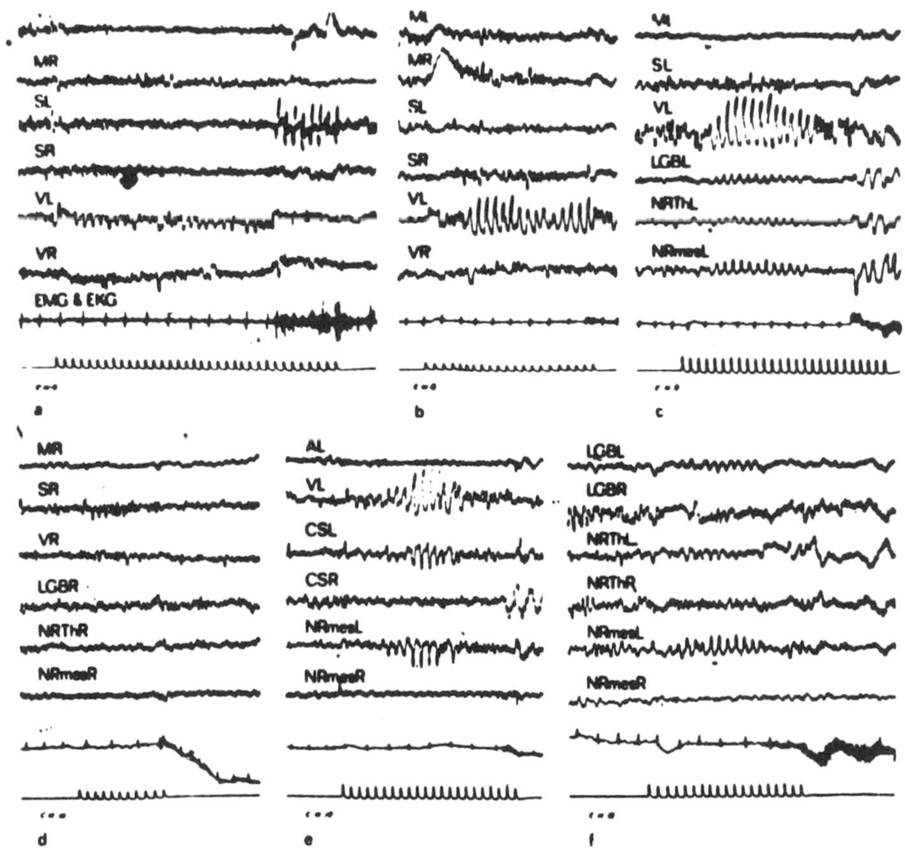

Fig. 17. Electrophysiological data from a split-brain cat. Sea text for details. (From Majkowski 1967)

since the purpose of the strategy is to permit comparisons between two different conditions. Furthermore, the double-label strategy does not by itself offer a solution to the problem of variability in regional metabolism across animals, because the distribution of 2-DG with one label under one condition cannot serve as a control for changes in distribution of 2-DG with the second label within the same animal, for exactly the same reasons.

Split-Brain Experimental Animals. More than 20 years ago, a technique was described by Sperry (1962) to separate the two hemispheres of the brain in cats, and to restrict the visual input to each hemisphere to that provided by the ipsilateral eye. Studies of learning in such split-brain cats, with the cerebral commissures and optic chiasm transected, established that visual information delivered only to one hemisphere remained localized on that side.

The split-brain cat was used for a particularly elegant electrophysiological study of learning by Majkowski (1967). He studied the labelled rhythms that appeared during training to a tracer-conditioned stimulus consisting of a flash at a specified repetition rate, delivered only to one hemisphere. Figure 17 shows data obtained from such an animal, during presentation of flicker only to the left eye while the right eye was covered by an opaque contact lens. Recordings from numerous brain regions on both the left and right sides demonstrate that visual input remains lateralized in this preparation. Majkowski used such split-brain animals to demonstrate that during generalization to a test stimulus at a different repetition rate, frequency-specific labelled rhythms at the frequency of the tracer stimulus used in conditioning appeared only on the trained side of the brain and therefore could not reflect unspecific influences.

This same strategy can be applied to functional mapping studies using double-labelled 2-DG to solve both of the problems identified above – the variance in glucose metabolism between different brain regions within the same condition caused by intrinsic regional differences, and the variance within the same brain regions between conditions caused by unspecific metabolic fluctuations, within the same animal. This permits each animal to provide all necessary control data, eliminating the need for separate control and experimental groups and potentially providing sufficient information about variance to permit data from a single animal to be evaluated statistically.

In general, the incorporation of a split-brain preparation into a double-labelled 2-DG functional mapping study of memory would entail using one "control" side of the brain to obtain an estimate of regional test–retest metabolic variability in uptake of 2-DG, during two successive time intervals in which a specific memory was not activated on that side. Those data would provide the baseline metabolic information necessary to evaluate the regional test–retest differences found on the other experimental side, comparing uptake during an experimental time interval in which a specific memory was activated on that side to uptake during a control time interval in which the specific memory was not activated.

Using the information from the control side, it should be possible to separate differences in regional uptake due to unspecific factors such as arousal from those due to activation of a specific function such as memory on the experimental side.

2.1 Strategy for Functional Mapping of Memory Using Double-Labelled 2-Deoxyglucose Radioautography in Split-Brain Cats

Our experimental design, derived from the above considerations, was as follows:

2.1.1 Experimental Treatment

Three split-brain cats were prepared, with transection of the cerebral commissural tracts and the optic chiasm using the methods described by Majkowski (1967), and permitted to recover from surgery for 4 weeks. Each cat was then trained to obtain food in a Yerkes box consisting of a starting chamber and a runway 1 m long, ending in two doors. Each door carried a card bearing a transparent geometric figure transilluminated from behind by a fluorescent tube. The positive cue was two concentric circles and the negative cue was a star. The transilluminated areas of the positive and negative cues were equated. The position of the positive and negative cues on the two doors was randomized according to a Gellerman schedule. Food, consisting of a bolus of horse meat approximately 1 g in size, could be obtained on each trial by pushing open the door bearing the positive cue to expose a food cup. The door bearing the negative cue was locked, and self-correction was not permitted. Training sessions of 40 trials occurred at the same scheduled time each day, including weekends. Trials were 1 min apart, with the animal replaced in the starting box after each trial. One hour after the end of each training session, the animals were permitted free access to food in their home cage for a period of 1 h. Water was always available in the cage.

After reaching criterion (90% correct discrimination), each cat was overtrained for an additional period of about 6 weeks. During this period, each cat was subjected to several procedures in preparation for the experiment. First, sessions were run with an opaque contact lens on one eye at a time. After brief attempts to dislodge the lens, each animal showed no further distress and performed as usual. These tests showed that each hemisphere was capable of performing the required discrimination at the criterion level or better. Next, the previously clear transparent cue figures were replaced by cue figures constructed from transparent green plastic. Two additional contact lenses had been constructed, one of the same green plastic and the other of red plastic. Further tests were carried out using these colored contact lenses. In the first test, the opaque lens was placed over one eye and the green lens over the other eye. Under this condition, each hemisphere was tested to establish whether discrimination of the green cue figures could be accomplished. In each cat, after a brief initial period of hesitation, the discriminations were performed at criterion level no matter which eye carried the green lens. In the final test, the opaque lens was placed over one eye and the red lens over the other. Under this condition, each cat performed at the random level, no matter which eye carried the red lens. These results were interpreted to prove that the green lens permitted the transilluminated green cue information to be adequately perceived by the receiving hemisphere, that the red lens effectively blocked reception of green cue information, and that each hemisphere contained an adequate memory trace.

The day before the 2-DG experiment, the cat was transproted from New York University to Brookhaven National Laboratory by car, together with the training apparatus. Several hours after arrival, at the regularly scheduled time, the animal was subjected to the usual behavioral session. Each cat performed with his customary accuracy during this session, with no visible signs of distress after the travel. After this test session, no further food was provided so as to ensure a high level of motivation on the next day.

On the following day, the 2-DG experiment was carried out. The green contact lens was placed over one eye and the red contact lens over the other eye. In two of the cats, the green lens was on the right eye, and in the third cat on the left eye. As soon as the contact lenses were in place, at the usually scheduled time, the cat received an intravenous injection of [^{14}C]-2-fluorodeoxyglucose (2FDG), in one forepaw. New cue cards were used, each bearing a transparent red triangle constructed from the same plastic as the red contact lens in addition to the transparent green circles or star. Sixty trials were performed, at the rate of one per minute, using the usual randomized schedule. During this initial uptake period, each cat performed at the criterion level or better, receiving approximately one-half the usual amount of food per trial. The time occupied by these trials exceeded the 45 min required for most of the circulating 2FDG to be taken up from the blood.

The animal was permitted to rest for 1 h after uptake of the first label. During this period, the radiochemistry group of Brookhaven National Laboratory completed the synthesis of an adequate amount of [^{18}F]-2FDG after production of the requisite [^{18}F] in the cyclotron of this research facility. After the interval of 1 h, the cat received an injection of 20–30 mCi of [^{18}F]-2FDG in the other forepaw. New cue cards were used, each bearing a transparent green triangle in addition to the transparent red triangle. No learned cue information was now available to either hemisphere. The red and green contact lenses remained in place. sixty trials were performed under these conitions, at the rate of one per minute. During this second uptake period, each cat performed at the random level, dividing its decisions almost equally between the two sides and displaying only temporary hesitation. All choices were reinforced with food, approximately one-half the usual amount. Each cat ran exactly the same distance in the same amount of time as under the previous condition and received the same amount of food.

As soon as the second uptake period was completed, each cat was sacrificed with an intravenous injection of a massive pentobarbital excess. When cardiac arrest occurred, the brain was removed, frozen in an alcohol bath containing liquid CO_2, and imbedded in methyl methacrylate. Serial sections 30 μ thick were prepared, using an LKB whole-body cryomicrotome. Alternate sections were air dried and placed on mammography film and placed in a deep freeze unit for radioautographs, and every tenth section was selected for conventional histological staining. Commercially prepared C-14 standards (Yonekura et al. 1983) were used. After 8 h, corresponding to approximately for [^{18}F] half-lives, the film was removed, yielding image II (second uptake period). After 12 h, when the [^{18}F] activity had decayed to less than 0.1% of its initial value, the sections were again placed on mammography film. After 15 days, the film was removed, yielding image I (first uptake period). The methods and systems used in these studies are presented in greater detail elsewhere (John 1976; Som et al. 1983).

This experimental procedure is summarized in Fig. 18. Figure 18A illustrates the experimental conditions under which [^{14}C]-2FDG uptake occurred, for a cat with the green lens on the right side. The control side of the brain is subject to multiple unspecific *influences*, including systemic variables such as blood pressure, heart rate, autonomic factors, arousal, motivation, movement, red patterned visual input, olfactory inputs from the experimenter and the apparatus, and the ingestion of food. The control side of the brain is subject to these same unspecific influences with green patterned visual input, plus the specific activation of the memory system mediating a previously learned visual pattern discrimination.

Figure 18B illustrates the experimental conditions under which [^{18}F]-2FDG uptake occurred, with unspecific influences affecting both sides of the brain equally. Red light input is equally present under both [^{14}C]-2FDG and [^{18}F]-2FDG uptake for the control side, while green light input is equally present under both conditions for the experimental side under both uptake conditions.

Thus, for each brain section on the control side, image I reflects [^{14}C]-2FDG uptake due to unspecific factors in the first session, while image II reflects [^{18}F]-2FDG uptake caused by unspecific factors in the second session, plus some contamination

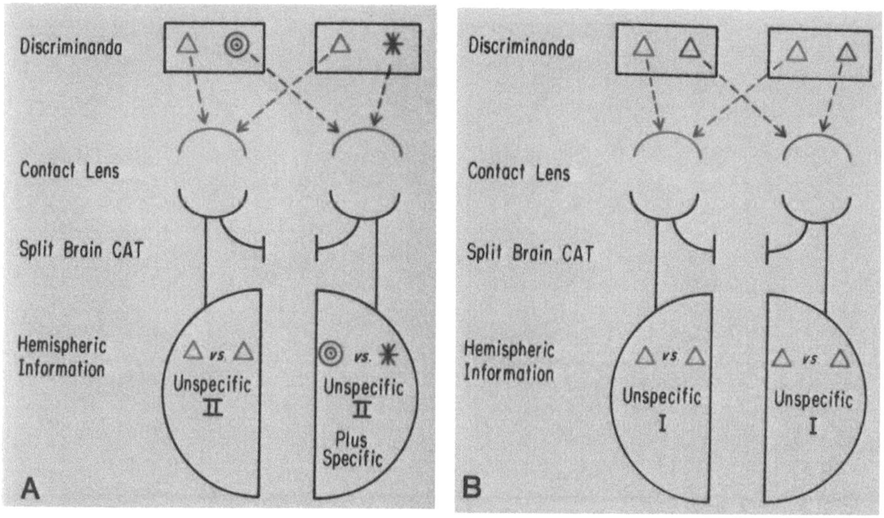

Fig. 18. A Diagram of experimental design showing input to reference hemisphere *(left)* and experimental hemisphere *(right)* for cat wearing a green contact lens on the right eye, during [^{14}C]-2DG uptake period (image II). B Input to hemispheres of same cat during [^{18}F]-2DG uptake period (image I). (Data from John et al. 1986)

caused by radioactive decay of the [^{14}C]-2FDG also present during the exposure. This contamination will be referred to as [^{14}C] "shine."

For each brain section on the experimental side, image I reflects [^{14}C]-2FDG uptake caused by unspecific factors *plus* activation of the specific memory of circle-star discrimination in the first session, while image II reflects [^{18}F]-2FDG uptake caused by unspecific factors during the second session, plus some contamination due to decay of the [^{14}C]-2FDG also present during the exposure ([^{14}C] "shine").

2.1.2 Image Processing

The goals of the image processing procedure were to quantify the 2-DG uptake on the control side and the experimental side in the two test periods, and to subtract the images of the [^{18}F] uptake in the second test period from that of the [^{14}C] in the first test period. On the control side, this would yield an estimate of the regional test–retest metabolic variability [unspecific − unspecific = replicability of metabolism; i.e., = metabolic "noise"]. On the experimental side, this would yield the distribution of the brain cells activated by previously learned cues, in addition to the metabolic noise [(specific memory + unspecific) − unspecific = memory + noise]. The noise variance would then be used to set statistical criteria to help identify brain regions mediating the specific memory.

To achieve these goals, the following steps were necessary for each section.

Quantification. Images I and II were digitized, using a video densitometer (Hamamatsu 512 × 512 buffer memory system with 8-bit AOC resolution, 100 × 100

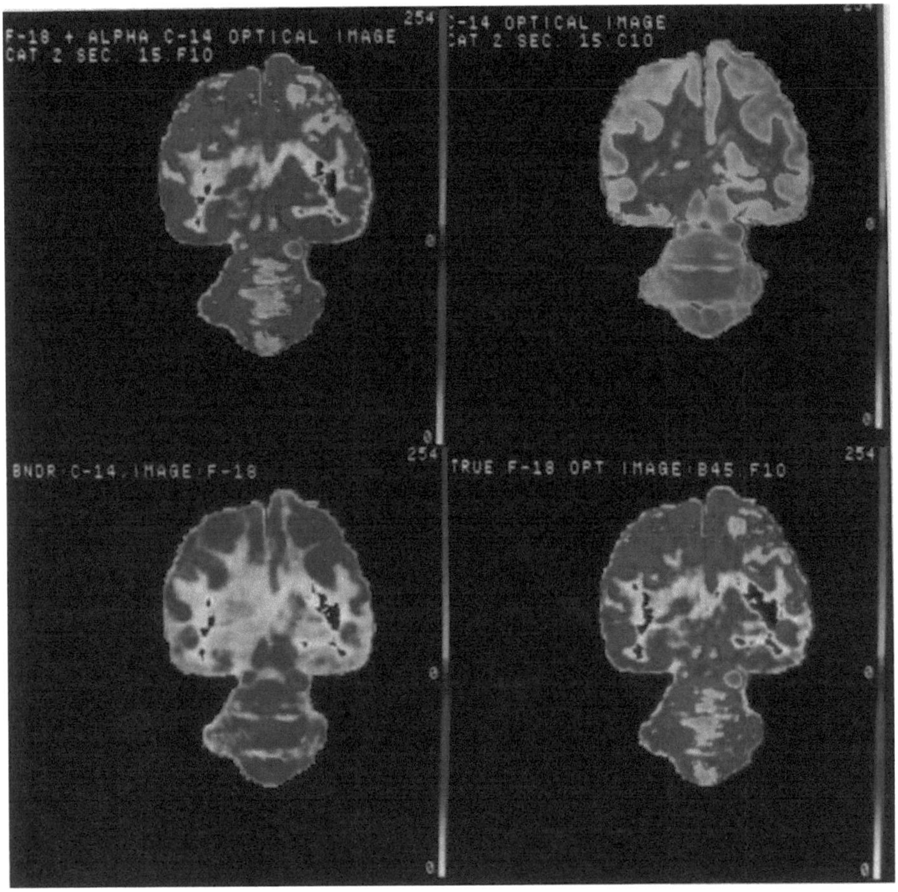

Fig. 19. Computer-generated images of same brain section as in Fig. 18 (15.10) corresponding to sequential stages in processing. *Color scale at right* confirms reproducibility of palette used for color coding of densitometric data. *Top left,* optical density image I, showing distribution of [^{18}F]-2DG uptake contaminated by detection of [^{14}C]-2DG also present in the tissue. *Top right,* optical density image II, showing distribution of [^{14}C]-2DG. *Bottom left,* boundary of image II *(pinke line)* encircles pale blue image I after registration of the two images. *Bottom right,* true [^{18}F]-2DG optical image after subtraction of an amount of image II proportional to the exposure ratio T1/T2. For further details, see text. (Data from John et al. 1986)

micron pixels with 64 images signal averaged). Film background was subtracted to produce an "optical density image." The first panel on Fig. 19 (top) shows the optical density image of image II ([^{18}F]-2FDG plus [^{14}C] "shine"), and the second panel shows the optical density image of image I ([^{14}C]-2FDG), for a typical section from cat 2, after background subtraction. The color coding on these images is a heat scale moving from white to red as optical density increases.

Registration. Images I and II had to be brought into precise registration. This was accomplished by an automatic registration algorithm at first, and then by superimposing the online of image I upon image II using visual correction. The automatic algorithm gave acceptable registration for about half the images, while visual correction was necessary for the remainder. The third panel in Fig. 19 shows the outline of the [^{14}C]-2FDG radioautograph as a pink line encircling image II.

Purification. The [^{14}C] "shine" had to be removed from image II in order to obtain the true image of [^{18}F]-2FDG uptake in the second test period. In order to make this correction, the optical density of each pixel in image I had to be multiplied by the exposure ratio factor $T2/T1$, where $T2$ was the short exposure time required to obtain the contaminated image II, and $T1$ was the long exposure time required to obtain image I. Since $T2/T1$ ([^{14}C]-2FDG) is precisely equal to the [^{14}C] "shine," this quantity was then subtracted from the optical density value of the corresponding pixels in image II, yielding the "true [^{18}F]-2FDG optical density image," seen in the bottom panel of Fig. 19.

Conversion. Optical density images of the anatomical distribution of a radioactively labelled 2-DG molecule are proportional to the glucose uptake of different brain regions in the section from which the radioautograph was obtained. However, the coefficient of proportionality that relates optical density to glucose uptake varies as a function of the capture efficiency of the photographic emulsion on the radioautographic film for the particles emitted when the radioactive tracer decays. Because the energy of emission is different for each radioisotope, calibration standards must be used to calculate a correction term for the different efficiency of the mammography film for [^{14}C] and [^{18}F] decay. Furthermore, because the exposure time required to obtain a distinct radioautograph with [^{18}F]-2FDG was long relative to the 2-h [^{18}F] half-life, additional corrections were necessary to take into account the interval between injection of the [^{18}F]-2FDG and completion of the exposure. The original optical density images were converted to "activity images" based on exposure times and the F-18 and C-14 standards. The top panel of Fig. 20 shows the optical density image for [^{14}C]-2FDG (left) and the true [^{18}F]-2FDG optical image (right). The second panel of Fig. 20 shows the computer-generated activity images of [^{14}C]-2FDG (left) and [^{18}F]-2FDG (right) uptake during the first and second test periods. The color scale on these images now reflects local [^{14}C] and [^{18}F] 2-FDG uptake in microcuries of radioactivity per pixel.

Normalization. Within each activity image, the relative [^{14}C]- and [^{18}F]-2-FDG uptake in micromoles per pixel (proportional to the microcuries of radioactivity) is reflected in the color scale. However, the [^{14}C]-2FDG and [^{18}F]-2FDG images from any section cannot yet be accurately compared to quantify the difference in local glucose metabolism in the two test periods, because the *specific activity* – i.e., the percentage of the total glucose pool that was radioactively labelled 2FDG – was different for the substances. In order to compensate for these differences so that quantitative comparison of local glucose utilization under the two test conditions can be accurately computed, the two activity images must be *normalized*. This normalization can be accomplished by equating the total activity across all pixels of image II to the total activity of image I. In common-sense terms, this compensates for the fact that the number of [^{18}F]-2FDG molecules available to the brain was different from the

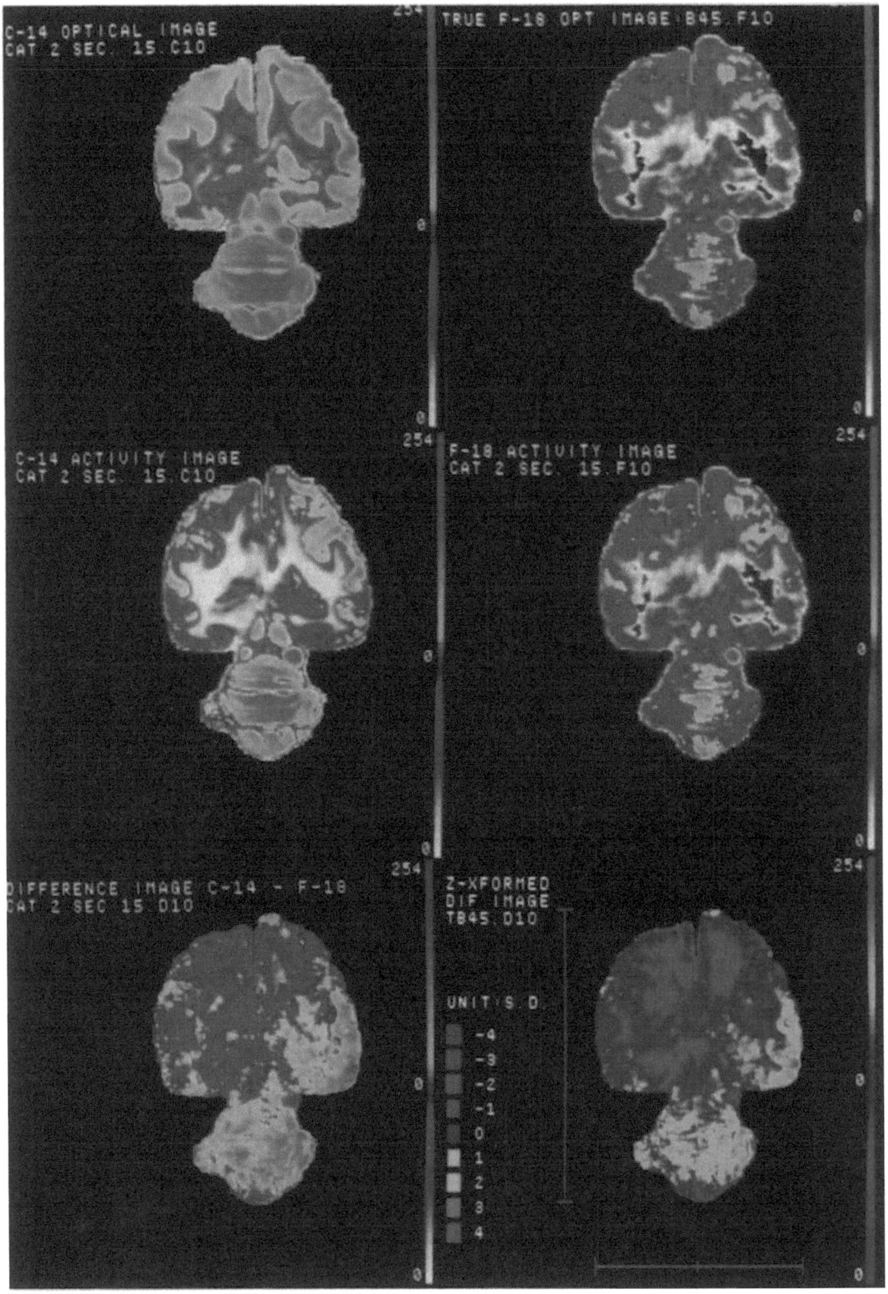

Fig. 20. Subsequent steps in processing are shown below, for the same section illustrated in Fig. 18B. *Top left,* optical image II, reflecting [^{14}C]-2DG uptake in the first session. *Top right,* corrected optical image I, showing true [^{18}F]-2DG uptake in the second session, after purification shown in Fig. 19. *Second row left,* activity image II, showing [^{14}C]-2DG uptake after correction for capture efficiency of emulsion. *Second row right,* activity image I, showing true [^{18}F]-2DG uptake after correction for emulsion capture efficiency and decay. *Third row left,* difference image obtained by subtracting the

number of [^{14}C]-2FDG molecules. The normalized activity images are not illustrated.

Subtraction. Once the original optical density images obtained in the two test periods have been brought into registration, converted to activity images, and normalized, comparison of local glucose utilization under the two test conditions could be validly made. Image II was subtracted from image I, yielding a "difference image," shown in the third panel of Fig. 20. The color scale in this difference image is proportional to the differences in local glucose utilization between the two test conditions. In the range from white to green, metabolism in the second test was higher than in the first. In the range from green to red, metabolism in the first test was higher than in the second.

According to our experimental design, the difference image for the control side (on the left in this animal) should reflect random metabolic variability, or "noise." The difference image on the experimental side should reflect the activation of a specific memory, in addition to metabolic "noise." Each pixel in this image has served as its own control. Inspection of the third panel of Fig. 20 shows that the right side of the brain, which was processing information related to the learned form discrimination in the first but not the second test, displayed a much greater difference in glucose utilization than the control side.

Z-Transformation. While the asymmetries in the third panel of Fig. 20 are obvious, the significances of differences between the two test conditions are difficult to evaluate because the color scale is still in microcuries. In order for the difference in any local region to be considered to reflect the metabolic consequences of activation of a specific memory, rather than random metabolic fluctuations or "noise," the noise variance must be taken into consideration. In order to accomplish this, the mean value and standard deviation across all pixels on the control side were computed from the difference image. These statistics were then used to z-transform each pixel in the difference image [z = (mean pixel difference minus individual pixel difference)/standard deviation of the mean].

It was now possible to compute the z-transformed difference image, shown in the bottom panel of Fig. 20. The color scale for the z-image ranges from -4 SD (pale blue) to $+4$ SD (red) and directly reflects the probability that the difference in glucose utilization observed in any pixel arises from random metabolic variance or from participation (excitatory or inhibitory) in the activation of a specific memory. It should be pointed out that a z-score of + or -4 SD corresponds to a probability much less than $P = 0.0001$.

---

true [^{18}F]-2DG image I from the [^{14}C]-2DG image II, after normalization for differences in specific activity accomplished by equating the total activity of the reference hemisphere in the two images. *Third row right,* Z-transformed difference image, obtained by Z-transformation of all pixels in the difference image relative to the mean and standard deviation of the reference side of the image above (third row). The *color coding* now reflects standard deviations of the difference. (Data from John et al. 1986)

2.2 Results of the Double-Label 2-FDG Study of Memory

The skeptical reader who questioned the need for the elaborate controls devised in this experiment to cope with the uncertainties in interpretation of single-label 2-DG radioautographic studies, caused by variations between different regional metabolic processes and fluctuations of metabolism arising from unspecific factors, is urged to compare the top left panel and the bottom panel of Fig. 20. The anatomical distribution of regions participating in a specific memory shown in the bottom panel of Fig. 20 is drastically different from that suggested in the top left panel.

It is not relevant to our purposes in this paper to describe in detail the anatomy of the neural structures affected by activation of a specific memory, identified in these studies. That anatomy will be presented elsewhere. For the present, suffice it to say that participating regions are extensively distributed throughout a wide range of brain structures. In cats 1 and 2, for which the experimental side was on the right, consistently greater glucose utilization was seen on the right side. In cat 3, for which the experimental side was on the left, the converse was true.

The steps in image processing illustrated in Fig. 19 were applied to serial sections from the top to the bottom of cats 2 and 3, at about 0.75-mm intervals (the brain of cat 1 was warped during freezing and not appropriate for thorough quantitative analysis). For each section, the number of pixels above and below the mean difference value of the control side was computed for both the control and experimental sides. Out of a total of 53 sections sampled in the two cats, 45 had a greater percentage of pixels above the mean difference than below on the experimental side rather than the control side.

The total number of pixels in the brains of cat 2 and 3 were somewhat different (13854192 vs. 11314728), possibly reflecting differences in age and weight between the two animals. If we assume that each brain contained about 2×10^9 neurons, the average pixel contained 144 neurons in cat 2 and 177 neurons in cat 3.

In cat 2, 5.61% more pixels were above the mean than below the mean on the experimental side rather than on the control side. In cat 3, 5.07% more pixels were above the mean than below the mean on the experimental side rather than the control side. Performing the appropriate calculations, we reach the conclusion that 111919705 neurons participated in the activation of a specific memory in cat 2, and 101537238 in cat 3, if we consider only significant increases in glucose utilization. If significant decreases were taken into account, these numbers might be almost twice as large. (Note: this computation should be repeated including only pixels whose z-scores corresponded to $P < 0.01$.)

2.3 Conclusions

The older electrophysiological data reviewed in the first portion of this paper failed to support notions that memory was mediated by dedicated discrete pathways established by learning and activated when the corresponding memory was retrieved. Those data showed involvement of large anatomically diffuse ensembles of neurons, with the readout of specific memories characterized by unique spatiotemporal patterns of nonrandomness. Information appeared to be represented by the statistical departures from randomness in large populations of neurons, with the activity of individual neurons important only insofar as it contributed to ensemble statistics. This model received support from studies of direct stimulation of different brain regions with electrical pulses, showing that discriminated learned behaviors could be selectively activated depending upon the temporal patterns of the electrical stimuli. Such findings were difficult to reconcile with memory models requiring the activation of specific neuronal pathways, even if such pathways were assumed to be anatomically distributed and extensively redundant.

The double-label 2-FDG studies reported here yielded results compatible with a probabilistic model and even more difficult to reconcile with a dedicated pathway model than the electrophysiological data. The conclusion that upwards of 100 million neurons are involved in mediation of a single two-form discrimination makes extremely unlikely the proposition that transmission across selected synaptic junctions reflects activation of a specific memory. Even if one were to concede that the 100 million participating neurons were activated selectively by the hypothesized junctions established during learning, the sheer volume of cells involved requires that all cells participate in numerous memories. "Labelled lines" restricted to mediation complex feature extractions, converging to percept or specific memory extractors, are no longer plausible, no matter how redundant or distributed such dedicated circuits might be.

These data better support notions of mass action, in which the spatiotemporal nonrandom behavior of huge ensembles of neural elements mediates the processing of information and the retrieval of memories. The observed phenomena do not fit with a computer-like model of the brain, with information stored in discrete registers no matter how many in number. A radically different model must be sought. In view of the enormous number of cells involved, the question of how the information represented in this swarm of neurons can be appreciated by the brain becomes of central theoretical interest. No conceivable neuron or neuronal set, no matter how diffuse its synaptic inputs, can evaluate the enormous amount of neural activity now shown to be involved in retrieval of even a simple discriminated response.

The explanation must be found in some property of the coherent neural activity demonstrated by these studies. When masses of hundreds of millions of cells enter synchronous discharge in a particular temporal pattern, the distribution of extracellular electrical charges must become extremely nonrandom and inhomogenous in space and time. Powerful currents must flow and powerful electrical fields must appear. We must consider the possibility that when negative entropy in the system reaches a certain level, those fields acquire a property which is not to be inferred from the properties of the individual neural elements in the system. Under certain conditions, negative entropy in matter may acquire the property of "consciousness;" that is, a unique sensitivity to its own spatiotemporal organization. Consciousness and memory in complex neural systems may be mediated by the physical properties of the system as a whole, and not depend upon any of the elements which contribute to the system. In previous speculations (Thatcher and John 1977) I have referred to this hypothetical system property as a "hyperneuron." This is simply a pseudo-biological term to cloak the heretical notion that what we consider the higher mental functions mediated by the brain, consciousness, self-awareness, and memory, are properties which are physical properties of matter rather than uniquely neural products. The unique property of the brain may simply be that because man has a brain, he can recognize that matter possesses these properties.

Acknowledgements. The 2-DG studies reported here were performed in collaboration with Drs. Bertrand Brill, Xiang Zhu, and Yuan Tang of Brookhaven National Laboratory; Dr. Kenji Ono, Nagasaki University; Dr. Ronal Young, University of the West Indies; Dr. Leslie Prichep, NYU; and Dr. Thomas Seitlin, University of Zürich.

References

Agranoff BW, Altenau LL (1977) A sequential double-labeled method for the measurement of radioactive 2-deoxyglucose phosphate distribution in the brain. Proc Int Soc Neurochem 6:513–516

Altenau LL, Agranoff BW (1978) A sequential double-labeled 2-deoxyglucose method for measuring regional cerebral metabolism. Brain Res 153:375–381

Bartlett F, John ER (1973) Equipotentiality quantified: the anatomical distribution of the engram. Science 181:764–767

Bartlett F, John ER, Shinokochi M, Kleinman D (1975) J Behav Biol 14:409–449

Galambos R, Sheatz GC (1962) An electroencephalograph study of classical conditioning. Am J Physiol 203:173–184

John ER (1954) Functional brain mapping using radioactive tracers. PhD thesis, University of Chicago

John ER (1967) Mechanisms of memory. Academic, New York

John ER (1972) Switchboard vs. statistical theories of learning and memory. Science 177:850–864

John ER (1976) A modell of consciousness. In: Schwartz GE, Shapiro D (eds) Consciousness and self-regulation. Plenum, New York, pp 1–50

John ER, Killam KF (1959) Electrophysiological correlates of avoidance conditioning in the cat. J Pharmacol Exp Ther 125:252–274

John ER, Killam KF (1960) Electrophysiological correlates of differential approach-avoidance conditioning in the cat. J Nerv Ment Dis 131:183–201

John ER, Kleinman D (1975) Stimulus generalization between differentiated visual, auditory, and central stimuli. J Neurophysiol 38:1015–1034

John ER, Morgades PP (1969a) Neural correlates of conditioned responses studied with multiple chronically implanted moving microelectrodes. Exp Neurol 23:412–425

John ER, Morgades PP (1969b) Patterns and anatomical distribution of evoked potentials and multiple unit activity elicited by conditioned stimuli in trained cats. Comm Behav Biol 3:181–207

John ER, Shimokochi M, Bartlett F (1969) Neural readout from memory during generalization. Science 164:1519–1521

John ER, Bartlett F, Shimokochi M, Kleinman D (1973) Neural readout from memory. J Neurophysiol 36:893–924

John ER, Bartlett F, Shimokochi M, Kleinman D (1975) Behav Biol 14:247–282

Kennedy C, des Rosiers MH, Jekle JW, Reivich M, Sharpes F, Sokoloff L (1975) Mapping of functional neural pathways by autoradiographic survey of local metabolic rate with [^{14}C]-deoxyglucose. Science 187:850–853

Kleinman D, John ER (1975) Contradiction of auditory and visual information by brain stimulation. Science 187:271–273

Majkowski J (1967) Electrophysiological studies of learning in split brain cats. Electroencephalogr Clin Neurophysiol 23:521–531

Ramos A, Schwartz E, John ER (1976) Stable and plastic unit discharge patterns during behavioral generalization. Science 192:393–396

Ruchkin DS, John ER (1966) Evoked potential correlates of generalization. Science 153:209–211

Sokoloff L, Reivick M, Kennedy C, des Rosiers MH, Pattak CS, Pettigrew KD, Sukurada O, Shinohara H (1977) The [^{14}C] deoxyglucose method for the measurement of local cerebral glucose utilization: theory, procedure, and normal values in the conscious and anesthetized albino rat. J Neurochem 28:897–916

Som P, Yonekura Y, Oster ZH, Meyer MA, Pelleteri ML, Fowler JS, MacGregor RR et al. (1983) Quantitative autoradiography with radiopharmaceuticals. II. Applications in radiopharmaceutical research. J Nucl Med 24:238–244

Sperry RW (1962) Some general aspects of cerebral integration. In: Mountcastle VB (ed) Interhemispheric relations and cerebral dominance. Johns Hopkins Press, Baltimore

Thatcher R, John ER (1977) Foundations of cognitive processes. Erlbaum, Hillsdale

Yonekurs Y, Brill AB, Som P, Bernett GW (1983) Quantitative autoradiography with radiopharmaceuticals. I. Digital filmanalysis system by videodensitometry: concise communication. J Nucl Med 24:231–237

Recent Advances in Neurocognitive Pattern Analysis

A. S. Gevins

1 Introduction and Methods

We use the generic term "neurocognitive pattern (NCP) analysis" to refer to procedures we have been developing to extract task-related spatiotemporal patterns from the unrelated "noise" of the brain. There have been three generations of NCP analysis. The first measured background EEG spectral intensities while people performed complex tasks, such as arithmetic problems lasting up to 1 min. These patterns had sufficient specificity to identify the type of task (Gevins et al. 1979a, b). However, when tasks were controlled for stimulus-, response-, and performance-related factors, they had identical, spatially diffuse EEG spectral scalp distributions (Fig. 1; Gevins et al. 1979a, c). This study suggested that previous (and the most current) reports of EEG hemispheric lateralization may have confounded electrical activity related to limb and eye movements, stimulus properties, and task difficulty with those of mental activity per se (Gevins et al. 1980). The second generation measured cross-correlations between 91 pairwise combinations of 15 electrodes recorded during performance of simple tasks. These split-second tasks, controlled so that only the type of judgment varied, were associated with complex, rapidly shifting neurocognitive patterns (Gevins et al. 1981). By using NCP analysis to extract differences between similar spatial tasks, rapidly shifting *focal* patterns were extracted (Fig. 2; Gevins et al. 1983, 1985). From these results, it is clear that a split-second temporal resolution is needed to isolate the rapidly shifting neurocognitive processes associated with successive information processing stages.

A third generation has been developed that operates on up to 64 channels recorded during a controlled sequence of stimuli in which a person prepares for, and executes, perceptual judgment and motor control tasks, and receives performance feedback (Gevins et al. 1987, submitted). This generation is also being applied to intracerebral recordings from primates. Because of the large size of the single-trial data sets (up to 150 megabytes for each person), two passes through the data are required to complete the analysis. The first pass selects channels, intervals, and trials with task-related information to reduce the amount of data, and then applies current source density or spatial deconvolution transforms to reduce volume conduction distortion. The second pass measures "functional interdependencies" (cross-covariance and time lag) between channels on enhanced averages obtained from the reduced data set. NCP analysis is implemented in the ADIEEG-IV analysis system (Fig. 3).

Springer Series in Brain Dynamics 1
Edited by Erol Başar

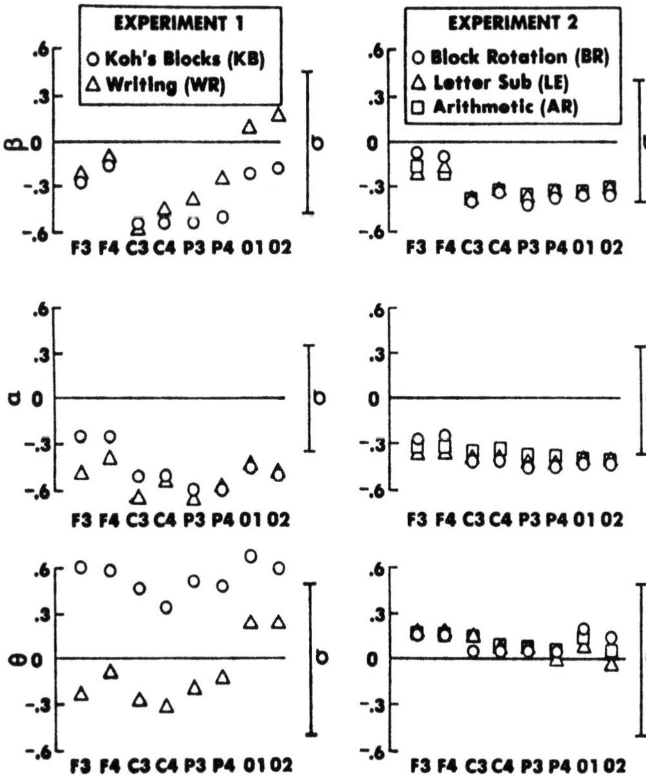

Fig. 1. Results of experiments designed to assess EEG correlates of higher cognitive functions. *Left:* tasks of experiment 1 were 1 min long and involved limb movements and uncontrolled differences in stimulus characteristics and performance-related factors. *Right:* tasks of experiment 2 were less than 15 s long and required no motion of the limbs; stimulus characteristics and performance-related factors were also relatively controlled. The graphs display means over all subjects of standard scores of EEG spectral intensities (expressed as changes from visual fixation values for clarity of display) recorded during performance of two tasks in experiment 1 and three tasks in experiment 2. *Upper, middle, and lower sets of graphs* are for spectral intensities in the theta, alpha, and beta bands. The *abscissas* show scalp electrode placements: *F3,* left frontal; *F4,* right frontal; *C3,* left central; *C4,* right central; *P3,* left parietal; *P4,* right parietal; *01,* left occipital; and *02,* right occipital. Standard deviations, which differed only slightly between electrode placements, are indicated on the *right* (I). Although there are prominent EEG differences between the uncontrolled tasks of experiment 1, EEG differences between the relatively controlled tasks of experiment 2 were lacking. Each of the controlled tasks is, however, associated with a remarkably similar bilateral reduction in alpha and beta band spectral intensity over occipital, parietal, and central regions. There is no evidence in these results that naturalistic, linguistic, and spatial cognitive tasks are associated with differentially lateralized EEG spectral intensity patterns. (From Gevins et al. 1979a)

Sixty-four-channel EEG Recording Technique and Automated Artifact Rejection. Recording capacity has been expanded to 64 channels to provide uniform scalp coverage with an interelectrode distance of about 3.25 cm on an adult head with a 10 cm radius (Fig. 4). Improved automated artifact rejection algorithms based on wave morphology, spectral, and topographic criteria are being developed for preliminary data screening (Gevins et al. 1975, 1977). Two eye movement artifact removal tech-

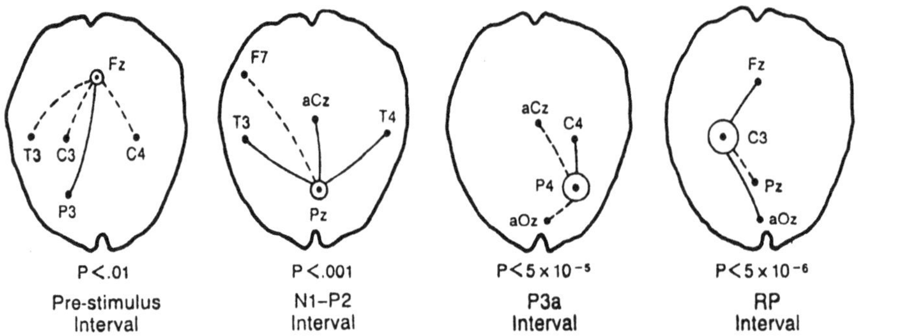

Fig. 2. Spatial brain potential differences between two split-second tasks requiring a spatial judgment are shown. A movement was required in one task, while the other required withholding the response. The most significant differing areas, their significance level, and the most prominent correlations with other electrodes are shown. A *solid line* between two electrodes indicated that the correlations were higher in the move task, while a *dotted line* indicates higher no-move task correlations. The appearance of very localized cognitive activity can be created by examining differences between two similar split-second tasks. Note how the lateralization shifts from right to left in less than one-tenth of a second. ———, Higher move correlations; ----, higher no-move correlations. (From Gevins et al. 1985)

niques were implemented: one based on amplitude subtraction and the other on spectral subtraction. Statistical pattern classification techniques were used to evaluate their performance. Although visual comparison showed both methods to be effective, the objective evaluation technique suggested that the frequency-dependent system identification method may be more effective for eye blink removal (Bonham 1985).

Digitization of Electrode Positions. Before and after a recording session, the position of each electrode is measured in three dimensions using a three-dimensional digitizer. Correction to scalp positions is made by a least-squares-fit multiple linear regression, which yields the general ellipsoid surface best fitted to the set of digitized positions. The digitized coordinates are then translated to a coordinate system centered on this ellipsoid.

Magnetic Resonance Imaging for Determining Positions of Electrodes and Cortical Structures. Magnetic resonance images (MRIs) are made of each subject. These are high-resolution (0.8 mm pixel) three-axis cross-section images of soft tissues 1 cm apart over the whole volume of the head. The pictures are digitized to give coordinate surfaces for scalp, outer and inner skull surfaces, and cerebral surface, including loci of major fissures. Registration of electrode positions and MRI measurements provides correlation of electrode positions and cortical areas, allowing comparison of functional neuroanatomy across people for localized sensory and motor processes. The direct measurement of thicknesses of scalp and skull provide accurate information for use in spatial deconvolution (see below).

Trial Selection Using Pattern Recognition. Most methods of event-related potential (ERP) estimation assume that task-related signals are present in every trial and they

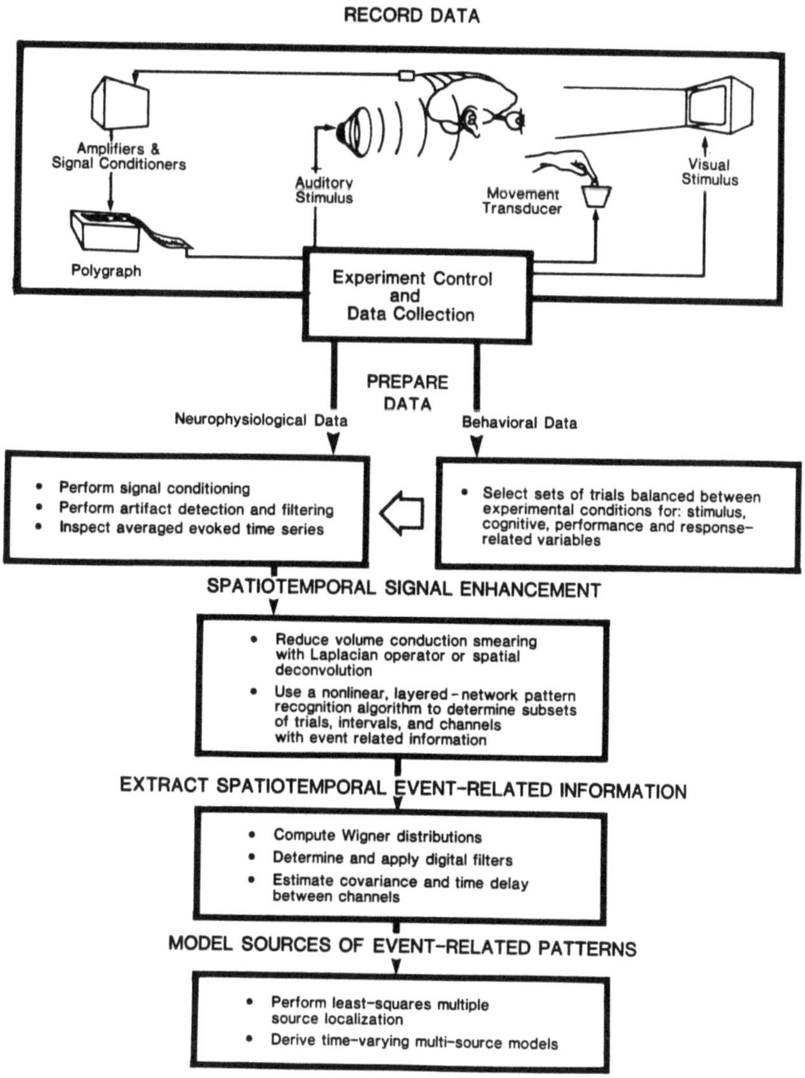

Fig. 3. ADIEEG-IV system for quantification of event-related brain signals. Separate subsystems perform on-line experimental control and data collection, data selection and evaluation, signal processing, and pattern recognition. Current capacity is 128 A/D channels. Spherical-head spatial deconvolution modules have been implemented, and source-modeling algorithms are being developed. Digital data tapes from other laboratories are converted into the ADIEEG data format using gateway programs; they are then processed using the same program modules as data collected in the EEG Systems Laboratory. The system is currently implemented on three computers (Masscomps) which are connected by high-speed buses

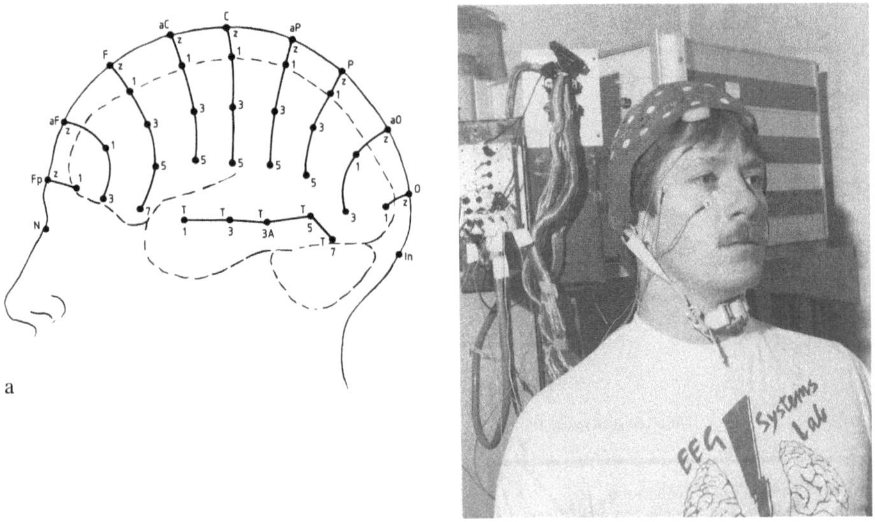

Fig. 4. (a) Expanded 10–20 system of electrode position nomenclature. Additional coronal rows of electrodes interpolated between the international 10–20 system coronal rows have the letter *a* for "anterior" added to the designation for the next row posterior, e.g., aPz for anterior parietal midline electrode. With 64 electrodes, the average distance between electrodes is about 3.25 cm. (b) Participant wearing 64-channel EEG recording cap

also have inherent assumptions about the statistical properties of signal and noise (see reviews in Gevins 1984; McGillem and Aunon 1987). A method for ERP estimation without the first assumption and with a relaxed second assumption has been developed (Gevins et al. 1986; Fig. 5). To do this, separate averages are formed from trials which are correctly or incorrectly classified by a statistical pattern recognition procedure. In the averages of correctly classified trials, the ERP peaks are enhanced in comparison with the original averages. The averages of incorrectly classified trials resemble the background EEG.

Wigner (Time–Frequency) Distributions. The ERP waveform is a function of time and does not provide explicit frequency information. Power spectra of ERP waveforms provide frequency information about obscure time-dependent phenomena. A view of the spectrum as it changed over time would give a new view of the evolution of different frequency components of the ERP. A simple but ineffective approach would be to compute the spectrum over highly overlapped windows of the average ERP. A preferred method is to compute a general function of time and frequency, called the Wigner distribution, which approximates the instantaneous energy for a given time and frequency. In practice, the "purified" ERPs (obtained using the procedure above) show strong enough energy "peaks" in the Wigner distribution to make very simple interpretations of the time and frequency locations of signal energy valid (Morgan and Gevins 1986; Fig. 6). The Wigner distributions are being used to determine digital filter characteristics that produce optimal time-frequency resolution for a given data set.

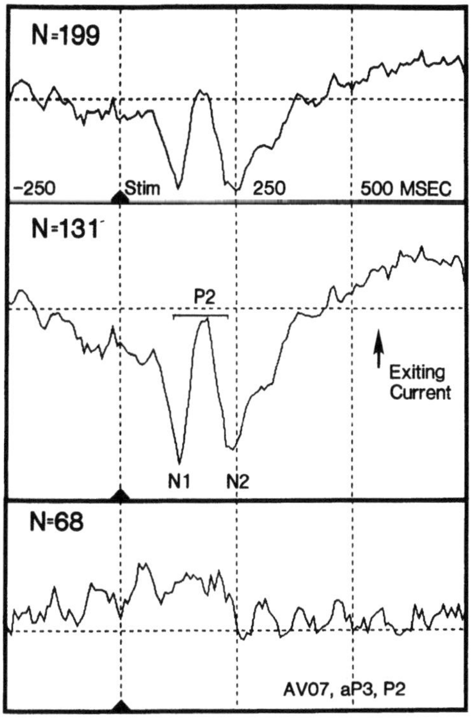

Fig. 5. Use of statistical pattern recognition analysis to remove trials without consistent task-related signals from a set of single-trial ERPs (actually event-related current source densities). This results in an average ERP with a higher signal-to-noise ratio obtained from fewer trials. Unlike optimal filtering methods, a priori hypothesized models of the structure and statistical properties of signal and noise components are not required. *Top,* original average ERP formed from 199 presentations of a visual numeric stimulus requiring an index finger response. *Middle,* Average of 131 trials with consistent task-related signals in the P2 interval. Trials were selected from the original set of 199 by applying a pattern recognition algorithm to distinguish a 125 ms, P2 time series segment from a precue "baseline" segment. Note the greatly increased size of the event-related peaks and lower-frequency wave forms. *Bottom,* Average of 68 trials which did not have consistent task-related signals in the P2 interval. Note the relative lack of event-related activity and the dominance of ongoing EEG alpha waves. (From Gevins 1984)

Minimizing Experimenter Delusion. In order to form controlled data sets that vary only according to a chosen neurophysiological hypothesis, the total set of artifact-free trials from each recording is submitted to an interactive program which displays the means, *t* tests, and histogram distributions of about 50 behavioral and other variables. Data sets can be quickly inspected for significant differences in variables not related to the hypothesis and pruned of outlier trials until balanced. Variables balanced in this manner include stimulus parameters, response onset and intermovement times, response movement force, velocity, acceleration and duration, error and adaptive performance measure, and indices of eye movement, muscle activity, and "arousal."

EEG Spatial Signal Enhancement. Electrical potentials generated by sources in the brain are volume conducted through brain, cerebrospinal fluid, skull, and scalp to

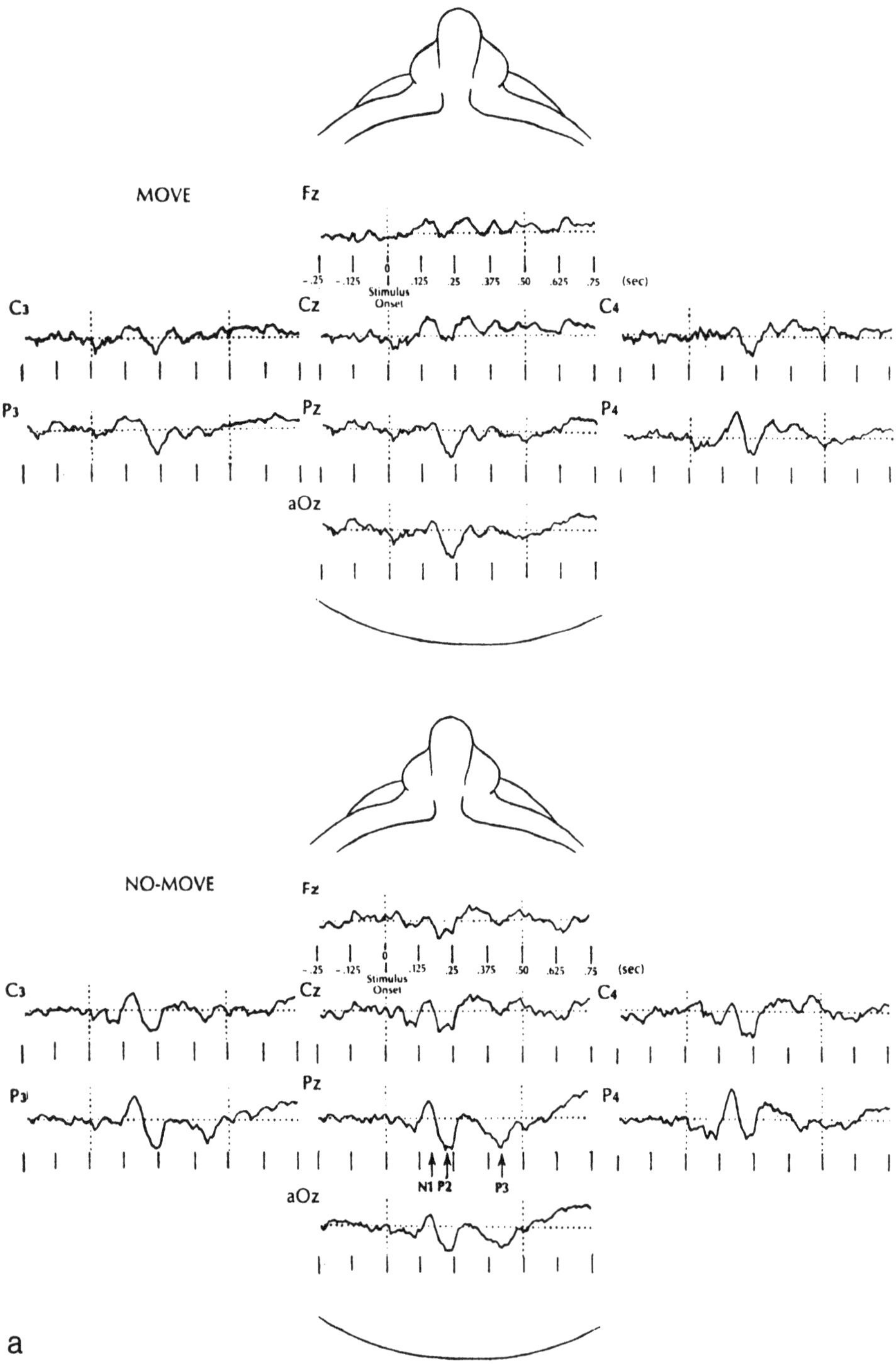

Fig. 6a, b. Two representations of eight average ERP channels for "move" and "no-move" visuospatial tasks. Looking down onto the top of the head, the nose is at the top of each set of eight channels. (a) Average time series of 40 no-move and 37 move trials. These trials were selected by a statistical pattern recognition algorithm as having consistent task-related signals which differed between the move and no-move tasks. Three of the most commonly studied ERP peaks are indicated on the *Pz* channel of the no-move task. Of these, the *P3* peak is larger in the infrequently occurring no-move

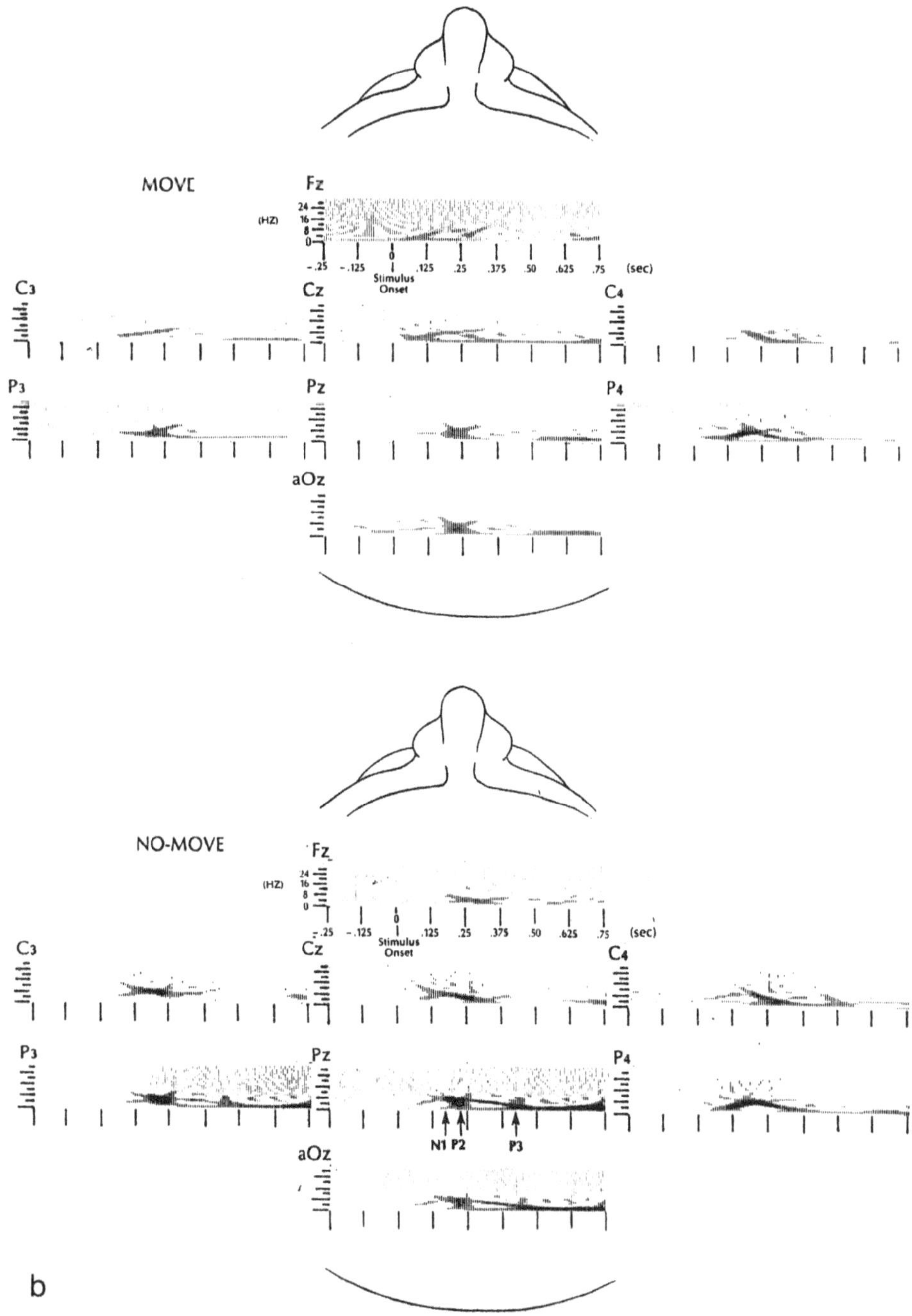

b

task. (b) The pseudo-Wigner distribution of the analytic signal of the same data. This representation shows that the event-related processes are rapidly changing in both time and frequency. The first moment along the *time axis* for each frequency is the group delay, while the first moment along the *frequency axis* at each time is the instantaneous frequency. There is also a build-up in energy after the stimulus. (From Gevins 1984)

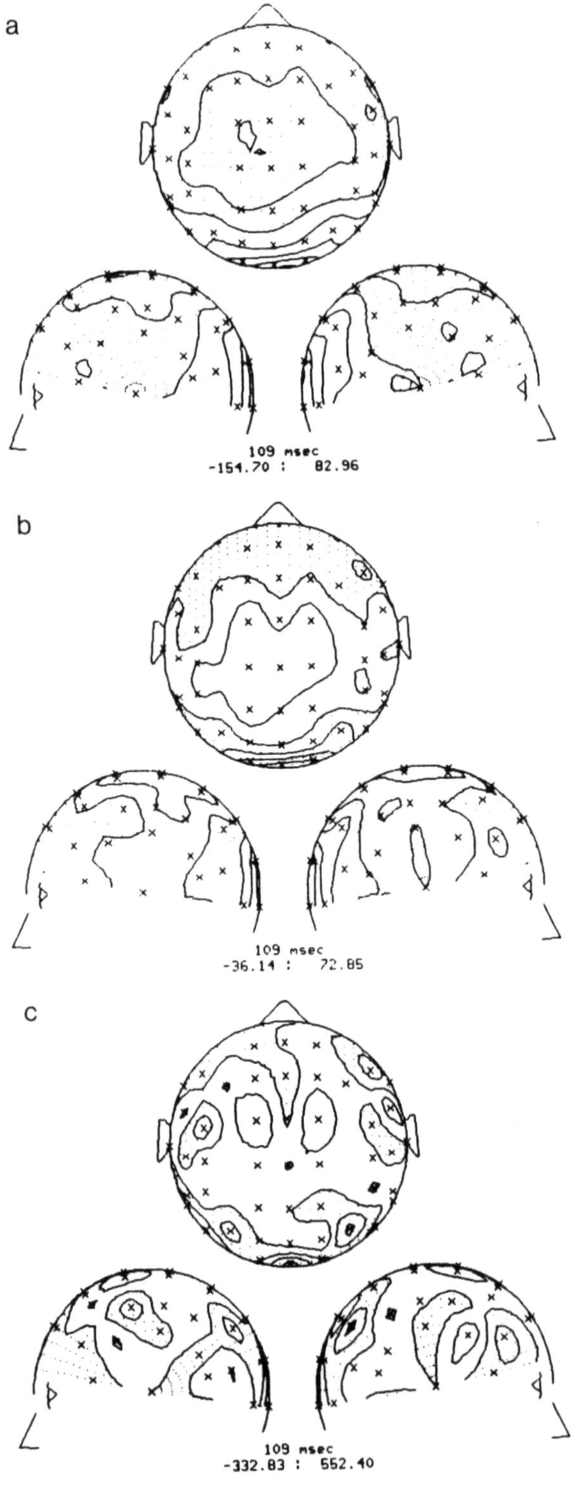

Fig. 7a–c. Contour plots of event-related waveforms from 50 channels at 109 ms after a visual stimulus. There is increased spatial detail with the current source density transform and a further increase with application of the spatial deconvolution. (a) Isopotentials from common average reference. (b) Isocurrent contours from current source density. (c) Iso-dipole strengths from deconvolution. (From Doyle and Gevins 1986)

the recording electrodes. Because of this, potentials from a localized source are spread over a considerable area of scalp, so that the potential measured at a scalp site represents the summation of signals from many sources over much of the brain. (We have estimated the "point spread" for a radial dipole in the cortex to be about 2.5 cm.) This spatial low-pass filtering makes source localization difficult, even for cortical sources, and causes the potentials from local sources to be mixed with those from more distant generators. By modeling the tissues between brain and scalp surface as surfaces with different resistivities, we can perform a deconvolution of the potential to just above the level of the brain surface without imposing assumptions as to the actual (cortical or subcortical) source locations (Doyle and Gevins 1986).

A simple form of the deconvolution was tested on a 50-channel recording obtained during an auditory-visual task (Doyle and Gevins 1986). The digitized electrode positions were used to implement the current source density transformation and a form of the spatial deconvolution modeling in terms of a distribution of radial dipoles over the cortex in a four-sphere model of the head (cerebrum, cerebrospinal fluid, skull, and scalp). In order to assess the reduction in interelectrode correlation caused by volume conduction, the interelectrode correlation was calculated over the stimulus-locked visual ERP for each pair of channels for the common average reference, current source density, and deconvolved ERPs. Then, using the measured electrode positions and the four-sphere model of the head, the point spread was calculated for each pair of electrodes. A linear regression of correlation versus point spread reveals the degree of association to be highest for the common average derivation. The same regression for the current source density and deconvolved derivations shows the degree of association to be more reduced for the deconvolution than for the current source density (Fig. 7). The advantage of the deconvolution increases with decreasing interelectrode distance. More realistic head models are under development for deconvolution and source localization.

Source Localization. Determination of the actual neuronal sources of potentials and fields may be done in the context of the same models as the deconvolution approach: a concentric-spherical-shell model of the tissues is used to represent the tissues; the boundary value problem is solved to express the scalp potentials (fields) in terms of cortical sources (Fender 1987; Nunez 1981; Plonsey 1969). In this case, only a few sources are postulated and their parameters (positions, orientations, strengths) are adjusted to minimize the (squared) discrepancy between predicted and measured potentials (fields). (Alternatively, the measured geometry of the tissues may be modeled in detail.)

Any given distribution of potential field at the scalp may be produced by an infinite number of different source configurations. To resolve this ambiguity, additional information is needed about the character, orientation, and number of sources. The convenient approach of modeling sources as one or two equivalent current dipoles has been successfully applied to sensory ERPs (review in Fender 1987), and event-related magnetic fields (ERFs) (reviews in Williamson et al. 1983; Weinberg et al. 1985; Williamson and Kaufman 1987). However, this approach seems unrealistic for higher cognitive processes which involve the integration of activity from a number of systems. In order to solve for a greater number of sources, more independent measures are required. Some benefit can be obtained by increasing the number of EEG channels, by using multiple time points (Scherg and von Cramon 1985), and by combining EEG, magnetoencephalogram (MEG), and MRI data. It is hoped that this will provide sufficient data to produce highly constrained fits for multiple sources.

The results of deconvolution and source localization must be independently validated using primate intracerebral models, intracerebral stimulation data from neurosurgery patients, and three-dimensional metabolic measures.

2 Comparison Between NCP Analysis, Conventional Averaged ERP Techniques, and Topographic Potential Maps

It is a great testimony to the ingenuity of psychophysiologists that so much has been learned about the timing of neurocognitive processes using very modest recording equipment and analysis techniques. Because of this, it is *certain* that when psychophysiologists are equipped with more advanced recording methods and more powerful analytic tools, they will make rapid advances in understanding human higher brain functions.

At the EEG Systems Laboratory, we are continuously striving to develop better methods and research tools. Our experiments and analyses are based on the body of information gained from ERP research and have the same underlying goal: to resolve spatially and temporally overlapping, task-related neural processes. Most ERP experiments have been concerned with measurement of the amplitude and/or latency of one to a few components associated with a single stimulus or response registration in a few channels in an epoch under 1 s. We are attempting to extend these limits and have made some modest progress. Because neurocognitive processes are complex, we are concerned with spatiotemporal task-related activity recorded by many (currently up to 64) scalp electrodes in many (currently about 25) time intervals spanning a 5-s period extending from before a cue, through stimulus and response, to presentation of feedback. In order to quantify time dependencies between channels, we have developed the method of event-related covariance. Feature extraction and hypothesis testing are performed as a single process which determines the differences in signal properties between the conditions of an experiment, or their differences with respect to a "baseline". Constraints are used to facilitate a neuroanatomical and neurophysiological interpretation.

These methods should be distinguished from currently popular interpolated color displays of the voltage of individual time points of 16–20 channels of averaged ERPs, or the difference between the voltages and a set of normative data. We use extensive signal processing and pattern recognition algorithms to reduce volume conduction effects and to extract minute event-related signals from unrelated background noise of the brain, compute between-channel interdependency patterns, and display their scalp distribution in three-dimensional perspective graphics using the head and brain (Gevins 1980, 1987a, b; Gevins and Morgan 1986; Gevins et al. 1987, submitted).

These aspects of our analysis may enable us to resolve previously unseen event-related signals from the overriding unrelated background activity of the brain, leading to a better understanding of mass neural processes of human goal-oriented behaviors. However, we must caution that although we have obtained several promising results, "the jury is still out". If results in the next several years prove that these new methods are worthwhile and that they successfully measure salient aspects of "functional interdependency", it will be possible to optimize and standardize our methods and analysis for application by other laboratories.

3 Results of Application of NCP Analysis to Primate Intracerebral Data

Analyses of pilot intracerebral primate data (recorded at C. Rebert's laboratory at Stanford Research Institute) revealed rapidly shifting correlation patterns between hippocampus, substantia nigra, premotor cortex, ventroanterior nucleus of the thalamus, and the midbrain reticular formation which distinguished "go" and "no-go" visuomotor tasks (Bressler and Gevins 1985). Interdependencies of the hippocampus with other loci were prominent in the go condition (which leads to reward), but not in the no-go condition, during intervals from 240 to 578 ms after the cue. The patterns were characterized by delays of up to 72 ms or more between loci.

4 Results of Application of NCP Analysis to a Bimanual Visuomotor Task

Data of seven right-handed adult males were collected in an experiment designed to study spatiotemporal patterns of human neurocognitive activity during preparation and execution of precise right- and left-hand finger pressures (Gevins et al. 1987, submitted). Twenty-six EEG channels were recorded, from which 16 Laplacian ("current source density") derivations were computed. Detailed analyses were made of the 4-s visuomotor task from cued preparation, through poststimulus perceptual and cognitive processing and response execution, to the "updating" associated with feedback about performance accuracy. Of the procedures described above, the following were applied: statistical pattern recognition methods for selecting trials with consistent event-related signals with which to form the enhanced average; computation of Laplacian derivations; determination of band-pass filter characteristics from Wigner distributions; computation of cross-covariance functions between averaged time series within brief (187 or 375 ms) intervals centered on event-related waves; and use of three-dimensional perspective color graphics to display the pattern of significant interdependencies. Several significant results were obtained:

- Covariance patterns for movement-registered time series closely corresponded to prior functional neuroanatomical knowledge (Fig. 8), lending a first level of validation for the covariance patterns associated with higher-order cognitive activity. The anterior midline precentral electrode, overlying the premotor and supplementary motor cortex, was the focus of all movement-related patterns. The pattern for the motor potential clearly reflected the sharply focused current sources and sinks spanning the hand areas of motor cortex. Both the Laplacian derivation topography and the covariance patterns suggest distinct source generator con-

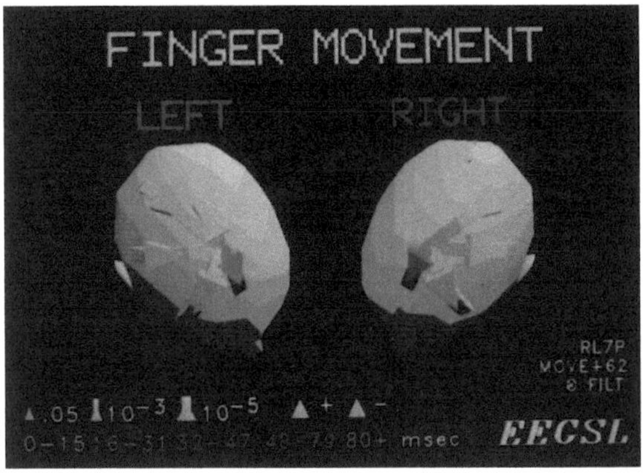

Fig. 8. Most significant (top standard deviation) covariance patterns at 62 ms after movement onset for right and left move trials for seven people, computed from theta band filtered data. Note that the anterior midline precentral (aCz) electrode is the focus of all covariances, 16–31 ms time delays between aCz and Fz. The patterns are distinctly lateralized according to responding hand. The sign of the aCz covariances is positive for lateral frontal, and negative for lateral central and anterior parietal electrodes. (From Gevins et al. 1987a)

figurations for the readiness potential, motor potential, and response after-potential.

– Differences were found in cue-registered covariance patterns which were *predictive* of trials that *subsequently* had accurate or inaccurate performance. Significant covariances involving electrodes overlying left frontal and appropriately lateralized parietal and motor areas characterized subsequently accurate performance of both hands. Neither poststimulus nor periresponse covariances were associated with performance accuracy. These results suggest that a spatially specific, neural preparatory set, composed of distinct left frontal and contralateral somesthetic-motor components, may be essential for accurate performance of certain types of visuomotor tasks (Gevins et al. 1987, submitted).

Thus, the new measures of "functional interdependency" that have been developed appear satisfactory. The results were clear-cut and consistent with neuropsychological models of the rapidly shifting cortical network accompanying expectancy, stimulus registration and feature extraction, response preparation and execution, and updating to feedback about response accuracy. Additionally, they appear to provide new information about these functions. Along with the pilot analyses of intracerebral recordings in a primate model described above, these results suggest that it is possible to characterize functional interdependencies of event-related processing between local neural areas by measuring the wave congruence and lag time of appropriately preprocessed low-frequency macropotentials. These results could, in principle, be understood in the context of a distributed network of specialized processing nodes. Unique determination of this network is a very difficult problem. This, and more detailed measurement and modeling of each node (Freeman 1987) is the focus of our current research.

5 Summary

We use the generic term "neurocognitive pattern (NCP) analysis" to refer to procedures being developed to extract spatiotemporal neurocognitive patterns from the unrelated "noise" of the brain. Recordings with up to 64 scalp channels during highly controlled tasks are now routine in our laboratory, as is the extended signal processing sequence required to extract minute neurocognitive signals from gigabyte sets of single trial data. More robust measures of the degree of "functional interdependency" between electrodes have been developed and applied to a bimanual visuomotor task recorded with 26 channels from seven people. The results were clear-cut and consistent with prior neuropsychological models of the rapidly shifting cortical network accompanying expectancy, stimulus registration and feature extraction, response preparation and execution, and "updating" to feedback about response accuracy. *Predictive* patterns have been identified distinguishing trials that subsequently had accurate or inaccurate performance. Along with pilot analyses of intracerebral recordings in a primate model, these results suggest that it is possible to characterize "functional interdependencies" of event-related processing between local neural areas by measuring the wave congruence and lag time of appropriately preprocessed low-frequency macropotentials. Many of these results cannot be explained by single equivalent current dipole source models, but could, in principle, be understood in the context of a distributed network of specialized processing nodes. However, unique determination of this network is a formidable problem, because event-related signals emitted by each node are obscured by unrelated brain activity and overlap both in time and in space when recorded at the scalp. We are attacking this problem with a number of technical developments: use of MRI to relate the EEG electrode positions to underlying cortical structures; use of spatial deconvolution to remove the "blurring" effect of conduction through skull and scalp; and incorporation of multichannel MEG data. In this way, we hope to determine more realistic source models.

Acknowledgements. The research of the EEG Systems Laboratory is sponsored by grants and contracts from the following agencies of the United States government: the Air Force Office of Scientific Research, the National Science Foundation, the National Institutes of Neurological and Communicative Diseases and Stroke.

References

Bonham B (1985) Comparison of frequency-dependent and frequency-independent methods for EOG artifact removal. Master's thesis, University of California

Bressler SL, Gevins AS (1985) Pattern analysis of primate brain functions: new pilot data. EEG Systems Laboratory Report, San Francisco

Doyle JC, Gevins AS (1986) Technical note: spatial filters for event-related brain potentials. EEG Systems Laboratory Technical Report, TR 86-001

Fender D (1987) Source localization. In: Gevins AS, Remond A (eds) Handbook of electroencephalography and clinical neurophysiology: methods of analysis of brain electrical and magnetic signals, vol 1. Elsevier, Amsterdam

Freeman W (1987) Analytic techniques used in the search for the physiological basis of the EEG. In: Gevins AS, Remond A (eds) Handbook of electroencephalography and clinical neurophysiology: methods of analysis of brain electrical and magnetic signals, vol 1. Elsevier, Amsterdam

Gevins AS (1980) Pattern recognition of human brain electrical potentials. IEEE Trans Patt Anal Mach Intell 2(5):383–404

Gevins AS (1984) Analysis of the electromagnetic signals of the human brain: milestones, obstacles and goals. IEEE Trans Biomed Eng 31 (12):833–850

Gevins AS (1987a) Correlation analysis. In: Gevins AS, Remond A (eds) Handbook of electroencephalography and clinical neurophysiology: methods of analysis of brain electrical and magnetic signals, vol 1. Elsevier, Amsterdam
Gevins AS (1987b) Statistical pattern recognition. In: Gevins AS, Remond A (eds) Handbook of electroencephalography and clinical neurophysiology: methods of analysis of brain electrical and magnetic signals, vol 1. Elsevier, Amsterdam
Gevins AS, Morgan NH (1986) Classifier-directed signal processing in brain research. IEEE Trans Biomed Eng 33(12): 1054–1068
Gevins AS, Yeager CL, Diamond SL, Spire JP, Zeitlin GM, Gevins AH (1975) Automated analysis of the electrical activity of the human brain: a progress report. IEEE Proc 63(10): 1382–1399
Gevins AS, Yeager CL, Zeitlin GM, Ancoli S, Dedon M (1977) On-line computer rejection of EEG artifact. Electroencephalogr Clin Neurophysiol 42: 267–274
Gevins AS, Zeitlin GM, Doyle JC, Yingling CD, Schaffer RE, Callaway E, Yeager CL (1979a) EEG correlates of higher cortical functions. Science 203: 665–668
Gevins AS, Zeitlin GM, Yingling CD, Doyle JC, Dedon MF, Schaffer RE, Roumasset JT, Yeager CL (1979b) EEG patterns during "cognitive" tasks. Part 1: Methodology and analysis of complex behaviors. Electroencephalogr Clin Neurophysiol 47: 693–703
Gevins AS, Zeitlin GM, Doyle JC, Schaffer RE, Callaway E (1979c) EEG patterns during "cognitive" tasks. Part 2: Analysis of controlled tasks. Electroencephalogr Clin Neurophysiol 47: 704–710
Gevins AS, Doyle JC, Schaffer RE, Callaway E, Yeager C (1980) Lateralized cognitive processes and the electroencephalogram. Science 207: 1005–1008
Gevins AS, Doyle JC, Cutillo BA, Schaffer RE, Tannehill RS, Ghannam JH, Gilcrease VA, Yeager CL (1981) Electrical potentials in human brain during cognition: new method reveals dynamic patterns of correlation. Science 213: 918–922
Gevins AS, Schaffer RE, Doyle JC, Cutillo BA, Tannehill RS, Bressler SL (1983) Shadows of thought: shifting lateralization of human brain electrical patterns of a brief visuomotor task. Science 220: 97–99
Gevins AS, Doyle JC, Cutillo BA, Schaffer RE, Tannehill RS, Bressler SL (1985) Neurocognitive pattern analysis of a visuomotor task: rapidly-shifting foci of evoked correlations between electrodes. Psychophysiology 22: 32–43
Gevins AS, Morgan NH, Bressler SL, Doyle JC, Cutillo BA (1986) Improved event-related potential estimation using statistical pattern recognition. Electroencephalogr Clin Neurophysiol 64: 177–186
Gevins AS, Morgan NH, Bressler SL, Cutillo BA, White RM, Illes J, Greer DS, Doyle JC, Zeitlin GM (1987) Human neuroelectric patterns predict performance accuracy. Science 235: 580–585
Gevins AS, Bressler SL, Morgan NH, Cutillo BA, White RM, Greer DS, Illes J (submitted) Event-related covariances during a bimanual visuomotor task, part I: Methods and analysis of stimulus- and response-locked data. Electroencephalogr Clin Neurophysiol
Gevins AS, Cutillo BA, Bressler SL, Morgan NH, White RM, Illes J, Greer DS (submitted) Event-related covariances during a bimanual visuomotor task, part II: Preparation and feedback. Electroencephalogr Clin Neurophysiol
McGillem C, Aunon J (1987) Analysis of event-related potentials. In: Gevins AS, Remond A (eds) Handbook of electroencephalography and clinical neurophysilogy: methods of analysis of brain electrical and magnetic signals, vol 1. Elsevier, Amsterdam
Morgan NH, Gevins AS (1986) Wigner distributions of human event-related brain potentials. IEEE Trans Biomed Eng 33(1): 66–70
Nunez PL (1981) Electric fields in the brain: the neurophysics of EEG. Oxford University Press, New York
Plonsey R (1969) Bioelectric phenomena. McGraw-Hill, New York
Scherg M, von Cramon D (1985) Two bilateral sources of the late AEP as identified by a spatio-temporal dipole model. Electroencephalogr Clin Neurophysiol 62: 32–44
Weinberg H, Stroink G, Katila T (eds) (1985) Biomagnetism: applications and theory. Permagon, New York
Williamson SJ, Kaufman L (1987) Analysis of neuromagnetic signals. In: Gevins AS, Remond A (eds) Handbook of electroencephalography and clinical neurophysiology: methods of analysis of brain electrical and magnetic signals, vol 1. Elsevier, Amsterdam
Williamson SJ, Romani GL, Kaufman L, Modena I (eds) (1983) Biomagnetism: an Interdisciplinary approach. Plenum, New York

Dynamic Changes in Steady-State Responses

R. Galambos and S. Makeig

1 Introduction

The research we report here began in 1980 when we rediscovered the so-called steady-state auditory evoked responses (SSRs) described some 20 years earlier (Chatrian et al. 1960; Geisler 1960; Galambos et al. 1981). Figure 1 shows such SSRs after their extraction by computer averaging from the scalp EEG of an adult who was receiving clicks monaurally through an earphone. Time-locked brain potentials appear at each rate (nominally 10/s in the top panel, 20 in the middle, and 40 in the bottom); at the 40-Hz rate, the individual response averages resemble single sine waves.

The question has been raised whether important relationships exist between evoked potentials such as those in Fig. 1 and the spontaneous EEG frequencies. As Başar points out (1980), the well-known alpha wave bursts appear spontaneously in the 10-Hz region but similar bursts can also be "evoked" by a single stimulus; furthermore, the brain readily follows or "resonates" when stimuli are applied at this rate. As for the 40-Hz region, Fig. 1 shows that the brain can be entrained at this frequency also, and, as Freeman describes elsewhere in this volume, several brain structures show both spontaneous and stimulus-initiated activity in the 40-Hz range. Thus the 40-Hz region, like the 10-Hz region (and 20, 60, and 70, as Başar points out) seems to display an unusual propensity for generating both spontaneous and driven responses.

Başar takes the position that brain research centered upon these frequency regions may well uncover significant new information. We have elected to present here some experiments that fit in with this idea – experiments that, it turns out, also delineate what may be a new class of dynamic processes that modulate transmission in the sensory pathways. We will also describe in some detail our quantitative methods for producing and analyzing SSRs like those in Fig. 1, since we have to date reported only briefly on this research (Galambos 1981, 1982; Makeig 1985).

2 Methods

2.1 Block Diagram

Figure 2 shows the instruments and procedures used by us (and others) to produce and analyze SSRs like those in Fig. 1. A train of computer-generated auditory signals (clicks or short-tone bursts at the rate and intensity specified in the program) are de-

Springer Series in Brain Dynamics 1
Edited by Erol Başar

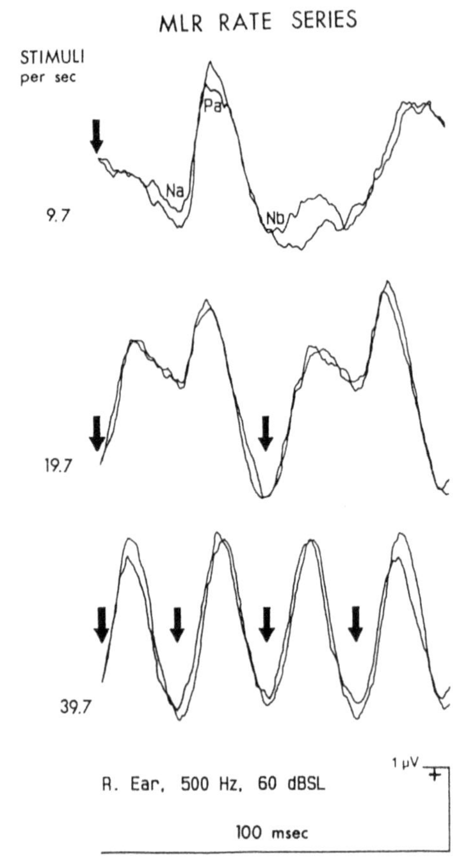

Fig. 1. Steady-state response (SSR) samples obtained at three stimulus rates suggesting evolution of the middle-latency response (*MLR*, 10/s) into the SSR at the highest rate (40 Hz) which closely resembles a sine wave. Electrode montage Cz-Oz

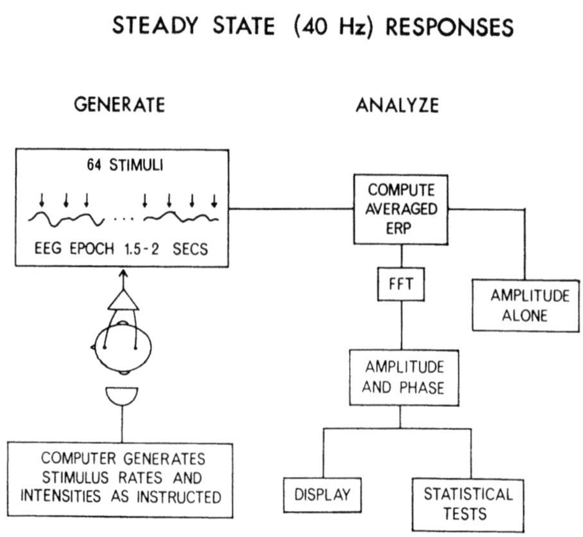

Fig. 2. Block diagram of apparatus for producing and analyzing SSRs

livered to a subject through earphones; these stimuli evoke a train of individual responses at the points indicated by the arrows above the EEG tracing. The EEG thus collected is then averaged and/or Fourier-transformed to determine the amplitude and phase of the EEG response at the same frequency as the rate of stimulation.

2.2 Response Analysis

Figure 3 plots the frequency distribution of the EEG amplitudes obtained during such an experiment in which the listener received 40 clicks/s. The plot reveals sharp amplitude peaks at the stimulus rate, 40 Hz, as well as at 60 Hz, the power-line frequency. Our procedure can be thought of as creating a unique electrophysiological event (the 40-Hz SSR) from which numbers representing its amplitude and phase are extracted.

Fourier analysis is conceptually an efficient method for fitting a sinusoid to a periodic signal such as the 40-Hz SSR (see Regan 1982 for a complete discussion). In our application, the computer matches such a best-fit sinusoid to an average (usually containing 64 responses, or about 2 s in real time) and records its amplitude and phase. We use peak-to-peak amplitude, and phase relative to signal onset as our measures. The meaning of "phase" in our case is illustrated in Fig. 4, where sinusoids simulating best fits to five responses are shown, each rotated 90° further relative to the stimulus onset; the numbers representing their amplitude and phase values appear in the top and bottom graphs, respectively, and in real experiments remain available for further statistical manipulations (Elberling 1979; Sayers et al. 1979; Stapells et al. 1984).

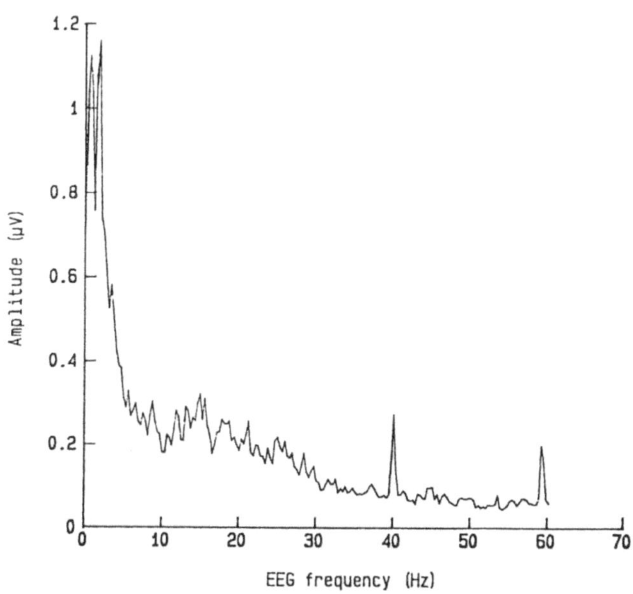

Fig. 3. The EEG spectrum obtained during a 15-min session in which monaural clicks were delivered throughout; note the amplitude peak at the stimulus rate, 40 Hz

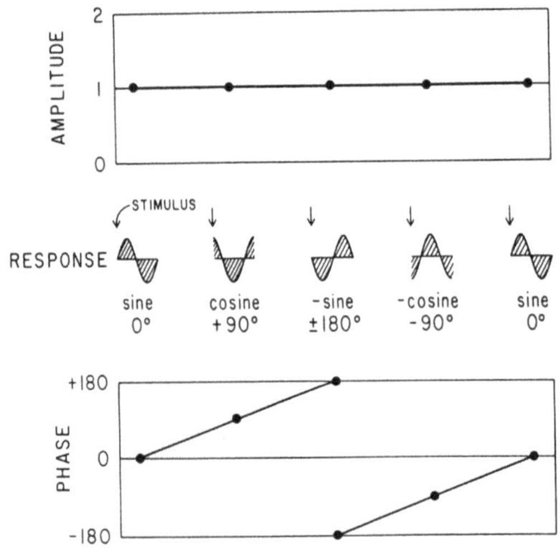

Fig. 4. Diagrammatic plots of the amplitude and phase numbers yielded by analysis of SSRs in the frequency domain

Two technical points should be made explicit. First, the stimulus rate must be an exact submultiple of the EEG sampling rate and locked to it under common clock control, for otherwise the response energy would spread into several frequency bins, and the moving average method described below would fail. Secondly, note that frequency-domain averaging can be carried out in two different ways. One can average in the time domain followed by computation of the amplitude and phase from a grand average or, equivalently, the real and imaginary parts of the phase plane vectors that represent each response can themselves be averaged and the result converted to amplitude and phase ("coherent averaging").

2.3 The Phase Plane

As shown on the left of Fig. 5, amplitude and phase of the computer-generated best-fit sinusoid can be represented in polar coordinates as a vector with its tip located at a particular spot in the phase plane. This vector tip can also be specified in Cartesian coordinates as distance along the horizontal (the so-called real) axis and the vertical (or "imaginary") axis; this so-called complex representation is what the discrete and fast Fourier transform (DFT and FFT) algorithms deliver, and, as described below, we use it to smooth or average a series of responses.

Another useful response measure is phase coherence, which estimates how closely the successive phase values cluster during an experimental session. The right half of Fig. 5 shows, for example, ten simulated sinusoidal best fits and their phase-plane vector representations. The ten vectors are not randomly distributed around the origin but are more or less clustered together, as will be the case when audible signals drive the nervous system. A statistical test of this phase aggregation is phase coherence or circular variance (Mardia 1972). Phase coherence varies between 1 (all phases identical) and 0 (phases equally distributed around the origin). Over much of its range, phase coherence is near-linearly correlated with the ratio of response amplitude to the amplitude of the background EEG at frequencies near the stimulus

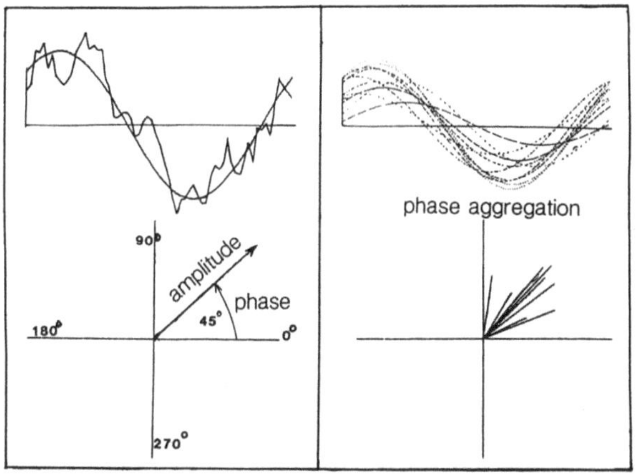

Fig. 5. Polar plots of response vectors. *Upper left,* the best-fit sine wave drawn through a simulated noisy SSR is represented *below* as a vector in polar coordinates. On the *right* ten simulated best fits are represented *below* as vectors clustered in one quadrant, a distribution typical of SSRs collected during suprathreshold stimulation

rate (Makeig 1985), and so it indexes both the presence and the robustness of a response.

2.4 EEG Background Estimation

A great advantage of the frequency-domain approach is that it can eliminate, if desired, all EEG activity not precisely at the stimulus rate. However, we frequently use the information available on spontaneous EEG amplitudes at frequencies near the stimulus rate because their average provides an excellent moment-to-moment estimate of what the spontaneous EEG amplitude at the stimulus rate would amount to if the stimulus were turned off.

2.5 Rate Series, Intensity Series

The way response amplitude and phase change as stimulus rate change is called a "rate series," and the way these vary as stimulus strength changes is called the "intensity series" (Galambos et al. 1981). Examples of both, including also the estimate of stimulus-rate EEG background, are seen in Fig. 6. Each of these graphs illustrates a fundamental SSR relationship.

The *rate series* data (Fig. 6A) come from two experiments on the same subject during which clicks were presented at the eight different rates shown on the abscissa. The solid curves show experimental data analyzed in the frequency domain, whereas the dotted (amplitude) curve comes from a study conducted 2 years earlier when only time-domain (peak-to-peak) analysis was available. Both methods produce similar amplitude curves, as would be expected (Stapells et al. 1984), and also give an indication of the stability of the response characteristics over time. Note that the response amplitude peaks in the 35–40 Hz region, which is normal for adults. The phase plot presents a second, different view of the orderly relationships in the data, one from which the time interval between stimulus delivery and the activation of the responding structure can be estimated (Van der Tweel and Verduyn Lunel 1965; Regan 1982). What has been termed the "apparent" or "implied latency" can be de-

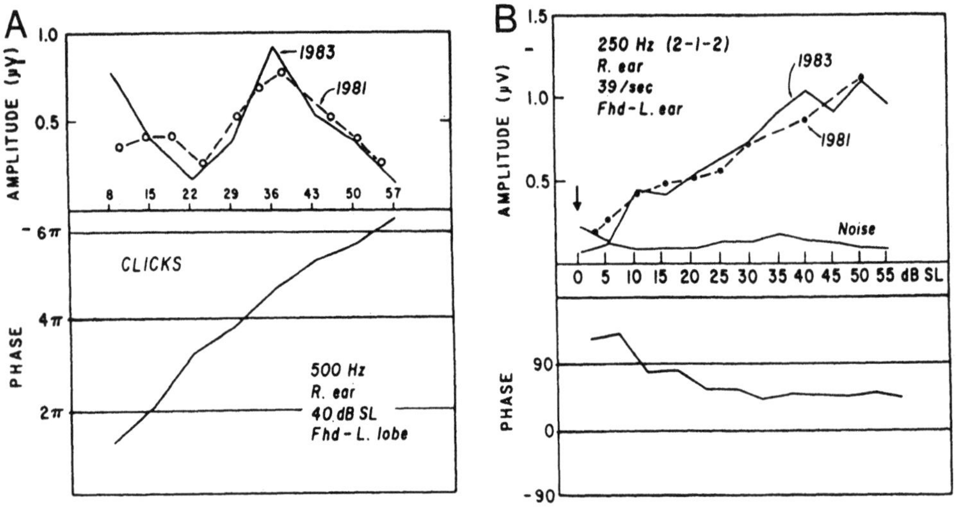

Fig. 6A, B. Amplitude and phase plots for SSR experiments in which stimulus parameters change. (A) The rate series; (B) the intensity series

rived from the slope of these phase plots; for visual data, this was estimated in 1965 to be about 60 ms and for the auditory data in Fig. 6A it is about 40 ms.

The *intensity series* in Fig. 6B plots data from another pair of experiments; this time SSRs were collected while 250-Hz tone bursts were being delivered at a constant rate but at the different intensities shown on the abscissa. (All our tone bursts are the so-called 2-1-2: two stimulus periods each for rise and for fall, one at plateau; total duration for the 250-Hz tone burst is therefore 20 ms.) The arrow marks the listener's behavioral threshold for these tone bursts. Again, response amplitudes established by time- and frequency-domain procedures are similar, and the phase plot once more indicates, separately and independently, the presence of an orderly relationship between the physiological responses and the stimuli that initiate them. Whether analyzed in the frequency domain or in the time domain, the SSR intensity series is excellent for threshold estimation, as shown here; the fact that analysis in the frequency domain so readily quantifies two important response features makes it the method of choice (Stapells et al. 1984).

2.6 The Moving Average

To assemble the data in Fig. 6, stimulus rate (Fig. 6A) and intensity (Fig. 6B) were randomly changed every few seconds, the responses being segregated, stored, and finally reconstituted as the grand averages shown. In the experiments of Fig. 6, this randomization procedure was selected deliberately in order to eliminate any time-dependent response fluctuations that might appear. If, instead of eliminating them, one chooses to examine these fluctuations, as we are about to do, the individual SSR averages are subjected to a moving-average procedure and displayed as in Fig. 7. To create this moving average, we move a rectangular averaging window of specified

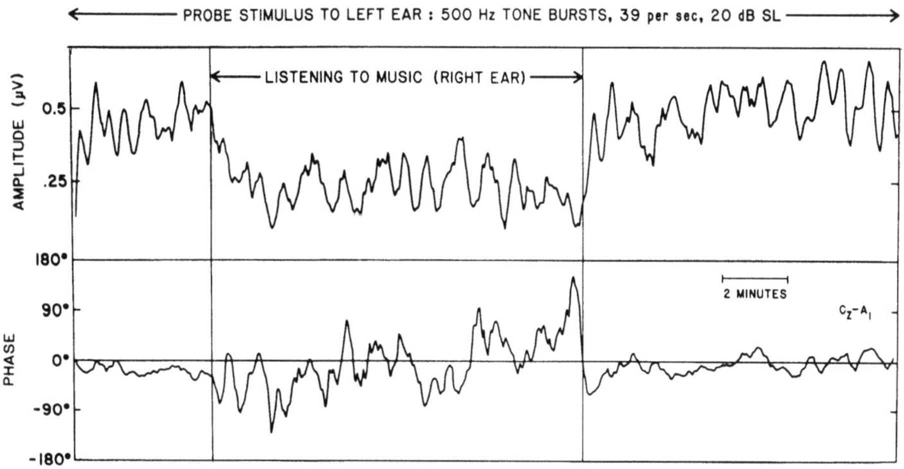

Fig. 7. Dynamic changes in SSR measures (amplitude and phase) during a 30-min session in which weak tone bursts were delivered into one ear throughout and a musical composition was played into the other ear for part of the time. Electrodes: Cz-left earlobe

length progressively through the approximately 2-s (64 stimulus) response epochs into which the EEG has been divided (see Fig. 2). In Fig. 7, a window length of 8 epochs (about 13 s) was used. Hence the first amplitude and phase points plotted in Fig. 7 give averages for epochs 1 through 8, the second for epochs 2 through 9, etc. Averaging windows representing about 15 s in real time remove the rapid fluctuations in the response (which are due largely to EEG activity unrelated to the stimulation) and reveal for examination the slow, time-dependent changes.

We can now turn to some experimental results obtained using the methods just described.

3 Results

3.1 The High-Rates Probe

The experiment of Fig. 7 was one of an exploratory series prompted by reports that the amount of 40-Hz EEG activity increases with attentive behavior in animals and man ("focused attention"; Sheer 1976; Flinn et al. 1977; Bouyer et al. 1981; Başar 1980), that it correlates positively with intelligence (Flinn et al. 1977), and that it decreases with sleep (Linden et al. 1985). The experimental plan involved delivering our brief tone bursts (or "probes") at a rate near 40 Hz to one ear for about 30 min with intermittent introduction into the other ear of tape-recorded music or story material intended to engage the subject's interest and attention. The hypothesis: listening intently may alter amplitude and/or phase of the probe response.

The musical composition played into the right ear in the experiment of Fig. 7 was the fourth movement of Beethoven's Third Symphony; subject instructions were to relax and enjoy the music as if in a concert hall. The results can be described as follows. SSRs are generated to the tone burst probes throughout, but their amplitude and phase fluctuate in two ways. First, the traces show numerous peaks and valleys, a count of which yields about 30, or close to 1 cycle/min. We use the term "minute rhythm" for these fluctuations; as we shall see, they actually display periods ranging from about 30 s to 2 min and are a constant feature of such recordings. Secondly, music onset initiates a sustained amplitude drop and alters the phase plot, and with music offset the premusic values for both are restored.

Numerous studies of the sort summarized in Fig. 7 have revealed that the high-rates SSR can be modulated in at least three ways: spontaneously on a near-minute time scale, as well as by a stimulus delivered to the contralateral ear, as seen in Fig. 7; and during the brain-state changes associated with drowsiness and sleep (Makeig 1985). In the experiments that follow, we have attempted to isolate and examine each of these factors by itself.

3.2 Sleep

Falling asleep does cause important changes in SSR measures, as demonstrated in Figs. 8 and 9 (and see also Brown and Shallop 1982; Klein 1983; Linden et al. 1985). The next time you are seized with an uncontrollable need to take a nap after lunch, remember Fig. 8; it illustrates what your SSRs might look like as you undergo the experience. The postprandial subject of Fig. 8, an adult male reclining comfortably on a couch, reported dozing off and waking up twice during a pilot study intended to chart the variability of his SSRs to the weak clicks presented at 39/s. This figure displays the raw data of the experiment; that is, 64 consecutive averages (each average is a single cycle in the figure) written out one after the other; in the figures shown elsewhere in this report, the Fourier-transform procedures have extracted the amplitude and phase information from averages such as these. In this instance, each epoch contains 128 rather than our usual 64 responses, which means it represents about 3.3 s in real time. The recordings show that response amplitudes drop by 50% or more with dozing and return abruptly with awakening.

Figure 9 similarly demonstrates awake-asleep brain changes monitored by SSRs, this time throughout an experiment lasting 30 min. Here the purpose of the study was

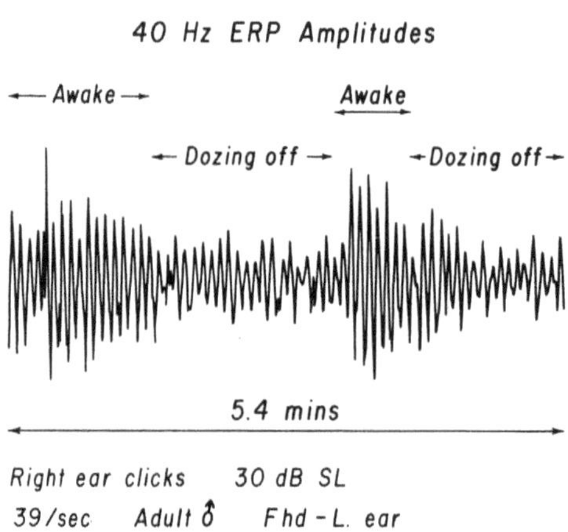

Fig. 8. Dynamic changes in SSR amplitudes during light sleep in a subject who was supposed to remain awake

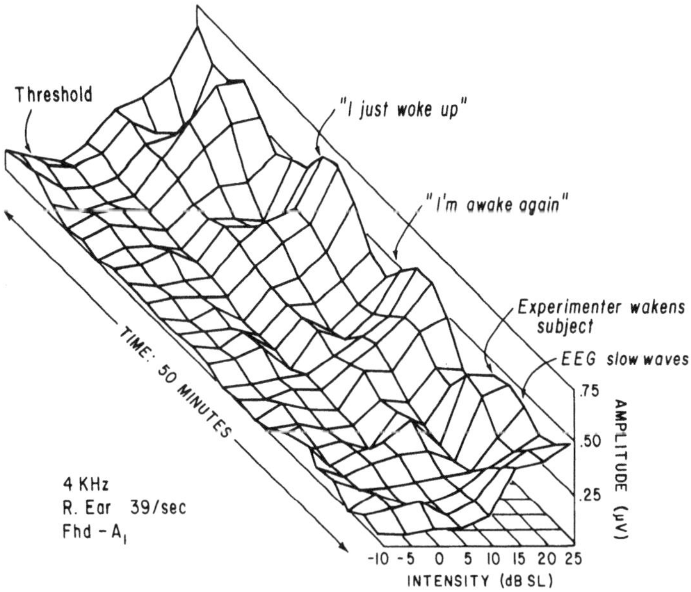

Fig. 9. SSR amplitude changes during a 50-min experiment in which the subject repeatedly dozed off. The data consist of 24 intensity series like the one shown in Fig. 6B; the artist has connected corresponding points to produce this three-dimensional display

to establish whether a change in SSR threshold occurs during dozing and sleep, and so a series of eight probe intensities (the intensity series; see Fig. 6B) was presented 24 times in succession; the figure plots the results in three dimensions, with response amplitude upward, time back to front, and signal strength left to right. The subject's verbal reports appear above the graph.

The plot can be described as an undulating surface that reveals the dynamic changes under way in the SSR brain generators. The 25-dB stimulus line shows four or five deep scallops with peaks at 10–15-min intervals, and these usually coincide with subject reports of reawakening. The 10-dB contour does not show these scallops, or at least not to the same degree. Also, the 24 lines charting the successive intensity series differ considerably from one another; some rise from threshold in a straight line (i.e., show a linear amplitude increase with intensity), while others show flat regions ("plateaus") or even actual reversals (amplitude drops as intensity rises). While some of these response irregularities may be due to our inability to sample intensities simultaneously, the maximum amplitude changes do approximate 50%, as in Fig. 8, and they correlate similarly with subject reports.

The experiments of Figs. 8 and 9 raise many questions for additional research to answer, but both agree that drops in mean amplitude seen during experiments like the one summarized in Fig. 7 could in fact be due to a change in the state of the subject on the sleep–wakefulness continuum.

3.3 Central Suppression

To test the possibility that, in the experiment of Fig. 7, the drop in mean response amplitude during the presentation of music was a masking phenomenon, we replaced

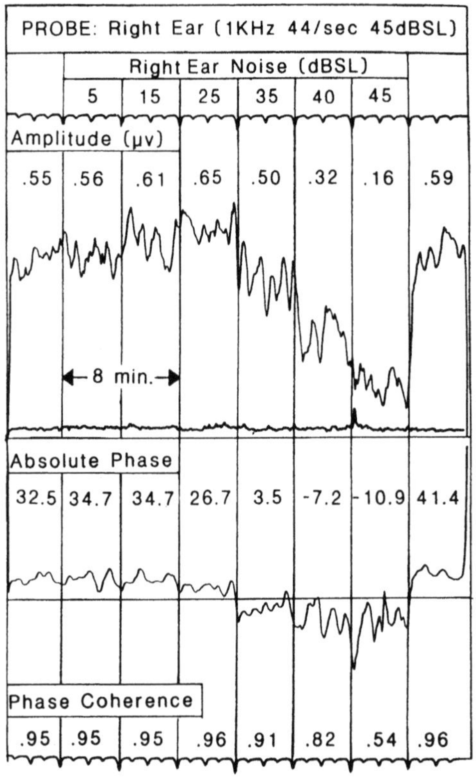

Fig. 10. SSR changes due to introduction of increasing levels of masking noise into the ear stimulated by the probe. All three response measures (amplitude, phase, and phase coherence) reflect the interaction of the stimuli in this direct masking situation

the music by noises controlled in amplitude, band-pass, and ear of delivery. The first example is Fig. 10, where increasing levels of narrow-band noise (1000-Hz center frequency, 18 dB per octave skirts) and the probe (a 1000-Hz tone burst 2−1−2) were both delivered to the same ear. This is a simple masking experiment; behaviorally, when both signals were presented together, both were heard except at the 5-dB noise level, where the noise was inaudible, and at the 45-dB noise level, where the probe was masked. Physiologically, SSR amplitudes seem to rise slightly at the 15- and 25-dB levels then drop steadily to their lowest values where masking is complete; the phase and phase coherence values also move, and in an orderly manner.

Figure 11 resembles Fig. 10 in all respects except that the noise is presented to the ear contralateral to the probe. Behaviorally, both signals were clearly audible throughout, and in separate tests the probe threshold remained unchanged – i.e., the probe was not masked – at contralateral noise levels up to 75 dB. Despite this behavioral evidence for no interaction between the signals, the physiological data resemble those seen in Fig. 10 in the following ways:

1. Amplitude is enhanced at the 5-, 15-, and 25-dB noise levels.
2. The absolute phase values change in the same direction, beginning in this case at the 15-dB noise level.
3. Changes appear in all three response measures at the 45- and 55-dB noise levels which, though considerably smaller in amount, resemble in kind and direction those seen in the ipsilateral masking case.

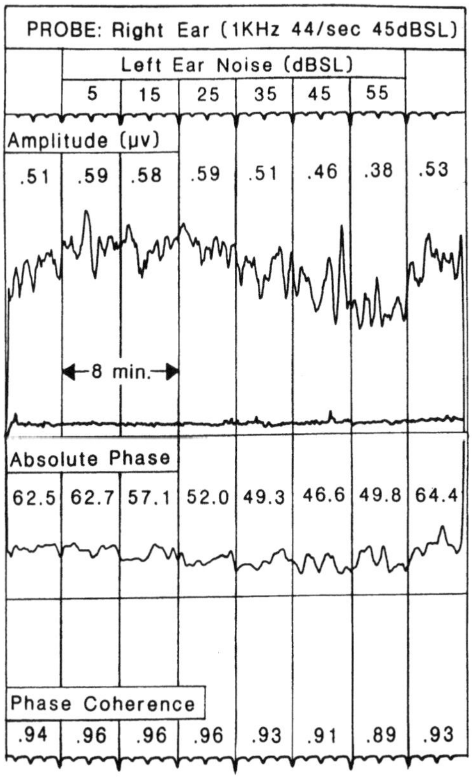

Fig. 11. Like Fig. 10 except the noise is delivered into the contralateral ear. The response measures suggest an interaction similar to that of Fig. 10 is taking place; the most intense noise (55 dB) is, however, still 20 dB below the level at which noise produces a threshold elevation in the opposite ear

These results strongly indicate that an interaction within the brain is what produces first enhancement and then suppression of the probe response as the noise delivered to the contralateral ear progressively rises in intensity. The possibility that sound waves physically interact at the basilar membrane level in the ear receiving the probe stimulus seems to be entirely excluded.

This contralateral suppression phenomenon has been explored in considerable detail using seven subjects each of whom received wide-band noise at three levels in one ear and 500-Hz tone bursts at three levels in the other. Every subject showed the same main effect seen in Fig. 11: with increase in noise intensity, probe response amplitude and phase coherence both dropped. At each of the three noise levels, the effect was usually greatest for the weakest probe, and progressively smaller as probe intensity rose.

Unexplained variability in the data calls for further study, but the main conclusion seems firm: contralateral noise at a level well below what is required for masking changes the way a probe activates the nervous system. The results further suggest that ipsilateral noise, in addition to interacting physically with a signal like our probe at the basilar membrane level, may be similar to contralateral noise in interacting with the probe response centrally. Both of these speculations disagree with conventional masking theory and obviously remain to be tested experimentally.

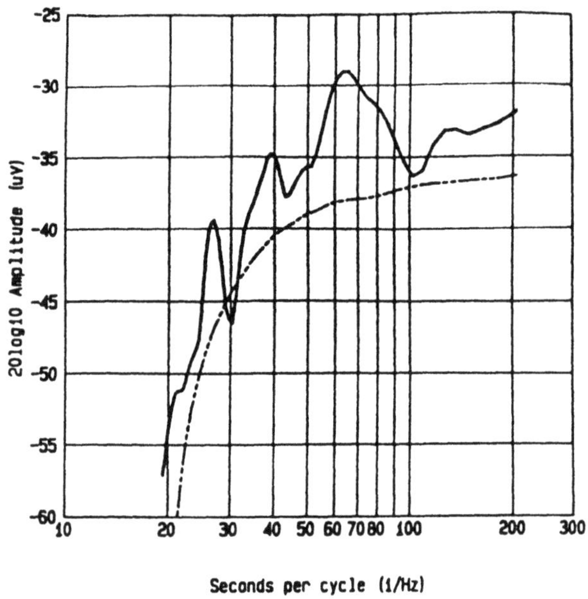

Fig. 12. Spectral analysis of a 15-min record containing typical minute rhythms *(solid curve)*, and a similar analysis of white noise *(dotted curve)*. The difference between them in the region of 60 s per cycle suggests the term "minute rhythm" is appropriate

3.4 Minute Rhythms

The distribution of minute-rhythm frequencies actually present in a recording like that of Fig. 7 can be seen in Fig. 12, which plots (solid curve) on loglog coordinates the spectrum of a typical 14-min (512 average) record made during right-ear stimulation by 500-Hz tone bursts. The spectrum was weighted by the spectral characteristic of the window filter used, which produces a null at 20-s periodicity. (This cutoff was chosen arbitrarily in order to resolve modulations in the 30–120-s range optimally). The dotted curve shows an expected value of the spectrum estimated by a bootstrap procedure in which 512 white-noise epochs were moving-window-averaged, Fourier-transformed, scaled, and plotted in the same manner as the responses. The difference between the dotted and solid curves shows relative response maxima near 60 s per cycle and at 28 and 40 s per cycle. This distribution of "minute" rhythms is typical, but when successive recordings receive this same analysis, the resulting distributions are never exactly the same. Evidently some centrally located system dynamically modulates the output level of the brain cells that mediate the SSRs, and in a way that itself varies through time.

We remark here parenthetically that SSRs driven by visual (checkerboard reversals) and tactile (vibratory) stimuli at rates in the 30–50-Hz range show minute-rhythm modulations that closely resemble the auditory examples illustrated here. Their relative amplitude fluctuations may in fact be larger, but the definitive study comparing minute rhythms in the three modalities remains to be done.

Every adult record examined during the past 4 years has shown minute rhythms, and this includes many collected during multichannel recording sessions. Figure 13 answers the question whether the rhythms at one scalp site are like those being recorded at the same time at another scalp site. In the experiment, the electrodes were placed symmetrically over left and right hemispheres (F3-T5 and F4-T6) across the

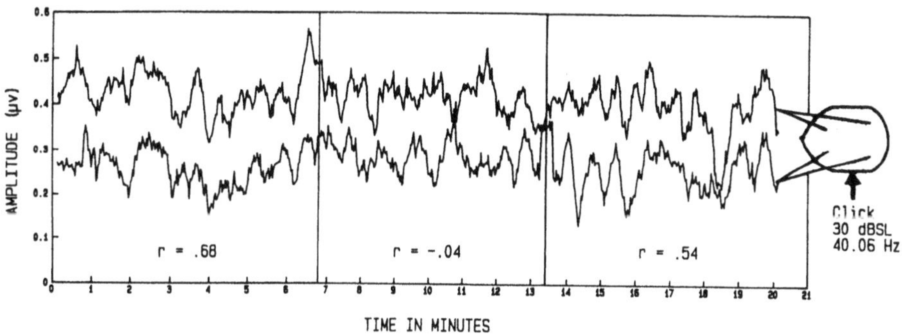

Fig. 13. Minute rhythms recorded over left and right hemispheres during 20 min of left ear stimulation with clicks. The correlation coefficients *(r)* show the rhythms synchronize quite differently in the successive 6–7-min periods

dipole Wood and Wolpaw (1982) identify as highly active during the first 100 ms post-stimulus. Clicks applied to the left ear produced SSRs on both sides (but larger on the right, a common finding in our studies), each showing typical minute rhythms. Even a cursory examination reveals that these rhythms are not exactly alike, and when a correlation coefficient is obtained for these successive 6-min samples (each 256 averages), the rhythms turn out to be sometimes entirely unrelated ($r = -0.04$) but usually reasonably in phase ($r = 0.68$ and 0.54). This gives a clear answer to the initial question; the rhythmicities recorded at different scalp sites vary widely. Apparently a mechanism for synchronizing the rhythms does exist but its level of operation changes through time. The numerous questions raised by this conclusion cannot be explored further here.

3.4.1 Concurrent Stimulation

When two (or more) stimuli are delivered at the same time, frequency-domain procedures allow the independent evaluation of each SSR produced. All that is required

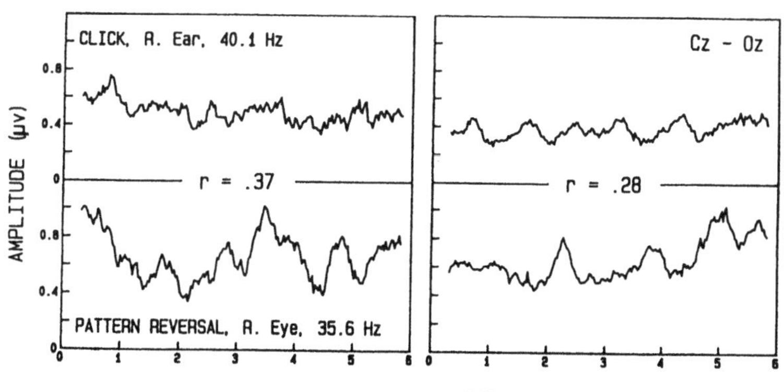

Fig. 14. Minute rhythms recorded at a single scalp location (Cz-Oz) during concurrent stimulation of an ear at one rate and an eye at another. The correlation between the two minute rhythms in this sample is 0.28

is the delivery of the stimuli at different rates; the resulting EEG spectrum will now contain peaks at the two stimulus rates (not just one, as in Fig. 3), each the candidate for the same complete analysis in the frequency domain. This was done in the experiment summarized in Fig. 14: stimuli were clicks at one rate, visual checkerboard alternations at another, with recording limited to a single scalp site. The display shows two minute-rhythm patterns, one for each stimulus modality, and these correlate 0.28. Identical runs on the same subject on the same day yielded these additional coefficients: 0.10, 0.37, 0.35, −0.09. To some, these results may suggest that the events responsible for minute rhythms are less likely to take place in the cortex – which the two sensory inputs presumably share in part at least – than at some subcortical location where the two pathways are separated. More experiments will, however, be required to settle the question.

3.4.2 Summary of the Minute-Rhythm Findings

1. Spectra of auditory SSR amplitude fluctuations show the minute-rhythm energy to be distributed unevenly between 20 and 120 s per cycle, with a peak frequently present in the region of 60 s per cycle. The term "minute rhythm" is therefore not entirely inappropriate, but should not be interpreted literally. Similar modulations of visual and tactile SSRs also occur. The process(es) responsible for these dynamic amplitude fluctuations in the three modalities remain to be identified.
2. Minute rhythms recorded simultaneously at different scalp locations are sometimes in phase and sometimes not. The process responsible for this partial and variable synchronization in the different brain regions is unknown.
3. Minute rhythms recorded at the same scalp site during concurrent stimulation of two end-organs similarly vary in synchrony and, again, the responsible mechanism(s) remain to be identified.

4 Discussion

4.1 SSR Generators

For more than 20 years, certain conceptual and analytic problems associated with evoked potential research have challenged members of this symposium – particularly Adey, Bullock, Freeman, John, and the senior author of this paper (Freeman 1962; Adey 1967; Bullock 1967; Galambos 1967; John 1967). Foremost among these problems is identifying the brain cells responsible for generating evoked potentials, including the SSRs considered here: what cells are involved, where are they located, and what rules control when they will generate the currents we record?

Drug studies have clearly shown that whereas cell groups such as the classical afferent pathways work about equally well in anesthetized and unanesthetized states (the human auditory brain stem response is such an example), others generate their currents only in unanesthetized animals, and still others only when experimentally specified external conditions are met (as both John and Freeman show elsewhere in this volume). Where are these labile cells that spring into action when the animal orients toward a novel sound (John 1967) or when a human listener perceives a word deliberately misplaced in a sentence (Kutas and Hillyard 1980)? What process swings a brain's electrical output over the range that extends from zero in anesthesia to high levels in concentrated attention?

Judging from where electrodes record evoked activities on the scalp and in the depths, these labile cell collections must be numerous, large in extent, and widespread throughout the forebrain. In animals, several regions contribute to SSRs resembling the potentials featured in our studies. Using ablation techniques in the guinea pig, Yoshida et al. (1984) have identified four: one in each hemisphere; one deep in the midline that remains active after bilateral removal of the cortex and inferior colliculi; and one present after decerebration. In man, strong evidence for a generator in each hemisphere comes from magnetic measurements (Romani et al. 1983), and a recent clinical study supports the claim that 40-Hz SSRs recorded at the scalp come from the midbrain, not the cortex (Spydell et al. 1985). Our present working hypothesis is that, as in the guinea pig, currents are generated in at least one deep (probably thalamic) midline site and at the cortical terminus of each afferent pathway. We assume that what the electrodes record, as theory predicts, is the algebraic sum of these at least.

4.2 SSR Versus Other Evoked Potential Procedures

In what way, one may ask, do these SSR experiments differ from those that have evolved during the past 20 years into the ones reported at this symposium by John, Freeman, and others? To answer this question, we will first list the essential features of the high-rates probe procedure we have developed, then compare it with the more conventional approaches, and finish with several generalizations extracted from the results reported in the previous section.

4.2.1 Characteristics of the High-Rates Probe Procedure

1. A stimulus in any modality probes the CNS continuously, at high rates near 40 Hz.
2. Every few seconds a new response average (the SSR) is computed or updated.
3. Frequency-domain analysis produces multiple response measures from each average.
4. Multidimensional data plots chart dynamic changes in CNS status.
5. Multichannel recording reveals for comparison the dynamic events at more than one cortical terminus of a given afferent pathway.
6. Concurrent stimulation within or across modalities, which produces two (or more) averages at the same cortical location, permits simultaneous evaluation of more than one afferent pathway at any electrode site.

4.2.2 Models

Evoked potential physiology has from the beginning been compared to testing procedures commonly used in engineering practice. Just as the geologist in search of oil explodes a dynamite charge on the ground and analyzes the echoes returning to his array of motion detectors, so the visual or auditory stimulus perturbs ongoing processes in EEG generators – starts some, stops others, changes phase-locking in still others – and thus creates the evoked potential. The high-rates probe procedure described here certainly fits this description, but perhaps the model of two people conversing in the question–answer format is even more apt. The probes represent a stream of identical questions that are answered every few seconds by the response average. Each answer is interesting in itself, but, when linked together (see Fig. 8), they reveal the even more interesting dynamic changes going on in the system that formulates the reply. Thanks to the frequency-domain procedures, every reply is decomposed into two independent components (amplitude, phase), each a complete story in itself and collectively a multidimensional description of a continuing interaction between stimulus events and the responding system.

With this model for reference, one detects few similarities between our probe procedure and Freeman's design. His EEG response frequencies are similar (40 Hz and above), but they are not stimulus driven, since the stimulus odor that initiates them has no frequency structure. John's animal conditioning experiments come closer. His stimuli ("tracers") resemble probes in being repetitive modulations of intensity; they recur, however, at relatively low rates (10 Hz or below) and are in fact conditioned stimuli to which the animal is trained to react. Our probes, by contrast, are weak signals which subjects are expected to ignore, but of course experiments could easily be designed in which probes become tracers in John's sense.

4.2.3 Modulators

This research seems to have identified at least three mechanisms, or processes, that control the output of the generators active during sensory stimulation. It may be heuristic to give these mechanisms a name – "modulators" – and to differentiate them on the basis of what they do. One class of modulators yields the minute rhythms, another the response changes associated with sleep, the third those events revealed in the central suppression studies. If we call these modulators "cyclic," "sleep-related" and "stimulus-linked," respectively, we not only describe the data observed but additionally suggests the time course of the action each exerts on the generators it controls.

Cyclic modulators introduce quasiperiodic fluctuations of varying size. Minute rhythms are the clearest example but whether still others will be found is left open. Cyclic modulators seem to be entirely under endogenous control, operate equally well on SSRs in any modality at any intensity, and persist in the presence of activity in the other modulator types. Their anatomical locations are uncertain: some evidence argues for cortical loci, some for their insertion into the individual afferent pathways at a subcortical level. Animal lesion studies and suitable human clinical cases will be required to resolve this problem.

The *sleep-related modulator,* identified by the changes seen during dozing and sleep, changes SSR amplitudes aperiodically, over wide ranges, and sometimes rapidly (as when someone suddenly wakens). Its unique linkage to sleep suggests an anatomical locus within CNS regions where changes in sleep and wakefulness are to be found. Like cyclic modulators it is entirely endogenous, but the two types of endogenous processes seem to proceed independently of each other and so may be located in different places in the brain.

The *stimulus-linked modulator* adjusts generator output to a new level whenever two auditory stimuli are presented at the same time. (This statement may also hold across modalities, but the evidence is still sparse.) This stimulus-dependent modulator is perhaps best classed as "exogenous," but whether its activities are affected by state changes such as sleep has not been studied. Locating it anatomically also invites further investigation; the interaural interactions already under electrophysiological study (Picton et al. 1981) suggest it lies above the brain stem level.

This introduction of the concept of modulators and their classification into three types attempts a synthesis of the information presently available. At the least, it may encourage the search for additional dynamic events complicating the processing of sensory information. At best, it could correctly characterize three major modulatory mechanisms of which perhaps only one (the sleep-related modulator) has heretofore been even suspected to exist. The most intriguing possibility is that these three modulators that so clearly introduce lability into SSRs may be related to the modulatory mechanism(s) that alter EP amplitudes in behavioral situations where, for instance, animals orient and people detect words misplaced in a sentence.

4.2.4 Behavior Correlations

Just as thinking is divided over whether the EEG and stimulus-locked potentials such as SSRs are related in any interesting or important way to the brain activities underlying behavior, so opinions are likely to differ on what, if anything, modulator activities have to do with behavior. Certain possibilities do exist, however, and we will discuss them briefly.

Alertness, Attention. It has been suggested in our laboratory that the minute rhythms might signal, or reveal, "microshifts" in alertness or attention, and it is intuitively obvious that "macroshifts" might well take place during the large movements in SSR measures that take place during dozing and awakening. Unfortunately, our tests of the first idea are so far inconclusive and we have as yet no information on the second. As for the stimulus-dependent modulator defined in the contralateral noise experiment, it seems unlikely it will turn out to be directly involved with alertness or attention.

Performance. The fact that correlation coefficients comparing minute rhythms at different scalp sites vary over time (Figs. 13 and 14) suggests the hypothesis that these coefficients may index speed, accuracy, or some other variable aspect of human performance. For instance, performance level might turn out to be highest (or lowest) whenever minute rhythms fall closely into phase at all scalp locations. This idea is reminiscent of John's "neurometrics" approach (John et al. 1977); to explore its possibilities would require application of the factor analysis and related statistical techniques developed by him, Adey, and Freeman, among others, for uncovering subtle and complex relationships among evoked potential and EEG events simultaneously recorded at many scalp sites. We have not yet implemented these procedures for application to our high-rates probe studies.

Perception. The central suppression results (Figs. 10, 11) identify physiologically an apparently new class of interaural interactions within the brain. We have therefore tried introspectively to detect an auditory experience associated uniquely with this alteration in the physiology. In the ipsilateral masking condition, the probe-response changes parallel, to a good first approximation, both the sensation of noise growing progressively in intensity and the partial masking of the signal going on at the same time. However, when the noise is introduced into the ear contralateral to the probe, the only obvious change is that the listener now perceives two sounds instead of only one. It is possible but we think not likely that the physiological changes index these perceptual differences only. If the physiological changes are related to "central masking" (Zwislocki 1973) or to loudness enhancement (Elmasian et al. 1980) we have not as yet detected any probe threshold or loudness change that would supply the link. We continue, therefore, to search for psychophysical or psychological experiments in which perception is found to be altered when the stimulus conditions described here are duplicated.

4.2.5 Summing Up

These experiments have not uncovered any relationship of the sort discussed in the Introduction between spontaneous and driven brain activity in the 40-Hz region. They have, however, disclosed three ways in which the streams of incoming sensory information are modulated by physiological activities endogenous to the nervous system. As yet behavioral correlations with the new physiological findings are meager.

5 Summary

Because a classical sensory pathway functions well even in deep anesthesia, some people assume that no important changes take place in the flow of sensory messages through the brain in the waking state. We present here a method that tests this assumption. It uses a steady stream of stimuli to challenge the human nervous system continuously for minutes or hours; every few seconds the potentials evoked by these stimuli are extracted from the EEG, averaged, analyzed in the frequency domain, and plotted. The results disclose that physiological activities modulate the stream of incoming sensory information in at least three different ways:

1. The response amplitude cycles up and down spontaneously every minute or so; these minute rhythms characterize the responses to auditory, visual, and tactile stimuli obtained from all adult subjects so far tested.
2. The response amplitude drops by 50% or more as a subject falls asleep and is promptly restored with awakening; this modulation, correlated to date only with sleep, is again found in all subjects and modalities.
3. The electrical response produced by tones delivered to one ear is modified by contralateral noise too weak to produce masking; the flow of impulses initiated by contralateral stimulation can either increase or decrease the response to the signal.

This account of the dynamic influences impressed upon sensory messages as they penetrate the nervous system is unfortunately still far more descriptive than it is analytic. Experiments are needed to show where in the nervous system these dynamic changes are introduced, what physiological mechanisms are involved, and whether the modulations index interesting alterations in psychological state.

Acknowledgements. This research has been supported by research grant NS17490 (R.G.) and a training grant (S.M.) from the National Institutes of Health; a dissertation research grant from the University of California, San Diego; the excellent technical assistance of Phyllis Galambos; and the advice and counsel of Dr. D. R. Stapells.

References

Adey WR (1967) Intrinsic organization of cerebral tissue in alerting, orienting and discriminative responses. In: Quarton GC, Melnechuk T, Schmitt FO (eds) The neurosciences: a study program. Rockefeller University Press, New York, pp 615–633

Başar E (1980) EEG-brain dynamics. Elsevier/North-Holland, Amsterdam

Bouyer JJ, Montaron MF, Rougeul A (1981) Fast frontal-parietal rhythms during combined focused attentive behaviour and immobility in cat: Cortical and thalamic localizations. Electroencephalogr Clin Neurophysiol 51: 244–252

Brown GD, Shallop JK (1982) A clinically useful 500 Hz evoked response. Nicolet Potentials 1 (5): 9–12

Bullock TH (1967) Signals and neuronal coding. In: Quarton GC, Melnechuk T, Schmitt FO (eds) The neurosciences: a study program. Rockefeller University Press, New York, pp 347–352

Chatrian GE, Petersen MC, Lazarte JA (1960) Responses to clicks from the human brain: Some depth electrographic observations. Electroencephalogr Clin Neurophysiol 12: 479–490

Elberling C (1979) Auditory electrophysiology: Spectral analysis of cochlear and brainstem evoked potentials. Scand Audiol 8: 57–65

Elmasian R, Galambos R, Bernheim A (1980) Loudness enhancement and decrement in four paradigms. J Acoust Soc Am 67: 601–607

Flinn JM, Kirsh AD, Flinn EA (1977) Correlations between intelligence and the frequency content of the evoked potential. Physiol Psychol 5: 11–15

Freeman WJ (1962) Phasic and long-term excitability changes in prepyriform cortex of cats. Exp Neurol 5: 500–512

Galambos R (1967) Brain correlates of learning. In: Quarton GC, Melnechuk T, Schmitt FO (eds) The neurosciences: a study program. Rockefeller University Press, New York, pp 637–643

Galambos R (1981) Addendum. Nicolet Potentials 1(4): 12

Galambos R (1982) Tactile and auditory stimuli repeated at high rates (30–50 per sec) produce similar event related potentials. Ann NY Acad Sci 388: 722–728

Galambos R, Makeig S, Talmachoff PJ (1981) A 40 Hz auditory potential recorded from the human scalp. Proc Nat Acad Sci USA 78: 2643–2647

Geisler CD (1960) Average responses to clicks in man recorded by scalp electrodes. MIT Research Laboratory of Electronics, Cambridge (Technical report 380)

John ER (1967) Electrophysiological studies of conditioning. In: Quarton GC, Melnechuk T, Schmitt FO (eds) The neurosciences: a study program. Rockefeller University Press, New York, pp 690–704

John ER, Karmel BZ, Corning WC, Easton P, Brown D, Ahn H, John M et al. (1977) Neurometrics. Science 196: 1393–1410

Klein AJ (1983) Properties of the brainstem response slow-wave component. I. Latency, amplitude and threshold sensitivity. Arch Otolaryngol 109: 6–12

Kutas M, Hillyard SA (1980) Reading senseless sentences: Brain potentials reflect semantic incongruity. Science 207: 203–205

Linden RD, Campbell KB, Hamel G, Picton TW (1985) Human auditory steady state potentials during sleep. Ear Hear 6: 167–174

Makeig S (1985) Studies in musical psychobiology. PhD dissertation, University of California at San Diego

Mardia KV (1972) Statistics of directional data. Academic, New York
Picton TW, Stapells DR, Campbell KB (1981) Auditory evoked potentials from the human cochlea and brainstem. J Otolaryngol [Suppl 9] 10:1–41
Regan D (1982) Comparison of transient and steady state methods. Ann NY Acad Sci 388:45–71
Romani GL, Williamson SJ, Kaufman L, Brenner D (1983) Characterization of the human auditory cortex by the neuromagnetic method. Exp Brain Res 47:381–393
Sayers BM, Beagley HA, Rhia J (1979) Pattern analysis of auditory evoked potentials. Audiology 18:1–16
Sheer DE (1976) Focused arousal and 40 Hz EEG. In: Bakker DJ (ed) The neurophysiology of learning disorders: theoretical approaches. University Park Press, Baltimore, p 71
Spydell JD, Pattee G, Goldie WD (1985) The 40 hertz auditory event-related potential: normal values and effects of lesions. Electroencephalogr Clin Neurophysiol 62:193–202
Stapells DR, Linden RD, Suffield JB, Hamel G, Picton TW (1984) Human auditory steady state potentials. Ear Hear 5:105–114
Van der Tweel LH, Verduyn Lunel HFE (1965) Human visual responses to sinusoidally modulated light. Electroencephalogr Clin Neurophysiol 18:587–598
Wood CC, Wolpaw JR (1982) Scalp distribution of human auditory evoked potentials. II. Evidence for overlapping sources and involvement of auditory cortex. Electroencephalogr Clin Neurophysiol 54:25–38
Yoshida M, Lowry LD, Liu JJC, Kaga K (1984) Auditory 40-Hz responses in the guinea pig. Am J Otolaryngol 5:404–410
Zwislocki JJ (1973) In search of physiological correlates of psychoacoustic characteristics. In: Moller AR (ed) Basic mechanisms in hearing. Academic, New York, pp 787–806

Cortical Structure and Electrogenesis

H. Petsche, H. Pockberger, and P. Rappelsberger

1 Introduction

Recording electrical brain potentials in the electroencephalogram (EEG) has become a major instrument in the clinical diagnosis of brain disorders. It has also contributed toward revealing both the nature and the degree of brain dysfunctions accompanying metabolic dysfunctions. Moreover, and particularly in the past decade, the EEG has also turned out to be a valuable aid for the detection of brain activities connected with specifically human brain functions, such as cognitive processes (for a review, see Giannitrapani 1985). In this latter context, the study of event-related potentials has turned out to be most valuable (Rockstroh et al. 1982). Spontaneous EEG activity is becoming more and more important for the analysis of cognitive processes (Duffy 1985), for which statistical processing of EEG data seems to be among the most effective methods at the present time (see Pockberger et al., in this volume).

All the methods mentioned above are based on the recording of field potentials, which are found throughout the brain and which can be recorded from the intact skull as the EEG. Since this kind of oscillation in potential represents a mass action of the nervous system (Freeman 1975) based on cooperative processes between neuronal elements and probably also glial cells, all attempts to explain the properties of the EEG by exploring the behavior of single nerve cells by intracellular recordings have failed up to now. The missing link between the abundant knowledge of the properties of the neuronal membrane and the phenomena observed by large electrodes, as in the EEG, may also be a major reason why the exploration of the nature of the EEG has been fairly neglected by the majority of neurophysiologists who use mainly intracellular recordings. Moreover, the cortical blueprint that is the structural base for its electrical properties is still far from being clear; the same holds true for the biochemistry of the cortex. Knowledge of the mode of action of neurotransmitters and neuromodulators is still increasing, as is the number of substances that are candidates for mediating the transmission of information from neuron to neuron. Furthermore, a description of electrical macroactivities requires systems of nonlinear partial differential equations, in contrast to the description of the electrical behaviour of a single neuron, the main properties of which may be described by the Hodgkin-Huxley equations.

One purpose of this paper is to demonstrate that a study of the properties of field potentials may yield some insights into the nature of brain processes, if certain requirements are fulfilled. Because of the immense number of unknown factors underlying field potentials, some of which are mentioned above, only descriptive state-

Springer Series in Brain Dynamics 1
Edited by Erol Başar

ments can be derived from these observations. However, if an appropriate scale of magnitude is chosen to study field potentials, several statements concerning the possible underlying structural properties can be made (Petsche et al. 1984). For this purpose we developed a microEEG method (Petsche et al. 1982), which is based on an appropriate strategy for comprehending the field potential continuum in a limited portion of the cortex. This strategy aims at: (a) using as many electrodes as possible; (b) recording from locations as close to one another as possible; and (c) recording from the cortical surface as well as from within the cortex.

It was evident at the onset that all of these requirements could not be met at one time and it was also clear that the number of electrodes would be limited by the particulars of the amplifier systems. In spite of these restrictions, however, we hoped to arrive at an idea of the continuous potential distribution in the portion of cortex under investigation by interpolation between the multiple recordings.

2 Methods and Material

For epicortical recordings, a square grid of 16 surface electrodes (4 × 4 at 2 mm distances; electrode diameter 0.1 mm) was mounted on the cortex of slightly anesthetized rabbits. This lissencephalic animal was chosen in order to obviate any geometrical complications. Simultaneous recordings from these 16 contacts, positioned with respect to the nasal bone, were recorded on analog tape and processed off line. The processing consisted, after digitation at 256 Hz, of plotting equipotential fields.

For intracortical recordings, a 16-fold semimicroelectrode was developed based on thin-film technology (Prohaska et al. 1979). This probe consists of 16 AgAgCl contacts of 10 × 10 μm^2 at 150 μm distances on a carrier needle inserted vertically into the cortex under microscopic control. The distances between the contacts are distributed so that recordings can be obtained from the total diameter of the rabbit's cortex. These 16 electrodes are connected to impedance transformers and the activities are amplified by EEG amplifiers (TC = 0.3 s, HF-filter setting ∞). After each experiment, silver is deposited from every contact for histological examination (Müller-Paschinger et al. 1979).

With this device, 16 simultaneous electrocorticographic recordings at 150 μV intervals can be obtained. By interpolation between these recordings, potential chronotopograms may be obtained; i.e., continuous representations of the oscillations of field potentials, continuous with respect to the cortical profile and time (Fig. 1).

Another great advantage of this device is the possibility of calculating current-source-density (CSD) profiles (Rappelsberger et al. 1981), which optimize localization of the places where currents are produced, a distinct improvement upon the commonly used potential-time traces. Morveover, errors caused by volume conduction are precluded by this method.

The usefulness of this method may be illustrated by the cortical response to a single stimulus of the nucleus lateralis thalami (Fig. 1); the multielectrode was in area praecentralis 1. The top of Fig. 1 shows the 16 traces of the response recorded in the conventional way. A short, complex initial event is followed by a slow wave, which

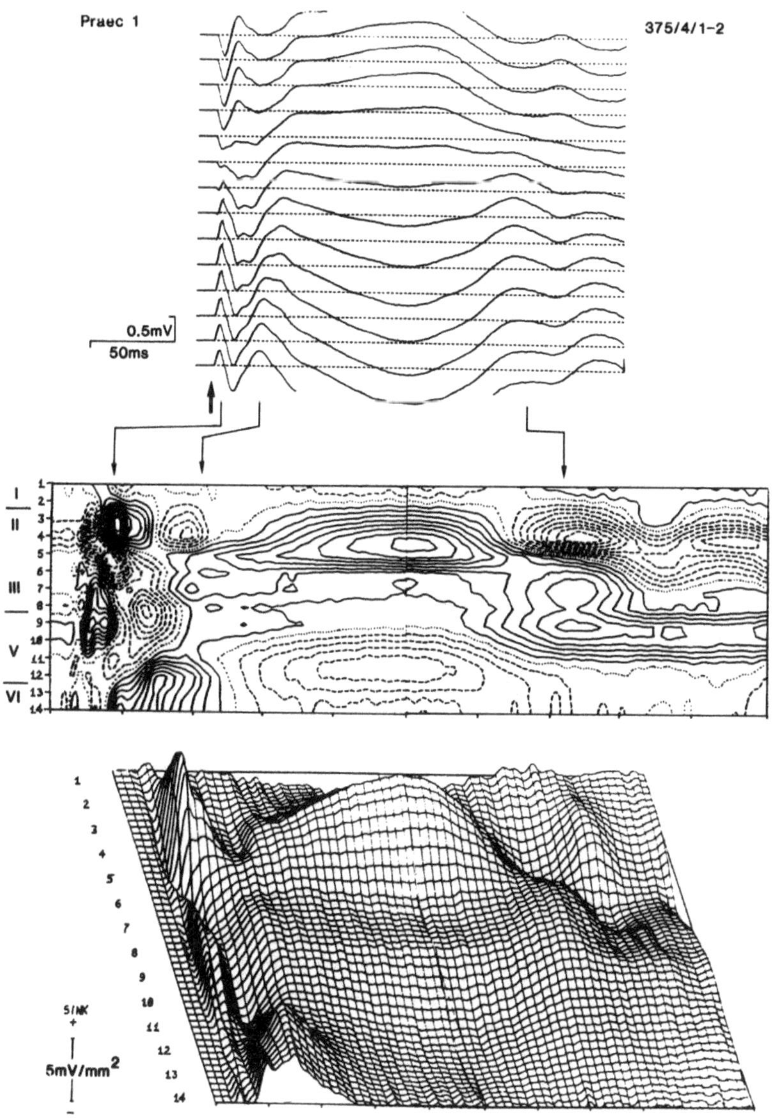

Fig. 1. *Top,* field potentials recorded with the 16-fold depth electrode from the area praecentralis of the rabbit (150-µm contact distances) after a single stimulus *(arrow)* to the nucleus lateralis thalami. *Middle,* CSD distribution of the same episode. *Solid lines,* sinks; *dashed lines,* sources; *abscissa,* time; *ordinate,* profile of the cortex; *Roman numerals,* layers. *Bottom,* the same episode, in three-dimensional representation. The field potentials turn out to correspond to a superimposition of numerous sinks and sources with different delays and at different locations within the cortex. This kind of presentation visualizes the pathways of cortical excitation and the reactions of neuronal substrates at different cortical levels

is negative near the surface and positive in the deep cortical layers. The diagram across the middle of Fig. 1 is the CSD configuration. The abscissa is the time and the ordinate is the profile of the cortex, with the layers indicated by Roman numerals. The density of the contour lines indicates CSD with sinks shown as full lines and sources as dashed lines. The lowermost diagram represents the same, three-dimensionally.

The event evoked by stimulation of the thalamus is fairly complex and seems to be composed of different sinks and sources at different cortical depths, beginning with a sink in layer Va. In contrast to this complicated configuration of the early electrical events, the late event may be understood as being caused by a single large dipole, the sink of which is in layer II, followed by oscillations of dipoles between layer II and III to Va, with the sinks in the lower layers.

Figure 1 was included in this paper simply to demonstrate that the method also permits studying cortical neuroanatomy in its functional aspects, e.g., electrohistology, by analyzing the sequence of electrical events at the different cortical levels. However, the figure may also serve as a basis for interpreting the configurations formed in the course of the brain's electrocortical activity.

The figure represents current sinks – i.e., locations of inward current – as families of full contour lines and sources as families of dashed lines. Both sinks and sources turn out to be more or less limited in their vertical extension but fairly variable in their horizontal extensions, i.e., time. According to the above considerations it is most likely that these sinks and sources are caused mainly by summated synaptic potentials, the contribution of soma discharges being almost negligible. It cannot be excluded, however, that glial processes may also contribute, particularly with regard to the slow processes (for a review see Somjen and Trachtenberg 1979). Since an overwhelming number of synapses terminate on dendrites, one may conclude that sinks and sources are mainly caused by a summation of de- and hyperpolarizations of dendrites (Mitzdorf and Singer 1979), with the contribution of soma discharges to sinks being minor. This idea is also supported by the observation that the location of maximal densities of current sinks and sources does not coincide with the cytoarchitectonic layers, which are defined by the different shapes and densities of neuron somata. However, those locations where the main sinks in Penicillin-induced "spikes" are most often observed do coincide with dendritic architectonics, as shown below.

Another point of view should be discussed in this context. When a nerve cell is locally depolarized, a passive outward current flows at other sites on the same neuron. Therefore, every active current sink produces passive current sources at other locations and vice versa. This also holds true for the sinks and sources in these diagrams. Nevertheless, because of the relatively low spatial resolution of the 16-fold electrode – the electrode distances were never smaller than 100 μm – these sinks and sources have to be considered as superimpositions of countless single events, probably mostly synaptic de- and hyperpolarizations. Therefore it is hardly possible to identify the source for every sink and vice versa. Moreover, it is also impossible to argue that any identified sink or source is an active event; i.e., that it is produced by an accumulation of excitatory or inhibitory postsynaptic potentials (EPSP or IPSP), respectively, or, instead, that it is a passive one. However, there may be some evidence that gives particular weight to one or the other argument. With regard to

Fig. 1, the initial brief sink in layer Va seems more likely to be caused by an accumulation of EPSPs as a response from pyramidal cells to the incoming excitation, than the extended accompanying source in layer II/III is likely to be caused by IPSPs. During this event, current is evidently drawn from the upper parts of the dendritic tree of layer V pyramidal cells during the excitation of the cells by the incoming thalamocortical shower of action potentials. However, it cannot be excluded that IPSPs may contribute to the source in layers II/III. The same holds true for the long-lasting afterswing, and the question remains whether the sink in layer II or the accompanying source in layer V is the active event. Therefore it is advisable to consider these pictures merely as a description of the spatiotemporal configuration of electrical events within the cortex. Another advantage of plotting CSD diagrams is that volume-conducted effects are eliminated by this method.

3 Ongoing Spontaneous EEG of the Rabbit

Since epi- and intracortical recordings can be made with the instruments described above only on animals immobilized in a stereotaxic frame, recordings on freely moving, awake rabbits could not be performed for ethical reasons.

In a slightly anesthetized rabbit (Nembutal), the electrocorticogram (ECoG) of the striate cortex consists of slow delta waves intermingled with single or grouped theta waves and waves of higher frequencies, with voltages rarely surpassing 500 μV. In analyzing these activities, the spatial resolution of the 16-electrode grid was too low to study the behavior of equipotential fields; these seem to be composed of events too poorly synchronized to be understood with interelectrode distance as large as 2 mm, quite in contrast to self-sustained seizure activities.

The intracortical properties, however, were studied by spectral analysis and by CSD analysis. The intracortical activity, shown at the top of Fig. 2, was recorded with 150 μm interelectrode distances. It has a pattern of slow waves, which is clearly different in the upper as contrasted with the lower cortical layers, these differences changing continuously along the cortical profile. As the power spectrum demonstrates (Fig. 3), maximal power is found at 2 Hz; this peak is lowest at contact 6. Moreover, higher-frequency bands are more pronounced in contacts beneath this level of minimal 2 Hz power (the power spectrum refers to a period of 60 s). A power profile at 2 Hz shows that there are two power maxima, one at contact 14 (layer VI) and another, lower one at contact 3 (layer II/III), separated one from the other by a power minimum at contact 6 (layer III). This is characteristic behavior for spontaneous activities and has been described in detail (Rappelsberger et al. 1982). The same holds true for coherence estimates: the least coherence with respect to the surface was found at contacts 6–7. It should be noted, however, that coherence with respect to the surface increases again with increasing depth to reach a maximum at contact 14, where the power maximum was found. There is also, as Fig. 3 shows, a phase reversal with respect to the contact next to the cortical surface. However, no significant phase shifts between the individual contacts were found. This analysis makes it evident that the preponderant frequency band of spontaneous activity may be explained by a dipole between layers II/III and layers V.

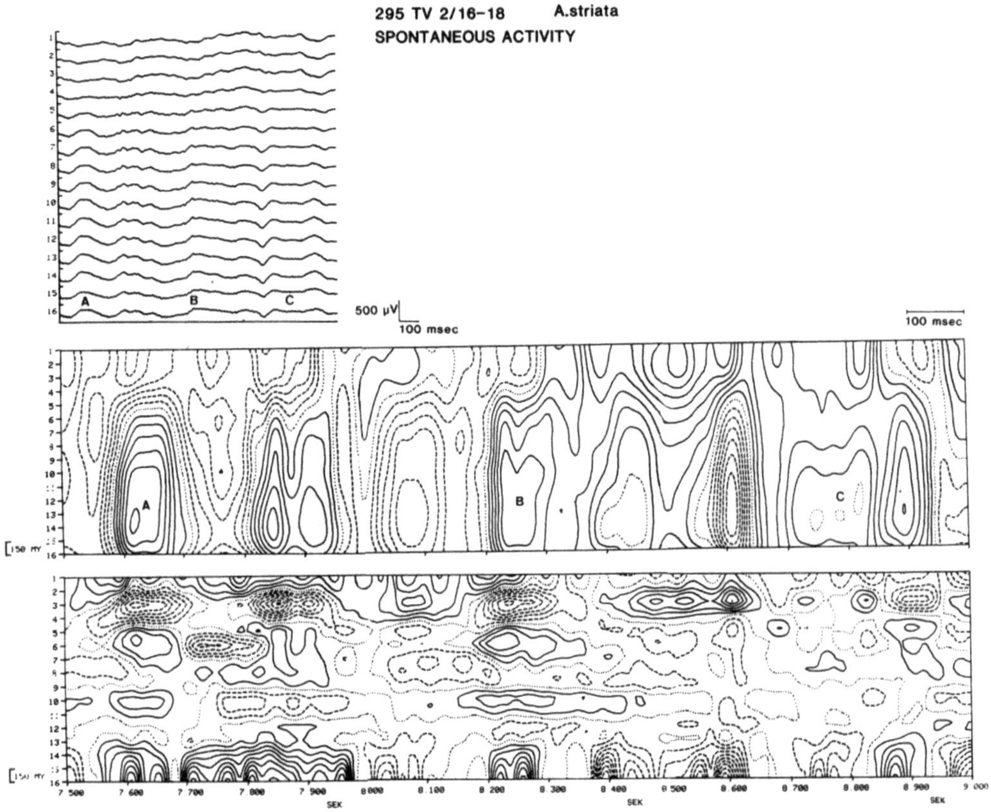

Fig. 2. Spontaneous activity of rabbit striate cortex. *Top,* a 1.5-s episode in conventional representation. *Middle,* the same episode as a potential chronotopogram with extended time scale. *Solid lines,* negativity; *dashed lines,* positivity. There is an approximate phase reversal between the deep and the most superficial cortical layers. *Bottom,* the same episode as a CSD diagram (see Fig. 1)

Figure 2 also represents a CSD analysis of spontaneous activity. At the top is a 1.5-s episode of spontaneous activity. In the middle, the same episode is represented chronotopographically. By interpolating between the 16 depth recordings at every 4 ms (corresponding to 256/s digitation rate), contour lines are obtained that connect points of equal potential. This pattern also documents that there is an approximate phase reversal of the main events between depth and surface.

The interpolation curves of the potential profiles were the basis for calculating CSD configurations which are represented at the bottom, also as contour lines. From this diagram, several conclusions may be drawn. Regarding sinks, there are several layers where sinks are most likely to appear, namely at contacts 3, 6, 10, and 15, located in layers II/III, IV, V, and VI. The sinks at contact 6 and at lower contacts often occur almost simultaneously, in contrast to the sinks and sources at contact 3, which seem to alternate with the former. The sinks often seem to consist of repetitive events.

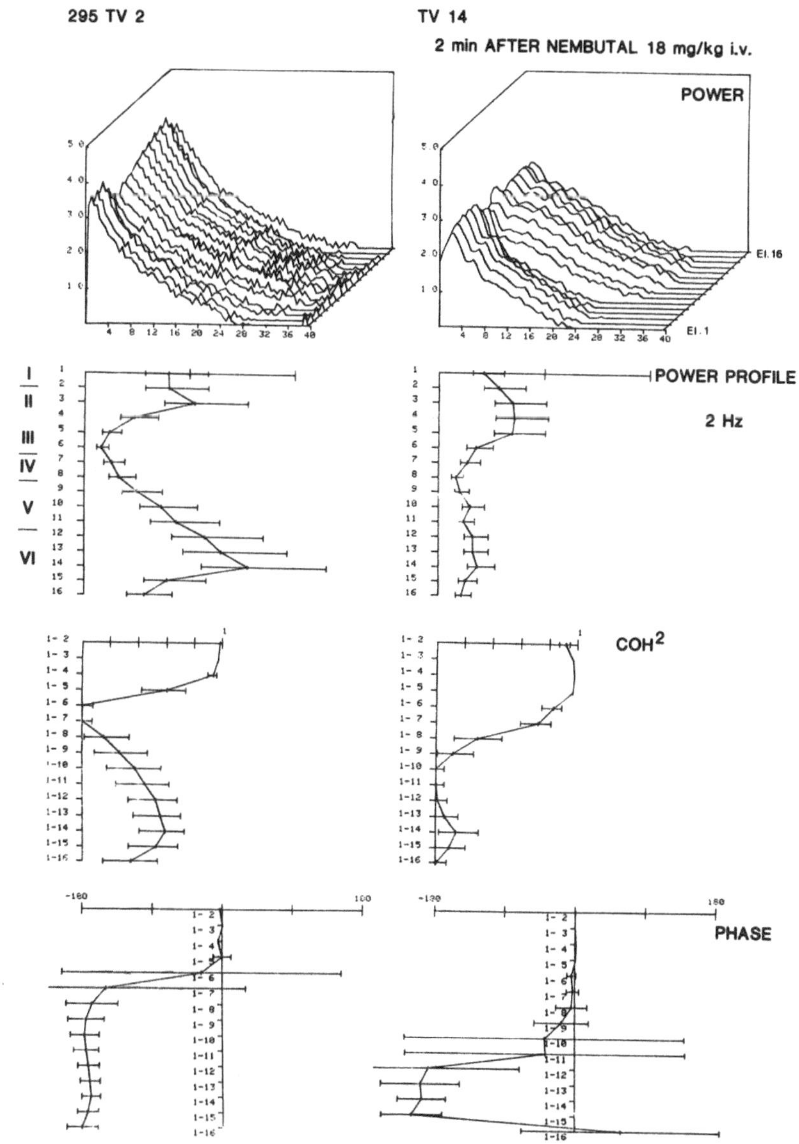

Fig. 3. Spectral analysis of *(left)* spontaneous activity and *(right)* activity 2 min after Nembutal (18 mg/kg i.v.). *First row,* families of power spectra (60-s episode) determined for all 16 contacts of the depth electrode (logarithmic scale). *Second row,* power profiles determined at 2 Hz (linear scale). *Third row,* squared coherence profiles, estimated with respect to contact 1. *Fourth row,* phase profiles. Under a high dosage of Nembutal, there is a considerable decrease of power, and the zone of minimum coherence, as well as the zone of phase reversal, shifts by 600 μm to deeper cortical levels

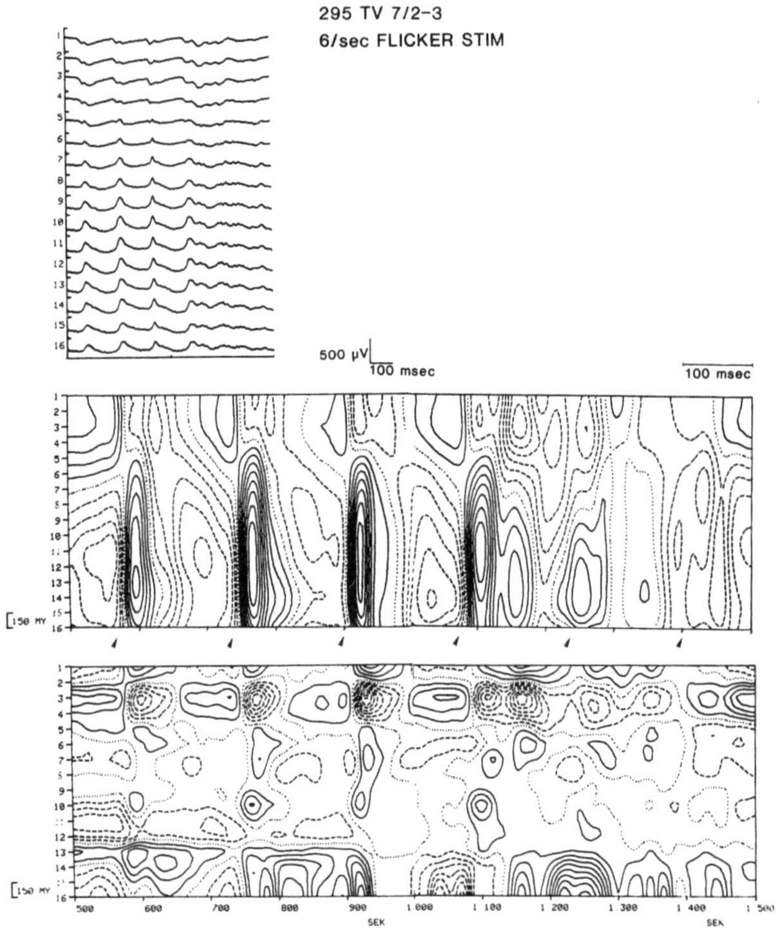

Fig. 4. Flicker stimulation of the contralateral eye (6 per second, *arrowheads*); same recording site as Fig. 2. The sequence of intracortical sinks begins in the deeper cortical layers

The impression that there is a time delay between sinks at contacts 14, 10, and 16, respectively, finds support when the animal is stimulates with a 6-Hz flicker (Fig. 4). This shows even more clearly that the sequence of infragranular sinks starts in the deepest cortical layers. This fact, as well as the fact that the flicker stimulus sometimes fails to elicit a distinct response, strengthens the assumption that the response to a flicker is but an amplification of the intrinsic tendency of the cortex to produce synchronized events. When the cortical propensity to synchronization is low (as for instance in Fig. 4, first stimuli), the flicker stimulus may beat time and synchronize the cortical events; but when the cortical propensity to synchronize is strong enough (last two stimuli), the flicker rhythm may fail, because the intrinsic cortical rhythm and the flicker rhythm may interfere, so that the regularity of the pattern decays. The most distinct dipoles during spontaneous and flicker-driven activity are found between contacts 3 and 6.

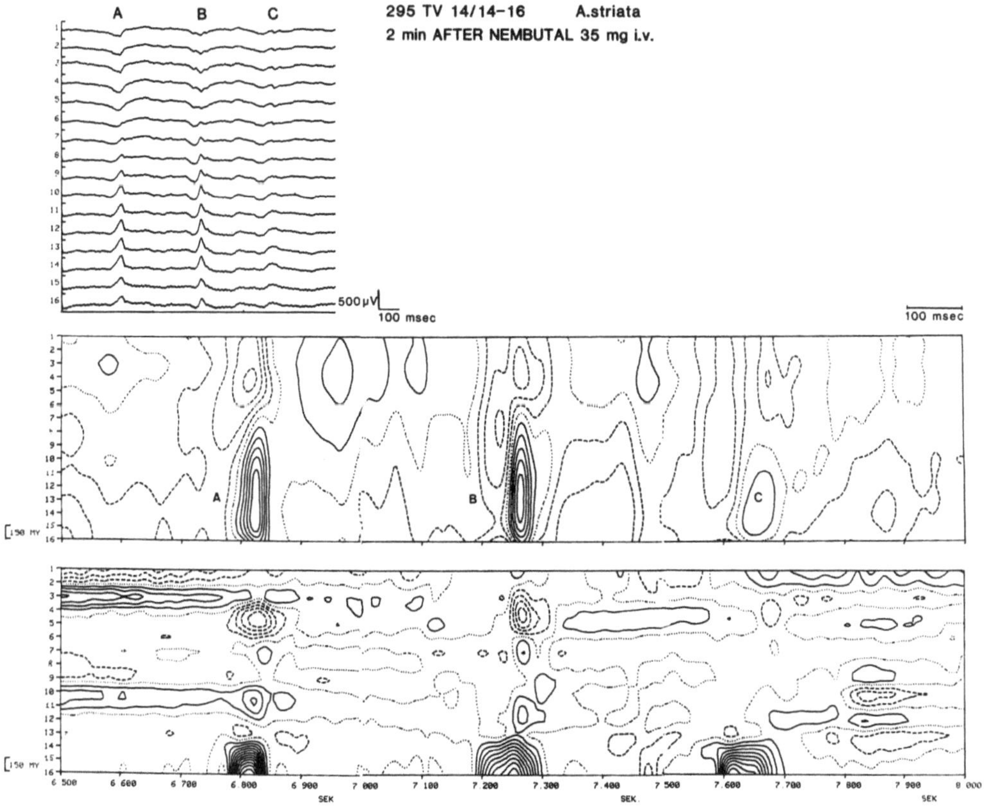

Fig. 5. Same experiment as in Figs. 2–4, but 2 min after Nembutal i.v. Spontaneous activity is largely reduced; the remaining single transients *(A–C)* exhibit a complex CSD configuration with an intensive sink in the deepest cortical layers

Figure 5 demonstrates how intracortical activity may change considerably under the influence of a barbiturate, Pentobarbital Sodium (Nembutal) given intravenously to the animal 2 min earlier. The intracortical field potentials now show the typical *tracé alternant,* known as a classical pattern also in humans, which appears immediately before the disappearance of the EEG. It appears as an arrhythmic sequence of short transients in an EEG with a very low voltage. In the potential chronotopogram at the center of Fig. 5, these transients appear as short, spike-like events, negative in the depth with a maximum at contact 14, with a roughly phase-reversed but lower transient near the surface. In the CSD configuration, these transients seem to be composed of infragranular sinks, with maximum intensity at contact 16 and accompanied by a source at contact 5. Apparently, layer VI cellular elements are mainly responsible for these transients.

The background activity has also changed considerably, apart from its generally large decrease in power, as Fig. 3 (right side) demonstrates. Now, the upper cortical layers produce, on the average, slightly more power than do the lower cortical layers, and the zone of phase reversal has shifted downward to contacts 10 and 11 and has also become broader. This becomes still more evident from coherence estimates with respect to the surface: the zero coherence zone has shifted from contacts 6–7 to contacts 10–12; there is almost no more coherence between surface and contact 14 than there was in the spontaneous EEG. These findings indicate that the populations of neurons responsible for the dipole behavior under Nembutal have changed dramatically. According to these findings, one may speculate that layer VI neurons seem to be most resistive to the action of Nembutal.

4 Self-sustained Activities (Seizures) Elicited by Penicillin

4.1 Penicillin "Spikes"

Since the mechanisms by which self-sustained activities arise may be studied best by observing an epileptic focus and its transition into a seizure, the development of a cortical focus after the application of penicillin onto the cortex was studied first (Pockberger et al. 1984a). Penicillin was chosen as an epileptogenic agent because the local brain edema accompanying its application is negligible. Furthermore, by restricting the epileptogenic action to a small area on the cortex, the role of midbrain structures in maintaining the oscillations that develop may be more clearly defined than in the case of intravenous application of an agent.

One of the first events to be seen is a sink in layer II/III, accompanied by sources above and below (Fig. 6). This sink increases in intensity and remains the event most characteristic of penicillin spikes throughout their existence. With increasing time and intensity, this layer II/III sink broadens and may oscillate. It is followed by another, lower-intensity sink of increasing duration, which is accompanied by sources above and below.

Very soon, however, another event becomes visible because of the penetration of penicillin to layer V (Fig. 6, right side): a short sink in layer Va, which always seems to trigger a large one in layer II/III. For the remaineder of the experiment, this sequence of a small triggering sink in layer Va and a larger one in II/III, followed by a longer, tail-like sink, remains the characteristic feature of the CSD pattern of penicillin spikes in both the motor and the striate cortex.

The nature of these sinks was substantiated by intracortical pressure applications of small (less than 1 IU) amounts of penicillin into different layers of the cortex (Pockberger et al. 1984b). It was shown that the structural basis of the main sink in the upper cortical layers must be neuronal elements in layers II/III. Layer IV cells are not essential for the appearance of this sink, because the same finding was made in the precentral cortex, where layer IV hardly exists. As for the triggering sink, layer Va was substantiated as the generator layer because tiny amounts of penicillin injected into layer Va also make the sink in layer II/III appear, but injections into layer Vb only induce sinks restricted to that level.

This is not the appropriate place for a detailed discussion of the mechanisms that lead to an epileptic seizure. Therefore only a few general remarks will be made. According to Prince (1985), it is most likely that a number of different mechanisms are involved in the synchronization of cellular burst discharges to form paroxysmal de-

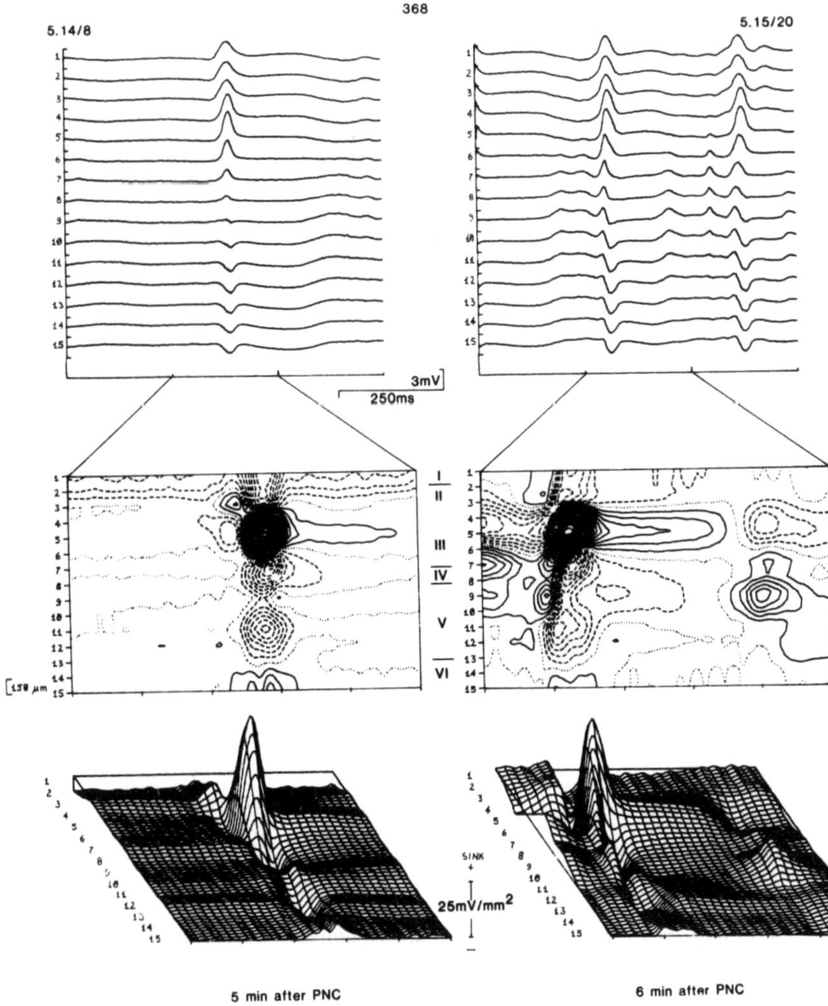

Fig. 6. Development of interictal spikes after the epicortical application of 1000 IU penicillin to the striate cortex. The main sink is fully developed in layer II/III after 5 min. One minute later, the CSD complex, characteristic for penicillin spikes, is fully developed: a triggering sink in layer Va, followed by the main sink in II/III with a long-lasting afterswing

polarization shifts (PDS) (Matsumoto and Ajmone-Marsan 1964) and thus to become visible as interictal spikes. The mechanisms include recurrent excitation by EPSPs, K^+ increase in the extracellular space, decrease of Ca^{2+}, release of ACh by the intense neuronal activity, the backfiring phenomenon, and also direct interaction with field effects, either by gap junctions or even without then. It is beyond the scope of this paper to discuss which of these mechanisms may contribute more and which less to the generation of seizures. That recurrent excitation by EPSPs seems to be of great importance in penicillin-induced seizures seems evident from the well-established fact that one of the most important actions of penicillin on the nervous system is an inhibition of gamma-aminobutyric acid (GABA)-releasing interneurons; thus, a disinhibition is the consequence (Krnjevic 1981).

There is increasing evidence that spike production in dendrites may also be important under pathological conditions (Wong and Prince 1979).

This point is emphasized here for the following reasons. In 1970, we found that seizure potentials recorded simultaneously from horizontally arranged double-microelectrodes were less well synchronized in the middle cortical layers than were the potentials recorded with the same device from either surface or deep cortical layers. This finding was later substantiated by coherence measurements (Petsche et al. 1975). Based on these observations, Fleischhauer et al. (1972) found, at the same time as did Peters and Walsh (1972), that the apical dendrites of pyramidal cells are arranged in bundles which show regional differences. The hypothesis that this organization may be involved in electrophysiological peculiarities was further supported by a study by Schmolke and Fleischhauer (1984), who compared the architectonics of bundles with cyto- and myeloarchitectonics. They found that these bundles extend to the same part of the cortex in which the main sinks are found during seizures, namely from Va to the upper part of layer II. Furthermore, the greatest number of dendrites per bundle is found in the upper two-thirds of this layer. The likelihood of the number of synaptic contacts being largest here was emphasized by Feldman (1975). It is in this layer that the most prominent sinks are found during penicillin spikes. There is, in addition, the probability that direct electrotonic interaction plays some role in the synchronization of discharges into interictal spikes, since Latz (1975) found that more than 37% of the close appositions of the dendrites showed no glial profiles inbetween them. Nonsynaptic mechanisms as an auxiliary means in the spreading of synchronization are made even more likely by the observation of dye-coupling between cortical neurons (Gutnick and Prince 1980).

4.2 Self-sustained Activities (Seizures)

Figure 7 demonstrates the beginning of a self-sustained seizure pattern. It always starts with a group of spikes which superimpose and cannot be clearly separated one from another; then comes a fairly regular ("tonic") activity, followed by more irregular groups of discharges ("clonic pattern") and variable patterns of oscillations, among which the most common are doublets, i.e., brief discharges consisting of two different transients. The voltage is usually greatest in the middle cortical layers. The lower part of Fig. 7 shows the current-source-sink pattern of the seizure shown above it, recorded three times as fast. At the very beginning, the source–sink pattern is very irregular and consists of a superimposition of sinks and sources of different intensities, durations, and locations in the cortex. During the tonic pattern, only a few source–sink configurations oscillate fairly regularly in different parts of the cortical levels, whereas, during the clonic pattern, sinks and sources change much more in shape and intensity and usually repeat at shorter intervals than during the tonic pattern.

A partial answer to the question of how this activity is initiated and maintained is given by surface recordings.

In Fig. 8, the last spike before and the spike initiating a seizure are shown on a larger time-scale in surface recordings. The two spikes differ from each other. The

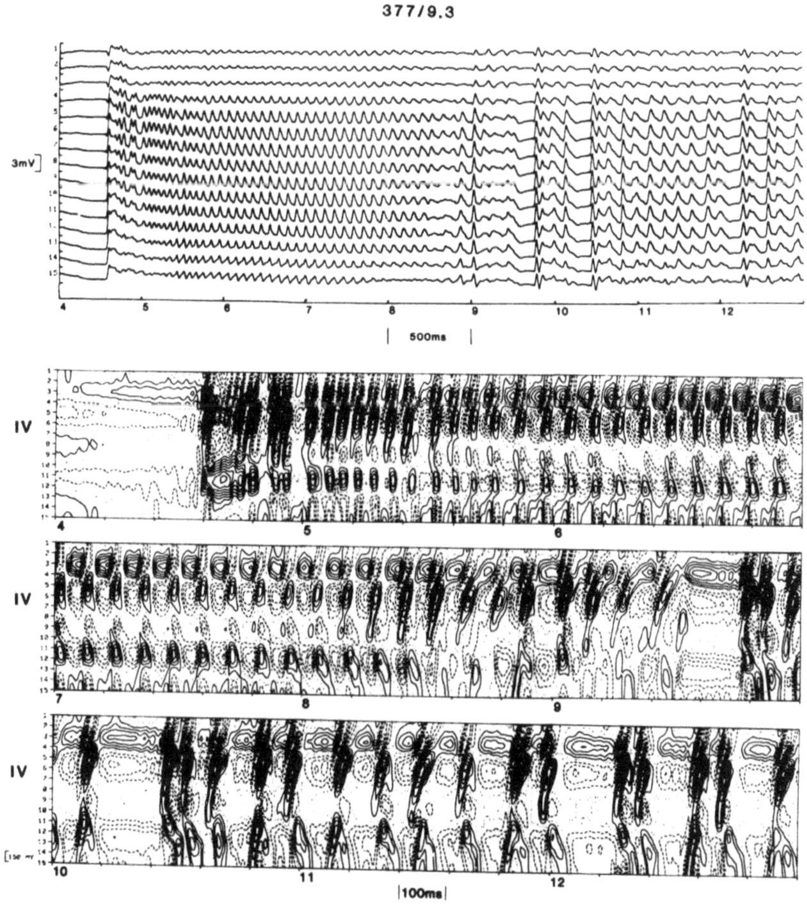

Fig. 7. The beginning of a self-sustained seizure after the application of 1000 IU penicillin to the striate cortex. *Top,* 6 s of conventional EEG recorded from the 16-fold depth electrode. *Bottom,* the same episode in CSD representation. Note the interaction of sinks and sources in different cortical layers which is more irregular in the clonic than in the tonic stage

first spike is lower than the second, and its positive prepotential is lower and also seems less well synchronized than the one initiating the seizure. Also, there are greater delays on the surface in the first spike than in the initiating spike.

More insight into the mechanisms is gained by plotting the equipotential fields of these two spikes (Fig. 9). In the last spike before the seizure, the rise time of the positive phase is 16 ms from zero level. In the spike initiating the seizure, the rise time is only 8 ms, and the negative equipotential field takes 12 ms longer to return to positivity in the spike initiating the seizure. Another difference is a shift in the location of the negative peak from contact 10 to contact 11 in spike A, whereas in spike B, the peak remains at contact 10, another suggestion for a better synchronization of underlying PDS.

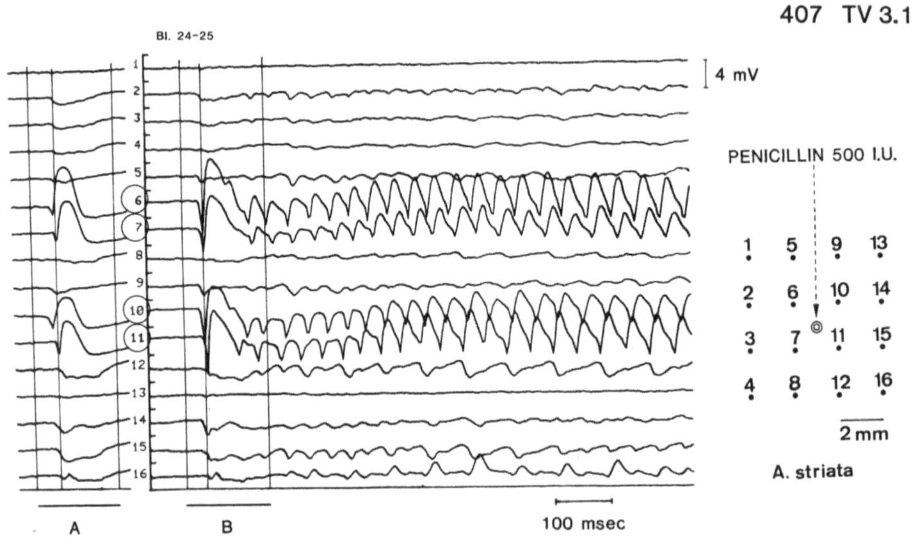

Fig. 8. Last interictal spike prior to a seizure (A) and first spike initiating the seizure (B), recorded from the surface of area striata. Note the difference in shape of these two events and also the steep potential gradients between the inner and the outer electrodes of the recording square. The *horizontal bars* relate to Fig. 9

The most remarkable difference, however, is found after these events. In spike A, a long-lasting positivity appears and extends even beyond the recording grid, whereas, during the beginning of a seizure, in spike B, very localized, steep potential fields appear and start revolving, in this case counter-clockwise; a positive potential field arises at electrode 6, reaches maximum voltage at electrode 7, and then shifts to electrode 11 and finally to electrode 10. This cycling goes on and on.

At later stages of the seizure, the frequency of oscillation, which at first is 26 Hz, subsides to about 9 Hz. Concurrently, the behavior of potential fields becomes more irregular. Moreover, the very small area initially involved – potential gradients of several mV/mm are found – expands, and doublets and clonic discharges appear. One characteristic of doublets is that the equipotential fields of the two single components show a different spatiotemporal behavior (Petsche et al. 1979). Clonic discharges, on the other hand, do not show any characteristic behavior but randomly change from one discharge to the next.

As for the mechanism of the development of the seizure, only a few general comments can be given here. The large voltage gradients between the site of application of penicillin and its surrounding suggest that some kind of inhibition is created by the seizure. Thise is supported by the findings of Prince and Wilder (1967) that in the neighborhood of penicillin foci, many more cells show IPSPs than under normal conditions. Possibly the slow positive-going shift of the field potentials surrounding the focus is caused by the surround inhibition, which may keep the pathological discharges localized for some time. This inhibition slowly declines as the seizure goes on, because the steep voltage gradient around the focus continuously diminishes with

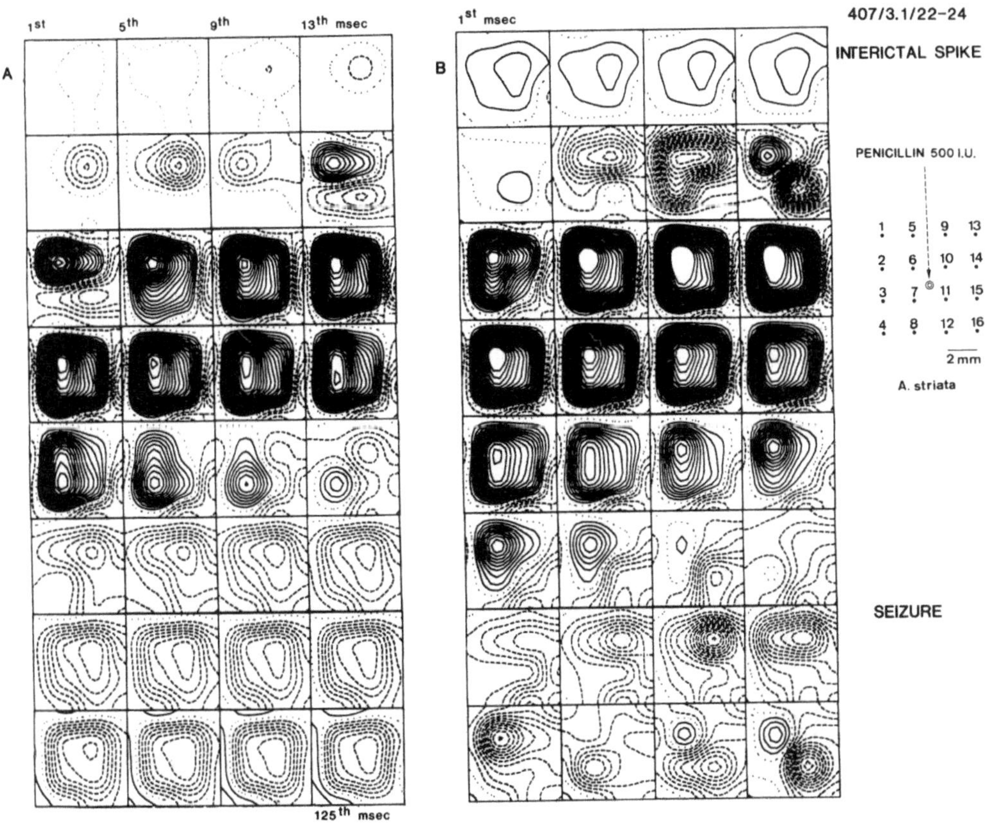

Fig. 9. The two interictal spikes (A) and (B) of Fig. 8 as uninterrupted sequences of equipotential fields at 4 ms distances, to be read from *top left* to *bottom right*. These two episodes correspond to the horizontal bars in Fig. 8. The tonic seizure succeeding spike B consists of almost circular and closely limited potential fields that start revolving anticlockwise at 26 Hz

the development of the seizure and with its decreasing frequency. A hypothesis attempting an explanation of these phenomena was proposed by Petsche (1983).

A strictly localized experimental cortical focus, which generates a seizure that slowly generalizes, is also an appropriate model for studying the eventual involvement of deeper brain structures, among them the thalamus, in the development and maintenance of self-sustained pathological activities. In the early days of electroencephalography, the thalamus was thought to play an important role in the genesis of epilepsy, an assumption that, however, was not substantiated when deep structures of the human brain were then studied. Our method seemed particularly suited to studying the question of thalamocortical interactions in seizures as, by using CSD analysis, errors due to volume conduction could be avoided. Since the eventual relationships between cortex and deep structures were expected to be nonlinear, Mars' method for the estimation of mutual information (Mars and Arragon 1982) was used (Rappelsberger et al. 1987).

The purpose of this paper has been to show that the study of field potentials, a domain badly neglected by neurophysiologists so far, can give results that lead to a better understanding of the mechanisms of spontaneous and self-sustained oscillatory electrical phenomena in the brain. An investigation of field potentials seems to be particularly productive when performing multiple simultaneous recordings in a small area of the brain while taking into account histological and intracellular findings.

Acknowledgements. This research was supported in part by the *Fonds zur Förderung der wissenschaftlichen Forschung in Österreich,* Project S 25/01. The authors wish to thank Mrs. S. Etlinger for linguistic advice and Mrs. E. Genner, Mrs. E. Trojan, and Mrs. G. Luger for experimental assistance and typing the manuscript.

References

Duffy F (1985) Topographic mapping of the brain. Butterworth, Stoneham

Feldman ML (1975) Serial thin sections of pyramidal apical dendrites in the cerebral cortex: spine topography and related observations. Anat Rec 181:354–355

Fleischhauer K, Petsche H, Wittkowski W (1972) Vertical bundles of dendrites in the neocortex. Z Anat Entwicklungsgesch 136:213–223

Freeman WJ (1975) Mass action in the nervous system. Academic, New York

Giannitrapani D (1985) The electrophysiology of intellectual function. Karger, Basel

Gutnick MJ, Prince DA (1981) Dye-coupling and possible electrotonic coupling in the guinea pig neocortical slice. Science 211:67–70

Krnjevic K (1981) Desensitization of GABA receptors. Adv Biochem Psychopharmacol 26:111–120

Latz H (1975) Quantitative Messungen an Dendritenbündeln in der Hirnrinde des Kaninchens. Thesis, University of Bonn

Mars NJT, Arragon GW (1982) Time delay estimation in nonlinear systems using average amount of mutual information analysis. Sign Proc 4:139–153

Matsumoto H, Ajmone-Marsan C (1964) Cortical cellular phenomena in experimental epilepsy interictal manifestations. Exp Neurol 9:286–304

Mitzdorf U, Singer W (1979) Excitatory synaptic ensemble properties in the visual cortex of the macaque monkey: a current source density analysis of electrically evoked potentials. J Comp Neurol 187:71–84

Müller-Paschinger IB, Prohaska O, Vollmer R, Petsche H (1979) The histological marking with multiple thin-film electrode probe for intracortical recordings. Electroencephalogr Clin Neurophysiol 47:627–628

Peters A, Walsh TM (1972) A study of the organization of apical dendrites in the somatic sensory cortex of the rat. J Comp Neurol 144:253–268

Petsche H (1983) EEG synchronization in seizures. Electroencephalogr Clin Neurophysiol [Suppl] 134:299–308

Petsche H, Prohaska O, Rappelsberger P, Vollmer R (1975) The possible role of dendrites in EEG synchronization. Adv Neurol 12:53–70

Petsche H, Rappelsberger P, Vollmer R, Lapins R (1979) Rhythmicity in seizure patterns – intracortical aspects. In: Speckmann EJ, Caspers H (eds) Origin of cerebral field potentials. Thieme, Stuttgart, pp 60–79

Petsche H, Pockberger H, Rappelsberger P (1982) The micro-EEG: methods and application to the analysis of the antiepileptic action of benzodiazepines. In: Herrmann WM (ed) EEG in drug research. Fischer, Stuttgart, pp 159–182

Petsche H, Pockberger H, Rappelsberger P (1984) On the search for the sources of the electroencephalogram. Neuroscience 11:1–27

Pockberger H, Rappelsberger P, Petsche H (1984a) Penicillin-induced epileptic phenomena in the rabbit's neocortex. I. The development of interictal spikes after epicortical application of penicillin. Brain Res 309:247–260

Pockberger H, Rappelsberger P, Petsche H (1984b) Penicillin-induced epileptic phenomena in the rabbit's neocortex. II. Laminar specific generation of interictal spikes after the application of penicillin to different cortical depths. Brain Res 309:261–269

Prince DA (1985) Physiological mechanisms of epilepsy. Epilepsia 26:S3–S14

Prince DA, Wilder BJ (1967) Control mechanisms in cortical epileptogenic foci. Arch Neurol Psychiatry 16:194–202

Prohaska O, Pacha F, Pfundner P, Petsche H (1979) A 16-fold semimicroelectrode for intracortical recordings of field potentials. Electroencephalogr Clin Neurophysiol 47:629–631

Rappelsberger P, Pockberger H, Petsche H (1981) Current source density: methods and application to simultaneously recorded field potentials of the rabbit's visual cortex. Pflugers Arch 389:159–170

Rappelsberger P, Pockberger H, Petsche H (1982) The contribution of the cortical layers to the generation of the EEG: field potential and current source density analyses in the rabbit's visual cortex. Electroencephalogr Clin Neurophysiol 53:254–269

Rappelsberger P, Pockberger H, Petsche H (1987) Evaluation of relationships between seizure potentials. In: Wieser HG, Elger C, Speckmann EJ, Engel P (eds) Methods of presurgical evaluation of epileptic patients. Springer, Berlin Heidelberg New York Tokyo, (in press)

Rockstroh B, Elbert T, Birbaumer N, Lutzenberger W (1982) Slow brain potentials and behaviour. Urban and Schwarzenberg, Baltimore

Schmolke C, Fleischhauer K (1984) Morphological characteristics of neocortical laminae when studied in tangential semithin sections through the visual cortex of the rabbit. Anat Embryol (Berl) 169:125–132

Somjen GG, Trachtenberg M (1979) Neuroglia as generator of extracellular current. In: Speckmann EJ, Caspers H (eds) Origin of cerebral field potentials. Thieme, Stuttgart, pp 21–32

Wong RKS, Prince DA (1979) Dendritic mechanisms underlying penicillin-induced epileptiform activity. Science 204:1228–1230

Evoked Potentials and Their Physiological Causes: An Access to Delocalized Cortical Activity

U. MITZDORF

1 Introduction

Event-related potentials and the phenomenologically closely related EEG provide noninvasive "on-line" access to normal and pathological CNS activity. Empirically, these signals have been shown to reflect the general state of the CNS and the central processing of afferent information, including higher cognitive processes (e.g., Freeman; John; Picton; in this volume).

However, these data are difficult to interpret neurophysiologically, because they are indirect and ambiguous reflections of the underlying neuronal activity. This is the main reason why basic neurophysiological research has concentrated on the investigation of single-unit properties (e.g., Hubel and Wiesel 1962). However, the single-unit approach not only contains the drawbacks of arbitrariness in selecting different cell types and unmanageability of the sample sizes needed for answering more complex questions, it also incorporates a conceptual shortcoming: the classical receptive fields of single cells reflect only spatially focused, localized activities. More subtle delocalized interaction phenomena are not accessible to single-unit studies. Moreover, these delocalized ensemble activities are likely to be more relevant expressions of central information processing (Fessard 1961; Katchalsky et al. 1974; Pribram 1971; Sperry 1969).

This chapter will describe an attempt to bridge the gap between the single-unit level and the mass-action level of event-related potentials and the EEG. Evoked potentials from the visual cortex of the cat have been analyzed and correlated with anatomical and physiological data, in order to identify their neuronal causes. The experimental results will demonstrate that evoked potentials contain aspects of the central processing of afferent information that are complementary to those accessible to conventional single-unit studies.

2 The Neuronal Causes of Event-Related Potentials

The relation between field potentials (profiles of event-related potentials or EEG waves) and cellular activity in the CNS is indicated in Fig. 1a: currents flow through the cell membranes during neuronal activities. If many similar elements of an anatomically ordered ensemble are activated simultaneously (ensemble activity), then these concomitant membrane currents can sum in the extracellular space to

Springer Series in Brain Dynamics 1
Edited by Erol Başar

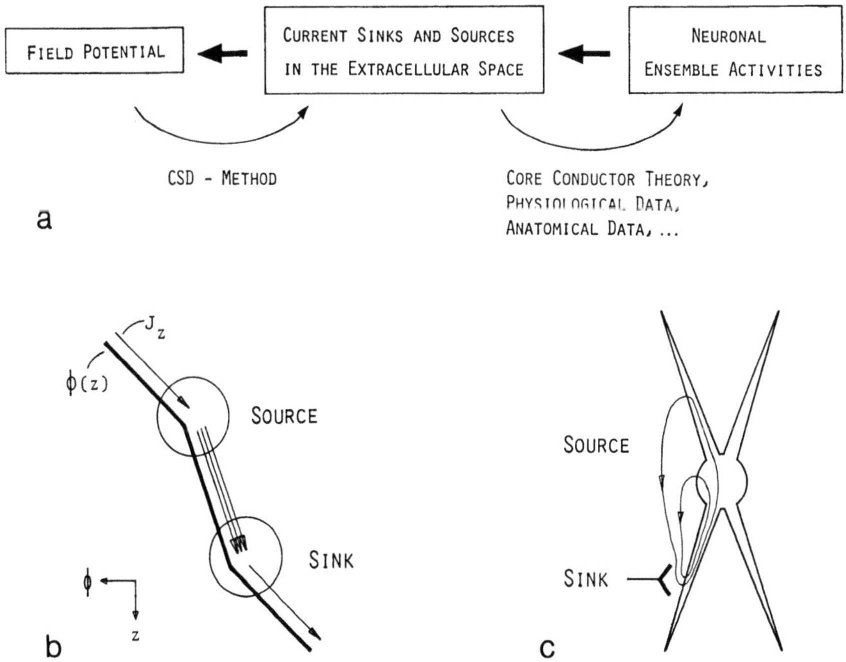

Fig. 1. (a) The *thick, straight arrows* represent the causal relations between field potentials, CSDs, and neuronal ensemble activities; the *curved arrows* represent the techniques required to get from the experimentally accessible field potential data down to the causative neuronal ensemble activities. (b) Illustration of the relation between the field potential $\phi(z)$ and the sinks and sources in the extracellular space. In a conducting medium, currents flow along potential gradients. The amplitude of the current (J_z) is proportional to the steepness of the potential slope. Changes in the potential slope ($\partial^2\phi/\partial z^2$) therefore imply changes in the amount of current flowing *(circled areas)*. A current increase indicates a source, a current decrease indicates a sink. As the extracellular space is a purely passive medium, sources and sinks must be caused by membrane currents that enter and leave, respectively, the extracellular space. (c) The physiological correlates of the sinks and sources in the extracellular space are membrane currents that flow during neuronal ensemble activities. Here, those membrane currents are indicated that flow during the excitatory synaptic activation of a cortical cell (synapse indicated by *chevron*). Active ionic current flows into the cell at the site of the synapse, and passive current of equal amplitude leaves the cell at proximal and more distant membrane sites. If many such excitatory synapses on many similar cells are activated simultaneously, then the membrane currents sum up in the extracellular space and cause a macroscopic sink in the depth region of these synapses, and a source above

macroscopic sinks (inward flowing membrane currents) and sources (outward flowing membrane currents). These sinks and sources are the physical causes of the field potentials. On the other hand, these sinks and sources provide the physiologically relevant information contained in the field potentials. Thus, they are the connecting link between field potentials and their physiological causes, the neuronal ensemble activities.

The sink and source distributions in the extracellular space can be calculated from the field potentials with the current-source-densitiy (CSD) method. These data are still ambiguous with respect to the neuronal activities that cause them. Therefore, additional physiological and anatomical cues are needed in order to identify these neuronal ensemble activities (for a detailed discussion of this subject, see Mitzdorf 1985).

Unfortunately, the locations of sinks and sources cannot be concluded from surface evoked potentials, but only from the spatial profiles of the potentials within the CNS. Therefore, straightforward analyses of evoked potentials via the current sink and source distributions as a link to the causal neuronal ensemble activities (Fig. 1a from left to right) are restricted to animal studies.

3 The CSD Method

The CSD method, invented by Pitts (1952), is based on the validity of Ohm's law in the extracellular space, and on the assumption that this extracellular space can be considered independently of the intracellular space. This latter assumption is plausible because the boundaries of the extracellular space, the cell membranes, have high resistances compared with the resistance of the extracellular space. Finally, the sinks and sources are the result of the volume averages of the membrane currents which flow into and out of the extracellular space. (The individual microscopic membrane currents that actually flow may be much larger than the fraction manifest in the macroscopic sinks and sources after the averaging procedure; but this averaging is inherent in the field potentials as well.)

These considerations are incorporated in Eq. 1.

$$\sum_{i=1}^{3} \left(\frac{\partial \sigma_{ii}}{\partial x_i} \cdot \frac{\partial \phi}{\partial x_i} + \sigma_{ii} \cdot \frac{\partial^2 \phi}{\partial x_i^2} \right) = -\mathrm{CSD} \tag{1}$$

This relates the CSD with the field potential ϕ and the conductivity tensor σ, in optimally oriented cartesian coordiantes x_i. If the outward membrane currents dominate in a volume element, the result is a current source (CSD > 0); if the inward currents dominate, the result is a current sink (CSD < 0). For detailed derivations of the CSD method, see Nicholson (1973), Nicholson and Freeman (1975), Mitzdorf (1985).

In the present study in the cat visual cortex, the simplified one-dimensional relation was applied (Eq. 2).

$$\sigma_z \cdot \frac{\partial^2 \phi}{\partial z^2} = -\mathrm{CSD} \tag{2}$$

It assumes homogeneity in the z-direction and translational invariance of ϕ in the two directions parallel to the laminar planes of the cortex. Experimentally, the profiles of ϕ were obtained by measurements at discrete equidistant depth locations, and the second spatial derivative was calculated according to a finite difference formula. This simplest form of the CSD method can easily be conceived intuitively; see Fig. 1b.

The one-dimensional CSD method makes possible the localization of the sinks and sources in the z-direction, but not in the x- and y-directions. The delocalization of the sinks and sources in these tangential directions, because of the inadequancy of the one-dimensional CSD method, can be estimated from the widths of retinotopically localized activation components (see Retinotopic and Nonretinotopic Activations and Fig. 4a). Taking into account the velocities of the focal stimuli and the magnification factors of the cat visual cortex (Bilge et al. 1967; Tusa et al. 1978, 1979), these widths correspond to about 1 mm in cortex (Mitzdorf 1987).

4 CSD Analysis of Evoked Potentials in the Cat Visual Cortex

Electrically and visually evoked field potentials from the two primary visual areas 17 and 18 of the cat cortex have been analyzed. In acute experiments (pentobarbital and/or N_2O anesthesia), the intracortical profiles of the potentials were recorded, either with one micropipette successively placed at different depths, or simultaneously in all cortical depths with a multiple electrode (an array of aligned pipettes). Usually, 20 single responses were averaged for the electrically evoked potentials and 100 or 200 single responses were averaged for the visually evoked potentials.

An example of a potential profile, evoked by electrical stimulation of the specific afferents, and the corresponding CSD profile are given in Fig. 2a and b. The cortical recording depths are indicated to the left of the potential profile, and the borders of the cortical laminae, as judged from anatomical and functional criteria (see Mitzdorf and Singer 1978), are indicated to the left of the CSD profile.

Even a cursory comparison of the two profiles demonstrates that the CSD distribution, in contrast to the original potential data, is well structured and reveals discrete events. The sinks of the various components have been marked by the letters a–f. A dipolar sink/source distribution in layers IV/III (a) is followed by two further dipoles in layers III/II (b and c). Symmetrical source/sink/source distributions are apparent with the sinks in lower layer IV (d) and in layer V (f). Component e in layer VI is dipolar at the start, but is later symmetrical. In several other profiles, two additional dipolar components, similar to the components a and b, are recognizable (Mitzdorf and Singer 1978). The indications of these sinks in Fig. 2b have been marked a′ and b′. These sinks and sources are the local generators of the evoked potential. The dipoles cause far-reaching potential components, but the symmetrical source/sink/source distributions cause only local contributions to the potential; compare the "far-fields" and "closed-fields" of Lorente de Nó (1947).

5 Interpretation of the Electrically Evoked CSDs

In order to identify the neuronal activity causing these various CSD components, the responses to stimuli applied at different sites along the specific afferents were compared. In addition, the effects of conditioning stimuli and of pharmacologically induced alterations of neuronal excitability on the CSDs were investigated.

Upon stimulation of the specific afferents at more distant sites, the CSDs in area 17 were delayed and more dissipated. According to their different latency increases upon stimulation of more distant sites, the early components could be attributed either to Y-type activity mediated by the fast-conducting afferents (component a and the early part of component e), or to X-type activity mediated by the more slowly

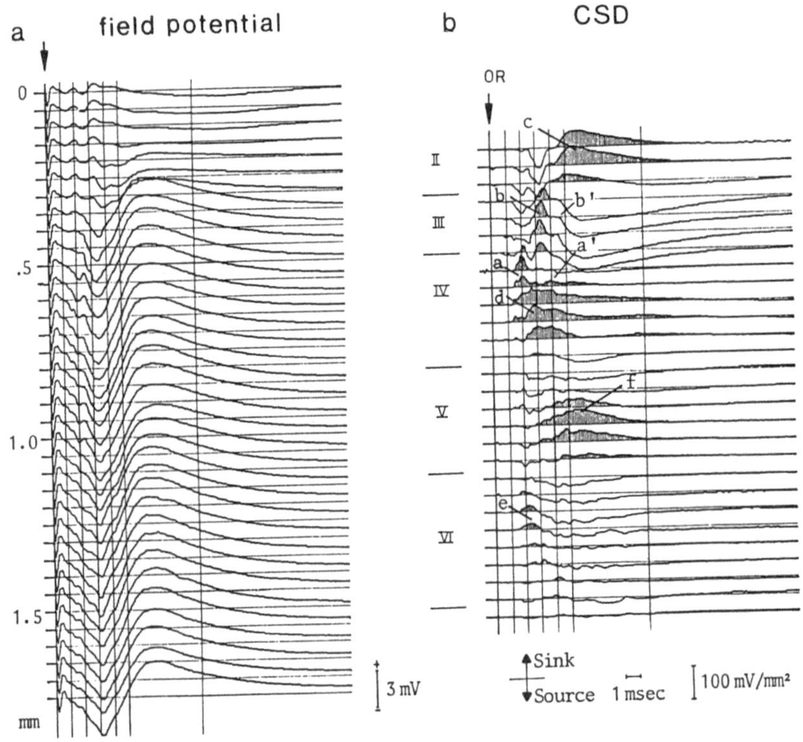

Fig. 2. (a) Field potential in the primary visual cortex (area 17) of the cat, evoked by electrical stimulation of the optic radiation *(arrow)*. The distance between adjacent recordings is 50 µm. The profile was obtained by successive recordings with one micropipette. (b) CSD distribution obtained from the potential profile in (a), according to Eq. 2, with a grid of 200 µm for the numerical differentiation (the conductivity is ignored in the calibration). The sinks, corresponding to active excitatory synaptic currents, are accentuated by *hatching*. At the *left margin*, the depth regions of the cortical laminae are indicated. Sinks *a, b, c* reflect mono-, di-, and trisynaptic Y-type activity; sinks *d* and *f* reflect mono-, di-, and trisynaptic X-type activity; sink *e* reflects Y-type and X-type monosynaptic activity. The deflections marked *a'* and *b'* indicate two further sinks which reflect di- and trisynaptic Y-type activity. (c) Schematic diagram of the successive intracortical excitatory relay stations, as well as the cell types involved, as suggested by the CSD in (b) and similar findings. Three main pathways, along which the afferent activity is relayed within the visual cortex, are shown (the *numbers* indicate whether the activations are mono-, di-, or trisynaptic, respectively). The first pathway transmits afferent Y-type activity from upper layer IV to layer III, and then to layer II. The second pathway relays afferent X-type activity from lower layer IV to layer V, where mainly lamina VI pyramidal cells are contacted (di- and trisynaptically). Along the third pathway, Y-type afferent activity is relayed within layer IV, and then also projected to layer III

conducting X-type afferents (component d). Furthermore, mono- and polysynaptic components could be distinguished by comparing the responses to double shock. When the second afferent volley arrives in the cortex, the cells are under strong inhibitory influence caused by the first activation. This inhibition does not affect the monosynaptic activation, but it prevents the generation of action potentials in many cells. Accordingly, the sinks and sources that were still present in the CSDs evoked

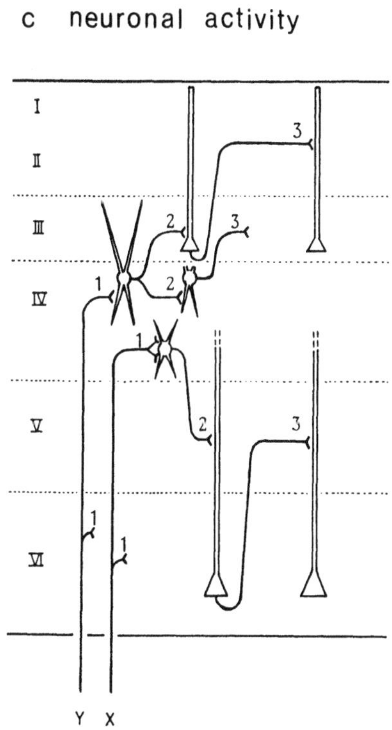

Fig. 2c

by the second of the double shocks were attributed to monosynaptic activity; those components that were drastically reduced or totally missing in the second CSDs were attributed to polysynaptic activity.

Picrotoxin and pentobarbital were applied systemically in order to investigate whether inhibitory synaptic activity contributes to the intracortical CSDs. These two drugs are known, respectively, to decrease and increase the strength of inhibitory synaptic activation. After the application of picrotoxin, all the sinks and sources were increased in amplitude. This effect was rather minor on the monosynaptic components, but was large on the polysynaptic components. Pentobarbital had the opposite effect: monosynaptic components were only slightly reduced, but polysynaptic components were strongly reduced or totally blocked. None of the sinks or sources was increased by pentobarbital. These results indicate that there are no discrete CSD components caused by inhibitory synaptic activation. The small drug effects on the monosynaptic components further indicate that these components are mainly caused by synaptic activity, and only to a small extent by action potentials. Large contributions of action potentials to the later components c and f can be ruled out, because single-unit discharges are rare at such latencies (Neumann 1979).

According to the arguments above, based on experimental data, the CSDs are essentially caused by excitatory synaptic ensemble activity. This conclusion is in good agreement with theoretical estimates of the relative contributions from excitatory and inhibitory synaptic activities and action potentials to CSDs (Mitzdorf 1985).

The membrane currents resulting from excitatory synaptic activation flow into the cells at the sites of the synapses, and leave the cells at proximal and more distant sites (see Fig. 1c). Therefore, the depths of the sinks indicate the laminar arrangements of the involved excitatory synapses; and the extent in depth of each sink and its corresponding source (or sources) indicates the region over which the activated cells extend their processes (see Fig. 1c and Mitzdorf 1985).

The successive intracortical excitatory relay stations, as well as the involved cell types suggested by the CSDs in area 17, are indicated in Fig. 2c. The monosynaptic sinks (a, d, and e in Fig. 2b) demonstrate that the afferent fibers terminate in laminae IV and VI. Likely candidates for the Y-type monosynaptic activation in layer IV (sink a) are stellate cells in upper layer IV and pyramidal cells of lower layer III (Lund et al. 1979); they are symbolically indicated in Fig. 2c by a large stellate cell which extends over laminae III and upper IV. As a minor source was seen in layer V, some deep pyramidal cells may also be contacted by these fibers in layer IV (not indicated in Fig. 2c). The X-type afferents mainly contact cells that do not extend into layer III; the most likely candidates are stellate cells in layer IV (Lund et al. 1979); but, because of a large source in layer V corresponding to sink d (see Mitzdorf and Singer 1978), deep pyramidal cells are involved as well. The polysynaptic sinks demonstrate three main pathways, along which the afferent activity is processed within the cortex. Along the first pathway (sinks a, b, c), afferent Y-type activity is projected to the supragranular layers. Along the second main pathway (sinks d and f), afferent activity is projected down to layer V. In area 17, the X-type activity is relayed along this pathway. Since the long-lasting sink f in layer V draws much current from below, mainly lamina VI pyramidal cells are activated along this pathway. The third pathway (sinks a, a′, b′; not well-recognizable in the CSD of Fig. 2b) has one relay within layer IV, and then projects Y-type activity to lamina III.

Thus, a further basic principle of topographic organization in cortex has been revealed. The results demonstrate that the afferent and intrinsic excitatory connections are arranged in a lamina-specific manner. The laminae in which the cell receives its inputs determine the types of input (Y-type or X-type; mono-, di-, or trisynaptic), as well as the degree of convergence of different inputs.

6 Similarity of Electrically and Visually Evoked and Nonspecific Cortical Activation Patterns

CSDs evoked by various types of visual stimuli and even CSDs evoked by nonspecific activations are qualitatively very similar to the CSDs evoked by electrical stimulation of the specific afferents. Figure 3 demonstrates this finding. As described above, early sinks are apparent in the input layers IV and VI (sinks a, d, e) in all CSDs; after a slight delay there follows a sink in layer III (b) and then, after considerable delay, the more dissipated sinks in layer II (c) and layer V (f).

The main differences between the activation patterns are in their time courses and their amplitudes. If the cortical activation process is initiated by strong electrical stimulation of the specific afferents, it occurs within 10 ms (Fig. 2b); if it is caused by abrupt visual stimuli, it needs about 200 ms (Fig. 3a, d, e); if it is caused by the movement of a pattern, it may need 1000 ms, or even more if the pattern is moved more

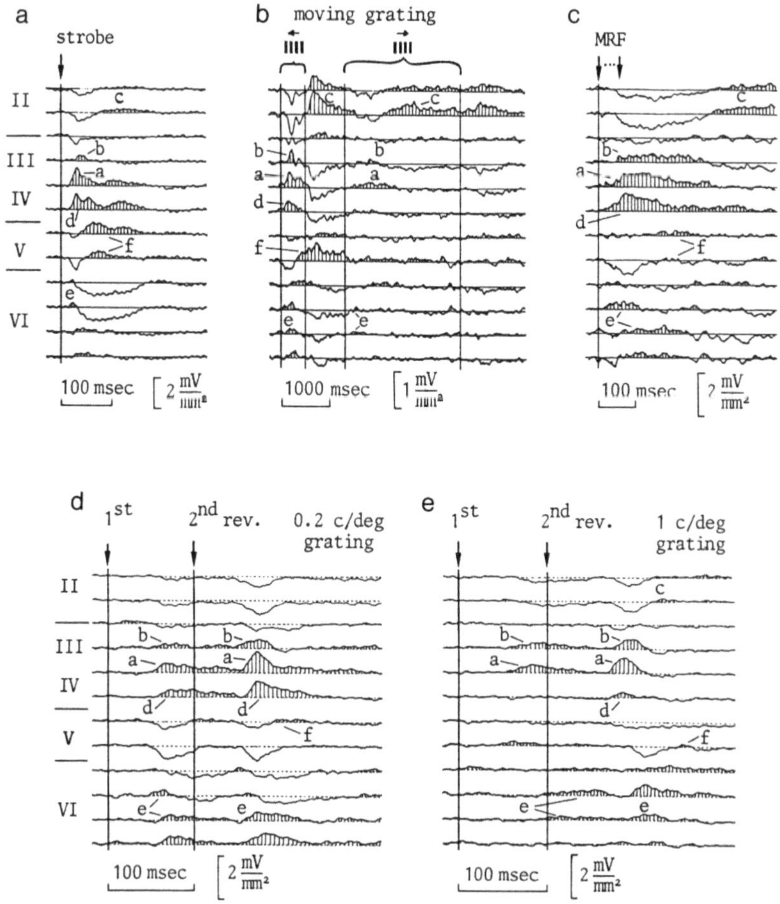

Fig. 3a–e. CSD distributions in area 17, recorded with a 16-fold multielectrode (intertip spacings 150 µm; differentiation grid 300 µm). The stimuli were: (a) strobe flash; (b) movement of an 0.5 c/deg grating, first to the left at a velocity of 10° per second, then to the right at a velocity of 2° per second (duration of movement is *bracketed*); (c) electrical activation of the mesencephalic reticular formation (a 60 ms train of five pulses); (d) two reversals of an 0.2 c/deg grating with an interstimulus interval of 100 ms; (e) two reversals of a 1 c/deg grating with an interstimulus interval of 100 ms. Conventions as in Fig. 2b. This figure demonstrates that the CSDs evoked by various different stimuli are qualitatively very similar. Note, however, the different time and amplitude scales. The CSDs d and e further demonstrate the gradual differences in responses to stimuli with few versus many contours, as well as the facilitatory effect of a conditioning stimulus

slowly (Fig. 3b). The faster this basic activation sequence proceeds, the larger the amplitudes of the sinks and sources.

The qualitative similarities of the sink and source distributions strongly suggest that all these CSDs reflect the same types of excitatory synaptic ensemble activities along the same intracortical pathways as the electrically evoked CSDs (for further arguments corroborating this conclusion, see Mitzdorf 1985 and 1987).

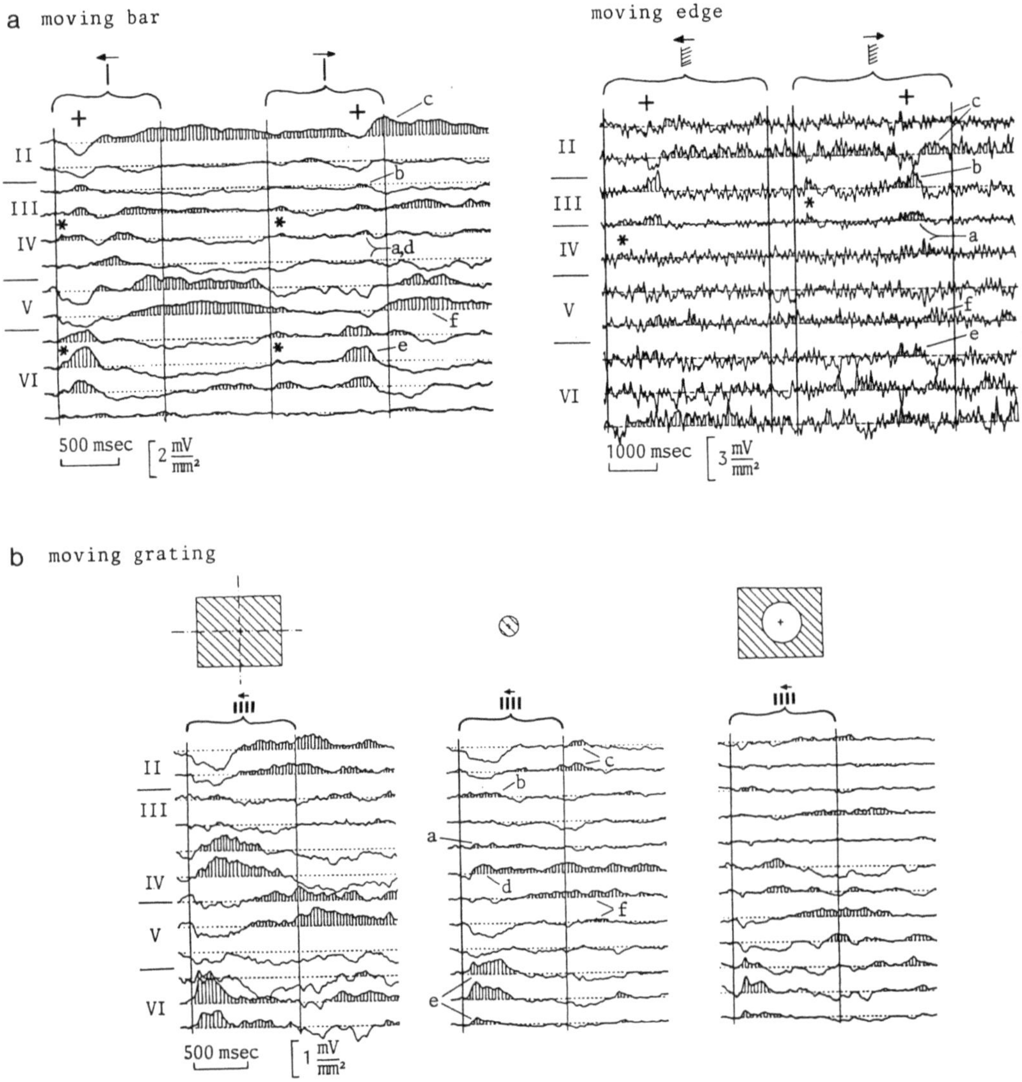

Fig. 4a, b

7 Stimulus-Specific and -Nonspecific Modulations of the Basic Pattern of Cortical Activation

Close inspection of the visually evoked CSD profiles reveals that the physical parameters of the visual stimuli are reflected in minor modulations of this basic activation process. As an example, the modulation caused by the amount of contour contained in an abrupt visual stimulus will be briefly described here (for further details, see Mitzdorf 1987). The responses to contour-rich stimuli (i.e., stimuli containing many

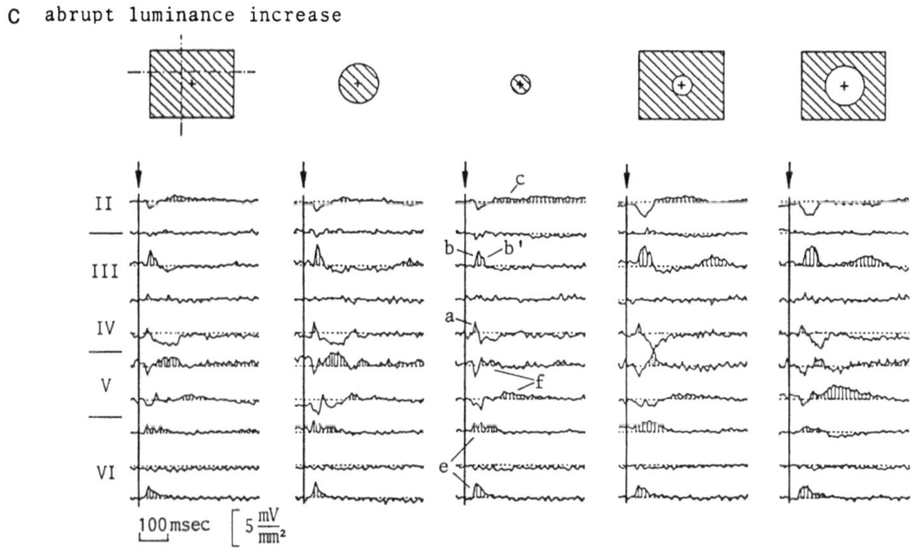

Fig. 4. (a) CSD responses to the movement of a single bar and of an edge. The times when the stimulus crosses the receptive field of the recording site are indicated by *crosses*. *Left CSD,* responses in area 17 to the back-and-forth movement of a bar at a velocity of 20° per second, recorded with a 16-fold multielectrode (intertip spacings 150 μm; differentiation grid 300 μm). The receptive field of the recording site was located within the central 3°. *Right CSD,* responses in area 18 to the back-and-forth movement of an edge at a velocity of 16° per second, recorded with a 12-fold multielectrode (intertip spacings 200 μm; differentiation grid 200 μm). The receptive field of the recording site was 12° eccentric. Note the retinotopically corresponding activations of the sinks in layers IV, III, and VI, and the responses to the onsets of movement (marked by *asterisks*). (b) CSDs evoked by an 0.2 c/deg grating, moving at a velocity of 13° per second, presented in a large area of the visual field (40° × 50°), in circular area in the central 12° only, and presented in the periphery, sparing the central area of 24° in diameter (area 17; 16-fold multielectrode; intertip spacings 150 μm; differentiation grid 300 μm; receptive field within the central 3°). Note the presence of sinks in layers III, IV, V, and VI in the right CSD, as well as the delayed onsets of the components in layers III, IV, and V. (c) CSDs evoked by Ganzfeld-"on" stimulation (area 18; 12-fold multielectrode; intertip spacings 200 μm; differentiation grid 200 μm; receptive field location 10° eccentric). The responses from *left* to *right* were evoked by presenting the stimulus in a large area of the visual field (40° × 50°), restricted to a region of 24° in diameter, centered on the receptive field, restricted to a region of 12° in diameter, centered on the receptive field, sparing the central 12°, and sparing the central 24° of the region around the receptive field. Note that the response to whole-field stimulation is not a linear summation of the responses to stimulation of subregions, and that the onset latencies of the responses are gradually more delayed the more that the region around the receptive field is spared by the stimulus

edges), as compared to stimuli that contain few or no contours, have longer latencies (compare Fig. 3d and e). The early sink in layer VI is delayed more than the early sink in layer IV, whereas it has the same or even a shorter onset latency when the stimulus contains few or no contours. The amplitudes of the sinks and sources are smaller in the responses to contour-rich stimuli. The amplitude of the late sink in layer II is usually reduced more than the amplitudes of the early sinks in layers IV/III.

This indicates that less afferent activity is projected from the input layer IV to the supragranular layers if the stimulus contains many contours.

Nonspecific factors influence the cortical CSD responses more than do the specific physical parameters of the stimuli. One of these factors is the general state of the CNS. For example, if the animal is very deeply anesthetized, then the amplitudes of the sinks and sources are rather small. The late phase of the activation process is reduced more than the early phase, indicating that less afferent activity is relayed intracortically from the input layers to the supra- and infragranular layers.

The response to a stimulus also depends strongly on the instantaneous state of activation in the cortex at the time when the afferent, evoked activity arrives. If the afferent activity coincides with the first phase of the basic pattern, i.e., with sinks in the input layers IV and VI (either evoked by a preceding stimulus, or spontaneously occurring), then the response is large (see Fig 3d and e: the response to the second stimulus is larger than the response to the first, identical stimulus). If the afferent activity arrives in the cortex during the late phase of the basic pattern, i.e., during the occurrence of sinks in the layers II and V, then the response is small. (Strong systematic interactions between ongoing activity and evoked responses have also been demonstrated by Başar 1980 and in this volume.)

8 Retinotopic and Nonretinotopic Activations

Examples of retinotopic and nonretinotopic activations of cortex are presented in Fig. 4. Figure 4a shows two CSDs which were evoked by to-and-fro movement of a single bar and of an edge over a large area of the visual field. These CSDs reveal that the cortical activation pattern, described above, is evoked when the bar or edge moves over the receptive field area of the cortical recording site. But the onset of movement, when the bar or edge is far away from the receptive field, also initiates this cortical activation pattern.

Such nonretinotopic activations are even more obvious in the two sets of responses shown in Fig. 4b and c. In these cases, the visual stimuli were presented either in a large area of the visual field or in small subregions which either corresponded retinotopically to the cortical recording site or which spared this retinotopically corresponding region of the visual field (see insets). Comparison of these responses shows that the cortical activation does not depend strongly on the site of stimulation. (The same conclusion has been drawn previously from evoked potential data; see Doty 1958; Vaughan and Gross 1969; Ebersole and Kaplan 1981.) Furthermore, the response to whole-field stimulation is obviously not the linear summation of the responses to stimulation of subregions (see Fig. 4c).

The CSD responses to retinotopically noncorresponding stimuli are slightly delayed in comparison with the CSDs evoked by retinotopically corresponding stimuli. However, the interrelations between most of the sinks in the different layers are very similar. These facts suggest that most of the nonretinotopically evoked activity is caused by retinotopic afferent activation, which is spread out laterally in cortex and then, at the terminal sites, activates the same interlaminar pathways as does the retinotopically evoked activity. The fibers of the two bands of Baillarger are likely anatomical substrates for the intracortical lateral spread of the nonretinotopically evoked afferent activity. (For differences in the degree of retinotopic precision of different types of afferents, as judged from CSDs, see Mitzdorf 1987).

9 Conclusions

9.1 Summary of Information from the CSDs

The CSD analysis of electrically and visually evoked potentials from the cat visual cortex revealed that these potentials reflect primarily excitatory synaptic activities and allowed tracing of the central processing of the afferent information. This information processing in the visual cortex is performed in one basic routine, which involves three main pathways with three successive synaptic relay steps each. The specific visual information is reflected only in slight modulations and in the speed of

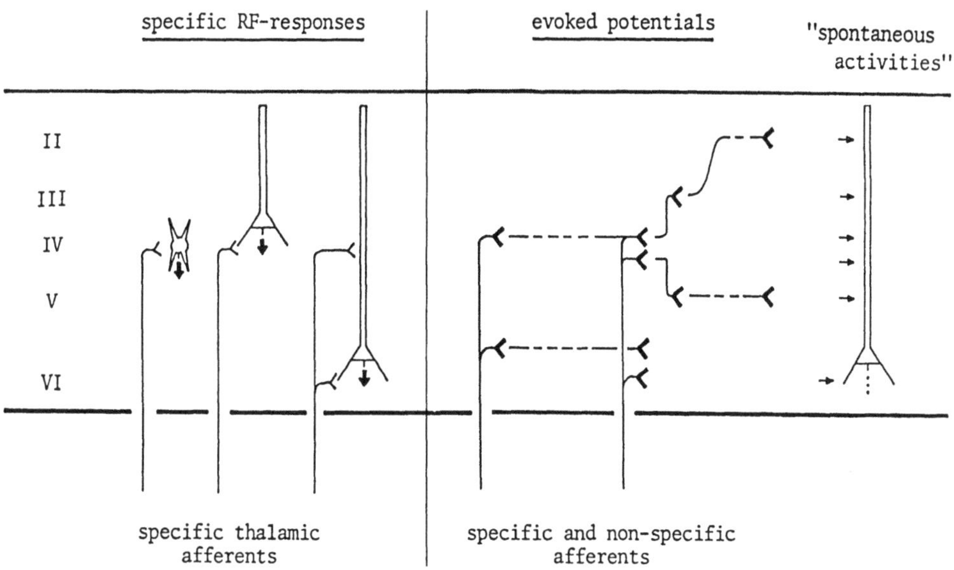

Fig. 5. Schematic diagram illustrating the complementary aspects of cortical information processing that are reflected in the specific receptive-field responses of cortical cells *(left part)* and in the evoked potentials and CSDs *(right part)*. The specific receptive-field responses of single units (indicated by the *thick arrows* in the *left part*) are essentially caused by direct monosynaptic activations by the specific thalamic afferents. Therefore, they reflect only the very first stage of the intracortical processing of afferent information. At this level, the retinotopic organization is still preserved. From there on, however, the afferent activity spreads laterally along tangential long-distance connections (*dashed lines* in layers IV and VI in the *right part*). The retinotopic organization is thereby overcome. From then on, more abstract features of the afferent information are processed. These higher stages of cortical information processing are also accessible by investigating CSDs or evoked potentials, because the CSDs are generated by mono- and polysynaptic excitatory mass actions; these are indicated by the *thick chevrons* in the *right part*. The *interconnections* indicate the three main pathways of the basic pattern of cortical activation. Since many (probably all) excitatory cells all over the cortical area participate in the delocalized higher-stage relays, the corresponding contributions from the individual cells may be sparse ("spontaneous activity", indicated by *dots* in the *rightmost part*)

Table 1. Information content of evoked potentials as compared to the specific receptive-field responses of single cortical cells

Receptive fields	Evoked potentials
Cell outputs (action potentials)	Excitatory synaptic cell inputs
Single cell properties	Mass actions
First intracortical stage of processing	Early and later stages of information processing
Still retinotopically organized	Retinotopic and nonretinotopic activities
Specific activity	Specific and nonspecific activity

activation of this sequence of components, while the intrinsic state of activation has the strongest influence on it.

A comparison of the data from the cat visual cortex with related mass-action data from other cortices (reviewed in Mitzdorf 1985) indicates that the finding from the cat visual cortex can be generalized. Essentially the same sequence of events is apparent in other neocortical areas in the cat and in other species in response to specific and nonspecific stimuli and in spontaneously occurring activity. It is therefore concluded that the activation sequence indicated in Fig. 2c and on the right of Fig. 5 represents a general, basic pattern of neocortical activation. Presumably, this pattern represents the physiological analogue to the uniform anatomical structure of neocortex.

9.2 Complementarity of Single-Unit Receptive-Field Properties and CSD Properties

The similarity of CSD responses to retinotopically corresponding and noncorresponding stimuli indicates that the retinotopic organization is already overcome at the first stage of intracortical processing. Thus, the primary evoked activity, as revealed by the CSD analysis of evoked potentials, is predominantly tangentially delocalized. According to single-unit studies, on the other hand, by far the most conspicuous and most prominent property of visually evoked activity in the primary visual areas is its retinotopic localization (Hubel and Wiesel 1962).

This discrepancy between single-unit results and the present CSD results is resolved if one assumes that the usual types of single-unit receptive-field studies reveal only the monosynaptic thalamocortical activations of the cells. Actually, two recent single-unit investigations do agree well with this somewhat heretical assumption (Malpeli 1983; Tanaka 1983). The consequential complementarity of information inherent in specific receptive-field responses and in evoked potentials is summarized in Table 1 and schematically illustrated in Fig. 5.

Acknowledgements. This work has been supported by grants from the Deutsche Forschungsgemeinschaft: Mi 141/4-1 and SFB 220/D3.

References

Başar E (1980) EEG-brain dynamics. Relation between EEG and brain evoked potentials. Elsevier/ North-Holland, Amsterdam
Bilge M, Bingle A, Seneviratne KN, Whitteridge D (1967) A map of the visual cortex in the cat. J Physiol (Lond) 191:116P–118P
Doty RW (1958) Potentials evoked in cat cerebral cortex by diffuse and punctiform photic stimuli. J Neurophysiol 21:437–464
Ebersole JS, Kaplan BJ (1981) Intracortical evoked potentials of cats elicited by punctate visual stimuli in receptive field peripheries. Brain Res 224:160–164
Fessard A (1961) The role of neuronal networks in sensory communications within the brain. In: Rosenblith WA (ed) Sensory communication. MIT Press, Cambridge, pp 585–606
Hubel DH, Wiesel TN (1962) Receptive fields, binocular interaction and functional architecture in the cat's visual cortex. J Physiol (Lond) 160:106–154
Katchalsky AK, Rowland V, Blumenthal R (1974) Dynamic patterns of brain cell assemblies. Neurosci Res Program Bull 12:1–187
Lorente de Nó R (1947) A study of nerve physiology. Rockefeller Institute of Medical Research, New York
Lund JS, Henry GH, MacQueen CL, Harvey AR (1979) Anatomical organization of the primary visual cortex (area 17) of the cat. A comparison with area 17 of the macaque monkey. J Comp Neurol 184:599–618
Malpeli JG (1983) Activity of cells in area 17 of the cat in absence of input from layer A of lateral geniculate nucleus. J Neurophysiol 49:595–610
Mitzdorf U (1985) Current source-density method and application in cat cerebral cortex: investigation of evoked potentials and EEG phenomena. Physiol Rev 65:37–100
Mitzdorf U (1987) Properties of the evoked potential generators: current source-density analysis of visually evoked potentials in the cat cortex. Int J Neurosci 33:33–59
Mitzdorf U, Singer W (1978) Prominent excitatory pathways in the cat visual cortex (A 17 and A 18): a current source density analysis of electrically evoked potentials. Exp Brain Res 33:371–394
Neumann G (1979) Intrinsic connectivity in area 18 of the cat. In: Freeman RD (ed) Developmental neurobiology of vision. Plenum, New York, pp 175–184
Nicholson C (1973) Theoretical analysis of field potentials in anisotropic ensembles of neuronal elements. IEEE Trans Biomed Eng 20:278–288
Nicholson C, Freeman JA (1975) Theory of current source-density analysis and determination of conductivity tensor for anuran cerebellum. J Neurophysiol 38:356–368
Pitts W (1952) Investigation on synaptic transmission. Transactions of the 9th Conference of the Josiah Macy Foundation, New York, pp 159–166
Pribram KH (1971) Languages of the brain. Prentice-Hall, Englewood Cliffs
Sperry RW (1969) A modified concept of consciousness. Psychol Rev 76:532–536
Tanaka K (1983) Cross-correlation analysis of geniculo-striate neuronal relationships in cats. J Neurophysiol 49:1303–1318
Tusa RJ, Palmer LA, Rosenquist AC (1978) The retinotopic organization of area 17 (striate cortex) in the cat. J Comp Neurol 177:213–236
Tusa RJ, Rosenquist AC, Palmer LA (1979) Retinotopic organization of areas 18 and 19 in the cat. J Comp Neurol 185:657–678
Vaughan HG, Gross CG (1969) Cortical responses to light in unanaesthetized monkeys and their alteration by visual system lesions. Exp Brain Res 8:19–36

II. Field Potentials and Nonlinear Dynamics: Do New Models Emerge?

Electromagnetic Field Interactions in the Brain

W. R. Adey

1 Introduction

For more than 100 years, structural and functional substrates of the organization of brain tissue have been based on considerations of connectivity as described in Waldeyer's neuronal doctrine and Sherrington's research on integrative action in the nervous system. The neuronal doctrine emphasizes signaling processes based on synaptic transmission. This in turn has focused attention on signal coding through nerve action potentials, conveyed along axonal paths from one cell to another.

1.1 Dendritic Slow Waves in Cerebral Neurons

Although the advent of intracellular recording in cerebral neurons revealed extensive and complex slow wave processes in many cells, originating primarily in dendrites (Creutzfeldt et al. 1966; Elul 1972; Fujita and Sato 1964; Jasper and Stefanis 1965), there has been relatively little attention to possible functional attributes of components of intraneuronal waves that appear in the pericellular environment as the far weaker electrochemical oscillations of the electroencephalogram (EEG).

More recently, studies in the hippocampal slice have confirmed propagation of rhythmic slow wave activity along the arch of hippocampal pyramidal cells in the absence of synaptic activation (Jeffery and Haas 1982; Snow and Dudek 1984; Taylor and Dudek 1984). Thus, in addition to direct functional contacts between dendrites of adjacent neurons, as described by Shepherd (1974), there are neuronal sensitivities to oscillating electric fields in pericellular fluid.

1.2 Functional Significance of Pericellular Slow Wave Fields: Sensitivities in Neural and Nonneural Systems

If these intrinsic electromagnetic fields in the tiny extracellular gutters between cells have functional significance, they must exert their influence on the membranes of cells which they enclose. From equilibrium considerations alone, there would be little reason to assign them this functional role, since they are typically six orders of magnitude smaller than the membrane potential gradient of 10^5 V/cm. Nevertheless, evidence for their functional role in brain tissue is strong. They can modulate cell firing patterns (Korn and Faber 1979); entrain EEG rhythms in rabbits (Takashima et al. 1979), cats (Bawin et al. 1973) and monkeys (Gavalas et al. 1970); alter neurotransmitter release (Kaczmarek and Adey 1974); and modulate behavioral states (Gavalas-Medici and Day-Magdaleno 1976; Smith 1984; Wever 1975).

Springer Series in Brain Dynamics 1
Edited by Erol Başar

These sensitivities have also been detected in nonneural cells, including bone (Bassett 1982; Luben et al. 1982; Luben and Cain 1984), liver (Byus et al. 1986), overian cells (Byus et al. 1986), pancreatic islets (Jolley et al. 1983), and lymphocytes (Byus et al. 1984; Lyle et al. 1983). We have therefore proposed that an intrinsic communication system between cells based on these weak electromagnetic (EM) influences may be a general biological property, allowing cells in tissue to "whisper together" (Young 1951).

2 Inward and Outward Signal Streams at Cell Membranes

An inward stream of signals is directed through the cell membrane to the fine tubes of the cytoskeleton, to intracellular organelles, including the nucleus, and to key enzyme systems. We have used three of these enzyme systems as markers of transduction of weak EM fields at the cell membrane surface with modulation of signals to the cell interior. These enzyme systems control messenger functions (Byus et al. 1984), metabolic energy production (Luben et al. 1982; Luben and Cain 1984), and the synthesis of essential chemical building blocks for cell growth and division (Byus et al. 1985).

2.1 Use of Imposed EM Fields as Modulators of Transmembrane Signals

We have found imposed EM fields to be powerful and highly specific tools in establishing the cell membrane as the site of detection and transductive coupling of oscillating EM Fields in the pericellular environment. Intracellular enzyme activity that is modulated by these fields provides sensitive molecular markers of both the sequence and the energetics of transmembrane coupling mechanisms. The findings emphasize physical aspects of the functional organization of these membrane events and suggest highly nonlinear, nonequilibrium processes in the first interactions of humoral stimuli at cell surface receptor sites (Adey 1984; Adey and Lawrence 1984; Lawrence and Adey 1982).

2.2 The Inward Signal Stream Through Cell Membranes

This inward signal stream is the result of a complex sequence of events. Binding of such humoral stimulating molecules as neurotransmitters, hormones, and antibodies at their specific surface receptor sites elicits a ripple effect extending along the membrane surface. It is manifested in altered ionic binding to the cell surface glycocalyx (Bawin and Adey 1976; Bawin et al. 1975, 1978a; Lin-Liu and Adey 1982), and in a concurrent ripple in membrane-related enzymes that serve both receptor and enzyme functions (Nishizuka 1983, 1984). Transduction of weak electrochemical stimuli at cell surface receptor sites leads to transmission of signals to the cell interior along coupling proteins (Luben et al. 1982; Luben and Cain 1984). This may involve nonlinear vibrational modes in the spines of helical proteins that span the membrane from the surface to the cell interior (Adey and Lawrence 1984; Lawrence and Adey 1982).

2.3 The Outward Signal Stream Through Cell Membranes

There is also an outward signal stream at cell membranes, involved in organization of the cell surface mosaic mediating allogeneic cytotoxicity (Lyle et al. 1983) and in secretion of neurotransmitters (Kaczmarek and Adey 1974), hormones (Jolley et al. 1983), connective tissue elements (Luben et al. 1982), and bone (Fitzsimmons et al. 1984). All these processes are calcium dependent and all have been found sensitive to one or more types of imposed EM field, including the spatial arrangement of the surface receptor mosaic (Lin-Liu et al. 1984).

2.4 Pathophysiology of Transmembrane Signaling

From these studies, there is the prospect that we may expect to identify major physical and chemical elements in the pattern of inward signals, and to distinguish between normal and abnormal signal streams. We have identified the cell membrane as a prime site of many EM field interactions (Adey 1983). From similar studies, there may be an opportunity to identify inward signal streams characteristic of cancerous cells and thus to identify an important role for the cell membrane in cancer promotion, one that is distinct from functions of the cell nucleus. We will examine aspects of the sequence and the energetics of cell membrane transductive coupling, with emphasis on transmembrane signals that reach intracellular enzyme systems.

3 A Three-Stage Model of Cell Membrane Transductive Coupling of EM Fields and Humoral Stimuli

Experimental findings are consistent with the "fluid mosaic" model of cell membranes (Singer and Nicolson 1972). External protrusions of intramembranous protein particles (IMPs) (Fig. 1) "floating" in the lipid bilayer have amino sugar (sialic acid) polyanionic terminals. They form a huge negatively charged sheet that attracts hydrogen and calcium ions in a "counterion" layer.

There is a minimal sequence of three steps in transductive coupling (Adey 1984), and each is calcium dependent: (a) cell surface glycoproteins that are stranded protrusions from intramembranous helical proteins (IMPs) sense the first weak electrochemical events associated with binding of neurohumoral molecules, hormones, and antibodies; (b) transmembrane portions of IMPs signal these events to the cell interior; (c) internally, there is coupling of this signal to intracellular enzyme systems and to the cytoskeleton (and thus to the nucleus and to other organelles).

3.1 Stage 1: Cooperative Modification of Calcium Binding with Amplification of Initial Signals and Modulation of Electrical Impedance in Cerebral Extracellular Space

Initial cell surface events appear to involve modulation of calcium binding to the numerous negative charges on the surface glycoprotein sheet, presumably in the plane of the membrane surface. A longitudinal spread would be consistent with the direction of flow of extracellular currents associated with physiological activity and

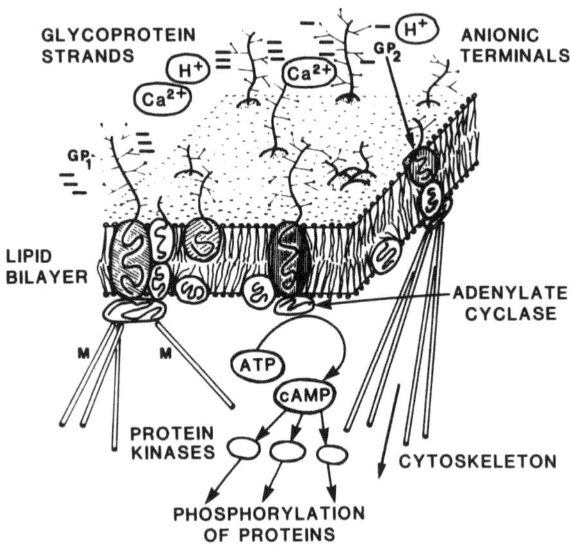

Fig. 1. Fluid mosaic model of cell membrane offers a structural basis for tissue interactions with EM fields. Intramembranous particles (IMPs) in the lipid bilayer have external protruding glycoprotein strands, negatively charged on their amino sugar terminals. They form receptor sites for antibodies, neurotransmitters, and hormones. They attract calcium ions. Stimulating molecules and EM fields alter surface calcium binding in the first step of transmembrane signal coupling. In stage 2, transmembrane signals pass along IMPs which act as coupling proteins to the interior. Stage 3 modulates intracellular enzyme activity. (Modified from Singer and Nicolson 1972)

from imposed EM fields. It would also be consistent with spreading calcium-dependent enzymatic activation from a single molecular locus proposed by Nishizuka (1984) and discussed below.

3.2 Calcium Efflux from Brain Tissue in Response to Intrinsic and Imposed EM Fields

In brain tissue, two distinct and contrasting patterns of calcium efflux occur in response to differing types of imposed fields. In isolated chick cerebral hemispheres, radio frequency (RF) fields at intensities around 1.0 mW/cm^2 (tissue levels of the order of EEG electric gradients) and with sinusoidal amplitude modulation from 3 to 35 Hz produced a "tuning curve" of increased calcium efflux, with a maximum increase at 16 Hz and smaller increments at higher and lower frequencies (Bawin et al. 1975). Unmodulated fields had no effect. Essential aspects of these studies in isolated cerebral tissue have been confirmed in awake cats at tissue gradients around 0.3 V/cm (Adey et al. 1982), again in the same range of electric gradients as the EEG at cellular dimensions (Elul 1962). Far weaker sinusoidal electric fields in the same low-frequency range (calculated levels in isolated chick cerebral tissue six orders of magnitude lower) also produced a tuning curve of modified calcium efflux, essen-

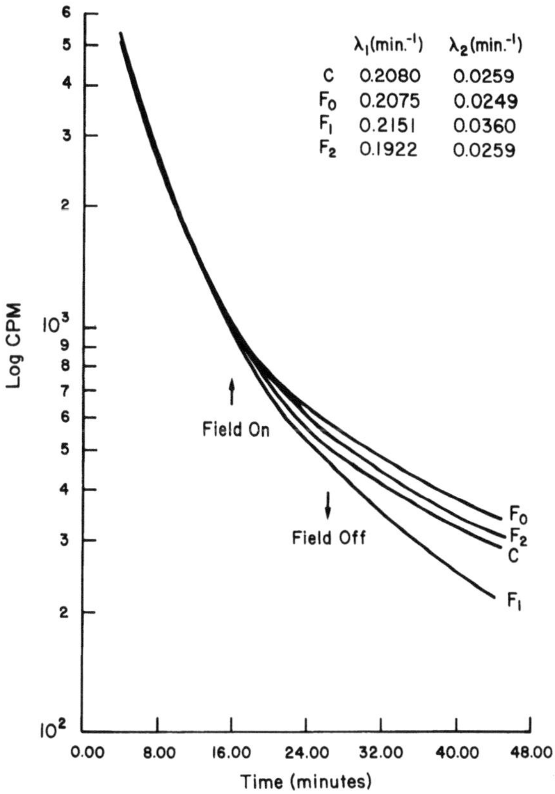

Fig. 2. Computer fitting of $^{45}Ca^{2+}$ data composites from rat cerebral synaptosomes exposed to "athermal" microwave fields (450 MHz, 0.5 mW/cm^2, sinusoidally modulated at 16 Hz). *C*, control; F_o, unmodulated 450-MHz field; F_1, same field with 16-Hz modulation; F_2, same field with 60-Hz modulation. (From Lin-Liu and Adey 1982)

tially as a mirror image of that from the stronger RF fields, with a decrease rather than an increase (Bawin and Adey 1976).

As a finer focus on the structural and functional basis of this field-sensitive calcium efflux in brain tissue, we have examined this relationship in synaptosome fractions (Lin-Liu and Adey 1982). Cerebral synaptosomes, typically 0.7 μm in diameter, retain the synaptic junction and the adjoining postsynaptic membrane. They have characteristics typical of brain chemical synapses. Calcium efflux was studied in synaptosomes preloaded with $^{45}Ca^{2+}$, using a continuous perfusion technique in a calcium-free physiological medium. A 450 MHz 0.5 mW/cm^2 field, sinusoidally modulated at 16 Hz, increased the rate constant of the calcium efflux by 38%. Unmodulated fields and fields modulated at 60 Hz were without effect (Fig. 2). This field-induced change was distinguishable from $CaCl_2$-stimulated efflux, which is most probably derived from intracellular sites.

Thus, these tiny elements of cerebral nerve fiber terminals also exhibited a frequency-selective calcium efflux in response to field exposure. The data support a model of field interaction with calcium at cell membrane surface sites. These responses are "windowed" with respect to field frequency and also to field intensity (Bawin et al. 1978b; Blackman et al. 1979; Dutta et al. 1984). This is the strongest single line of evidence for the essential nonlinearity of these effects.

3.3 Cooperative Models of Field Interactions at Cell Surface Ion Binding Sites

We have proposed the following model for these highly cooperative interactions (Adey 1981b). From intracellular sources, anionic charge sites on terminals of protruding glycoprotein strands are raised to energy levels substantially above ground state, forming "patches" or domains with coherent states between neighboring charge sites. Weak triggers at the boundaries of these coherent domains, such as oscillating EM fields or proton tunneling, may initiate a domino effect, with the release of much more energy than in the initial triggering events. Modulation of membrane surface calcium binding is thus an amplifying step in the transductive sequence, and is sensitive to imposed EM fields.

3.4 Combined Steady Magnetic and Oscillating Electric Fields; Effects on Cerebral Calcium Binding

Recent studies by Blackman et al. (1985) have shown that there are strong interactions between the earth's magnetic field and a weak imposed low frequency EM field (40 V/m peak-to-peak in air, estimated tissue components 10^{-7} V/cm) in determining calcium efflux from chick cerebral tissue. For example, halving the local geomagnetic field with a Helmholtz coil rendered a previously effective 15 Hz field ineffective; and doubling the geomagnetic field caused an ineffective 30 Hz signal to become effective.

A sensitivity of the pineal gland to the orientation of the head with respect to the earth's magnetic field has been reported by Semm (1983). In pigeons, guinea pigs, and rats, about 20% of pineal cells respond to changes in both direction and intensity of the earth's magnetic field. The peptide hormone melatonin secreted by the pineal powerfully influences the body's circadian rhythms. During the night, experimental inversion of the horizontal component of the earth's magnetic field significantly decreased secretion of melatonin and activity of its synthesizing enzymes (Welker et al. 1983).

3.5 Neurobehavioral Correlates of Cerebral Electrical Impedance and the Role of Calcium Ions in the Extracellular Space

As discussed above, most current flow in cerebral tissue occurs along membrane surfaces, with extracellular fluid as a preferred pathway. There are impedance "transients" accompanying alerting, orienting, and visual discriminative responses (Adey et al. 1966) (Fig. 3). They exhibit differential characteristics in different brain regions. Longlasting but reversible impedance changes occur with anesthetic and psychotropic drugs. We have shown that this impedance relates to extracellular calcium levels (Nicholson 1965).

We have hypothesized that "electrical impedance changes accompanying physiological responses may arise in perineuronal fluid with a substantial macromolecular content and calcium ions may modulate perineuronal conductivity" (Adey 1966). Macromolecular strands derived from intramembranous proteins lie in the perineuronal space and may modulate conductance as a function of cerebral tissue state.

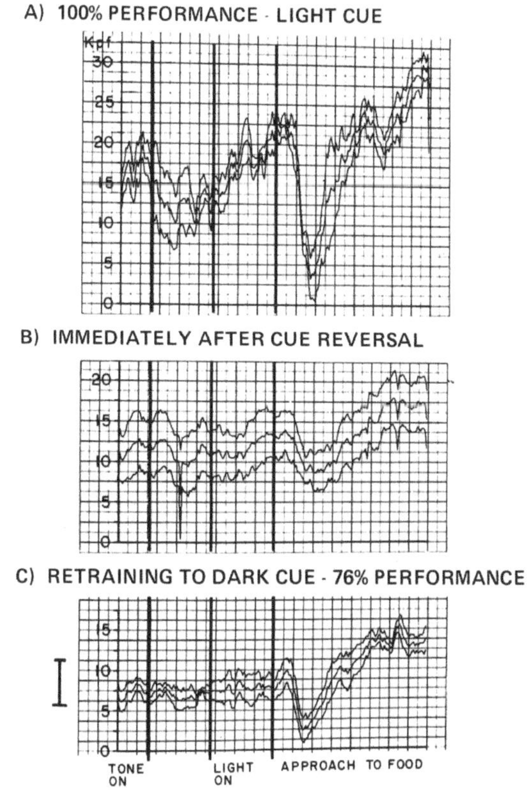

Fig. 3A–C. Hippocampal impedance measurements over 5-day periods at various levels of training, with successive presentations of alerting, orienting, and discriminative stimuli. In each *graph*, the *middle trace* indicates mean, with *upper* and *lower traces* showing one standard deviation. Calibration = 50 p, (baseline at 11100 pf). Variability was low at 100% performance (A), increased after cue reversal (B), but decreased again after retraining (C). (From Adey et al. 1966)

4 Coupling – Stage 2: An Enzymatic Marker of Signaling Along Proteinaceous Molecules Spanning the Plasma Membrane

In the following sections of this paper, evidence will be presented for sensitivities to imposed EM fields in nonneural as well as neural cells, with strong evidence that there may be an intrinsic communication system in tissue as a general biological property. For example, these sensitivities have been detected in bone, liver, ovary, skin, and lymphocytes. They appear organized around the cell's ability to detect electrochemical signals generated by other cells in its immediate environment.

In studies with EM fields imposed on bone cells, Luben has identified aspects of the role of intramembranous proteins in conveying signals from hormone receptor sites on the membrane surface to the cell interior (Luben et al. 1982; Luben and Cain 1984). These studies first examined effects of pulsed low-frequency magnetic fields on stimulation of adenylate cyclase by parathyroid hormone (PTH). In bone cells, PTH binds to specific receptor sites in membrane surface glycoproteins; adenylate cyclase is located on the internal surface of the membrane. The receptor site outside and the catalytic subunit inside are coupled by the N-protein. With induced pericellular gradients of only 1–3 mV/cm, one-millionth of the gradient of the membrane potential, stimulation of adenylate cyclase by PTH was inhibited by about 90%

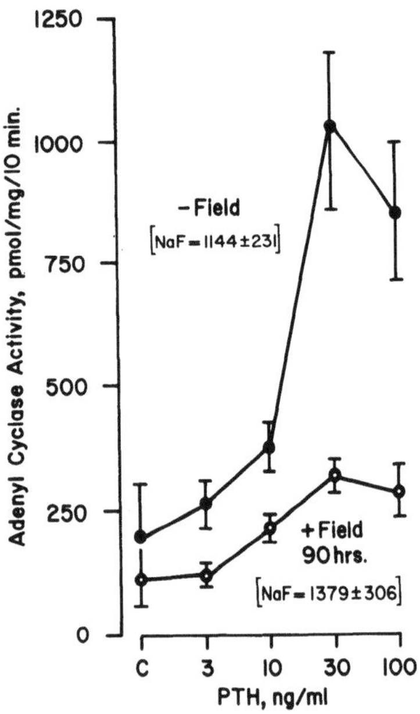

Fig. 4. Adenylate cyclase activity in bone cells stimulated with parathyroid hormone (PTH), with and without an imposed 72-Hz pulsed magnetic field. (From Luben et al. 1982)

(Fig. 4). However, this inhibition did not relate to inactivation of the adenylate cyclase. In further experiments with NaF activation, the enzyme showed full activity with and without field exposure. Nor did the field interfere with binding of PTH to its receptor site, since studies with I-125-labeled PTH showed the same levels of binding to these bone cells in control and field-exposed cultures. By exclusion, the evidence thus points strongly to events involving a protein that couples between the receptor and the adenylate cyclase, such as the N-protein, as the probable site of an important EM field action.

Further evidence that the cell membrane is indeed the prime site of interaction with these fields came from their effects on collagen synthesis by bone cells stimulated either with PTH or with vitamin D_3. This did not occur in the presence of the pulsed magnetic field. By contrast, vitamin D_3 has a primary site of action within the cell, possibly at the nucleus. Field exposure did not influence the inhibition of collagen synthesis by vitamin D_3. We therefore conclude that evidence from both adenylate cyclase and collagen synthesis studies points to the cell membrane as a prime site of field interaction.

5 Coupling – Stage 3; Protein Kinase and Ornithine Decarboxylase Activity as Markers of EM Field Transduction at Cell Membranes

By their phosphorylating actions, protein kinases constitute a major intracellular messenger system. By its role in the synthesis of polyamines, ornithine decarboxylase is essential for cell growth. We have shown that both groups of enzymes are sensitive to imposed EM fields.

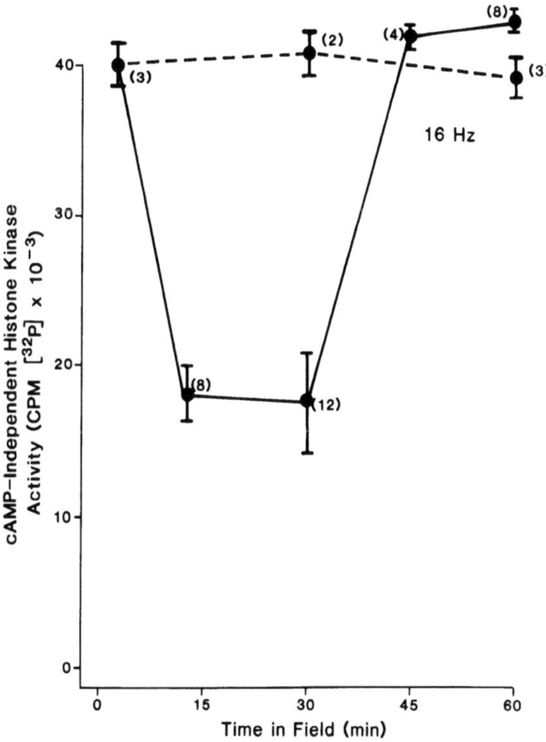

Fig. 5. Transient decrease in protein kinase activity at 15 and 30 min in human lymphocytes exposed to 450-MHz field (1.0 mW/cm^2 peak envelope power), sinusoidally modulated at 16 Hz. ●----●, Control; ●——●, exposed. (From Byus et al. 1984)

5.1 Time and Frequency Windows in Lymphocyte Kinase Responses

Protein kinases may be grouped in two broad classes, the cyclic adenosine monophosphate (cAMP)-dependent and the cAMP-independent protein kinases. The latter are activated by signals arising in cell membranes (as discussed below in relation to actions of cancer-promoting phorbol esters) that do not involve the cAMP pathway. We have examined responses of cAMP-independent protein kinases in cultured human lymphocytes exposed to a weak 450 MHz microwave field (Byus et al. 1984).

In cultures with approximately 50% T and 50% B cells, cAMP-independent protein kinase activity was sharply modified. The interaction showed "windowing" with respect to exposure duration and to modulation frequency. Activity fell to less than 50% of control levels after 15–30 min exposure, but despite continuing exposure, returned to control levels by 45 and 60 min (Fig. 5). Reduced enzyme activity occurred at modulation frequencies between 16 and 60 Hz, but not at higher and lower frequencies. Unmodulated fields were without effect.

5.2 Windowed Sensitivity of Lymphocyte Cytotoxicity to Low-Frequency Modulation of Microwave Fields

The cytolytic capacity of allogeneic T lymphocytes targeted against lymphoma cell cultures also showed sensitivity to modulation frequency when exposed to the same 450 MHz fields used in the protein kinase studies above. Cytolysis was reduced by

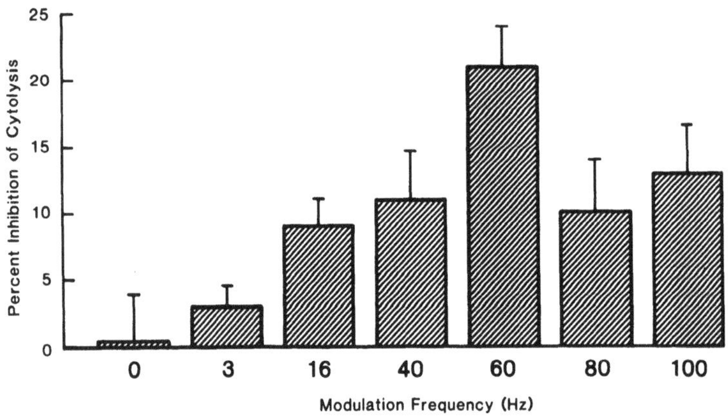

Fig. 6. Inhibition of cytotoxicity of allogeneic T lymphocytes by exposure to a 450-MHz 1.5 mW/cm^2 field as sinusoidal amplitude modulation was varied between 0 and 100 Hz. (From Lyle et al. 1983)

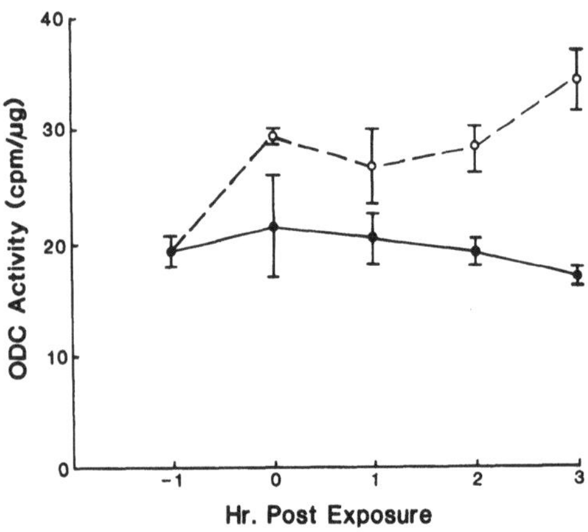

Fig. 7. Ornithine decarboxylase activity in Reuber H35 hepatoma cells before (●) and after (○) exposure to a 450-MHz field (1.0 mW/cm^2 peak envelope power; 1-h exposure) amplitude modulated at 16 Hz. (From Byus et al. 1986)

about 20% with 60 Hz modulation, with smaller interactions at higher and lower frequencies (Lyle et al. 1983) (Fig. 6). Here, too, unmodulated fields exerted no effect.

5.3 Ornithine Decarboxylase Responses to EM Fields in Cultured Liver, Ovary, and Bone Cells

Ornithine decarboxylase occurs in all cells and is essential for normal and abnormal cell growth. Clinically, its activity in cultures of suspected cancer cells (e.g., in human prostatic cancer) has proved a reliable index of malignancy. Pathways for its activation are not well defined, but our studies indicate that binding of phorbol esters at membrane receptor sites acts as an inducer (Byus et al. 1986).

Cultures of ovary and liver cells were exposed for 1 h to the same 450 MHz fields used above, sinusoidally modulated at 16 Hz. Enzyme activity was increased by 50% or more in the 3-h test period after exposure (Fig. 7). Increased ornithine decarboxylase activity has also been observed in cultured bone cells following exposure to 15-Hz pulsed magnetic fields (Cain et al. 1985).

6 Cancer-Promoting Phorbol Esters in the Pericellular Environment and Activities of Protein Kinases and Ornithine Decarboxylase

As noted above, an important cAMP-independent protein kinase that functions both as a receptor and an enzyme occurs widely in cell membranes and is in highest concentration in brain tissue (Nishizuka 1983, 1984). This enzyme, phosphatidyl serine kinase (kinase C), is calcium dependent and is normally activated by diacylglycerol formed from inositol phospholipids by the action of cell surface stimuli.

Nishizuka has described a striking sequence of response in the interactions of kinase C with diacylglycerol. A single molecule of diacylglycerol can activate one molecule of kinase C. Thereafter there is a spreading domino or brushfire effect that activates all kinase C molecules around the whole membrane surface. Nishizuka points out that "these findings provide an entirely new concept of receptor function." He also emphasizes the significance of a similar spreading brushfire of altered calcium binding along the membrane surface, which we have described as occurring concurrently with actions of humoral stimuli and EM fields at cell-surface receptor sites (Adey 1981a, 1983).

6.1 Kinase C as a Receptor for Cancer-Promoting Phorbol Esters; Synergy of Phorbol Esters and EM Fields

Nishizuka has also shown that kinase C is a specific receptor for phorbol esters, a class of tricyclic ring compounds widely used as promoting agents in experimental cancer models. They bind strongly to kinase C, activating it irreversibly. Kinase C is cAMP independent, and we have identified kinase enzymes in this group as sensitive to weak pericellular EM oscillations, as described above (Byus et al. 1984)

Our studies now show that induction of ornithine decarboxylase also follows stimulation of liver and ovary cells with a phorbol ester (tetradecanoyl phorbol acetate, TPA) and that this response is sharply enhanced by imposed 450 MHz fields, sinusoidally modulated at 16 Hz (Fig. 8). These findings offer evidence for a synergic action between environmental EM fields and the effects of cancer-promoting substances. However, it is not known whether this activation of ornithine decarboxylase occurs as a response to kinase C or through other pathways.

Kinase C is found in highest concentration in brain tissue, but until recently there has been no evidence for its involvement in the control of neuronal excitability. Studies in Kaczmarek's laboratory (DeRiemer et al. 1985) now show that activation of endogenous protein kinase C by TPA, or intracellular injection of the purefied enzyme, enhances the voltage-sensitive calcium current in bag-cell neurones of Aplysia.

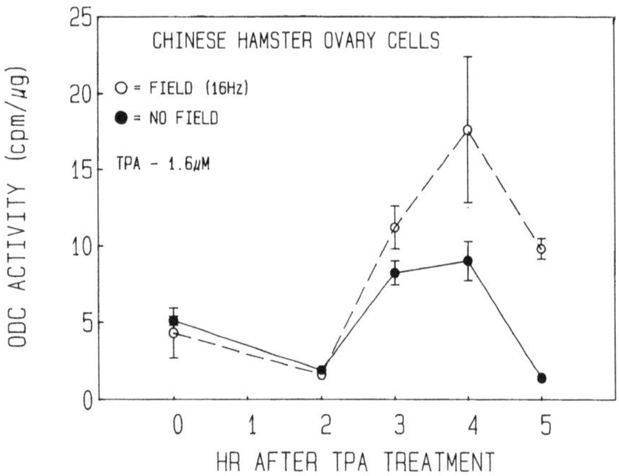

Fig. 8. Ornithine decarboxlase activity in hamster ovary cells stimulated by a phorbol ester (*TPA*), with and without concurrent exposure to the same 450-MHz field as in Fig. 7. (From Byus et al. 1986)

7 Emergent Concepts of Cell Membrane Transductive Coupling; Implications in Brain Transactional Mechanisms

It was a stated goal of the 1985 conference Dynamics of Sensory and Cognitive Processing of the Brain to encourage speculation on "how the brain works;" and more specifically, to seek broader syntheses than hitherto possible on mechanistic substrates of transaction, storage, and recall in cerebral tissue. Experimental evidence presented here offers strong support for concepts of communication between adjacent cells in cerebral tissue based on slow electrochemical oscillations.

In cerebral tissue, these oscillations are a dominant feature of dendritic activity. Their extracellular components are many orders of magnitude less than the electric gradient of the cell membrane potential. Development of intraneuronal recording in the 1950s focused attention on the membrane potential and changes in it associated with synaptic excitation. With membrane potential gradients of the order of 10^5 V/cm and with synaptic potentials changing this gradient by 10^3 V/cm, generally accepted views on so-called "field potentials" in pericellular fluid denied them a functional role. However, more recent studies utilizing imposed fields that mimic aspects of intensity and frequency characteristics of natural bioelectric activity have confirmed sensitivities of cultured cells, isolated tissues, and intact organisms to fields at pericellular intensities in the range 10^{-1}–10^{-7} V/cm (Adey 1981a; Adey and Lawrence 1984).

We have presented evidence that cell membranes play a key role in detecting, transforming, and transmitting signals of chemical and electrical stimuli from the cell surface to the interior. Initial studies focused on these phenomena in brain tissue, but our more recent findings have disclosed a much wider occurrence in noneural tissues, suggesting that they may be a substrate for intercellular communication as a general biological property.

In the continuing search for mechanisms underlying these interactions, it has been possible to trace the major events in cell membrane transductive coupling that culminate in modulation of intracellular enzyme activity. Weak extracellular electric gradients oscillating at low frequencies have been shown to influence the functions of adenylate cyclase, protein kinases, and ornithine decarboxylase. Independent replication and mutual consistency of many of the key findings have served to establish this experimental record as a highly credible body of evidence. Statements that these findings are "sensational," or "attributable to experimental error" (Haken 1983), therefore appear to be inappropriate.

Imposed EM fields have proved unique tools in this developing awareness of the profound importance of the pericellular microenvironment. First, there is strong evidence that cell membranes are powerful amplifiers of weak electrochemical events in their immediative vicinity. Secondly, virtually all these sensitivities appear to involve natural or imposed field frequencies below 100 Hz, a spectral span that we have named the "biological spectrum." Thirdly, there are changes in conductance in the pericellular environment, possibly mediated by calcium ions that are closely correlated with higher nervous activity in cerebral structures.

7.1 Physical Bases of Low-Level, Low-Frequency Interactions with Biological Macromolecules

No known mechanisms explain low-frequency bioeffects on the basis of direct interactions with component dipoles of molecular systems oscillating at these low frequencies. Therefore, a structural and functional basis must reside in properties of molecular systems. It is also necessary to elucidate the windowed character of these responses with respect to these low frequencies, and to take account of similar windowing with respect to intensity. We conclude that these interactions are not only nonequilibrium in character; they are also highly nonlinear (Adey 1981a, b; 1984; Lawrence and Adey 1982).

Low-frequency sensitivities have been modeled in terms of Lotka-Volterra (predator-prey) processes involving slow shifts in energy states of coherent populations of fixed charges on cell surface glycoproteins (Fröhlich 1975); in limit-cycle behavior of calcium ions binding to cell surface macromolecules (Kaczmarek 1976); in Einstein-Bose phase transitions in populations of electric dipoles in the cell surface glycocalyx (Grodsky 1976); in chaotic behavior of pseudorhythmic molecular oscillations (Kaiser 1984); and in cyclotron oscillations of calcium ions exhibiting coherent states at the cell membrane surface (Polk 1984). A precise understanding of these low-frequency sensitivities awaits future research.

7.2 Transmembrane Signaling Along Helical Proteins by Soliton Waves; Quasiparticles and New Concepts of the Organization of Matter

We have speculated that the linear macromolecules spanning the membrane play a key role in signal coupling to the cell interior, and that this coupling is a direct consequence of the initial longitudinal events on the membrane surface discussed above. Surface events may initiate a dispersive process, in which solitary waves of the Davydov type may move as phonons down the length of the long spirals of the helical protein molecules spanning the membrane (Davydov 1979). They offer a means of signaling or of energy transport (Lawrence and Adey 1982).

Solitons may be considered as traveling "packets" of a vibrational state, forming quasiparticles that pass along the triple spines of the protein molecule. They are relatively uninfluenced by random vibration of particles through which they pass, or even by other solitons. These concepts suggest a fundamentally new aspect to the organization of matter at molecular and atomic levels, and there is an intense search to detect solitons in physical and biological systems. Atoms in solitons are not necessarily randomly located, as classic statistical mechanics of matter has postulated, and in this "clustering" or graininess, first perceptions indicate some unique properties of these quasiparticles.

These concepts are as new to the physical sciences as they are new in their applications to biology and medicine. They bespeak a fundamentally new aspect in the organization of matter. To the biologist, now aware of highly nonlinear phenomena in cell membrane transduction of hormonal, neurohumoral, and immunological stimuli, these models from the frontiers of the physics of matter offer a unique opportunity to move beyond accepted equilibrium models in cellular and subcellular systems.

References

Adey WR (1966) Intrinsic organization of cerebral tissue in alerting, orienting and discriminative responses. In: Quarton GC, Melnechuk T, Schmitt FO (eds) The neurosciences: a study program. Rockefeller University Press, New York, pp 615–633

Adey WR (1981a) Tissue interactions with nonionizing electromagnetic fields. Physiol Rev 61:435–514

Adey WR (1981b) Ionic nonequilibrium phenomena in tissue interactions with nonionizing electromagnetic fields. Am Chem Soc Symp Ser 157:271–297

Adey WR (1983) Molecular aspects of cell membranes as substrates for interaction with electromagnetic fields. In: Başar E, Flohr H, Haken H, Mandell AJ (eds) Synergetics of the brain. Springer, Berlin Heidelberg New York, pp 201–211

Adey WR (1984) Nonlinear, nonequilibrium aspects of electromagnetic field interactions at cell membranes. In: Adey WR, Lawrence AF (eds) Nonlinear electrodynamics in biological systems. Plenum, New York, pp 3–22

Adey WR, Lawrence AF (eds) (1984) Nonlinear electrodynamics in biological systems. Plenum, New York

Adey WR, Kado RT, McIlwain JT, Walter DO (1966) The role of neuronal elements in regional impedance changes in alerting, orienting and discriminative responses. Exp Neurol 15:490–510

Adey WR, Bawin SM, Lawrence AF (1982) Effects of weak amplitude-modulated microwave fields on calcium efflux from awake cat cerebral cortex. Bioelectromagnetics 3:295–309

Bassett CAL (1982) Pulsing electromagnetic fields: a new method to modify cell behavior in calcified and noncalcified tissues. Calcif Tissue Int 34:1–8

Bawin SM, Adey WR (1976) Sensitivity of calcium binding in cerebral tissue to weak environmental oscillating low frequency electric fields. Proc Natl Acad Sci USA 73:1999–2003

Bawin SM, Gavalas-Medici RJ, Adey WR (1973) Effects of modulated VHF fields on specific brain rhythms in cats. Brain Res 58:365–384

Bawin SM, Kaczmarek LK, Adey WR (1975) Effects of modulated VHF fields on the central nervous system. Ann NY Acad Sci 247:74–80

Bawin SM, Adey WR, Sabbot IM (1978a) Ionic factors in release of $^{45}Ca^{2+}$ from chicken cerebral tissue by electromagnetic fields. Proc Natl Acad Sci USA 75:6314–6318

Bawin SM, Sheppard AR, Adey WR (1978b) Possible mechanisms of weak electromagnetic field coupling in brain tissue. Bioelectrochem Bioenergetics 5:76–76

Blackman CF, Elder JA, Weil CM, Benane SG, Eichinger DC, House DE (1979) Induction of calcium ion efflux from brain tissue by radio frequency radiation; effects of modulation frequency and field strength. Radio Sci 14:93–98

Blackman CF, Benane SG, Rabinowitz JR, House DE, Joines WT (1985) A role for the magnetic field in radiation-induced efflux of calcium ions from brain tissue in vitro. Bioelectromagnetics 6 (4)

Byus CV, Lundak RL, Fletcher RM, Adey WR (1984) Alterations in protein kinase activity following exposure of cultured lymphocytes to modulated microwave fields. Bioelectromagnetics 5:34–351

Byus CV, Kartun K, Peiper S, Adey WR (1986) Ornithine decarboxylase activity in liver cells is enhanced by low-level amplitude-modulated microwave fields. (To be published)

Cain CD, Luben RA, Donato NJ, Byus CV, Adey WR (1985) Pulsed electromagnetic field effects on responses to parathyroid hormone in primary bone cells (Abstr). Annual Meeting of the Bioelectromagnetics Society, San Francisco, p 8

Creutzfeldt OD, Watanabe S, Lux HD (1966) Relations between EEG phenomena and potentials of single cortical cells. I. Evoked responses after thalamic and epicortical stimulation. Electroencephalogr Clin Neurophysiol 20:1–18

Davydov AS (1979) Solitons in physical systems. Phys Scripta 20:387–394

DeRiemer SA, Strong JA, Albert KA, Greengard P, Kaczmarek LK (1985) Enhancement of calcium current in aplysia neurons by phorbol ester and kinase C. Nature 313:313–316

Dutta SK, Subramoniam A, Ghosh B, Parshad R (1984) Microwave radiation-induced calcium ion efflux from human neuroblastoma cells in culture. Bioelectromagnetics 5:71–78

Elul R (1962) Dipoles of spontaneous activity in the cerebral cortex. Exp Neurol 6:285–289

Elul R (1972) The genesis of the EEG. Int Rev Neurobiol 15:227–272

Fitzsimmons R, Farley J, Adey R, Baylink D (1984) Bone formation is increased after short term exposure to very low amplitude electric fields in citro (Abstr) J Cell Biol 99:422a

Fröhlich H (1975) The extraordinary dielectric properties of biological materials and the properties of enzymes. Proc Natl Acad Sci USA 72:4211–4215

Fujita Y, Sato T (1964) Intracellular records from hippocampal pyramidal cells in rabbit during theta rhythm activity. J Neurophysiol 27:1012–1025

Gavalas RJ, Walter DO, Hammer J, Adey WR (1970) Effect of low-level, low-frequency electric fields on behavior in Macaca nemestrina. Brain Res 18:491–501

Gavalas-Medici R, Day-Magdaleno SR (1976) Extremely low frequency weak electric fields affect schedule-controlled behavior in monkeys. Nature 261:256–258

Grodsky IT (1976) Neuronal membrane: a physical synthesis. Math Biosci 28:191–219

Haken H (1983) Synposis and introduction. In: Başar E, Flohr H, Haken H, Mandell AJ (eds) Synergetics of the brain. Springer, Berlin Heidelberg New York, pp 3–25

Jasper H, Stefanis C (1965) Intracellular oscillatory rhythms in pyramidal tract neurons in the cat. Electroencephalogr Clin Neurophysiol 18:541–553

Jefferys JGR, Haas HL (1982) Synchronized bursting of CA1 hippocampal pyramidal cells in the absence of synaptic transmission. Nature 300:448–450

Jolley WB, Hinshaw DB, Knierim K, Hinshaw DB (1983) Magnetic field effects on calcium efflux and insulin secretion in isolated rabbit islets of Langerhans. Bioelectromagnetics 4:103–105

Kaczmarek LK (1976) Frequency sensitive biochemical reactions. Biophys Chem 4:249–252

Kaczmarek LK, Adey WR (1974) Weak electric gradients change ionic and transmitter fluxes in cortex. Brain Res 66:537–540

Kaiser F (1984) Entrainment-quasiperiodicity-chaos-collapse: bifurcation routes of externally driven self-sustained oscillating systems. In: Adey WR, Lawrence AF (eds) Nonlinear electrodynamics in biological systems. Plenum, New York, pp 393–412

Korn H, Faber DS (1979) Electrical interactions between vertebrate neurons: field effects and electrotonic coupling. In: Schmitt FO, Worden FG (eds) The neurosciences: fourth study program. MIT Press, Cambridge, pp 333–358

Lawrence AF, Adey WR (1982) Nonlinear wave mechanisms in interactions between excitable tissue and electromagnetic fields. Neurol Res 4:115–153

Lin-Liu S, Adey WR (1982) Low frequency amplitude modulated microwave fields change calcium efflux rates from synaptosomes. Bioelectromagnetics 3:309–322

Lin-Liu S, Adey WR, Poo M-M (1984) Migration of cell surface concanavalin A receptors in pulsed electric fields. Biophys J 45:1211–1218

Luben RA, Cain CD (1984) Use of bone cell hormone response systems to investigate bioelectromagnetic effects on membranes in vitro. In: Adey WR, Lawrence AF (eds) Nonlinear electrodynamics on biological systems. Plenum, New York, pp 23–33

Luben RA, Cain CD, Chen M-Y, Rosen DM, Adey WR (1982) Effects of electromagnetic stimuli on bone and bone cells in vitro: inhibition of responses to parathyroid hormone by low-energy, low-frequency fields. Proc Natl Acad Sci USA 79:4180–4184
Lyle DB, Schechter P, Adey WR, Lundak RL (1983) Suppression of T lymphocyte cytotoxicity following exposure to sinusoidally amplitude-modulated fields. Bioelectromagnetics 4:281–292
Nicholson PW (1965) Specific impedance of cerebral white matter. Exp Neurol 13:386–401
Nishizuka Y (1983) Calcium, phospholipid and transmembrane signalling. Philos Trans R Soc London [Biol] 302:101–112
Nishizuka Y (1984) The role of protein kinase C in cell surface transduction and tumor promotion. Nature 308:693–697
Polk C (1984) Time varying magnetic fields and DNA synthesis: magnitude of forces due to magnetic fields on surface-bound counterions (Abstr) Proceedings of the 6th Annual Meeting of the Bioelectromagnetics Society, Atlanta, p 77
Semm P (1983) Neurobiological investigations on the magnetic sensitivity of the pineal gland in rodents and pigeons. Comp Biochem Physiol 76A:683–689
Shepherd GM (1974) The synaptic organization of the brain. Oxford University Press, Oxford
Singer SJ, Nicolson GL (1972) The fluid mosaic model of the cell membrane. Science 175:720–7321
Smith RF (1984) Core temperature as a behavioral indicant of the rat's reaction to low frequency magnetic fields. PhD Thesis, University of Kansas, Lawrence
Snow RW, Dudek FE (1984) Electrical fields directly contribute to action potential synchronization during convulsant-induced epileptiform bursts. Brain Res 323:114–118
Takashima S, Onoral B, Schwan HP (1979) Effects of modulated RF energy on the EEG of mammalian brains. Radiat Environ Biophys 16:15–27
Taylor CP, Dudek FE (1984) Excitation of hippocampal pyramidal cells by an electrical field effect. J Neurophysiol 52:126–142
Welker HA, Semm P, Willig RP, Wiltschko W, Vollrath L (1983) Effects of an artificial magnetic field on serotonin-N-acetyltransferase activity and melatonin content of the rat pineal gland. Exp Brain Res 50:426–432
Wever R (1975) The circadian multi-oscillatory system of man. Int J Chronobiol 3:19–55
Young JZ (1951) Doubt and certainty in science. Oxford University Press, New York

Global Contributions to Cortical Dynamics: Theoretical and Experimental Evidence for Standing Wave Phenomena

P. L. Nunez

1 Local and Global Theories of Cortical Dynamics

A number of characteristic frequencies observed in either spontaneous EEG or late evoked potentials appear to owe their origins to resonant phenomena in the cortex (Freeman 1975; Başar 1980; Nunez 1981a). One can distinguish between local circuit models of cortical dynamics (Beurle 1956; Griffith 1963; Wilson and Cowan 1972; Lopes da Silva et al. 1974; Freeman 1975; Başar 1980; van Rotterdam et al. 1982; Wright and Kydd 1984a; Zhadin 1984), global models (Nunez 1974a; 1981a, c, 1985; Katznelson 1981) and attempts to integrate the two viewpoints (Nunez 1981a, b; Ingber 1982, 1984, 1985). It is shown here that a relatively simple global model predicts many of the salient characteristics of EEG with minimal assumptions about unknown physiological parameters, and independent of local circuit effects that may be dominant during cognitive processing.

In order to illustrate the distinction between local and global effects with minimal mathematics, it is instructive to consider the dispersion relation relating temporal frequency ω to wavenumber (or spatial frequency k) and characteristic velocity in the wave medium v (Eq. 1).

$$\omega^2 = \omega_0^2 + v^2 k^2 \tag{1}$$

When the local contribution ω_0 is zero, Eq. 1 is the well-known relation between frequency and spatial wavelength (equal to $2\pi/k$) for electromagnetic radiation in a vacuum, sound waves, or waves in stretched string. Electromagnetic waves in transmission lines exhibit the local contribution ω_0 caused by the product of inductance and capacitance per unit length in addition to the global contribution vk. In a transmission line or string of length L, with the field variable (potential or string displacement) held fixed at the ends, standing waves occur with discrete wavenumbers,

$$k = \pi l / L, \quad 1 = 1, 2, 3 \ldots \tag{2}$$

and corresponding resonant (or normal mode) frequencies ω_l, given by the substitution of Eq. 2 into Eq. 1. The spatial extent and location of the external input (the manner in which a guitar string is plucked, for example) determines how the total energy in the wave is distributed between the fundamental frequency ω_1 and the overtones ω_l, $l = 2, 3, 4, \ldots$. These ideas may be extended to more complicated geometries. For example, if one were to postulate that cortical waves are nondispersive ($\omega_0 = 0$), that the characteristic velocity (v) of the cortex is known, and that the cortical surface of one hemisphere is shaped like a prolate spheroid, discrete resonant fre-

Springer Series in Brain Dynamics 1
Edited by Erol Başar

quencies are predicted that depend on the size and eccentricity of the prolate spheroid. For an average brain with $v = 7$ m/s, the resonant frequencies in Hz are 10, 11.6 ($l = 1$); 17.2, 18.1, 21.2 ($l = 2$); 24.1, 24.8, 27.2, 30.6 ($l = 3$), etc. Of course, we have no a priori reason to expect "brainwaves" to be nondispersive ($\omega = v k$); the dispersion relations, if they exist, must be derived from cortical anatomy and physiology. To the extent that our analogy to simple physical systems is valid, cortical dynamics determine the dispersion relation. Cortical size, shape, and inhomogeneity/isotropy determine the spatial shapes of the eigenfunctions (analogous the sine functions with wavenumbers k_l for one-dimensional systems), specific afferent input to the cortex corresponds to plucking the guitar string or applying an external pulse to the transmission line, and diffuse afferent input changes cortical resonant properties.

2 A Cortical Dispersion Relation

Whereas mammalian cortex consists of distinct layers, it exhibits a large-scale homogeneity/isotropy in directions parallel to its surface. A number of experiments, including the famous work of Hubel and Wiesel (1962), suggest that the functional unit of the cortex is, to some degree, the cortical column rather than the single neuron in that much redundancy of neural firing patterns within columns is observed. The interconnections between cortical neurons are distinctly separated into two types: the short-range intracortical fibers having an average length of less than 1 mm and the long-range corticocortical (association) fibers having an apparent average length of several centimeters (Braitenberg 1978). The latter axons, which form most of the white matter, number of the order of 10^{10}; that is, nearly every cortical pyramidal cell sends an axon into the white matter which re-enters the cortex at some distant location. A significant number have lengths of 20 cm or more. The velocity of action-potential propagation in these corticocortical fibers is approximately 6–9 m/s. Many intracortical fibers are known to terminate with inhibitory synapses; others may end in excitatory synapses. The corticocortical fibers are believed to be exclusively excitatory (Szentagothai 1978).

In humans, only a few percent of the fibers entering the underside of a cortical column originate in the thalamus and other midbrain structures; the remaining are corticocortical fibers (Braitenberg 1978). It then appears that, to a first approximation, the cortex may be regarded as a self-contained system in which afferent input from the midbrain acts as a driving "force" and/or changes the threshold for neural firings. Even in the absence of supporting EEG data, the fact that finite velocities occur for signal propagation along the closed cortical surface would suggest the existence of standing waves of electric field with frequencies dependent on cortical dynamics (brain wave equations) and cortical boundary conditions. Note that these waves can have nothing to do with the macroscopic version of Maxwell's equations; their existence is ultimately dependent on the peculiar, nonlinear membrane properties that allow for the propagation of action potentials.

Surface EEG reflects the space-averaged neural activity of at least several square centimeters of cortical surface; thus, it is suggested that the number of synaptic and action-potential firings in columns of cortex be followed with a macroscopic theory that avoids much of the mathematical complexity of local circuit models. The columns are connected by both long- and short-range fibers. Delays caused by both the finite velocity of action potential and the rise times of postsynaptic potentials are included in the most general version of the theory (Nunez 1981a, b, c). Solutions of somewhat simplified brain wave equations have recently been obtained for waves on the surface of a homogeneous sphere (Katznelson 1981). In the interest of both brevity and clarity, the simplest one-dimensional version with negligible local delays is outlined here. The number, size, and length distribution of the long-range corticocortical fibers is only partly known; it is likely that these connections are both inhomogeneous and anisotropic. It is assumed here that white matter contains N overlapping excitatory fiber systems, where the number of connections between cortical locations separated by distance x falls off as $\exp(-\lambda_n x)$, $n = 1, N$. For large N, this assumption can be made to fit nearly all distributions. The dispersion relation relating complex frequency $p = j\omega + \gamma$ to wavenumber k is obtained as a solution of an integral equation (Nunez 1981a, c):

$$1 - \sum_{n=1}^{N} A_n \int_0^\infty \frac{\lambda_n^2 v^2 + p\lambda_n v}{(\lambda_n v + p)^2 + k^2 v^2} f_n(v)\,dv = 0 \tag{3}$$

The sum is over the N excitatory fiber systems and $f_n(v)$ is the velocity distribution function for action potential propagation in each system. To examine the general character of the dispersion relation, let each fiber system carry action potentials with fixed velocity v so that $f_n(v) = \delta(v - v_n)$. Also, $|p/v_n|$ is on the order of $(20\pi\,\mathrm{s}^{-1})/(6\text{–}9\,\mathrm{m/s}) \sim 0.1\,\mathrm{cm}^{-1}$ for scalp EEG; thus, all intracortical and short corticocortical systems have $\lambda_n \gg |p/v_n|$. If all systems but one ($n = 1$) satisfy this condition, the dispersion relation reduces to

$$1 - B\frac{\lambda_1^2 v_1^2 + p\lambda_1 v_1}{(\lambda_1 v_1 + p)^2 + k^2 v_1^2} = 0 \tag{4}$$

where the nondimensional parameter,

$$B \cong \frac{Q_+ \rho_1}{Q_- \rho_-} \tag{5}$$

For example, if a cortical pyramidal cell requires a transmembrane potential change of 50 mV at the soma to fire an action potential, and each excitatory postsynaptic potential contributes 1 mV, $Q_+ = 1/50$ and Q_- is similarly defined for inhibitory synapses. The number density of long-range fibers with excitatory synapses and total number density of inhibitory synapses are given by ρ_1 and ρ_-, respectively. Physiological data suggest that B is of order 1, at least for most brain states. Input from the brainstem reticular formation is well known to regulate sleep/waking; we cannot stay awake without this diffuse and apparently inhibitory influence on the cortex. Thus, it is suggested here that the threshold parameter B is a macroscopic descriptor of physiological state, with increases in B occurring as it becomes more easy for the "average" cortical neuron to fire an action potential. Equation 4 has the following solutions for the frequency (ω_l) and temporal damping (γ) of "brain waves,"

$$\omega_l^2 = v^2\left(k_l^2 - \lambda^2\frac{B^2}{4}\right), \quad l = 1, \infty \tag{6}$$

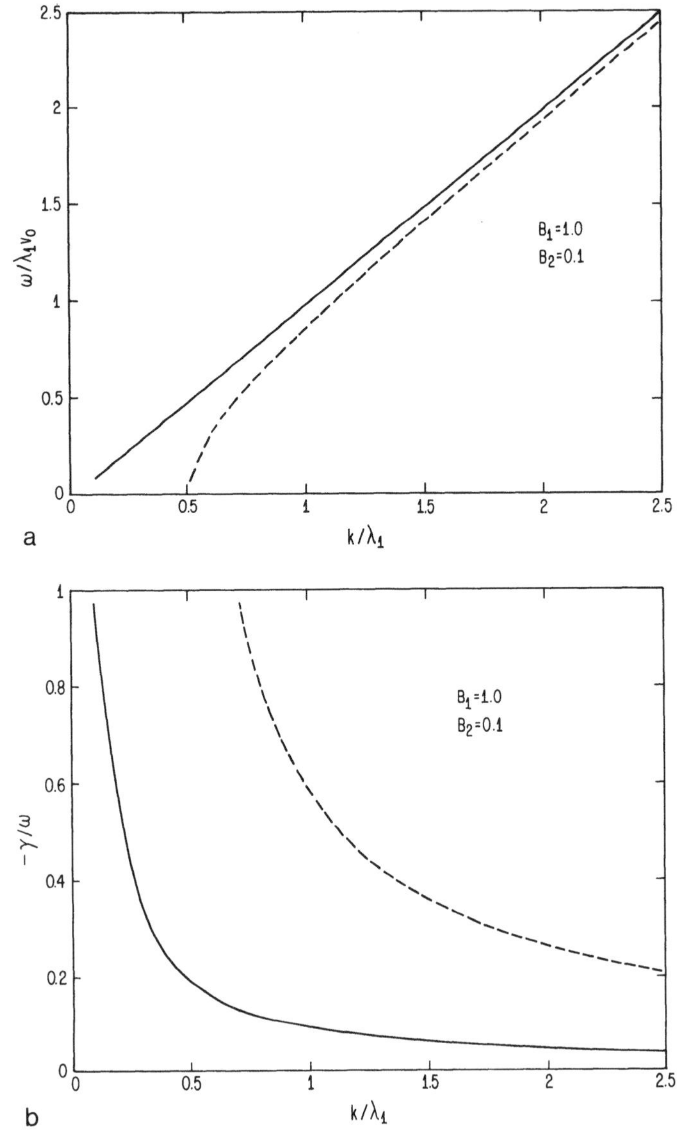

Fig. 1. (a) Two sets of corticocortical fiber systems ($\lambda_1/\lambda_2 = 10$) yield two branches of the dispersion relation corresponding to two kinds of "brain waves" indicated by *solid* and *dashed lines.* (b) Unequal damping of the two wave phenomena is shown

$$\gamma = -\lambda v\ (1 - B/2) \tag{7}$$

where the 1 subscript has been dropped for clarity and the subscript l has been added to indicate that only discrete frequencies are expected in a finite medium. Thus, as B approaches zero, the waves are nondispersive, $\omega_l = v k_l$ and the higher modes tend to have less relative damping; that is, $|\gamma/\omega| \sim \lambda k_l$. In an anisotropic medium, waves will tend to propagate along the directions of the longest fibers, which largely determine the dispersion relation. If the longest fibers cut across fissures, their effective lengths

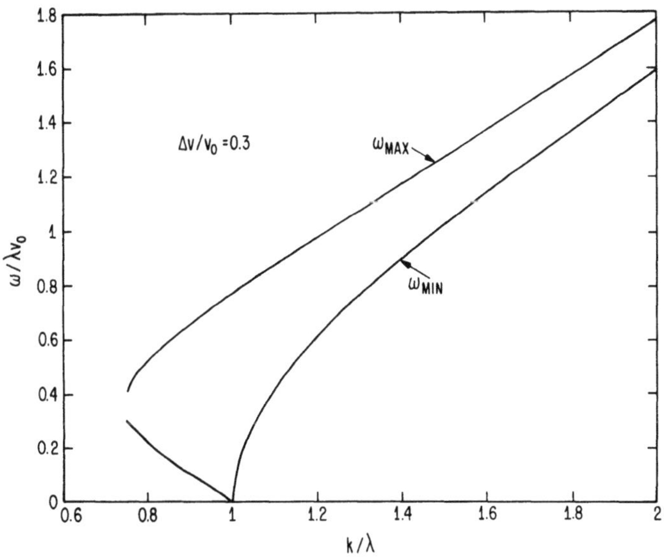

Fig. 2. Maximum and minimum frequencies (ω) allowed in brain with distributed action potential velocities in corticocortical fibers subject to the condition of weak to moderate damping for any value of the threshold parameter B

and propagation velocities are doubled and halved, respectively; they then yield the same delay between cortical regions separated by x, the surface distance measured in and out of fissures and sulci.

Multiple long-range fiber systems, corresponding to the inclusion of terms $n > 1$ in Eq. 3, result in polynominal equations of higher order for the complex frequency p. The case of two corticocortical systems is illustrated in Fig. 1, which shows frequencies and damping corresponding to two branches of the dispersion relation (both branches depend on both fiber systems). The physiological correlate of this mathematical result is that more than one type of brain wave can exist simultaneously in the cortex in a manner analogous to the occurrence of optical and acoustical waves in crystals or plasmas. In relatively simple cases waves have an independent existence, but the individual identities may be lost if sufficiently strong coupling effects occur. Corticocortical fibers carry action potentials with velocities distributed according to the functions $f_n(v)$ in Eq. 3 rather than the single velocity used to obtain Eqs. 6 and 7. Diameter histograms of white matter fibers suggest that the dominant $f_n(v)$ may be approximated by a Gaussian with peak velocity (v_0) in the 6–9 m/s range. The half-width of the distribution $\Delta v/v_0$ is approximately of the order of 0.2–0.4 (Katznelson 1981). Figure 2 represents a summary of a large number of numerical solutions of Eq. 3. The effect of the spread in velocity distribution is to confine the range of moderately to weakly damped frequencies that correspond to a given wavenumber for any positive value of the threshold parameter B, a very restrictive condition for the higher modes. For example, when $k/\lambda = 1$, Eqs. 6 and 7 indicate that the condition of moderate damping $-1 \leq \gamma/\omega \leq 0$ yields an allowed frequency range $0 \leq \omega/\lambda v_0 \leq 1$, whereas Fig. 2 yields the slightly more restrictive frequency range $0 \leq \omega/\lambda v_0 \leq 0.76$. However, when $k/\lambda = 2$, Eqs. 6 and 7 show an allowed frequency range $0 \leq \omega/\lambda v_0 \leq 1.73$; whereas Fig. 2 yields the highly restrictive range $1.60 \leq \omega/\lambda v_0 \leq 1.76$, due to the effects of spread in the velocity distribution function. This frequency restriction occurs in addition to that determined by boundary conditions.

The effects of local delays on the dispersion relation were approximated in an earlier version of the global wave theory (Nunez 1981a). If rise times of postsynaptic potentials are approximately equal to membrane time constants, that is, of the order of $\tau = 8$ ms, the principal effect of local delays is to restrict allowed frequencies to lower ranges. Local delays can be expected to influence strongly the dispersion relation only when $\omega\tau \sim 1$; that is,

$$f = \omega/2\pi \cong 1/2\pi\tau \cong 20\,\text{Hz} \tag{8}$$

which could partly account for the paucity of EEG power above this frequency. However, neural potentials are known to exhibit much longer time scales, due partly to nonlinear membrane properties that could invalidate the estimates made in this paragraph. Another view is that interactions across neural hierarchies must be considered from higher levels of organization down as well as lower levels up. That is, delays observed at the local level may result from interactions at many higher levels, rather than local delays causing global delays, an issue considered further in the next section.

3 Experimental Support of the Theory

A large fraction of scalp EEG rhythms are spatially coherent over large regions of the cortex. (It has been shown that the amplitude of spatially incoherent cortical activity is often too small to be measured on the scalp, except in the case of some evoked potentials, where averaging is required.) In this case, the magnitude of the scalp EEG may be roughly proportional to the threshold parameter B. The range of amplitudes for nonepileptic EEG between the alert, nonalpha state and deep sleep is at least a factor of 10 or 20, which is perhaps a reasonable first guess for the range of B. For small B, the frequency of each mode l is approximately independent of B. However, as B approaches $2k_l/\lambda$, there is a sharp drop in the frequency of the k_l mode in the manner suggestive of the transition from the awake to the sleeping state. Furthermore, higher modes are reduced to a lesser extent by increasing B. This observation, together with arguments presented in a following section, suggest that nondelta activity observed during sleep may be composed of a number of these higher modes.

If the threshold parameter B is small, the frequency of the lowest expected mode may be estimated from Eq. 6. The longest "circumference" (in and out of fissures and sulci) of one hemisphere is $L \sim 100$ cm. With propagation velocities in the 6–9 m/s range (or doubled for fibers cutting across fissures), standing waves in a closed strip have discrete wavenumbers determined by continuity of both the potential and its derivatives,

$$k_l = \frac{2l\pi}{L}, \quad l = 1, \infty$$

$$k_1 \sim 0.06\,\mathrm{cm}^{-1}$$

$$f_1 = \frac{\omega_1}{2\pi} \sim 6\text{–}18\ \mathrm{Hz} \tag{9}$$

Thus, either of the two lowest modes appear to be possible candidates for the alpha rhythm, because of uncertainty in the parameter estimates. With the assumption of an effective length (twice the actual length) of 10–50 cm for the longest corticocortical fibers, Eqs. 6 and 7 yield damping estimates for the case where B approaches zero; that is,

$$\left|\frac{\gamma}{\omega_1}\right| \sim \frac{\lambda}{k_1} \sim 0.3\text{–}1.7$$

$$\left|\frac{\gamma}{\omega_2}\right| \sim \frac{\lambda}{k_2} \sim 0.16\text{–}0.8 \tag{10}$$

with damping somewhat reduced as B is increased. Because of uncertainty of homogeneity/isotropy of the long-range connections, detailed investigation of boundary conditions does not appear warranted at this stage, but some qualitative connections to EEG are evident. For example, lack of symmetry in a brain-like shape can be expected to cause more splitting of the modes of each l index. The splitting of the $l=1$ or $l=2$ modes may account for the double-peaked alpha rhythm observed in many subjects (Nunez 1981a). Even in subjects with a single sharp alpha peak of width ~1 Hz, the spatial distribution of frequency components just below the peak significantly differs from that just above the peak, suggesting the occurrence of multiple modes near 10 Hz in all subjects. The question of the paucity of higher modes in EEG data was addressed in the previous section.

An interesting similarity between physical wave phenomena and the pattern visual evoked potential concerns the latter's resonance-like response at particular spatial frequencies in the pattern, corresponding to particular spatial behavior of afferent input to the visual cortex (Teyler et al. 1978). In the macaque, this spatial frequency selectivity is observed in visual cortex but not in the lateral geniculate body, suggesting a cortical rather than thalamic resonance (DeValois et al. 1977). In humans, spatial frequency selectivity is observed at 200 ms latency but not at 100 ms (White et al. 1983), suggesting that the longer latency is required for global wave phenomena to be organized in the cortex.

The existence of a dispersion relation has been confirmed by means of frequency-wavenumber spectral analysis of the alpha rhythm (Nunez 1974b, 1981a). While these experiments were limited by poor spatial resolution, all subjects of the studies showed the expected increase in wavenumber Δk across the alpha band Δw. A very rough estimate of the group velocity ($d\omega/dk$) indicated an approximate range of 4–20 m/s, in qualitative agreement with the dispersion relation Eq. 6. The relative contributions of free and forced oscillations to this frequency-wavenumber shift are unknown. The recorded wavenumber spectra appeared roughly consistent with the lower modes $l=1$ or 2.

In an exhaustive study of phase and coherence of scalp EEG in 189 subjects, the average anterior/posterior phase shift of alpha rhythm translates to a phase velocity in the 7–16 m/s range (Thatcher et al. 1986). Long-range coherence estimates indicate that corticocortical connections are anisotropic; they are consistent with anterior/

posterior standing wave patterns. This study also showed a strong negative correlation between high average coherence and scores on IQ tests, a result consistent with the idea that cognitive events disrupt standing wave patterns. In a comprehensive series of studies of event-related potentials using sophisticated methods to obtain spatial patterns of correlation, quantitative evidence was obtained that supports the idea that cognitive events involve parallel activity in many cortical columns which are integrated in rapidly shifting patterns of focal activity (Gevins et al. 1985).

Whereas the frequency of each mode is determined by the parameters (λ, v, B), for which only rough estimates can be made, it is generally true that if other factors are held constant, larger systems should produce lower frequencies. This prediction was verified in a study of alpha frequency vs. head size in an adult sample of 123 having peaked alpha rhythm (Nunez et al. 1977); that is, there is a weak (significance, $p = 0.02$) negative correlation between head size and alpha frequency. This study excluded children, if only because myelination of axons during maturation can be expected to change velocity distributions.

Since mammalian cortices are quite similar, one might inquire whether the size-frequency relationship can be applied to nonhumans. This idea is complicated by the fact that the ratio of corticocortical to afferent fibers is much higher in man than in other mammals, as well as by other anatomical differences. The large number of corticocortical connections has been cited as a major factor in making the human brain "human" (Braitenberg 1978). Nevertheless, rough estimates of expected lower modes can be obtained from linear scale factors, which are the cube roots of the ratios of human to nonhuman brain volumes. For the dog, this ratio is $(1500/70)^{1/3} \sim 2.8$; for the cat, the ratio is $(1500/30)^{1/3} \sim 3.6$. The dog does indeed exhibit a cortical EEG rhythm which is peaked near 28 Hz (Lopes da Silva et al. 1970a). Also, driving the visual system of the dog with sine wave-modulated light produces a resonance-like peak in the same frequency range (Lopes da Silva et al. 1970b), which appears analogous to a similar effect in humans near 10 Hz (van der Tweel et al. 1965). Under halothane anesthesia, the dog may produce rhythms which are also roughly twice the frequency of the human halothane rhythm (Nunez 1981a). But, these theoretical/experimental comparisons are clouded by the existence of a 12-Hz rhythm in the dog, which is recorded in some parts of the cortex and the thalamus, is attenuated by opening eyes, and is considered by some to be the analog of the human alpha rhythm (Lopes da Silva et al. 1973). Possible explanations of the 12-Hz dog rhythm are that it is analogous to a lower mode than the human alpha, or is more dominated by local delays, or a combination of both effects.

The cat produces a cortical rhythm near 40 Hz which is believed to be generated by local delays in the olfactory bulb (Freeman 1975). It is interesting to speculate that this apparent matching of local and global delays may not be fortuitous; perhaps the local delays are "learned" as the result of the influence of global modes. Other animal experiments are also of interest to this theory. Lesions of midbrain structures in cats are shown to change the damping, but not the dominant frequencies of the EEG, thereby supporting the idea that the cortex is the medium responsible for resonance (Wright and Kydd 1984b). Also, lesions in the cortex of cats indicate that EEG is altered only when the white matter is also lesioned (Gloor et al. 1977), a result consistent with the theoretical idea that the long-range connections may dominate normal modes.

While the oversimplifed version of the global wave theory outlined here appears to account at least partly for spontaneous EEG rhythms and some aspects of event-related potentials, cognitive processing can be expected to disrupt standing waves and to complicate spatiotemporal patterns. Also, some EEG phenomena - for example, the 3–8 Hz hippocampal theta rhythm - appear to be dominated by local effects and do not fit into global theories. A more comprehensive theory is required that would encompass many hierarchical levels of neural interaction. With this goal in mind, a formal theory of statistical mechanics of neocortical interaction has been under development for the past several years (Ingber 1982). The theory suggests that EEG rhythms may owe their origins to multiple mechanisms at multiple scales of interaction between cortical columns containing a few hundred to a few million neurons. In this view, local and global theories of cortical dynamics are not mutually exclusive, even when applied to the same EEG phenomena. Standing waves occur in the cortex as a limiting case (Ingber 1985) and a quantitative mechanism for short-term memory is described (Ingber 1984). We may be approaching a time when aspects of conscious experience can be realisticly derived from neuronal firing patterns.

References

Başar E (1980) EEG brain dynamics: relations between EEG and evoked potentials. Elsevier/North-Holland, New York

Beurle RL (1956) Properties of a mass of cells capable of regenerating pulses. Trans R Soc Lond [Biol] 240:55–94

Braitenberg V (1978) Cortical architectonics: general and areal. In: Brazier MAB, Petsche H (eds) Architechtonics of the cerebral cortex. Raven, New York

DeValois RL, Albrecht DG, Thorell LG (1977) Spatial tuning of LGN and cortical cells in monkey visual system. In: Spekreijse H, van der Tweel LH (eds) Spatial contrast. North-Holland, Amsterdam, pp 60–63

Freeman WJ (1975) Mass action in the nervous system. Academic, New York

Gevins AS, Doyle JC, Cutillo BA, Schaffer RE, Tannehill RS, Bressler SL (1985) Neurocognitive pattern analysis of a visuospatial task: rapidly-shifting foci of evoked correlations between electrodes. Psychophysiol 22:32–43

Gloor P, Ball G, Shaul N (1977) Brain lesions that produce delta waves in the EEG. Neurology (Minneap) 27:326–333

Griffith JS (1963) On the stability of brain-like structures. Biophys J 3:299–308

Hubel DH, Wiesel TN (1962) Receptive fields, binocular interaction, and functional architecture in the cat's visual cortex. J Physiol (Lond) 160:106–154

Ingber L (1982) Statistical mechanics of neocortical interactions. Basic formulation. Physica 5D:83–107

Ingber L (1984) Statistical mechanics of neocortical interactions. Derivation of short term memory capacity. Physiol Rev A29:3346–3358

Ingber L (1985) Statistical mechanics of neocortical interactions. EEG dispersion relations. IEEE Trans Biomed Eng 32:91–94

Katznelson RD (1981) Normal modes of the brain: neuroanatomical basis and a physiological theoretical model. In: Nunez PL (1981a) Electric fields of the brain: the neurophysics of EEG. Oxford University Press, New York

Lopes da Silva FH, van Rotterdam A, Storm van Leeuwen W, Tielew AM (1970a) Dynamic characteristics of visual evoked potentials in the dog. I. Cortical and subcortical potentials evoked by sine wave modulated light. Electroencephalogr Clin Neurophysiol 29:246–259

Lopes da Silva FH, van Rotterdam A, Storm van Leeuwen W, Tielew AM (1970b) Dynamic characteristics of visual evoked potentials in the dog. II. Beta frequency selectivity in evoked potentials and background activity. Electroencephalogr Clin Neurophysiol 29: 260–266
Lopes da Silva FH, van Lierop THMT, Schrijer CF, Storm van Leeuwen W (1973) Organization of thalamic and cortical alpha rhythms: spectra and coherences. Electroencephalogr Clin Neurophysiol 35: 627, 639
Lopes da Silva FH, Hoeks H, Smits H, Zetterberg LH (1974) Model of brain rhythmic activity. Kybernetik 15: 27–37
Nunez PL (1974a) The brain wave equation: a model for the EEG. Math Biosci 21: 279–297
Nunez PL (1974b) Wave-like properties of the alpha rhythm. IEEE Trans Biomed Eng 21: 473–482
Nunez PL (1981a) Electric fields of the brain: the neurophysics of EEG. Oxford University Press, New York
Nunez PL (1981b) A study of origins of the time dependencies of scalp EEG. I. Theoretical basis. IEEE Trans Biomed Eng 28: 271–280
Nunez PL (1981c) A study of the origins of the time dependencies of scalp EEG. II. Experimental support of theory. IEEE Trans Biomed Eng 28: 281–288
Nunez PL (1985) The evidence for standing waves in the brain. 14th International Conference on Medical and Biological Engeneering, 7th International Conference on Medical Physics, Espoo
Nunez PL, Reid L, Bickford RG (1977) The relationship of head size to alpha frequency with implications to a brain wave model. Electroencephalogr Clin Neurophysiol 44: 344–352
Szentagothai J (1978) The neural network of the cerebral cortex: a functional interpretation. Proc R Soc Lond 201: 219–248
Teyler CW, Apkarian P, Nakayama K (1978) Multiple spatial frequency tuning of electrical responses from human visual cortex. Exp Brain Res 33: 535–550
Thatcher RW, Krause PJ, Hrybyk M (1986) Corticocortical associations and EEG coherence: a two compartmental model. Electroencephalogr Clin Neurophysiol 64: 123–143
Van der Tweel LH, Verduyn Lunel HFE (1965) Human visual responses to sinusoidally modulated light. Electroencephalogr Clin Neurophysiol 18: 587–598
Van Rotterdam A, Lopes da Silva FH, van der Ende J, Viergever MA, Hermans AJ (1982) A model of the spatial-temporal characteristics of the alpha rhythm. Bull Math Biol 44: 283–305
White CT, White CL, Hintze RW (1983) Cortical and subcortical components of the pattern VEP. Int J Neurosci 19: 125–131
Wilson HR, Cowan JD (1972) Excitatory and inhibitory interactions in localized populations of model neurons. Biophysics 12: 1–24
Wright JJ, Kydd RR (1984a) A linear theory for global electrocortical activity and its control by the lateral hypothalamus. Biol Cybern 50: 75–82
Wright JJ, Kydd RR (1984b) A test for constant natural frequencies in electrocortical activity under lateral hypothalamic control. Biol Cybern 50: 83–88
Zhadin MN (1984) Rhythmic processes in the cerebral cortex. J Theor Biol 108: 565–595

Gross Electrocortical Activity as a Linear Wave Phenomenon, with Variable Temporal Damping

J. J. Wright, R. R. Kydd, and G. J. Lees

1 Introduction

Electrocortical waves may be defined as potentials arising at a recording site, generated by current flows in a surrounding conductive medium. The currents are themselves generated by distributed electromotive forces on the brain surface. The distributed sources are slow potentials in dendrites, now known as "neuronal waves". The recorded signal represents coherent fields of dendritic activity, extracted against a background of incoherent activity (Elul 1972). Since these fields of coherence appear to be closely correlated with cognitive states, it is necessary to find a physically meaningful mathematical model of this activity, as a step towards explaining the overall processing of information in the brain.

Başar (1976) has described mathematical models as falling into "black box", "white box", and "grey box" categories. The model of electrocortical activity described here is a "grey box", which aims to bypass some of the following problems:

1. "White box" models of EEG activity have to isolate simplified properties of neuronal behaviour and coupling a priori (Beurle 1956; Karawahara 1980; Lopes da Silva et al. 1974; Pringle 1951; Nakagawa and Ohashi 1980; Wilson and Cowan 1972, 1973; Zhadin 1974). Since real neurones are very complicated, there is difficulty in isolating the most significant properties for modeling. Often, critical tests are hard to formulate and apply.
2. "Black box" analysis of the EEG is equally difficult. Real EEG waves are nonstationary and complicated (Walter and Brazier 1968; Walter and Adey 1968). Many different types of equation can mathematically characterise the wave forms over a given epoch, particularly if some of the parameters of description are allowed to be time varying, but this does not determine the *class* of wave phenomenon involved. There are also incompletely defined transforming processes intervening between the surface dendritic voltages and the recorded signal (Elul 1972; Nunez 1981; Walter and Adey 1968).

Because of these difficulties, we have formulated a model of electrocortical activity which made *minimal* physiological assumptions, and allowed maximum freedom for complicated and nonlinear interactions between neurones. However, the model is sufficiently constrained that a definite class of wave phenomena are predicted. Thus we hoped to establish a physically correct parameterisation of EEG waves, and also a framework within which detailed cellular models could then be developed. In this line of reasoning we were following arguments similar to those advanced by Freeman (1975), Başar (1980) and Nunez (1981).

Springer Series in Brain Dynamics 1
Edited by Erol Başar

The essence of our conclusions is that EEG waves are *linear waves*; that is, waves which obey a superposition principle, meaning that separate waves pass through each other by adding together at each point. As a consequence, in conditions where the wave medium is bounded, or closed upon itself, such waves can form "standing waves" from their bidirectional passage. At frequencies at which this occurs, there is said to be a resonant mode of the system. Millions of such modes must be possible in the brain, but their frequencies are limited to a few bands. Further, control of these wave patterns depends both on the strength of excitation of the system, and the rates at which the excited patterns die away. These propositions are discussed below.

2 Outline of the Model

The conceptions upon which the model is constructed are diagrammed in Fig. 1. Details are given in Wright and Kydd (1984a) and Wright et al. (1985c).

2.1 A Generalized Description of Cortical Dendritic Activity Considering All Cortical and Subcortical Interactions

We begin by considering the dendrites of the cerebral cortex as a mass of voltage (or current) sources of arbitrarily small size. The point voltage of each of these elements varies over time about a mean value, and can thus be fitted to the general, inhomogeneous differential equation,

$$\ddot{x}(t) + D(t)\dot{x}(t) + N^2(t)x(t) = \text{input signals},$$

where $x(t)$ is the membrane voltage at time t, and $D(t)$ and $N(t)$ are first considered as free parameters.

Next, notice that neuronal elements ordered into many classes of closed loops exist in this system, within cortical columns, via cortico-cortical reciprocal connections, cortico-subcortical pathways, etc. These nonlinear loops are cross-coupled to a massive extent. Thus, without loss of generality, we can represent these as quasi-additively coupled, as shown by the set of equations

$$\ddot{x}_1 + D_1(t)\dot{x}_1 + N_1^2(t)x_1 = K_2^1(t)x_2 + \cdots + K_n^1(t)x_n \quad \text{to}$$

$$\ddot{x}_n + D_n(t)\dot{x}_n + N_n^2(t)x_n = K_1^n(t)x_1 + \cdots + K_{n-1}^n(t)x_{n-1}$$

where $K_j^i(t)$ are again free parameters representing a wealth of nonlinear interactions.

2.2 Restrictive Assumptions

The parameters of the above generalised description can be constrained in the following way, by imposing physiological interpretations on the parameters:

1. Each $N_i(t)$ principally represents system perturbation about a dominant cycle time for a specific circuit, created by closed cycles of activity in pathways of nearly fixed conduction time, but subject to outside interference and nonlinearities

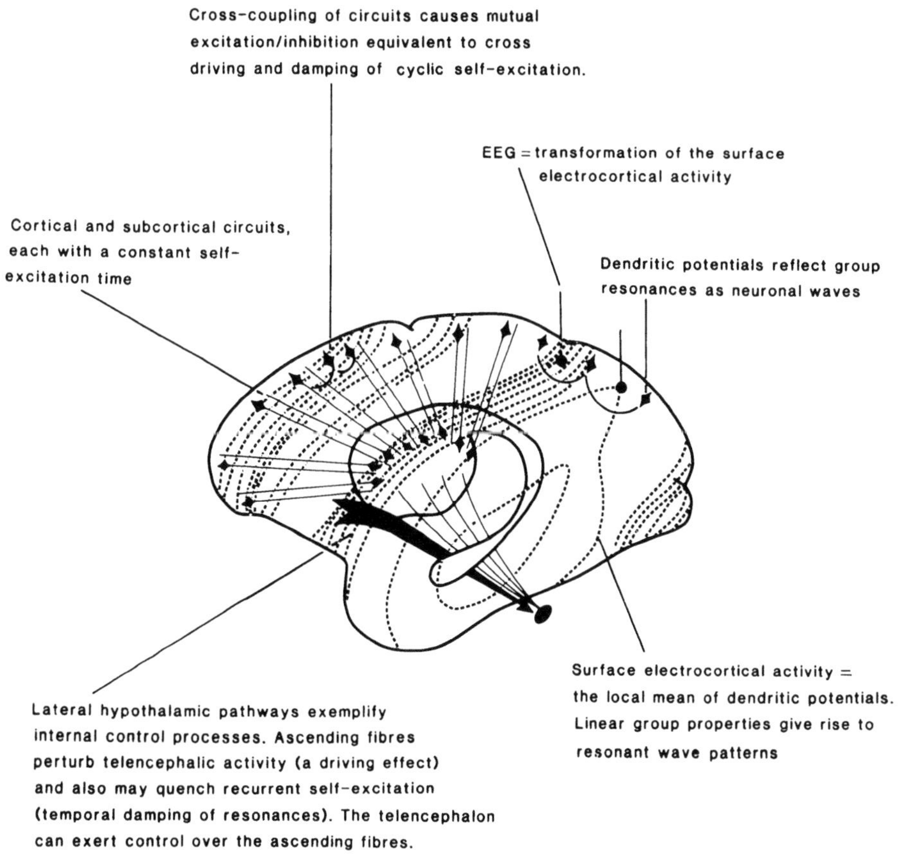

Fig. 1. A schematic representation of the brain as a "grey box". (After Wright et al. 1985a)

2. Each $K^i_j(t)$ represents a coupling imposed by fixed structural features, perturbed by complex nonlinearities and interactions
3. The parameters $D_i(t)$ can reflect multiplicative interactions between the coupled loops in accord with the following general relationship,

$$D_1(t) = F^1_j(t)\,x_1 + \cdots + F^1_n(t)\,x_n \quad \text{to}$$
$$D_n(t) = F^n_1(t)\,x_1 + \cdots + F^n_n(t)\,x_n\,.$$

   The free parameters $F^i_j(t)$ used to represent this multiplicative interaction are subject to the same type of considerations as $N_i(t)$ and $K^i_j(t)$.
4. Thus, each $N_i(t)$, $K^i_j(t)$, $F^i_j(t)$ can be assumed to have a time-invariant mean $\overline{N_j}$, $\overline{K^i_j}$ or $\overline{F^i_j}$, with unspecified variances. All $N_i(t)$, $K^i_j(t)$, $F^i_j(t)$ may also be assumed to be *stochastically independent in the large,* because of their complexity, and also because they are largely influenced by *local* interactions. However, the anatomical orderliness of the brain, with its many classes of similar repeating circuits and couplings, implies that all $\overline{N_i}$, $\overline{K^i_j}$, and $\overline{F^i_j}$ must be clustered about certain further center values $\overline{\overline{N_i}}$, $\overline{\overline{K^i_j}}$, and $\overline{\overline{F^i_j}}$.

2.3 Mathematical Development and Consequences

2.3.1 System Properties

The above coupled second-order inhomogeneous differential equations can be transferred to first-order state variables, z_i, $i=1 \ldots 2n$. The system can then be represented *at any instant* by the matrix relations,

$$\dot{\mathfrak{z}} = \mathcal{A}(t)\,\mathfrak{z}$$

and

$$\mathfrak{D} = \mathcal{B}(t)\,\mathrm{x},$$

where $\mathcal{A}$, $\mathcal{B}$ are matrices representing the instantaneous couplings and natural frequencies of the elements, $\mathfrak{z}$ is the state-variable eigenvector, $\mathfrak{D}$ the vector of instantaneous damping coefficients, and x the vector representing position in phase space.

In steady state conditions, diagonalisation of $\mathcal{A}(t)$, using the central limit theorem, reveals that $\mathcal{A}$ and $\mathcal{B}$ are effectively invariant with time, and identical to a similar linear system of many cross-coupled oscillators that form "families" of repeating subsystems. Thus:

1. The system is a linear wave medium. Since the system has closed boundary conditions, each resonant frequency is associated with an invariant standing-wave pattern.
2. The frequencies of the resonant modes are clustered about some smaller number of central values.
3. A multitude of steady states are possible for systems in this class, which means that electrocortical waves are effectively waves in a linear system with time-varying damping for each of many invariant resonant modes.

Transient nonlinear effects are to be expected during some or all of the transitions between the wealth of possible steady states, particularly in circumstances in which the instantaneous coupling parameters transiently become strongly stochastically dependent.

2.3.2 Input and Output Relations

It can be further shown that input signals and active cell firing mean that the system is *driven by white noise*, so that, overall, a process of balance is being struck between driving signals and dissipative effects attributable to the temporal damping of the resonances.

The recorded output signal is equated with some linear transform of the coherent component of the overall electrocortical activity. Thus input and output relations are consistent with the findings of Elul (1972), and the power spectrum of the recorded signal, $V^2(\omega)$, is given by

$$V^2(\omega) = |A(\omega)|^2 \cdot |\Sigma x_i(\omega)|^2,$$

where $|A(\omega)|$ is the transformation mediating between the surface signal and the recorded output.

2.4 Role of a Selected Pathway Within the System – the Lateral Hypothalamus

To develop critical tests for the theory, we considered the effects of a unilateral lesion in some critical pathway upon the symmetry of left and right electrocortical signals. For this purpose, the lateral hypothalamus was of interest. This pathway contains fibre systems that include ascending inhibitory neurones, particularly the dopaminergic and noradrenergic groups, running from brainstem to telencephalic sites (Graybiel and Ragsdale 1979; Lindvall and Bjorklund 1974). Reciprocal fibre systems exert descending controls from telencephalon to brainstem, suggesting that the lateral hypothalamus is an important mediator of telencephalon-brainstem interactions. Lesions or stimulations of the pathway and of the specific ascending fibres exert profound effects upon motivation and attention (as evidenced by the phenomena of sensorimotor neglect and intracranial self-stimulation) (Ljungberg and Ungerstedt 1976; Marshall et al. 1971; Olds and Forbes 1981; Wise 1982), supporting the assumption from structure.

Within the terms of the theory it is to be expected that damage to ascending lateral hypothalamic fibres should exert two main effects:

1. The release of telencephalic resonant modes from a damping input
2. A change in the strength of active and noise-like driving signals within the telencephalon

This follows from the essentially closed and prolific connections within the telencephalon. Thus, most terms $N_i(t)$, $K_j^i(t)$ in the $\mathscr{A}$ matrix are telencephalic parameters, and these principally determine the resonant modes. But the wide diffusion of brainstem pathways to the telencephalon implies that many terms in the $\mathscr{B}$ matrix reflect these diffuse inputs. Similarly, such diffuse inputs would be expected to perturb the levels of depolarisation of many cells in the telencephalon, thus leading to increased rates of firing and to increased strength of driving noise. (Put another way, *impulses* in the pathway perturb the higher system; sustained input quenches afterdischarge.)

3 Experimental Tests of Theory

Tests were evolved to avoid many difficulties implicit in interpreting standard systems analysis procedures. Our basic experimental method was to record epochs of electrocortical activity from symmetrical left and right sites in experimental rats, and to make these recordings before and 2–4 days after unilateral lesion of ascending lateral hypothalamic fibres. This permitted several different procedures to be applied, all aimed at testing the internal consistency of the theory when theoretical equations fitted to certain aspects of the experimental data were used to predict other, independent aspects of the experimental findings.

Using selective methods, these lesions were *either* of the dopaminergic mesostriatal and mesocortical systems, *or* the noradrenergic neurones of the locus coeruleus, *or* gross electrolytic lesions of the entire lateral hypothalamus. Controls with no specific damage were also studied. Recordings were obtained in quasi-steady-state conditions of rest.

3.1 Values Obtained Experimentally

Fourier analysis of left and right channels, before and after lesion, yielded the sets of Fourier components $a(\omega)$, $b(\omega)$ for each channel/epoch. These yield the power spectrum

$$V^2(\omega) = a^2(\omega) + b^2(\omega)$$

and phase

$$\phi(\omega) = \tan^{-1}\left[\frac{b(\omega)}{a(\omega)}\right].$$

Thus, *experimentally* we could obtain V^2_{LA}, V^2_{LB}, V^2_{CA}, and V^2_{CB}, the power spectra of each channel/epoch. The subscripts LA, LB, CA and CB indicate lesion and control spectra, before and after lesion. From these we could directly calculate the *ratio change in power attributable to the lesion*, as

$$G^2(\omega) = \frac{V^2_{\mathrm{LA}}}{V^2_{\mathrm{LB}}} \Big/ \frac{V^2_{\mathrm{CA}}}{V^2_{\mathrm{CB}}}(\omega),$$

and also the *relative shift in phase attributable to the lesion* is given by

$$\Delta\phi(\omega) = \left\{\tan^{-1}\left[\frac{b(\omega)}{a(\omega)}\right]_{\mathrm{LA}} - \tan^{-1}\left[\frac{b(\omega)}{a(\omega)}\right]_{\mathrm{CA}}\right\} - \left\{\tan^{-1}\left[\frac{b(\omega)}{a(\omega)}\right]_{\mathrm{LB}} - \tan^{-1}\left[\frac{b(\omega)}{a(\omega)}\right]_{\mathrm{CB}}\right\}$$

A priori, there is no necessary relation between power and phase changes.

3.2 Theoretical Relations Expected

It can be shown within the model that in idealised conditions of true steady state, with complete symmetry between left and right electrocortical signals before lesions, then

$$G^2(\omega) = K\,\frac{|A(\omega)|^2_{\mathrm{LA}}\left[\sum^{n}\frac{M_i^2-\omega^2}{(M_i^2-\omega^2)^2+\mathscr{D}^2_{i\mathrm{LA}}\omega^2}\right]^2+\left[\sum^{n}\frac{-\mathscr{D}_{i\mathrm{LA}}\,\omega}{(M_i^2-\omega^2)^2+\mathscr{D}^2_{i\mathrm{LA}}\omega^2}\right]^2}{|A(\omega)|^2_{\mathrm{CA}}\left[\sum^{n}\frac{M_i^2-\omega^2}{(M_i^2-\omega^2)^2+\mathscr{D}^2_{i\mathrm{CA}}\omega^2}\right]^2+\left[\sum^{n}\frac{-\mathscr{D}_{i\mathrm{CA}}\,\omega}{(M_i^2-\omega^2)^2+\mathscr{D}^2_{i\mathrm{CA}}\omega^2}\right]^2}$$

where K represents the lesion/control ratio of power in noise-like driving signals, M_i are natural frequencies of the resonant modes, and $\mathscr{D}_{i\mathrm{LA}}$, $\mathscr{D}_{i\mathrm{CA}}$ are the appropriate damping coefficients of the modes on each side, following lesion. Notice that $|A(\omega)|^2$ cancels with the assumed equality of $|A(\omega)|_{\mathrm{LA}}$, $|A(\omega)|_{\mathrm{CA}}$.

We also know that for any epoch and channel

$$|A(\omega)|^2 = V^2(\omega)/|x_0(\omega)|^2 \times \left\{\left[\sum^{n}\frac{M_i^2-\omega^2}{(M_i^2-\omega^2)^2+\mathscr{D}_i^2\omega^2}\right]^2+\left[\sum^{n}\frac{-\mathscr{D}_i\,\omega}{(M_i^2-\omega^2)^2+\mathscr{D}_i^2\omega^2}\right]^2\right\}$$

where $|x_0(\omega)|$ is a scaling factor.

Thus $|A(\omega)|$ can be obtained from $V^2(\omega)$, the direct power spectra, and parameters obtained by curve fitting the *ratio* change in power.

Likewise

$$\Delta\phi(\omega) = \tan^{-1}\left|\frac{\sum^{n} \frac{-\mathscr{D}_{i\,\mathrm{LA}}\,\omega}{(M_i^2-\omega^2)^2+\mathscr{D}_{i\,\mathrm{LA}}^2\,\omega^2}}{\sum^{n} \frac{M_i^2-\omega^2}{(M_i^2-\omega^2)^2+\mathscr{D}_{i\,\mathrm{LA}}^2\,\omega^2}}\right| - \tan^{-1}\left|\frac{\sum^{n} \frac{-\mathscr{D}_{i\,\mathrm{CA}}\,\omega}{(M_i^2-\omega^2)^2+\mathscr{D}_{i\,\mathrm{CA}}^2\,\omega^2}}{\sum^{n} \frac{M_i^2-\omega^2}{(M_i^2-\omega^2)+\mathscr{D}_{i\,\mathrm{CA}}^2\,\omega^2}}\right|$$

which is composed of parameters obtained from $G^2(\omega)$, the ratio change in power.

4 Specific Predictions and Tests

From the above equations, critical tests are possible by curve fitting *ratio change in power*, obtaining approximate parameters, and thus predicting other experimentally determinable relations. This must be done using an approximate low-order model. For convenience, we chose a model order describing the five principle resonant modes. These are the predictions to be tested:

1. Curve fitting the *ratio power change* from different experimental animals will yield values for the principle resonant mode frequencies, clustered about the frequencies of the major cerebral rhythms. There is no a priori reason why the theoretical function would find these values in ratio power changes.
2. With animals showing progressive degrees of lesion of a critical ascending system (e.g. mesotelencephalic dopaminergic fibres), curve fitting will reveal a progressive *decrease* in damping associated with a fall in the power of driving signals, on the side of the lesion (compared to the control hemisphere).
3. The parameters obtained from curve fitting $G^2(\omega)$ should predict the values of $\Delta\phi(\omega)$ with a significant correlation.
4. Values of $|A(\omega)|_{\mathrm{LA}}$ and $|A(\omega)|_{\mathrm{CA}}$ calculated from the *absolute* power spectra, and fitted parameters from the *relative* power, should be essentially equal.
5. Similar results in all of these respects should be obtained with all three classes of lesion (dopaminergic, noradrenergic and nonspecific).

5 Results

Detailed results have been presented elsewhere (Wright and Kydd 1984a, b, c; Wright et al. 1984, 1985 a, b, c). Here, we have summarised results relevant to the predictions made.

5.1 Values of Natural Frequencies Found by Curve Fitting the Ratio Power Change Attributable to Unilateral Lesion

Table 1 gives the centre values and distribution of the natural frequencies found in this way, from all three classes of unilateral lesion. Results are significantly clustered

Table 1. Summary of the group median natural frequencies and the mean absolute deviations about the cluster centres, under three conditions of unilateral lesion. Kolmogorov-Smirnov two-sample tests reveal no significant differences between distributions, although each distribution is nonrandom

Class of unilateral lesion used to induce left/right EEG asymmetry	Cluster medians for natural frequencies estimated from ratio changes in power spectra	Corresponding mean absolute deviations about each cluster median
Noradrenergic	2.81,7.67, 12.06, 16.31, 25.09	0.78, 0.77, 1.07, 0.73, 2.43
Dopaminergic	2.33, 7.1, 10.77, 19.65, 25.01	0.85, 0.55, 1.96, 1.16, 1.08
Lateral hypothalamic electrolytic	3.89, 7.03, 10.73, 18.6, 24.57	0.75, 0.69, 0.91, 1.11, 1.22

Table 2. Changes in parameters of relative power obtained from animals with varying degree of unilateral damage

Class of unilateral lesion	Ratio power of driving signals, lesion/control	Ratio of mean damping coefficients, lesion/control
Severe mesotelencephalic dopaminergic damage ($n = 5$, median damage 27.5/33)	0.33	0.46
Moderate mesotelencephalic dopaminergic damage ($n = 3$, median damage 22.5/33)	0.41	0.51
Mild mesotelencephalic dopaminergic damage ($n = 5$, median damage 19/33)	0.63	≈ 1
No dopaminergic damage ($n = 4$, needle passage damage only)	0.97	1.5

about frequencies close to those of the great cerebral rhythms ($p < 0.01$, to $p < 0.05$), and the type of lesion has no significant effect on the results.

5.2 Graded Changes in Relative Left and Right Damping, and Strength of Noise-Like Driving Signals, with Increasing Degree of Unilateral Lesion

Table 2 lists the results which show the effects predicted. The association between damage, asymmetry of left and right damping, and decrease in K (the left/right ratio of driving power) is highly significant ($p < 0.002$).

5.3 Correlation of Relative Phase Shift Attributable to Lesion, to That Predicted from Relative Amplitude

This test has been performed for animal groups with severe and mild unilateral dopaminergic lesion, and for unilateral locus coeruleus lesion. Prediction is met at a high level of significance in each instance, but it is clear that a very rough fit of phase is obtained. An analysis of the errors shows that they are of the types expected from limited model order, imperfect maintainance of steady state, and limited left–right hemisphere coherence. One case is shown in Fig. 2.

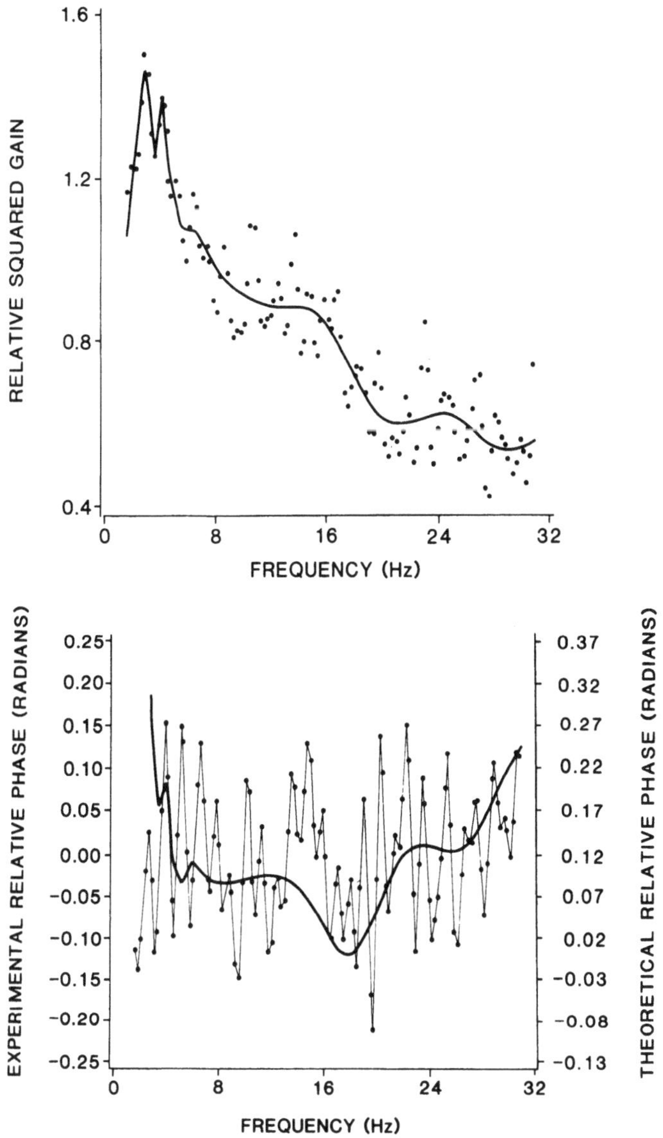

Fig. 2. *Top*, relative amplitude changes following mild dopaminergic lesion to unilateral substantia nigra pars compacta. Data *(dots)* fitted by theoretical equation. *Bottom*, superimposed plots of experimentally determined shift in phase attributable to lesion *(connected dots)* and that predicted *(line)* from the amplitude changes. Several classes of approximation error explain the displaced baseline and inverse correlation at low frequences of the spectrum. Data averaged from three animals

5.4 Equality of Retrospective Calculations of Left and Right Electrode Transfer Characteristics

Left and right electrode transfer characteristics for a single lesion class are shown in Fig. 3. Autocorrelated errors (also equal in degree left and right) have been introduced by limitations of model order. The fitted straight line is introduced to give an

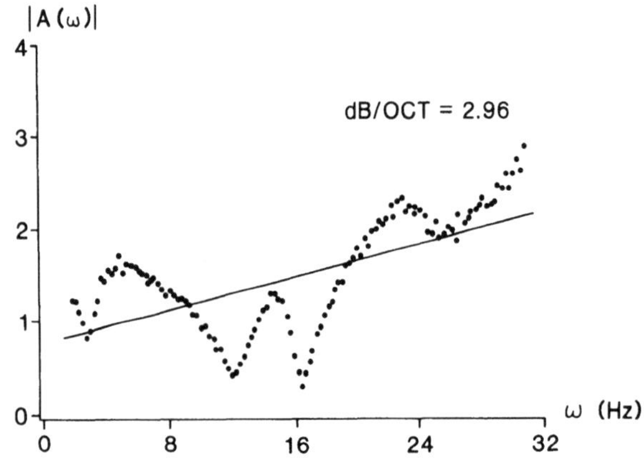

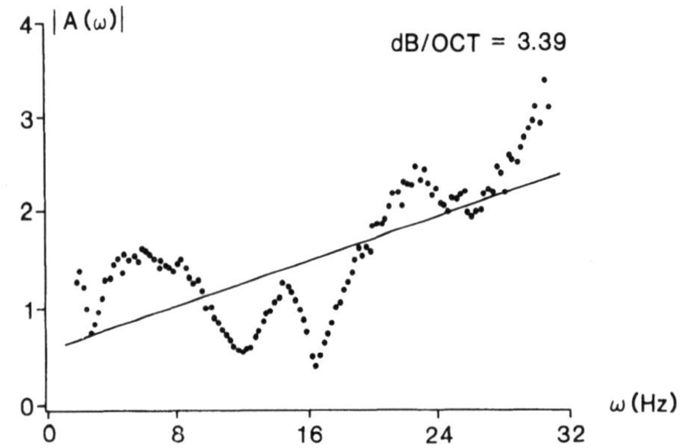

Fig. 3. Lesion *(top)* and non-lesion *(bottom)* electrode transfer characteristics, back-calculated from parameters of relative power and raw absolute power spectra. Findings are from animals with unilateral lesions of the dorsal noradrenergic bundle

Table 3. Linear trends in estimated electrode transfer characteristics

Type of unilateral lesion	Linear trend on side of lesion (dB gain per octave)	Linear trend on side opposite lesion (dB gain per octave)	Difference in gain per octave (%)
Gross electrolytic lesion, lateral hypothalamus ($n = 8$)	0.98	1.12	2
Severe mesotelencephalic dopaminergic damage ($n = 5$)	1.03	1.21	2
Moderate mesotelencephalic dopaminergic damage ($n = 3$)	2.93	4.06	12
Mild mesotelencephalic dopaminergic damage ($n = 5$)	1.26	1.63	4
Dorsal noradrenergic bundle damage ($n = 7$)	2.96	3.39	5

estimate of mean linear trend, in db/octave. Table 3 shows this mean linear trend over a variety of other cases, which are shown to all be roughly comparable, and repeatedly equal left and right.

5.5 Similarity of Results for All Classes of Lesion Affecting Unilateral Ascending Fibres

The above results show this equality in all important features.

6 Conclusion

The results of all the tests support the hypothesis, despite the unavoidable errors and approximations involved in the estimates. The agreement of findings using a variety of fibre systems, each projecting to different telencephalic sites, gives reason to believe the model is not unique for a specific control system.

While the experimental limitations do not permit a conclusion of absolute wave linearity, they show that the system is approximately linear. The theory suggests that wave actions will be linear for all stationary periods of EEG, with nonlinear transitions between steady states; and thus, positive results under the extreme limitations and simplifications needed to apply the tests argue for a fairly considerable inherent linearity in steady states. Given this approximate physical description of the nature of electrocortical waves, the following features are worthy of further consideration:

1. During periods of comparative stationarity, multichannel recordings (from as many cortical sites as possible) could be analysed by linear techniques, to obtain dispersion relations, and hence wave equations of the telencephalic surface. Suitable experimental methods appear to include those of Lehmann (1971, 1984). These findings would greatly constrain the possible forms of integro-differential equations used in "white box" modeling (see Nunez 1981).
2. A format for "black box" modeling is also implicit. Using techniques arising from autoregression analysis, different steady states of whole brain can be characterised and represented in a phase space which is parameterised by dimensions equivalent to the damping coefficients of the resonant modes (see the matrix equations outlined earlier). A beginning in analysis along these lines has been made by Franaszczuk and Blinowska (1986), who have shown that autoregressive analysis applied to EEG is consistent with a variable-temporal-damping, constant-resonant-mode interpretation. Each steady state may be equated to an attractor basin within the phase space (see the chapters by Babloyantz and by Başar and Röschke in this volume).
3. Transitions between steady states may be viewed as a Markov process described by movement from one attractor state to another. Such a process may have the capacity to control information processing at the cellular level over widespread areas, and also to organise sequential stepping and looping of brain state transitions, akin to a general computational programme (Wright et al. 1985c).
4. The physical interpretation of evoked potentials (at least of the late components) is clear for such a system: evoked potentials represent impulse functions of a

linear system, with the proviso that the "impulse" is poorly defined, and the damping coefficients of the modes may be time-varying over the epoch of recording. Thus the present "grey box" model does appear to offer a link between "black box" and "white box" methods.

7 Summary

A theory of electrocortical wave activity has been described and critical tests for the theory outlined. The results support the theory's validity as a first approximation.

The theory is provisional, in that detailed specifications of cell-to-cell couplings are not given. Instead, a general mathematical treatment for masses of linked circuits of neurone-like elements is developed. Diffuse fibre projections from brainstem to telencephalon are included.

Assumptions constraining this general description arise from consideration of the circuit conduction times, the extreme complexity of cross-couplings, and the anatomical orderliness of the system. These properties imply several important consequences for gross electrocortical waves:

1. The waves are physically equivalent to standing wave patterns on a closed surface. The millions of possible standing waves occur at frequencies clustered about certain centre values.
2. Ascending brainstem pathways (in particular, catecholamine pathways) exert effects combining noise-like driving of the resonant wave patterns and modulation of the degree of damping of each standing wave.

These findings may permit analysis of global brain function, using techniques of linear systems theory, to determine basic system properties, including wave equations, and the distribution of attractors within phase space.

References

Adey WR (1974) On line analysis and pattern – recognition techniques for the electroencephalogram. International symposium on signal analysis and pattern recognition in biomedical engineering

Başar E (1976) Biophysical and physiological systems analysis. Addison-Wesley, Reading

Başar E (1980) EEG-Brain dynamics. Relation between EEG and brain evoked potentials. Elsevier North-Holland, Amsterdam

Beurle RL (1956) Properties of a mass of cells capable of regenerating pulses. Trans Roy Soc Lond 240:55–94

Elul R (1972) The genesis of the EEG. Int Rev Neurobiol 15:227–272

Franaszczuk PJ, Blinowska KJ (1985) Linear model of brain electrical activity – EEG as a superposition of damped oscillatory modes. Biol Cybern 53:19–25

Freeman WJ (1975) Mass action in the nervous system. Academic, New York

Graybiel AM, Ragsdale CW Jr (1979) Fiber connections of the basal ganglia. Prog Brain Res 51: 239–301

Karawahara T (1980) Coupled van der Pol oscillators – model of excitatory and inhibitory neural interactions. Biol Cybern 39:37–43

Lehmann D (1971) Multichannel topography of human EEG fields. Electroencephalogr Clin Neurophysiol 31:439–449
Lehmann D (1984) EEG assessment of brain activity: spatial aspects segmentation and imaging. Psychophysiol 1:267–276
Lindvall O, Bjorklund A (1974) The organisation of the ascending catecholamine neurone systems in the rat brain. Acta Physiol Scand [Suppl 412]
Ljungberg T, Ungerstedt U (1976) Sensory inattention produced by 6-hydroxydopamine-induced degeneration of ascending dopaminergic neurones in the brain. Exp Neurol 53:585–600
Lopes da Silva FH, Hoeks A, Smits H, Zetterberg LH (1974) Model of brain rhythmic activity. The alpha rhythm of the thalamus. Kybernetik 15:27–37
Marshall JF, Turner BM, Teitlebaum P (1971) Sensory neglect produced by lateral hypothalamic damage. Science 174:523–525
Nakagawa T, Ohashi A (1980) A spatio-temporal fitter approach to synchronous brain activities. Biol Cybern 36:33–39
Nunez PL (1981) Electric fields of the brain: the neurophysics of EEG. Oxford University Press, New York
Olds ME, Forbes JL (1981) The central basis of motivation: Intracranial self-stimulation studies. Ann Rev Psychol 32:523–574
Pringle JWS (1951) On the parallel between learning and evolution. Behavior 3:174–215
Walter DO, Adey WR (1968) Is the brain linear? In: Iberall A, Reswick JB (eds) Technical and biological problems of control. A cybernetic view. Proceedings of the International Federation of Automatic Control, Vervan
Walter DO, Brazier MAB (1968) Advances in EEG analysis. Electroencephalogr Clin Neurophysiol [Suppl 27]
Wilson HR, Cowan JD (1972) Excitatory and inhibitory interaction in localized populations of model neurons. Biophys J 21:1–24
Wilson HR, Cowan JD (1973) A mathematical theory of the functional dynamics of cortical and thalamic nervous tissue. Kybernetik 13:55–80
Wise RA (1982) Neuroleptics and operant behaviour: The anhedonia hypothesis. Behav Brain Sci 5:39–87
Wright JJ, Kydd RR (1984a) A linear theory for global electrocortical activity, and its control by the lateral hypothalamus. Biol Cybern 50:75–82
Wright JJ, Kydd RR (1984b) A test for constant natural frequencies in electrocortical activity under lateral hypothalamic control. Biol Cybern 50:83–88
Wright JJ, Kydd RR (1984c) Inference of a stable dispersion relation for electrocortical activity controlled by the lateral hypothalamus. Biol Cybern 50:89–94
Wright JJ, Kydd RR, Lees GJ (1984) Amplitude and phase relations of electrocortical waves regulated by transhypothalamic dopaminergic neurones: A test for a linear theory. Biol Cybern 50: 273–283
Wright JJ, Kydd RR, Lees GJ (1985a) Quantitation of a mass action of dopaminergic neurones regulating temporal damping of linear electrocortical waves. Biol Cybern 52:281–290
Wright JJ, Kydd RR, Lees GJ (1985b) Contributions of noradrenergic neurones of the locus coeruleus to the temporal damping of linear electrocortical waves. Biol Cybern 52:351–356
Wright JJ, Kydd RR, Lees GJ (1985c) State-changes in the brain viewed as linear steady states and non-linear transitions between steady states. Biol Cybern 53:11–17
Zhadin MN (1984) Rhythmic processes in the cerebral cortex. J Theor Biol 108:565–595

Chaotic Dynamics in Brain Activity

A. Babloyantz

1 Introduction

The aim of this paper is to report on a new attempt at characterizing the electroencephalogram (EEG), which is based on recent progress in the theory of nonlinear dynamical systems (Brandstäter et al. 1983; Nicolis and Nicolis 1984, 1986; Babloyantz et al. 1985). The method is independent of any modeling of brain activity. It relies solely on the analysis of data obtained from a single-variable time series. From such a "one-dimensional" view of the system, one reconstructs the $\{X_k\}$ (where $k = 1, \ldots, n$) variables necessary for the description of systems dynamics. With the help of these variables, phase-space trajectories are drawn. Provided that the dynamics of the system can be reduced to a set of deterministic laws, the system reaches in time a state of permanent regime. This fact is reflected by the convergence of families of phase trajectories toward a subset of the phase space. This invariant subset is called an "attractor."

Thus, from an analysis of the EEG considered as a time series, it is possible to answer the following questions:

1. Is it possible to identify attractors for various stages of brain activity? In other words, can the salient features of neuronal activities be described by deterministic dynamics?
2. If attractors exist, what is their Hausdorff dimension D? This quantity gives a means of classifying attractors, and the dynamics they portray, as periodic, quasiperiodic, or chaotic.
3. What is the minimum number of variables necessary for the description of a given EEG activity?

2 Phase Portraits

Let us assume that the dynamics of the brain activity is described by a set of $\{X_0(t), X_1(t), \ldots, X_{n-1}(t)\}$ variables satisfying a system of first-order differential equations. A differential equation of order n with a single variable X_0, accessible from experimental data, is equivalent to the original set. Now both X_0 and its derivatives, and therefore the ensemble of n variables, can be obtained from a single time series. However, it is more convenient to construct another set of variables $\{X_0(t),$

Springer Series in Brain Dynamics 1
Edited by Erol Başar

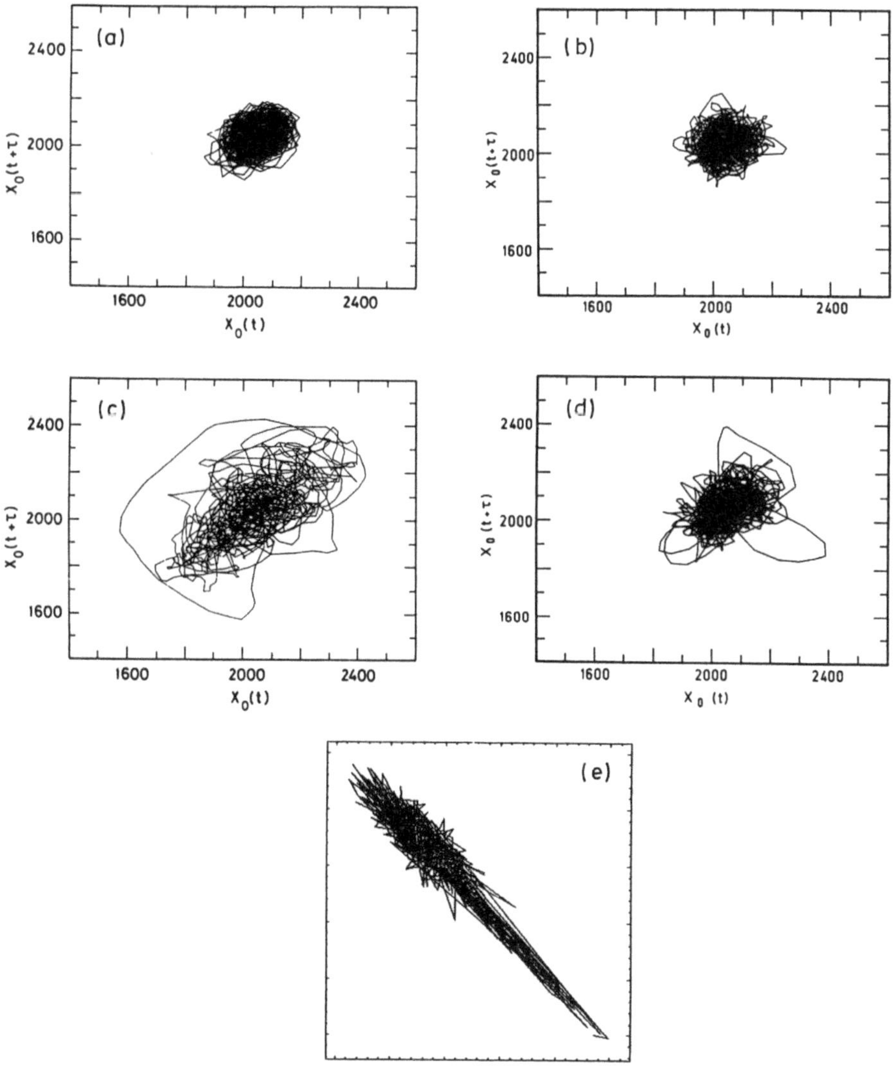

Fig. 1a–e. Two-dimensional phase portraits derived from the EEG of (a) an awake subject, (b) stage 2 sleep, (c) stage 4 sleep, (d, e) REM sleep. The time series $X_0(t)$ is made of $N = 4000$ equidistant points. The central EEG derivation C4-A1 according to the Jasper system was recorded with a PDP 11-44, 100 Hz for 40 s. The value of the shift from 1a to 1e is $\tau = 10\Delta t$

$X_0(t+\tau), \ldots, X_0[t+(n-1)\tau]\}$, which is topologically equivalent to the original set (Takens 1981). X may represent the electrical potential V recorded by the EEG. These variables are obtained by shifting the original time series by a fixed time lag τ ($\tau = m\Delta t$, where m is an integer and Δt is the interval between successive samplings).

These variables span a phase space, which allows the drawing of the phase portrait of the system or, more precisely, its projection into a low-dimensional subspace of the full phase space. With the help of the procedures cited above, we have constructed the phase-space portraits of various stages of sleep cycles.

The phase portrait of the awake subject is densely filled and occupies only a small portion of the phase space (Fig. 1a). The representative point undergoes deviations from some mean position in practically all directions. In sleep stage 2, already a larger portion of the phase space is visited and a tendency toward a privileged direction is seen (Fig. 1b). This tendency is amplified in sleep stage 4 and one sees preferential pathways, suggesting the existence of reproducible relationships between instantaneous values of the pertinent variables (Fig. 1c). This phase portrait is the largest and it exhibits a maximum "coherence," which diminishes again when rapid eye movement (REM) sleep sets in (Fig. 1d).

A universal attractor for different REM episodes of a single night and a given individual seems unlikely, as the REM episodes are associated with intense brain activity and generation of dreams. Figure 1e shows a second REM episode in the sleep cycle of the same individual whose EEG recording was used in Fig. 1d.

We must now determine whether these phase portraits represent chaotic attractors, in which case they could be characterized further by a number corresponding to their dimensionality as defined below.

3 The EEG Attractors

Let us consider an ensemble of points in an n-dimensional phase space. We cover the set with hypercubes of size ε. If $N(\varepsilon)$ is the minimum number of hypercubes necessary to cover the set, the Hausdorff, dimension D of the attractor is defined as (Bergé et al. 1984):

$$D = \frac{\ln N(\varepsilon)}{\ln(1/\varepsilon)} \tag{1}$$

where for small ε, $N(\varepsilon) \simeq \varepsilon^{-D}$.

From this definition, one easily verifies that if the set is reduced to a single point, then $D=0$. If the set of points represents a segment of line length L, then $N(\varepsilon) = L\varepsilon^{-1}$ and therefore $D=1$. In the case of a surface S, $N(\varepsilon) = S\varepsilon^{-2}$ and $D=2$. For these simple cases, the Hausdorff dimension coincides with the euclidean dimension. However, this is not so for a class of objects called "fractals," which may be illustrated by the following example. Let us consider a segment of unit length and remove the middle third of the segment. We repeat the same operation on the remaining segments. If the deletion is performed an infinite number of times, we obtain an infinite number of disconnected points called a "Cantor set." A simple calculation based on Eq. 1 shows that the Hausdorff dimension of this set is $D=0.63$, which is between 0 and 1. This number is the fractal dimension of the set.

The phase portraits of Fig. 1 belong to the family of fractal objects. However, the Hausdorff dimension of the attractor cannot be evaluated in a simple way from Eq. 1.

Chaotic attractors constructed from a time series can be characterized by another method proposed by Grassberger and Procaccia (1983a, b). Let $\{(X_0(t_1), \ldots,$

$X_0[t+(n-1)]\}$ represent the coordinates of a point $\vec{X}$ in the phase space of Fig. 1. Given an $\vec{X}_i$, we compute all distances $|\vec{X}_i - \vec{X}_j|$ from the $N-1$ remaining points of the data. This allows us to count the data points that are within a prescribed distance r from point $\vec{X}_i$ in the phase space. Repeating the process for all values of i, one arrives at the quantity

$$C(r) = \frac{1}{N^2} \sum_{i \neq j = 1}^{N} \theta(r - |\vec{X}_i - \vec{X}_j|), \quad (2)$$

where θ is the Heaviside function, $\theta(X) = 0$ if $X < 0$, and $\theta(X) = 1$ if $X > 0$.

The nonvanishing of $C(r)$ measures the extent to which the presence of a data point $\vec{X}_i$ affects the position of the other points. $C(r)$ may thus be referred to as the integral *correlation function* of the attractor.

Let us fix a small ε and use it as a yardstick for probing the structure of the attractor. If the latter is a line, clearly the number of data points within a distance r from a prescribed point should be proportional to r/ε. If the attractor is a surface, this number should be proportional to $(r/\varepsilon)^2$ and, more generally, if it is a d-dimensional manifold, the number should be proportional to $(r/\varepsilon)^d$. We therefore expect that for relatively small r, $C(r)$ should vary as

$$C(r) \sim r^d.$$

In other words, the dimensionality d of the attractor is given by the slope of $\log C(r)$ versus $\log r$ in a certain range of values of r:

$$\log C(r) = d \left|\log r\right| + C^\circ \quad (3)$$

The results cited above suggest the following algorithm:

1. Starting from a time series provided by the EEG, construct the correlation function, Eq. 2, by considering successively higher values of the dimensionality of the phase space.
2. Deduce the slope d near the origin according to Eq. 3 and see how the result changes as n is increased.
3. If the d versus n dependence is saturated beyond some relatively small n, the system represented by the time series should possess an attractor. The saturation value d is regarded as the dimensionality of the attractor represented by the time series. The value of n beyond which saturation is observed provides the minimum number of variables necessary to model the behavior represented by the attractor.

The procedure cited above has been applied to two sets of EEG data corresponding to stage 2 sleep of two individuals and stage 4 sleep of three individuals (Babloyantz et al. 1985). Figure 2 gives the $\log C$ versus $\log r$ dependence for $n = 2$ to $n = 7$ computed for stage 4 sleep. We observe the existence of a region over which this dependence is linear.

The slope of the curve $\log C(r)$ versus $\log r$ has been evaluated with extreme care. After determining the boundaries of the linear zone by visual inspection, we determine the slope of m first points in this segment by using the least-square method. The operation is repeated all along the linear region by sliding m one point further. The computation is repeated for increasing values of m. If the region is linear, all these

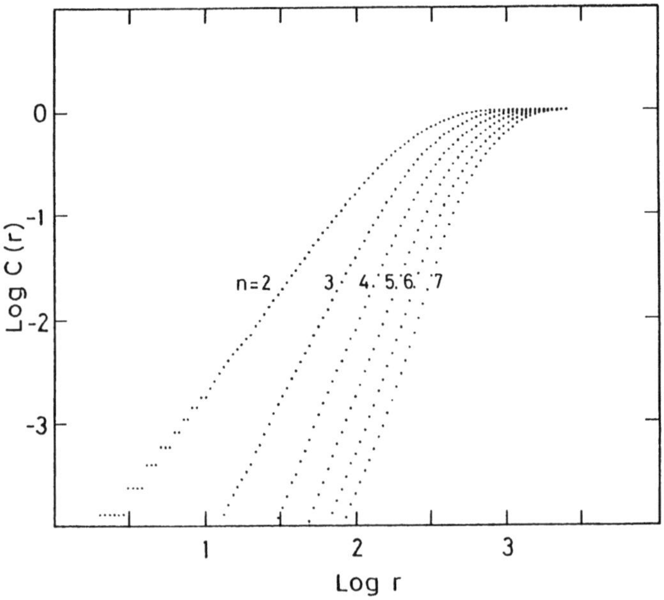

Fig. 2. Dependence of the integral correlation function $C(r)$ on the distance r for stage 4 sleep. Parameter values as in Fig. 1

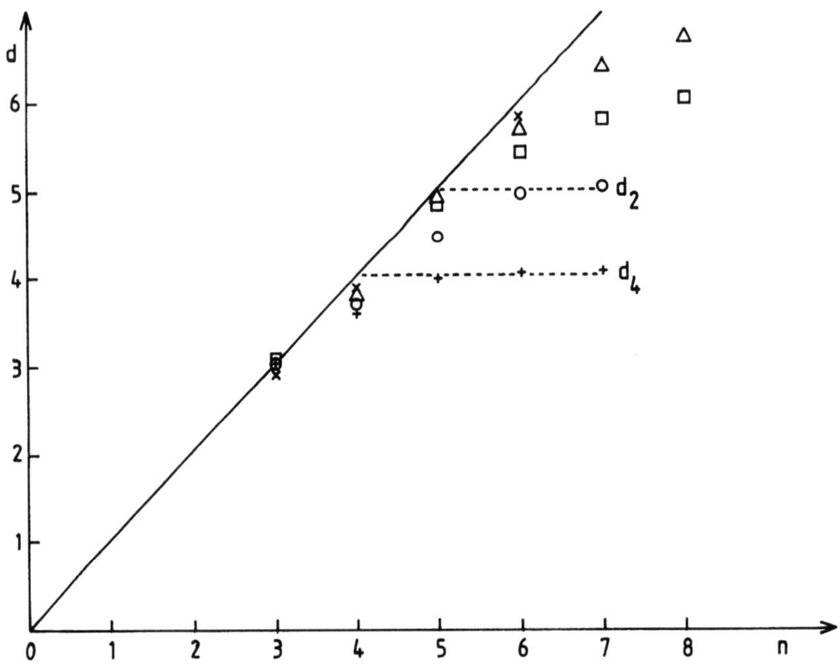

Fig. 3. Dependence of dimensionality d on the number of phase space variables n for a white noise signal (x), the EEG attractor of an awake subject (Δ), stage 2 sleep (o), stage 4 sleep (+), and REM sleep (□), for the same number of data points as in Fig. 1

operations must yield the same value of the slope (within acceptable error boundaries).

Although in principle every value of time lag τ is acceptable for the resurrection of the system's dynamics, in practice, for a given time series, only a well-defined range of τ gives satisfactory results.

Figure 3 shows saturation curves describing the dependence of the slope d versus the dimension n of the embedding phase space computed for the awake state and for several stages of the sleep cycle. They are compared with the behavior obtained from a random process such as Gaussian white noise. There is no saturation for the awake state; however, we are far from random noise. A satisfactory saturation exists for stage 2 sleep. In this case, we find $d = 5.03 + 0.07$ and $d = 5.05 + 0.1$. Saturation curves for stage 4 sleep show $d = 4.08 \pm 0.05$, $d = 4.05 + 0.05$, and $d = 4.37 + 0.1$. For the REM state, saturation is again poor.

4 Time-Dependent Property

The chaotic nature of the attractor can also be assessed with the help of a time-dependent property. Although in the presence of a chaotic attractor all trajectories converge toward a subset of the phase space, inside the attractor, two neighboring trajectories may diverge. This fact reflects the extreme sensitivity of chaotic dynamics to the initial conditions. The rate of the divergence of the trajectories in time may be assessed from a time series (Wolf et al. 1985). The Lyapunov exponents λ_i are the average of these individual evaluations over a large number of trials. A negative Lyapunov exponent indicates an exponential approach of the initial conditions on the attractor; on the contrary, a positive λ_i expresses the exponential divergence on an otherwise stable attractor. Thus, a positive Lyapunov exponent indicates the presence of chaotic dynamics.

Using the Fortran code described by Wolf et al. (1985), we evaluated the largest positive Lyapunov exponent λ for stage 2 and stage 4 of deep sleep. For stage 2, we find a positive value of λ_2 between 0.4 and 0.8. For stage 4, we find also a positive number $0.3 < \lambda_4 < 0.6$. The inverse of this quantity gives the limits of predictability of the long-term behavior of the system.

5 Conclusions

We have shown that from a routine EEG recording, the dynamics of brain activity could be reconstructed. The fact that chaotic attractors could be identified for two stages of normal brain activity indicates the presence of deterministic dynamics of a complex nature. This property should be related to the ability of the brain to generate and process information.

Unlike periodic phenomena, which are characterized by a limited number of frequencies, chaotic dynamics show a broadband spectrum. Thus, chaotic dynamics increase the resonance capacity of the brain. In other words, although globally a chaotic attractor shows asymptotic stability, there is an internal instability reflected by the presence of positive Lyapunov exponents. This results in a great sensitivity to the initial conditions and, thus, an extremely rich response to external input.

The topological properties of the attractors and their quantification by means of dimensionality analysis may be an appropriate tool in the classification of brain activity and, thus, a possible diagnostic tool. For example, various forms of epileptic seizures could be classified according to their degree of coherence (Babloyantz and Destexhe 1986). Moreover, the determination of the minimum number of variables necessary for the description of epileptic attractors is a valuable clue for model construction.

References

Babloyantz A, Destexhe A (1986) Low-dimensional chaos in an instance of epilepsy. Proc Natl Acad Sci USA 83:3513–3517

Babloyantz A, Nicolis C, Salazar M (1985) Evidence of chaotic dynamics of brain activity during the sleep cycle. Phys Lett [A] 111:152–156

Berge P, Pomeau Y, Vidal C (1984) L'ordre dans le chaos: vers une approche déterministe de la turbulence. Hermann, Paris

Brandstäter A, Swift J, Swinney HL, Wolf A (1983) Low-dimensional chaos in a hydrodynamic system. Phys Rev Lett 51:1442–1445

Grassberger P, Procaccia I (1983a) Characterization of strange attractors. Phys Rev Lett 50:346–349

Grassberger P, Procaccia I (1983b) Measuring the strangeness of strange attractors. Physica [D] 9:189–208

Nicolis C, Nicolis G (1984) Is there a climatic attractor? Nature 311:529–532

Nicolis C, Nicolis G (1986) Reconstruction of the dynamics of the climatic system from time-series data. Proc Natl Acad Sci USA 83:536–540

Takens F (1981) Detecting strange attractors in turbulence. In: Rand DA, Young LS (eds) Dynamical systems and turbulence, Warwick 1980. Springer, Berlin Heidelberg New York pp 366–381 (Lecture notes in mathematics, vol 898)

Wolf A, Swift JB, Swinney HL, Vastano JA (1985) Determining Lyapunov exponents from a time series. Physica [D] 16:285–317

The EEG is Not a Simple Noise: Strange Attractors in Intracranial Structures

J. RÖSCHKE and E. BAŞAR

1 Introduction

Since Berger's (1929) first description of the electrical activity of the brain, several approaches have been undertaken in order to correlate the activity at neuronal levels with the origin of the electroencephalogram (EEG). Creutzfeldt (1974) pointed out that the spontaneous electrical activity of the CNS and sensory evoked potentials are highly correlated to intracellularly measured postsynaptic potentials (EPSPs and IPSPs). Ramos et al. (1976) postulated that it is impossible to specify any general causal or predictable relationship between the waveform of an evoked potential and the firing pattern of a neuron. Some authors take the view that the spontaneous EEG activity is an expression of the incessant, irregular background neural firing. Do we have the right to consider the spontaneous activity of the brain as a background noise in the sense of ideal communication theory? Or rather, is the EEG a most important fluctuation, which controls the sensory evoked and event-related potentials? We have written elsewhere that the spontaneous activity plays an active role in the signals transmitted through various structure and recorded at various sites in the brain and that the EEG should not be considered as a noisy signal. Especially, we have assumed that regular patterns of the EEG reflect coherent states of the brain during which cognitive and sensory inputs are processed (Başar 1980, 1983a, b).

The main goal of the present paper is to show that the brain's spontaneous activity is not a simple noise, but is an active signal probably reflecting causal responses from hidden events and sources during sensory and cognitive processing in the brain.

There are several difficulties to overcome in order to describe well-defined states of spontaneous activity and evoked potentials. Although conventional methods of system theory, such as power spectral analysis, have been very useful for analyzing the brain waves as a first approach, the highly nonlinear character of the brain's dynamic behavior led us first to use phase portrait analysis, analogies with laser theory, and the Duffing equation (Başar 1980, 1983a, b). Our preliminary results from an analysis of the EEG in phase space allowed us to speculate on the existence of strange attractors in the EEG (Başar 1983a, b).

In our newest approach we used the algorithm of Grassberger and Procaccia (1983), similar to the analysis of Babloyantz and Nicolis (1985). The EEG signal was embedded into phase space and we computed the dimension of the attractors of the acoustical cortex (GEA), the hippocampus (HI), and the reticular formation (RF) of the cat brain during slow-wave sleep stage (SWS).

Springer Series in Brain Dynamics 1
Edited by Erol Başar

Our preliminary results showed that the field potentials (or EEG) in intracranial structures do not reflect the behavior of a simple noisy signal, as has been shown for the human EEG derived from scalp electrodes (Babloyantz and Nicolis 1985). However, field potentials from various brain structures have properties of strange attractors, indicating the presence of a chaotic system. The most important result in the present paper is the existence of differentiated dimensionality in functionally independent brain structures. This in turn makes the future application of this method most useful for the differentiated analysis of brain states and function. In this study we also discuss what is meant by "attractors" and "strange attractors", and our belief that the use of such concepts in EEG research may lead to basic trends (see also the Epilogue).

2 The Mathematical Procedure and the Concept of "Attractor"

In order to describe periodic, aperiodic, or even chaotic behavior of nonlinear systems arbitrarily with more degrees of freedom, several approaches have been applied. Lorenz (1963) applied concepts of nonlinear dynamics to the convection phenomenon of hydrodynamics in order to describe atmospheric turbulence (Navier-Stokes equation). He demonstrated the possibility that the unpredictable or chaotic behavior observed in such an infinite-dimensional system might be caused by a three-dimensional (deterministic) dynamical system.

In order to understand these arguments, we have to consider some recent tools from the theory of nonlinear dynamical systems. The description of systems behavior (in our case the EEG from different brain structures) must be analyzed not only in the time domain or frequency domain, but also in the *phase space*. In general, a phase space is identified with a topological manifold. An n-dimensional phase space is spanned by a set of n independent linear vectors. This requirement is generally sufficient. There are several possibilities for defining a phase space. We consider a proposal of Takens (1981) and span a ten-dimensional phase space by $x(t)$, $x(t+\tau)$, $\ldots$, $x(t+9\tau)$, where τ means a fixed time increment. Every instantaneous state of a system is therefore represented by a set $(x_1, \ldots, x_n)$, which defines a point in the phase space. The sequence of such states (or points) over the time scale defines a curve in the phase space, called a "trajectory". As time increases, the trajectories either penetrate the entire phase space or they converge to a lower-dimensional subset. In this latter case, the set to which the trajectories converge is called an "attractor". Figure 1 shows some simple (converging) attractors and a noise that does not converge in a two-dimensional phase space.

In relation to the topological dimension of the remaining attractor, one can deduce various properties of the investigated system. If the dimension of an *attractor* is a noninteger, called a "fractal", the attractor is a "strange attractor" and can be identified with the properties of deterministic chaos. A characteristic phenomenon of deterministic chaos is a sensitive dependence on initial conditions. Similar causes do not produce similar effects. This is a very extensive statement, which apparently damages the causality principle of natural philosophy. However, by examining the properties of a strange attractor more precisely, one finds that a strange attractor

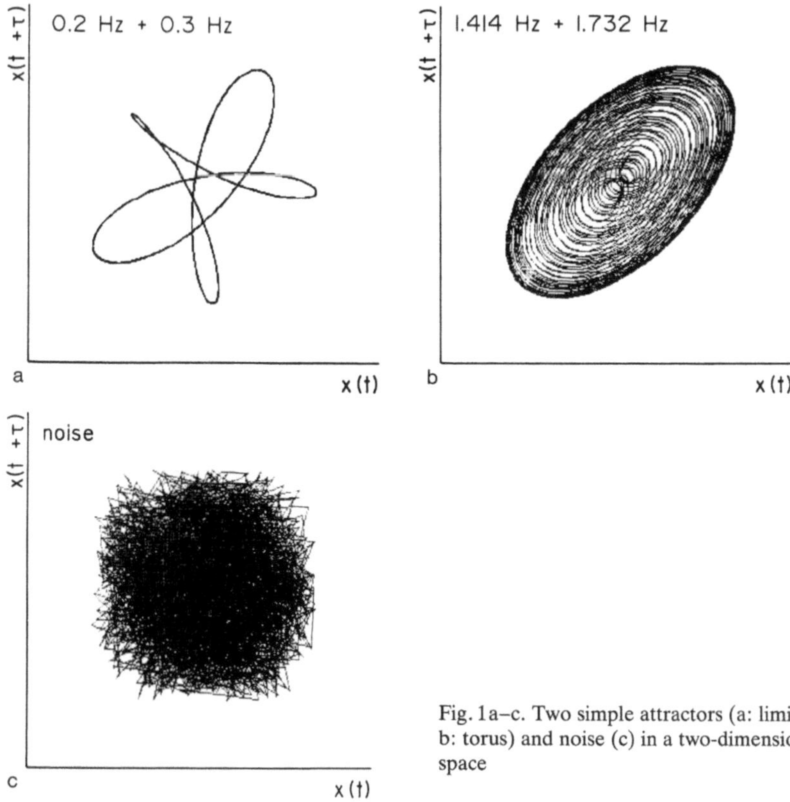

Fig. 1a–c. Two simple attractors (a: limit cycle; b: torus) and noise (c) in a two-dimensional phase space

may have a strong conformity, called „self-similarity", which is an invariance with respect to scaling.

Self-similar objects possess a fractal dimension (Schroeder 1986). What is a "fractal dimension"? One of the oldest notions of dimension is that of a topological dimension D_T. For a point, $D_T = 0$; for a line, $D_T = 1$, and for a plane, $D_T = 2$. A first generalization is the *Hausdorff dimension or fractal dimension* D_F. For simple sets, for example a limit cycle or a torus, the fractal dimension D_F is an integer and is equal to the topological dimension D_T. For a n-dimensional phase space, let $N(\varepsilon)$ be the number of n-dimensional balls (or cubes) of radius ε required to cover an attractor. Then the fractal dimension D_F is defined as

$$D_F = \lim_{\varepsilon \to 0} \frac{\log N(\varepsilon)}{|\log \varepsilon|}$$

The classical example of a set whose fractal dimension exceeds its topological dimension [such sets are called "fractals" by Mandelbrot (1977)] is Cantor's set (Fig. 2). If one chooses $\varepsilon = \left(\frac{1}{3}\right)^n$, then $N(\varepsilon) = 2^n$, and it follows that

$$D_F = \lim_{\varepsilon \to 0} \frac{\log N(\varepsilon)}{|\log \varepsilon|} = \lim_{n \to \infty} \frac{\log 2^n}{\log 3^n} = \frac{\log 2}{\log 3} = 0.630 \ldots$$

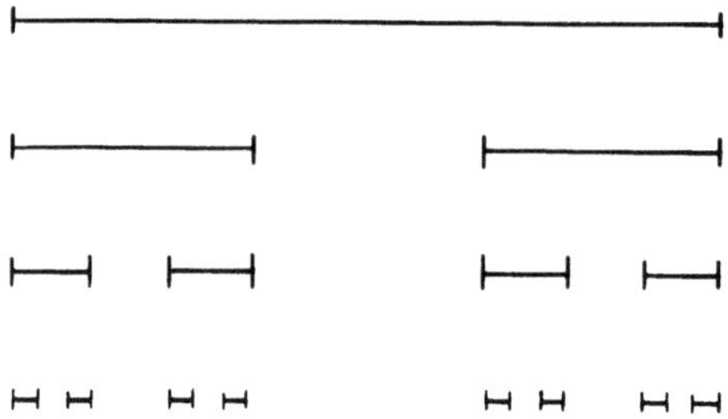

Fig. 2. Cantor's set

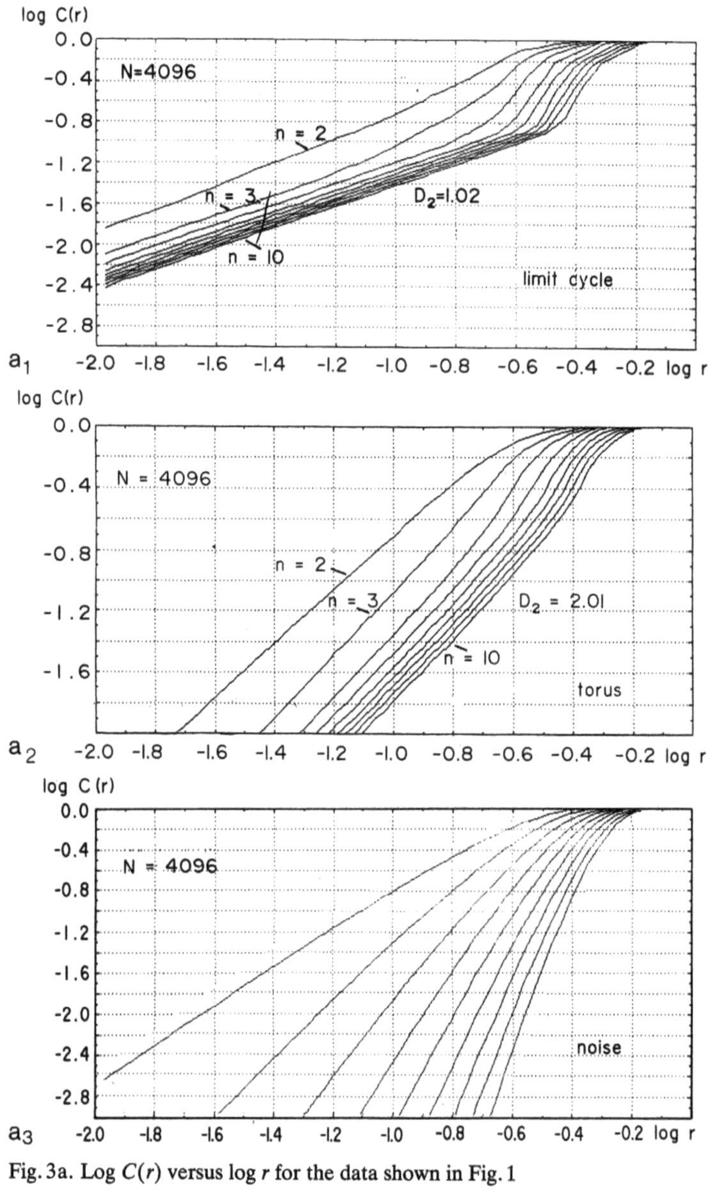

Fig. 3a. Log $C(r)$ versus log r for the data shown in Fig. 1

Note the self-similarity of Cantor's set. For example, the fractal dimension of the Lorenz attractor is $D_F = 2.03$. A generalization of the fractal dimension is introduced in information theory. The Renyi information of order q is defined as

$$I_q = \frac{1}{1-7} \log \sum_{i=1}^{N(R)} p_i^q \qquad (q \neq 1)$$

$$I_q = -\sum_{i=1}^{N(R)} p_i \log p_i \qquad (q = 1)$$

Let p_i be the probability that an arbitrary point (of an attractor) falls into cube i with-radius R and let $N(R)$ be the number of nonempty cubes. The generalized dimensions D_q of order q are given by

$$\mathrm{D_q} = \lim_{R \to 0} \frac{I_q(R)}{\log(1/R)}$$

For $q = 0$ we find $D_0 = D_F$. D_1 is called the "information dimension" and D_2 is called the "correlation dimension". It is the case that

$$D_0 \geqslant D_1 \geqslant D_2 \geqslant \ldots$$

In practice, the correlation dimension D_2 is the generalized dimension easiest to estimate from attractors generated by experimental data (Grassberger 1984, personal communication), because

$$I_2 = -\log \sum_{i=1}^{N(R)} p_i^2 = -\log C(R)$$

where $C(R)$ is a measure of the probability that two arbitrary points $\vec{x}$, $\vec{y}$ will be separated by distance R. $C(R)$ is called the "correlation integral" and can be easily computed:

$$C(R) = \lim_{N \to \infty} \frac{1}{N^2} \sum_{\substack{i,j=1 \\ i \neq j}}^{N} \theta(R - |\vec{x} - \vec{y}|)$$

where θ is the Heavyside function.

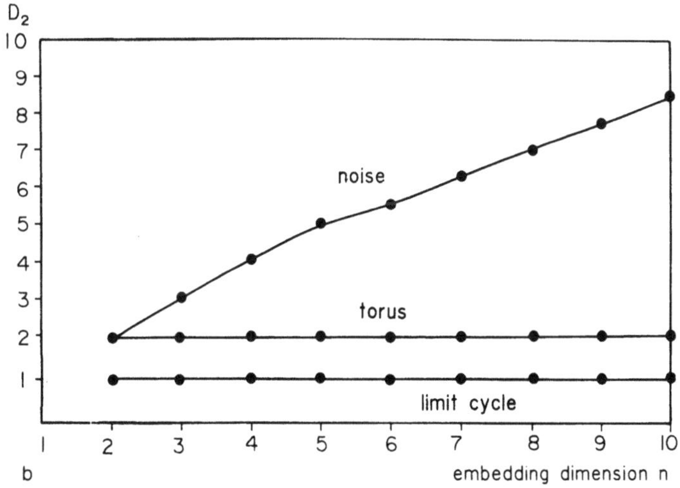

Fig. 3b. The correlation dimension D_2 versus the embedding dimension n for the limit cycle, the torus, and noise shown in Fig. 1

It then follows that

$$D_2 = \lim_{R \to 0} \frac{\log C(R)}{\log R}$$

or

$$C(R) \sim R^{D_2}$$

The main point is that $C(R)$ behaves as a power of R for small R. This means that it is possible to find a measure for the dimensionality of an attractor by evaluating $C(R)$ and plotting $\log C(R)$ versus $\log R$. For the Lorenz attractor, Grassberger and Procaccia (1983) found $D_2 = 2.05$.

Figure 3 shows the computation of the correlation dimension for the examples presented in Fig. 1. It is obvious that a noise signal has no attractor. By plotting the slope of the curves versus the embedding dimension, no saturation can be observed. On the contrary, there exist attractors of $D_2 = 1.00$ for a limit cycle and $D_2 = 2.00$ for quasi-periodic flow.

3 Experimental Procedure and Results

In order to analyze the dimensionality of field potentials, five cats with chronically implanted electrodes were studied. The chronic electrodes were implanted in the GEA, HI, and RF. In total, 15 experimental trials during SWS activity were evaluated. The intracranial EEG signals were digitized by a 12-bit AD converter and stored in the memory of an HP 1000-F computer. The sampling frequency was $f_s = 100$ Hz for all trials.

Dimensions of the EEG signals were evaluated over a time period of about 20 s ($N = 2048$) and 40 s ($N = 4096$). Details of the software have been described elsewhere (Röschke 1986). The phase space was constructed by using the time-delayed coordinates proposed by Takens. Theoretically, the evaluation of the dimension of the attractors should be independent of the arbitrary but fixed time increment τ. In practice, this independence is not generally valid. Investigations of low-pass filtered noise (non-deterministic signals) have shown that $\mathrm{d}D_2/\mathrm{d}n$ depends on the time increment τ. It is evident that in the case of nondeterministic signals, $\mathrm{d}D_2/\mathrm{d}n \neq 0$ is observed for every embedding dimension.

However, by evaluating signals from a deterministic system, it is observed as a rule that the dimension D_2 of an attractor does converge towards a saturation value. In this case, $\mathrm{d}D_2/\mathrm{d}n = 0$ and this convergence is independent of the choice of τ in a given interval $\tau_1 < \tau < \tau_2$. Some recent investigations (Fraser 1985; Holzfuss 1985) have assumed that the best choice of τ corresponds to a minimum of the "mutual information" between two measurements, but these investigations have not yet been properly concluded.

Figure 4 shows the two-dimensional phase space representation of the EEG signal for the three investigated brain structures in the case of the cat named Toni. The time period used for evaluation of these curves was about 40 s and the time delay to construct the phase space was about 60 ms. Figure 5 shows the plot of $\log C(r)$ versus $\log r$. In all the computations, the EEG signal was embedded in a ten-dimensional

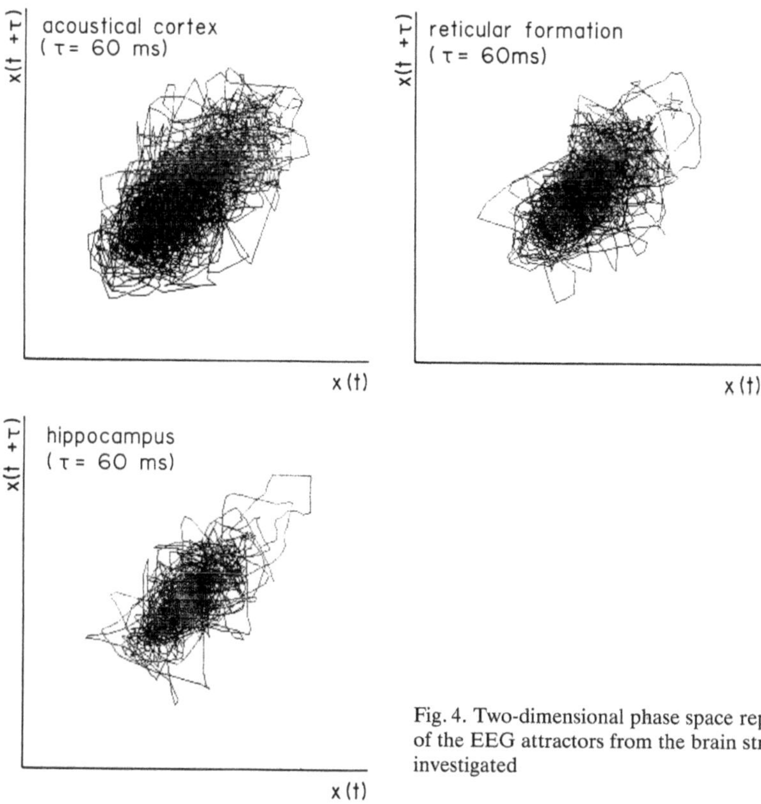

Fig. 4. Two-dimensional phase space representation of the EEG attractors from the brain structures investigated

phase space. Figure 6 shows the convergence of D_2 (slope of the curves from Fig. 5) towards the saturation value as a function of the embedding dimension n. Especially in this case, the dimensions of the attractors had the following values:

$D_{\mathrm{GEA}} = 5.00 \pm 0.10$
$D_{\mathrm{RF}} = 4.25 \pm 0.07$
$D_{\mathrm{HI}} = 4.32 \pm 0.07$

Table 1 presents the correlation dimension D_2 computed for all of the experimental data from five cats and 15 experiments.

1. For both $N = 2048$ and $N = 4096$, one cannot determine an unambiguous conformity. Both the results from a single cat and the dimensions from a single region vary within the range of acceptable limits.
2. In 86% of the investigated trials, D_{GEA} presents the maximal dimension. By taking into account all the evaluated data, the following mean values have been obtained:
   $D_{\mathrm{GEA}} = 5.06 \pm 0.31$
   $D_{\mathrm{RF}} = 4.58 \pm 0.38$
   $D_{\mathrm{HI}} = 4.37 \pm 0.36$

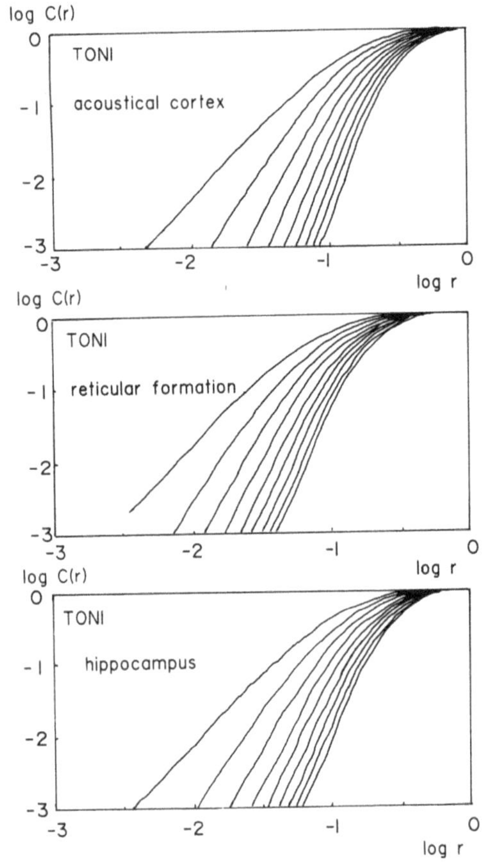

Fig. 5. Log $C(r)$ versus log r for the cat named Toni

In other words, the data confirm the following important relation:

$D_{\mathrm{GEA}} > D_{\mathrm{RF}} > D_{\mathrm{HI}}$

4 Discussion

The concept of dynamic patterns to use in understanding bioelectric phenomena was proposed by Katchalsky et al. (1974). A fundamental problem in the physical as well as in the biological sciences is the origin of a dynamic pattern. In physical science this problem can be attacked at vulnerable points; i.e., in systems that are simple enough to permit analysis both in physical and mathematical terms, dynamic patterns refer to those patterns that arise and are maintained by the dissipation or consumption of energy, such as *traveling or standing waves* generated in the air, on the surface of water, or by a vibrating violin string. They can be contrasted to static patterns, such as a stack of nesting chains, a crystal, or a virus capsid.

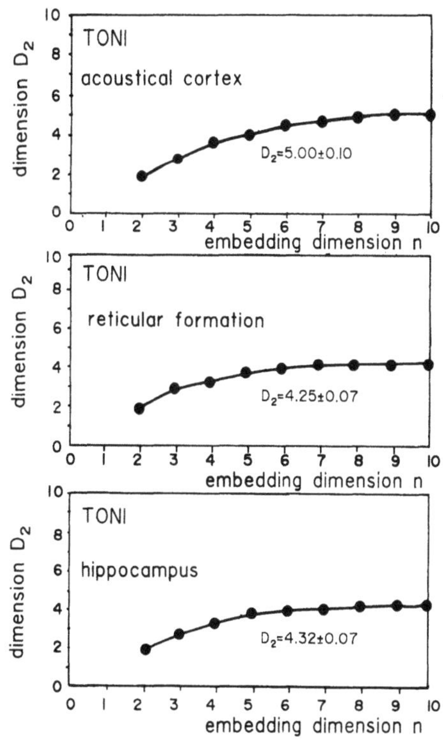

Fig. 6. The convergence of the slope D_2 of the curves (from Fig. 5) towards the saturation value as a function of the embedding dimension n

Table 1. Correlation dimension D_2 computed for all of the experimental data from five cats and 15 experiments

$N = 2048$	TONI	LENA	DESY	SARA	ROMY	LENA	SARA	D_2
GEA	5.00	4.35	4.66	4.93	5.00	5.00	–	4.82 ± 0.27
RF	4.25	4.62	4.30	4.24	5.22	4.60	4.05	4.48 ± 0.39
HI	4.32	4.35	4.00	3.80	5.00	4.15	4.85	4.35 ± 0.44

$N = 4096$	TONI	LENA	ROMY	DESY	SARA	TONI	LENA	ROMY	D_2
GEA	5.30	5.00	5.17	5.30	5.10	5.00	5.60	5.45	5.24 ± 0.21
RF	4.20	4.35	4.75	5.15	4.76	4.20	5.00	5.00	4.68 ± 0.38
HI	4.40	4.28	4.44	4.66	4.00	3.90	4.61	4.76	4.38 ± 0.31

According to Katchalsky et al. (1974), the central question is: how does uniform matter, obeying physical principles, i.e., laws of conservation of momentum, matter, and energy, spontaneously develop regular patterns? In other words, how is it that a set of isotropic causes can give rise to anisotropic dynamic effects? This appears to be the root problem of morphogenesis; extending from it are the more widely encountered problems of how preexisting static structures influence dynamic patterns.

In the book by Katchalsky et al. (1974), some dynamic patterns observed in geology, meteorology, and astrophysics are also described; for example, dynamic patterns on a large scale in clouds and the solar coronasphere. According to traditional Newtonian mechanics, if certain things are known about a system – all of the forces acting on it, its position, and the velocity of its particles – it is possible to describe, in theory, all of its future states. However, the current that started with the pioneering work of H. Poincaré at the turn of this century has made clear that the predictability of even classical deterministic systems can be quite limited. Simple nonlinear systems which are just as deterministic as the motions of the planets can behave in a manner so erratic as to prelude predictability past a short time (Shaw 1981). The existence of these "chaotic" systems raises both practical and conceptual questions.

A simple example is described by Hooper (1983): "Suppose you are sitting beside a waterfall watching a cascade of white water flow regularly over jagged rocks, when suddenly a jet of cold water splashes you in the face. The rocks have not moved, nothing has disrupted the water, and presumably no evil sprites inhibit the waterfall. So why does the water suddenly "decide" to splash you?" Physicists studying fluid turbulence have wondered about this kind of thing for several hundred years, and only recently have they arrived at some conclusions that seem to solve the problem at least in part: the waterfall's sudden random splashes do not come from some inperceptible jiggle, but from the inner dynamics of the system itself. Behind the chaotic flow of turbulent fluids or the shifting cloud formations that shape the weather lies an abstract descriptor which the physicists call a "strange attractor." What is an "attractor" and what makes it "strange"? We shall try to describe it again by using the simple explanation of Hooper (1983). Suppose one puts water in a pan, shakes it up, and then stops shaking it; after a time it will stop whirling and come to rest. The state of rest – the equilibrium state – can be described mathematically as a "fixed point," which is the simplest kind of attractor.

Let us now imagine the periodic movement of a metronome or a pendulum swinging from left to right and back again. From the viewpoint of geometry, this motion is said to remain within a fixed cycle forever. This is the second kind of attractor, the *limit cycle*. All of the various types of limit cycles share one important characteristic: *regular, predictable motion*. The third variety, the *strange attractor*, is *irregular, unpredictable*, or simply *strange*. For example, when a heated or moving fluid moves from a smooth, or laminar, flow to wild turbulence, it switches to a strange attractor.

Chaotic behavior in deterministic systems usually occurs through a *transition* from an orderly state when an external parameter is changed. In studies of these systems, particular attention has been devoted to the question of the route by which the chaotic state is approached. An increasing body of experimental evidence supports the belief that apparently random behavior observed in a wide variety of physical systems is caused by underlying deterministic dynamics of a low-dimensional chaotic (strange) attractor. The behavior exhibited by a chaotic attractor is predictable on short time scales and unpredictable (random) on long time scales.

The unpredictability, and so the attractor's degree of chaos, is effectively measured by the parameter "dimension". Dimension is important to dynamics because it provides a precise way of speaking of the number of independent variables inherent in a motion. For a dissipative dynamical system, trajectories that do not diverge to infinity approach an attractor (Farmer 1982).

The dimension of an attractor may be much less than the dimension of the phase space that it sits in. In other words, once transients die out, the number of independent variables to the motion is much less than the number of independent variables required to specify an arbitrary initial condition. With the help of the concept of dimension it is possible to discuss this precisely. For example, if the attractor is a fixed point, there is no variation in the final space position; the dimension is zero. If the attractor is a limit cycle (pendulum) the phase space varies along a curve; the dimension is one. Similarly, for quasi-periodic motion with n incommensurate frequencies, motion is restricted to an n-dimensional torus (dimension of chaotic attractors).

Noise is a common phenomenon in systems with many degrees of freedom. Under the influence of noise, observables show irregular behavior in the time and broad-band Fourier spectra. There is an important difference between a noise signal and chaotic fluctuations resulting from the motion of a larger number of system dimensions. The noise signal does not have a finite dimension, whereas chaotic systems with differential equations show finite dimensionality. This difference can be shown by the evaluation of the dimension.

The field potentials of the cat brain showed almost stable mean values of 5.06, 4.58, 4.37 for the GEA, the HI, and the RF, respectively. In other words, various structures of the brain indicate the existence of various chaotic attractors with fractal dimensions. The signals measured in these different structures do not reflect properties of noise signals, but reflect behavior of strange attractors of quasi-low dimension. The measured dimensions in these various structures are seemingly different attractors which might be functionally uncoupled; in other words, even during the SWS stage, where a state of hypersynchrony is observed in all of the various brain structures, the attractors show significant differences. These differences are stable and statistically relevant. We want especially to report that the dimension of the GEA is significantly higher than that of the HI and the RF. This is a kind of differentiated behavior that cannot be observed by the analysis of power spectra (see Fig. 7). In simple words, the cat GEA seems to show a more complex behavior with a greater degree of freedom than do such structures as the RF and HI. There is *no limit cycle behavior* in the studied structures (see also the Epilogue of this book).

Our computations, which are not yet finished and may be theoretically imperfect, showed that the spontaneous activity of the cat cortex depicted a dimension of around 8–9. The acoustical evoked potentials during the waking stage showed much lower dimensionality than did the spontaneous EEG during the same waking state: the dimension of the evoked potential usually varied between 3.5 and 5.

In our earlier studies, we pointed out the possibility of modeling globally the evoked potentials of the brain with the Duffing equation, assuming that nonlinear forced oscillations of various brain structures could be described with solutions of the Duffing equation or similar equations (Başar 1980, 1983b). We argued further that neural populations of the brain may be regarded as a large number of coupled oscillators, each comprising millions of neurons. If n oscillators are left uncoupled, their attractor will be an n-dimensional torus, with n independent frequencies. If the oscillators are coupled, however, the dimension will be reduced. It is possible, for example, that they will entrain at a single frequency (limit cycle). In a condition of slightly less entrainment, the attractor might have three independent frequencies, i.e., be a three-dimensional torus. One might think that with even less entrainment, the attractor could be a four-dimensional torus. Our preliminary analysis of the brain's

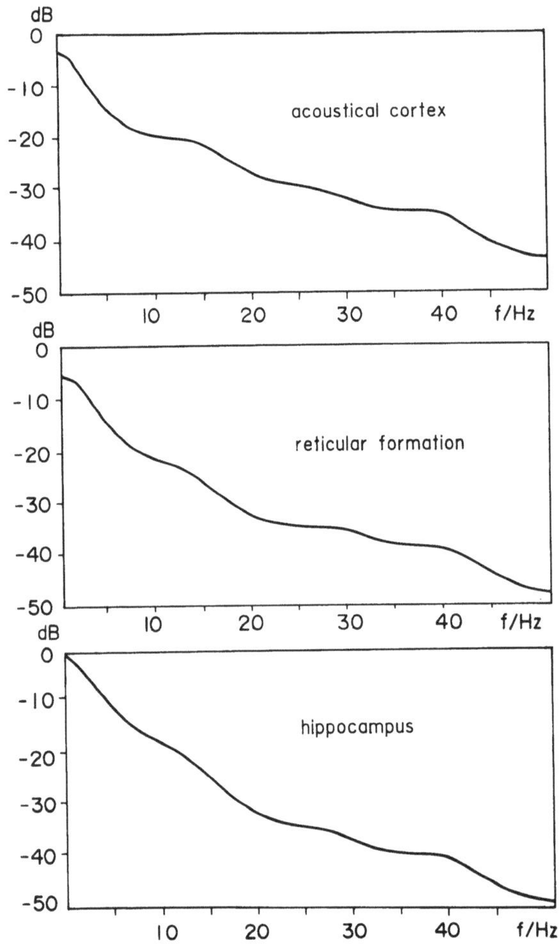

Fig. 7. Power spectra of various structures of the cat brain during the slow-wave sleep stage

spontaneous activity had already pointed out the possibility of correlating the EEG with the behavior of a strange attractor (Başar 1983b). The present results now show that the new algorithm of Grassberger and Procaccia offers an index of the degrees of freedom in the spontaneous activity of the brain. The results presented in this study have a number of new implications.

Three important applications are straightforward:

1. If the fractal dimension is a more precise indicator of state changes of the brain than are power spectra, it can be used in addition to power spectra to describe the state changes in every dimension of EEG analysis: for example, in pharmacoencephalography and in studies of the evolution of the EEG by comparative analysis of the field potentials of vertebrates, low vertebrates, and invertebrates.
2. Even in the same brain there can exist structurally and functionally independent EEG dimensions. Can one structure of the brain undergo a transition, say from a low-dimensional to a high-dimensional level, while another structure shows the opposite behavior? It is difficult to interpret changes in entropy. The use of the

dimension concept seems to be an adequate method for describing ordered states and for describing dynamic transitions impossible to see in power spectral analysis.

3. It seems that various intracranial structures have various independent attractors. It is possible to describe the ensemble of attractors with a matrix configuration in which various substructures of the brain can occupy a defined place in the matrix. In this case the transition matrices could tell a lot about the multiplicity of attractors as well as about the coupling and decoupling of attractors. The description of multiple attractors could be very useful for the description of cognitive processes, for which sufficiently descriptive parameters are still missing.

The results of the present study constitute a step forward in our preliminary efforts to try to describe the evoked potentials of the brain and the EEG as forced nonlinear oscillations, for which we tried to correlate the processes involved in the Duffing equation. Haken (1985) pointed out interesting formal analogies between laser and brain. We also pointed out similar formal analogies in the generation of evoked potential patterns (Başar 1983b). However, at present, formal analogies with a laser can be considered only for a simple laser. By using fractal dimensions to describe the EEG and the evoked potential, we believe that we can find new descriptors to describe neural analogies with a multimodal laser.

References

Babloyantz A, Nicolis C, Salazar M (1985) Evidence of chaotic dynamics of brain activity during the sleep cycle. Phys Lett [A] 111:152–156

Başar E (1980) EEG-brain dynamics. Relation between EEG and brain evoked potentials. Elsevier/North-Holland, Amsterdam

Başar E (1983a) Toward a physical approach to integrative physiology. I. Brain dynamics and physical causality. Am J Physiol 245(4):R510–R533

Başar E (1983b) Synergetics of neuronal populations. In: Başar E, Flohr H, Haken H, Mandell AJ (eds) Synergetics of the brain. Springer, Berlin Heidelberg New York

Berger H (1929) Über das Elektroencephalogramm des Menschen. Arch Psychiatr Nervenkr 87: 527–570

Creutzfeldt OD (1974) The neuronal generation of the EEG. In: Renard A (ed) Handbook of electroencephalography and clinical neurophysiology. Elsevier, Amsterdam

Farmer JD (1982) Dimension, fractal, measures, and chaotic dynamics. In: Haken H (ed) Evolution of order and chaos. Springer, Berlin Heidelberg New York

Fraser AM (1985) Using mutual information to estimate metric entropy in dimensions and entropies in chaotic systems. In: Mayer-Kress G (ed) Dimensions and entropies in chaotic systems. Springer, Berlin Heidelberg New York Tokyo

Grassberger P, Procaccia I (1983) Measuring the strangeness of strange attractors. Physica [D] 9:183–208

Holzfuss J (1985) An approach to error-estimation in the application of dimension algorithms. In: Mayer-Kress G (ed) Dimensions and entropies in chaotic systems. Springer, Berlin Heidelberg New York Tokyo

Hooper J (1983) What lurks behind the wild forces of nature? Ask the connoisseurs of chaos. Omni 5:85–92

Katchalsky AK, Rowland W, Blumenthal R (1974) Dynamic patterns of brain cell assemblies. MIT Press, Cambridge

Lorenz EN (1963) Deterministic nonperiodic flow. Atmos. Sci. 20:130

Mandelbrot B (1977) Fractals, form, chance and dimension. Freeman, San Francisco
Ramos A, Schwartz E, John ER (1976) Evoked potential–unit relationship in behaving cats. Brain Res Bull 1:69–75
Röschke J (1986) Eine Analyse der nichtlinearen EEG-Dynamik. Dissertation, University of Göttingen
Schroeder MR (1985) Number theory in science and communications. Springer, Berlin Heidelberg New York Tokyo
Shaw R (1981) Strange attractors, chaotic behaviour, and information flow. Z Naturforsch 36a:80
Takens F (1981) Detecting strange attractors in turbulence. In: Rand A, Young LS (eds) Dynamical systems and turbulence, Warwick 1980. Springer, Berlin Heidelberg New York, pp 366–381 (Lecture notes in mathematics, vol 898)

III. New Scopes at the Cellular Level

Assessment of Cooperative Firing in Groups of Neurons: Special Concepts for Multiunit Recordings from the Visual System

R. Eckhorn and H. J. Reitboeck

1 Introduction

Progress in elucidating the cellular basis of visual perception has always depended on relating structure to function. At present, structure-function problems confront the field of cortical neurophysiology with the following types of questions: (a) what are the intrinsic dynamic operations in a local cortical module and what is its relevance for visual perception; (b) what are the principles of sensory processing within a single cortical area with its laminae, columns, and slabs; (c) what is the function of the distributed systems connecting the multiple visual areas? These problems are inherently population problems; i.e., to answer these questions, the *dynamic interactions of neuron groups* have to be studied. In our Marburg group we have developed (a) techniques for recording the spike trains from up to 19 single units; (b) computer-aided procedures for the simultaneous visual stimulation of several units; and (c) real-time correlation methods to assess cooperative firing in groups of neurons.

2 Recording Equipment

Figure 1 shows our fiber electrodes and electrode manipulator equipment (Reitboeck 1983a, b; Reitboeck and Werner 1983). A 7-channel and a 19-channel device have been constructed. They allow moving each electrode independently of the others, which has some major advantages over fixed electrode arrays, because (a) neurons with correlated activities can be searched for; (b) the electrodes can be positioned for optimal single-unit isolation; and (c) the intact dura of cats and monkeys can be penetrated by lowering the electrodes one by one. The electrode tip arrangements are interchangeable between concentric and linear arrays. With a specially developed high-temperature puller, both the electrode's metal core and its quartz insulation are drawn to fine tips.

3 Recording Procedure

After dura penetration, the positions and preferred directions as well as the ocularities of the multiunit fields are plotted for all electrodes on the projection

Springer Series in Brain Dynamics 1
Edited by Erol Başar

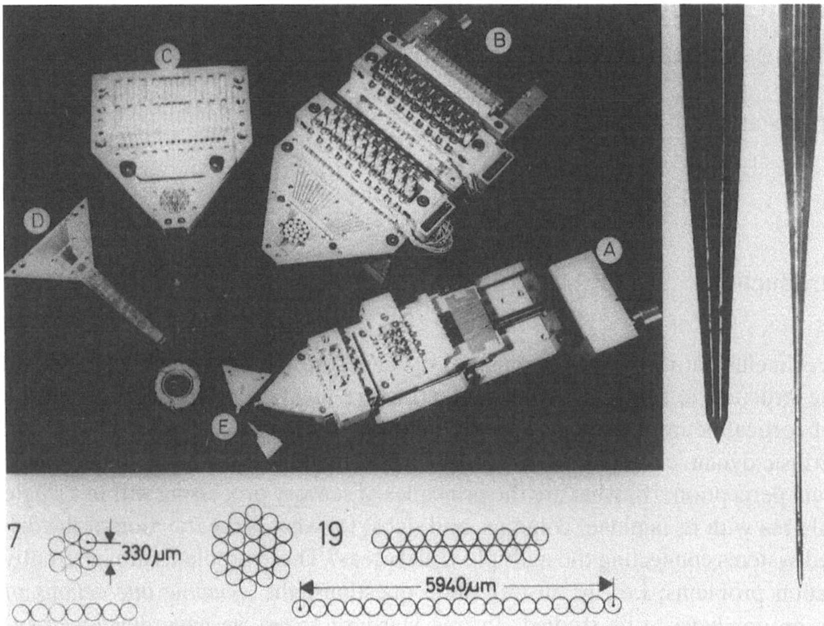

Fig. 1. Recording technique. *Upper left*, 7-channel (*A*) and 19-channel (*B*) microelectrode manipulators (Reitboeck 1983a). Exchangeable manipulator heads for 19- (*C, D*) and 7- (*E*) electrodes. Their geometric tip configurations are shown *below*. *Right,* fiber electrodes with 100 and 65 µm shafts with differently drawn tapers and ground tips

screen in front of the cat. Then single-unit activity of the desired cell type in a given layer is searched for with a single electrode. During electrode advancement, an appropriate visual stimulus is given. Having sufficient isolation of a single unit, we advance another electrode to a position where a second unit is isolated, etc.

Generally we can record single unit activity from up to 19 electrodes. In practice, however, it would be too time consuming to search always for 19 isolated units. Our restriction normally to four to six units is caused by conceptual problems with multi-unit data analysis that have not been solved by any group to date. Not the recording techniques but rather the insensitive, complicated, and time-consuming correlation procedures generally used are the main bottleneck for effective multiunit experiments.

4 Assessment of a Group's Stimulus–Response Correlations

A first estimate of the stimulus–response transfer properties is obtained with our method of deriving the receptive field cinematogram (RF-Cine) of visual units (Krause and Eckhorn 1983). Fig. 2 shows a schematic diagram of our method. A bright or dark disk of appropriate diameter moves randomly across the RF locations of all simultaneously recorded units. For clarity, only a single electrode and one

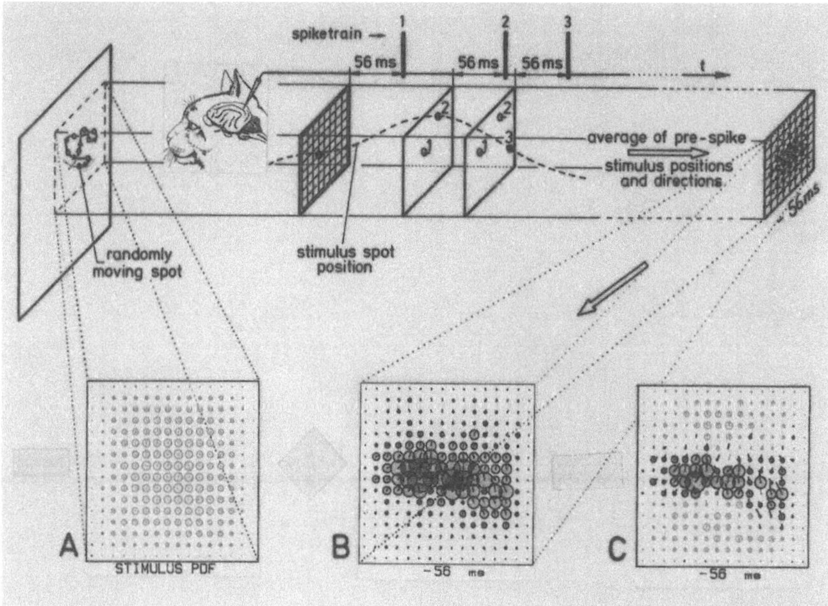

Fig. 2. Scheme of stimulation and data evaluation for receptive field cinematograms. *Left*, randomly moving disk stimulus on the RFs of the investigated units. *Center and right*, scheme for the evaluation of the average prespike stimulus ensemble. For simplicity, only a single electrode, spike train, and delay (– 56 ms) are shown

spike train are plotted. From the stimulus and response data of each unit, a computation is made of where, when, how often, and in what direction the disk has preceded a spike. By this procedure, probability-density distributions of the disk's positions and movement directions are obtained as preimpulse stimulus ensembles in visual space at different stimulus–response delays. The position probabilities are plotted as circles with proportional diameters. Taking the uncorrelated stimulus–response situation as reference, there are values above the statistical expectation, plotted in black, from where spikes were elicited with increased probability, and grey circles where the stimulus appeared less often than was expected. The dashes seen at the centers of the circles indicate the mean stimulus direction and, by their length, the probability of the mean direction at that particular position.

Figure 3 shows 12 consecutive stimulus distributions (frames) representing the same visual space at different prespike times. Such a sequence is called a RF-Cine and represents 2×256 first-order (linear) cross-correlations, one set for the stimulus position-to-response correlations and the other for the direction-to-response correlations. The typical elongated shape of a simple cell's excitatory subfield flanked by two inhibitory regions is seen in the frame at $\tau = 56$ ms before spike initiation. Also interesting are the facts that (a) the excitatory region continuously shifts its position from the uppter left to the lower right in a direction perpendicular to its orientation; and that (b) it shows significant metamorphosis of its shape (Krause et al. 1984). Such substantial *dynamic field changes*, which cannot be unraveled with common RF

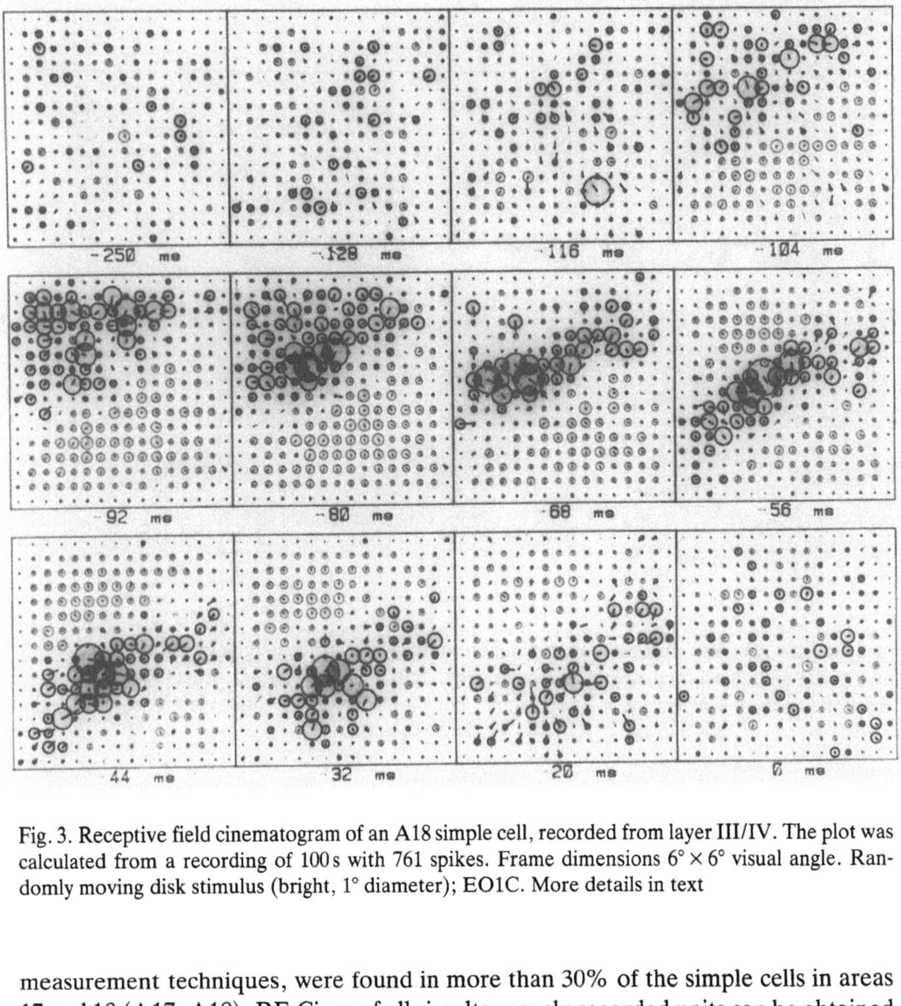

Fig. 3. Receptive field cinematogram of an A18 simple cell, recorded from layer III/IV. The plot was calculated from a recording of 100 s with 761 spikes. Frame dimensions 6° × 6° visual angle. Randomly moving disk stimulus (bright, 1° diameter); EO1C. More details in text

measurement techniques, were found in more than 30% of the simple cells in areas 17 and 18 (A17, A18). RF-Cines of all simultaneously recorded units can be obtained from a single 100–150 s experimental run.

5 Automatic Search for an Appropriate Group Stimulus

For the assessment of internal group connections with correlation techniques, the spikes on these connections have to occur during a 10–20 ms interaction window, the duration of which is given by the postsynaptic potentials (PSP) time constants and by delays in interneuronal spike transmission. A stimulus capable of eliciting such simultaneous discharges is here called an "appropriate group stimulus." Such a stimulus can be derived from a *group RF-Cine*, which can be obtained either by superimposing the single unit's RF-Cines (appropriate for similar overlapping RFs) or by calculating the pre-event stimulus history not for the single-unit spike events but for nearly simultaneous spikes in all neurons of the group. An *appropriate group stimulus* then is derived either by selecting the stimulus disk's most effective movement trajectories from the group RF-Cine or by directly using the forms and positions of excitatory areas in the frame sequence for the generation of a stimulus picture sequence. Both procedures can be carried out on-line and automatically with the aid of computers and a controlled picture generator (Habbel and Eckhorn 1985).

6 Cross-Correlation Analysis

One of our three real-time methods for the assessment of cooperative firing is cross-correlation. In Fig. 4, typical visual cortex interactions are illustrated by cross-correlograms. The lower plots have a time scale of ± 1 s while the upper ones are expanded from the lower central ± 100 ms. In A, the correlations are shown between two simple cells from A17 and A18, respectively, with overlapping RFs having nearly the same RF properties. The random-dot pattern stimulus was gated during its movement course. The start/stop movement cycle generated a strongly locked response in phase, as can be seen by the broad peaks in the lower part of Fig. 4A. The narrower center peak, however, is caused by the random structure of the dot pattern moving across the fields. A time-expanded version of this peak (upper Fig. 4A) shows a delay of 12 ms between the A17 and A18 cells, which is confirmed by the small 4 ms peak probably caused by a direct excitatory connection. In Fig. 4B, the correlograms are shown for a pair of A18 complex cells. They were stimulated with a continuously and randomly moving disk. The slightly leftward-shifted broad center peak is caused by the common stimulation, which caused different response delays

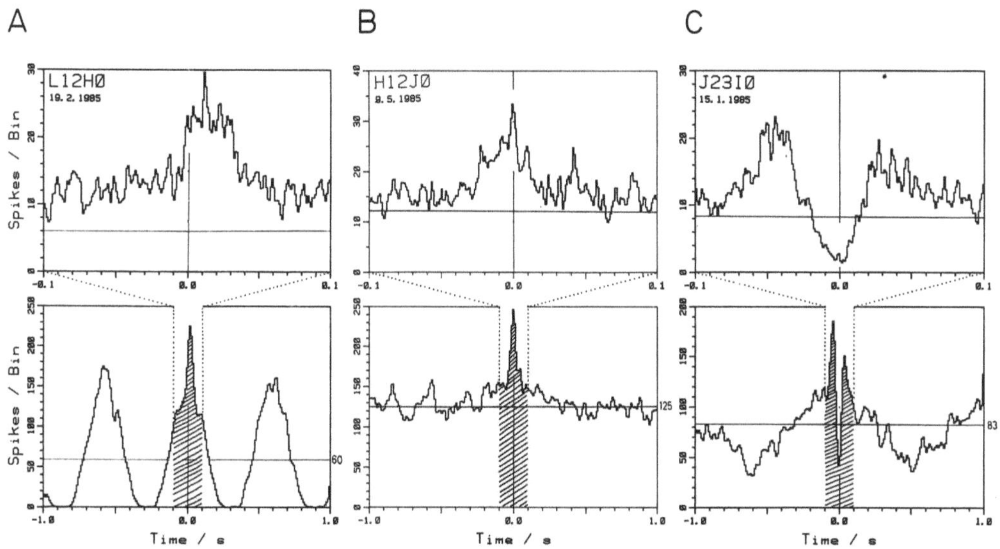

Fig. 4A–C. Cross-correlograms of three pairs of visual cortex neurons with overlapping RFs. (A) Two simple cells, A17 and A18. (B) Two complex cells, A18. (C) Simple and complex cell A18. More details in text

in both units. The time-expanded version also shows a common synaptic input, as can be read from the narrow centered peak. Inhibitory interactions or common antagonistic input was found less frequently and was in most cases accomplished by excitation. In Fig. 4C, an example of this is shown for a simple/complex pair in A18 stimulated with a random-dot pattern gated in its movement path.

7 Dynamic Changes in Correlated Activities

A major problem with the cross-correlation is its low sensitivity to inhibitory interactions (Aertsen and Gerstein 1985) and its failure to comprise *dynamic coupling changes*. For periodic stimuli, we improved a method of Gerstein and Perkel (1972) that reveals the periodic coupling changes in correlation histograms (Schneider et al. 1983). Such a dynamic coupling measure has the advantage over time averages that the varying types as well as the different instances of high and low correlation can be related to their causes (special stimulus phases, sudden changes in attentiveness, etc.). Figure 5 shows correlation histograms of an A18 simple/complex cell pair stimulated under four different conditions. A central correlation peak ($\tau = 0$ ms) and two smaller structures at delays of about $\pm$ 180 ms can be seen with changing amplitudes at different stimulus phases.

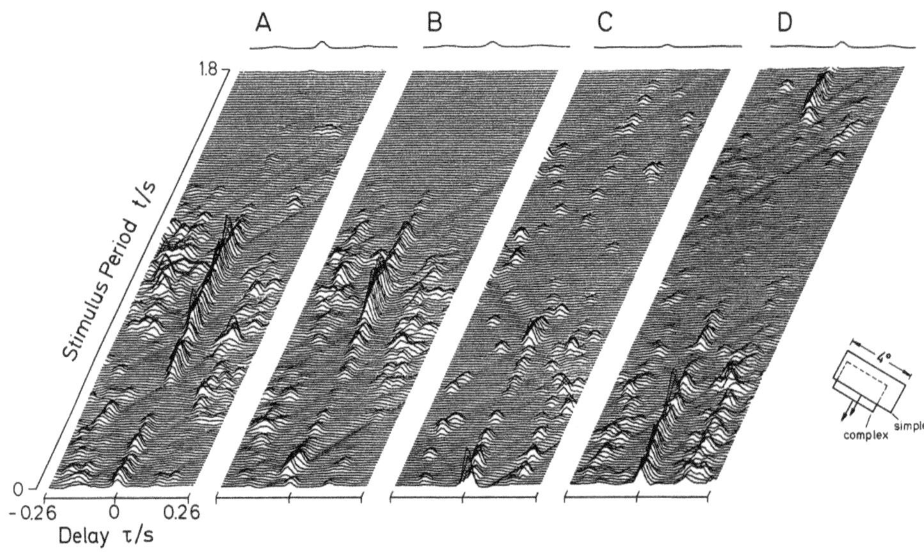

Fig. 5A–D. Correlation histograms of A18 complex/simple-cell pair under four stimulus conditions. (A) Constant velocity movement of bar in preferred and opposite direction, 2 s cycle, left eye. (B) Same as A but right eye. (C) Circular movement of stimulus, 2 s cycle, right eye. (D) Same as C but left eye. Electrode separation 330 µm; complex cell in 700 µm, simple cell in 775 µm depth from surface. $N = 106$ stimulus cycles. Scheme of RFs on the *right*. At the *top* the cross-correlograms are plotted on the same scale

Both cross-correlograms and correlation histograms comprise only pair interactions. For groups larger than about six neurons, new methods are needed.

8 A "Neuron-Like" Network for Real-Time Correlation

The brain has no direct knowledge of physical stimuli, but extracts sensory information from the spatiotemporal patterns of the afferent neural sensory signals. This is accomplished by a network of interacting neurons. It, therefore, seemed promising for us also to assess the spatially and temporally correlated activities of neuron groups with a "neural network." A correlator was developed that has some simple properties derives from interconnected neurons. In Fig. 6, the first two channels of the correlator are shown diagrammatically (Schneider and Eckhorn 1984; Eckhorn et al. 1986). Each recorded spike train (a) is converted to smooth PSP (c) by leaky integrators (b). The subsequent correlation procedure is performed by multiplications of all possible combinations of PSP signals; that is, all pair-wise, triple, quadruple, etc., interactions are evaluated (f). This evaluation is greatly simplified by the use of an analog-to-stochastic converter (d), which encodes the PSPs into stochastic impulse sequences (e). Multiplications are simply carried out by AND gates. The participation of the single neurons in the joint activity of the group are derived as weighted running averages (i) of the multiplicator-signals (g) with a second set of leaky integrators (h). These coupling values are displayed via LED bars (j). That is, only one actual coupling strength is displayed per recorded spike train. This restriction is of great value for real-time assessments of correlated group activities, because the experimenter is not "overloaded" by controlling outputs from all combinations of connections. It is sufficient during the experiment to know whether the recorded group activity is worthy of being stored in the computer so that more detailed analyses can be obtained afterwards off-line.

Figure 7 shows two examples of correlator output signals. In B the actual correlation values of two complex units are shown. They are moderately phase-locked to the cyclic movement of a random bar pattern. This is better seen in the correlation cycle histogram below (Bb), which is the ensemble average of the original signals

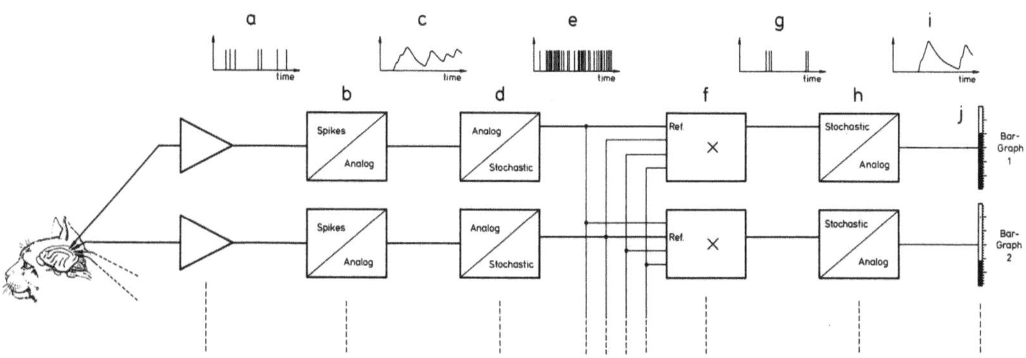

Fig. 6. Scheme of real-time correlator with "neuronal properties"

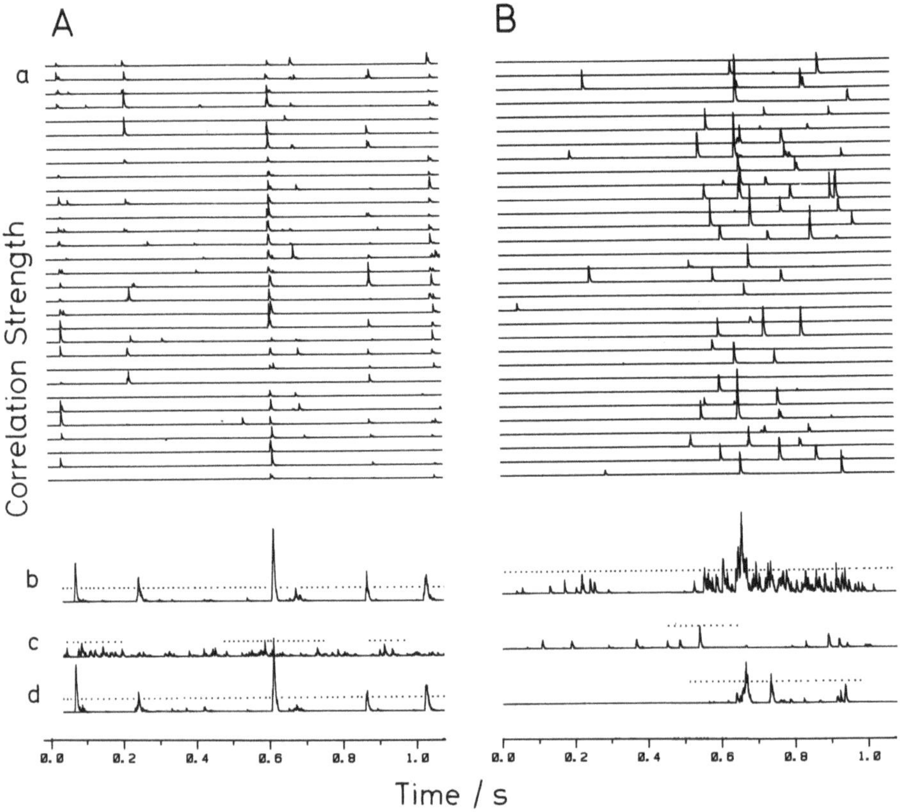

Fig. 7. A, B. Actual correlator output and correlation cycle histograms for two pairs of visual cortex neurons with overlapping RFs. Stimulus: constant velocity movement of random bar pattern against (first half cycle) and in optimal direction (second half cycle). (A) Two simple units. (B) Two complex units. *a*, Signals at one correlator output in response to 31 successive stimulus periods. *b*, Correlation cycle histogram, the ensemble average of a. *c*, Correlation cycle histogram obtained after interval shuffling of original spike trains; *dotted line*, maximum of c. *d*, Correlation cycle histogram calculated after shifting one spike train by one stimulus period

above. In A, the correlated activities of two simple units are shown in response to the same stimulus. A bar with appropriate orientation travelling across the RFs always excited the units nearly simultaneously. Here, sharp phase-locking to the stimulus pattern is present, which can be seen more clearly in the correlation histogram below. Especially interesting are the comparisons of traces (Fig. 7b) with the lowermost traces (Fig. 7d). The latter were calculated with one spike train shifted by one stimulus period ("shift predictor"). In the case of the two simple units (Fig. 7A), the original correlation cycle histogram and its shift predictor are nearly identical; i.e., all correlations were caused by the common stimulus. On the contrary, the shift predictor calculated for the two complex units (Fig. 7Bd) shows only smaller and much less frequent events; i.e., the numerous correlations seen above (Fig. 7Bb) are mainly caused by synaptic interactions. These results were confirmed by cross-correlations.

Summarizing our preliminary results from the cat's primary and secondary cortex, we can say that the correlations that could be ascribed to horizontal intracortical connections are normally weak. Synergistic activities were found at distances of up to 2 mm, but only when the FRs overlapped and when the cell type and the orientational and directional specificities were nearly identical. This finding is supported by the preliminary data of Gilbert et al. (1985), who made pair recordings in cat and monkey striate cortex. More frequent and stronger correlations were found when the electrode tips were confined to a small vertical column of about 200 μm width. Correlated activities from the same type of cell were synergistic because of common excitatory input or mutual excitation. Inhibition was seldom found, and only in conjunction with excitation when both cells were of different types.

Acknowledgements. The recording equipment was developed mainly by H. J. Reitboeck and built by P. Muth and W. Gerber. Electrode improvements and fabrication were by U. Thomas. W. Lenz and H. Wagner assisted in the experimental preparations and W. Adamczak, C. Habbel, S. Hansch, F. Krause, H. Lohmann, and J. Schneider were involved in the experiments and in the development of our concepts. S. van Rennings typed the manuscript. The authors are grateful to all these co-workers. This work was supported by DFG grants Re 547/1-1,2 and VW I/35695.

References

Aertsen A, Gerstein G (1985) Evaluation of neuronal connectivity: sensitivity of cross-correlation. Brain Res 340: 341–354

Eckhorn R, Habbel C, Krause F, Lohmann H, Reitboeck HJ, Schneider J (1984) Multiunit recordings from the visual system require special concepts. Neled [Suppl] 18: 164

Eckhorn R, Schneider J, Keidel R (1986) Real-time covariance computer for cell assemblies is based on neuronal principles. J Neurosci Methods 18: 371–383

Gerstein G, Perkel D (1972) Mutual temporal relationships among neuronal spike trains. Statistical techniques for display and analysis. Biophys J 12: 453–473

Gilbert C, Wiesel T, Ts'o D (1985) Clustered intrinsic connections and functional architecture of the visual system. Invest Ophthalmol Vis Sci [Suppl] 26: 133

Habbel C, Eckhorn R (1985) Automatic search for effective stimuli for units of the cat's visual cortex. Neled [Suppl] 22: 203

Krause F, Eckhorn R (1983) Receptive fields for motion stimuli determined for different types of cat visual neurons. Neled [Suppl] 14: 209

Krause F, Eckhorn R, Habbel C (1984) Two types of direction selective mechanisms in simple units revealed by Receptive Field Cinematograms. Eur J Physiol [Suppl] 402: R51

Reitboeck HJ (1983a) Fiber microelectrodes for electrophysiological recordings. J Neurosci Methods 8: 249–262

Reitboeck HJ (1983b) A 19-channel matrix drive with individually controllable fiber microelectrodes for neurophysiological applications. IEEE Trans Syst Man Cybern 13: 676–683

Reitboeck HJ, Werner G (1983) Multi-electrode recording system for the study of spatio-temporal activity patterns of neurons in the central nervous system. Experientia 39: 339–341

Schneider J, Eckhorn R (1984) Real time correlator for neuronal assemblies. Neled [Suppl] 18: 412

Schneider J, Eckhorn R, Reitboeck HJ (1983) Evaluation of neuronal coupling dynamics. Biol Cybern 46: 129–134

Generation of Fast and Slow Field Potentials of the Central Nervous System – Studied in Model Epilepsies

E.-J. SPECKMANN and J. WALDEN

1 Introduction

This chapter deals with the potentials detectable in the space surrounding cellular elements of central nervous structures. Such potentials, which can in part also be recorded from outside the central nervous system as for example, the electroencephalogram, sensory evoked potentials, contingent negative variatons, etc., are generally called "field potentials."

The first section describes the elementary mechanisms underlying the generation of field potentials. The second and third sections are devoted to cortical field potentials during focal and generalized tonic-clonic seizure activity elicited in animal experiments. Such model epilepsies have the advantage that the bioelectrical activity of a single cell is representative of a larger population of elements, which facilitates the analysis of field potential generation (Caspers and Speckmann 1970; Creutzfeldt and Houchin 1974; Speckmann and Caspers 1979a; Lopes da Silva and van Rotterdam 1982; Speckmann and Elger 1982, 1984; Caspers et al. 1984; Speckmann et al. 1984).

2 Elementary Mechanisms Underlying the Generation of Field Potentials

In this chapter, neurons and glial cells will be considered as the generative structures of field potentials in the central nervous system. Other structures – such as, for example, meninges and the blood-brain barrier – can be neglected as generators in this context (de Robertis and Carrea 1965; Palay and Chan-Palay 1977; Caspers et al. 1984).

In *neurons*, postsynaptic potentials are thought to play a predominant role in the generation of field potentials. With the initiation of an excitatory postsynaptic potential, a net inflow of cations occurs at the membrane region located under the synapse (Eccles 1964; Hubbard et al. 1969; Shepherd 1974). That inflow induces a depolarization of the subsynaptic membrane. The process described is associated with the development of a potential gradient along the neuronal membrane in the intracellular and extracellular space. This potential gradient forces ions to move along the nerve cell membrane. The flow of cations is directed to the subsynaptic region in the extracellular space and is inverse in the intracellular space. With the generation of inhibitory postsynaptic potentials, an inflow of anions or an outflow of cations appears, which hyperpolarizes the membrane in the subsynaptic region as compared with the surrounding segments of the membrane. Thus, a potential gradient along the cell

Springer Series in Brain Dynamics 1
Edited by Erol Başar

membrane is established as in the case of the excitatory postsynaptic potentials, but with an opposite polarity. This potential gradient again induces an ionic current. Under these conditions, the flux of cations is directed from the subsynaptic membrane to the surrounding membrane regions in the extracellular space, but has an inverse direction in the intracellular space (Hubbard et al. 1969; Creutzfeldt and Houchin 1974; Lopes da Silva and van Rotterdam 1982; Speckmann and Elger 1982; Speckmann et al. 1984).

In *glial cells*, the membrane potential is determined mainly by the extracellular K^+ concentration (K^+ activity). When the extracellular K^+ concentration increases and then re-decreases, the glial cells depolarize and repolarize, respectively. The dependence of the membrane potential of glial cells on the extracellular K^+ concentration causes a functional linkage between glial and neuronal structures, as neuronal discharges are accompanied by a potassium outflow. When the K^+ concentration is elevated only locally, a potential gradient develops between the depolarized area and the adjacent regions. This leads to an intra- and extracellular current flow as described above. Because glial cells have been found to have widespread processes and to be connected with each other, potential gradients and current flows of considerable spatial extent may develop (Kuffler and Nicholls 1966; Kuffler et al. 1966; Orkand et al. 1966; Somjen 1973, 1975; Somjen and Trachtenberg 1979).

As described above, membrane potential changes in neurons and glial cells are accompanied by primary transmembraneous ionic currents, which may lead to secondary ionic currents along the cell membranes in the extra- and intracellular space. Field potential generation is caused by the current component flowing through the extracellular space. The mechanisms concerned are shown in Fig. 1. Figure 1A shows a long neuronal element perpendicularly oriented within a structure of the central nervous system; e.g., within the cerebral cortex. One end of the element is lying close to the surface of the cortex and is contacted at an excitatory synapse by an afferent fiber. The bioelectric activity of the model tissue in Fig. 1 is explored with intracellular and extracellular electrodes. The activity in the afferent fiber is monitored by the microelectrode ME1. The membrane potential of the perpendicular neuronal element is recorded by the microelectrodes ME2 and ME3 inserted into the two ends of the unit. The field potentials are picked up by the extracellular electrodes E1 and E2, one located at the surface of the cortex and the other within the cortex, near the deeper end of the neuronal element.

When an action potential appears in the afferent fiber (Fig. 1A1), the excitatory synapse is activated. That leads (a) to a net inflow of cations; (b) to a potential gradient along the neuronal element; and thus (c) to a current flow in the intracellular and extracellular space. Because of the movement of positive ions in the intracellular space, an excitatory postsynaptic potential can be recorded from all parts of the model neuron (Fig. 1A2, 3). As a consequence of the ionic movements in the extracellular space, field potentials are built up. At the surface electrode E1, the net inflow of positive charge into the neuron gives rise to a negative field potential (Fig. 1A4) and at the depth electrode E2 – metaphorically speaking – the approach of positive charges leads to a positive field potential (Fig. 1A5). A reversal of polarity of the field potentials occurs between the electrodes E1 and E2.

When an inhibitory synapse located at the deeper end of the neuron is activated, an extracellular current flow and consequent establishment of field potentials take

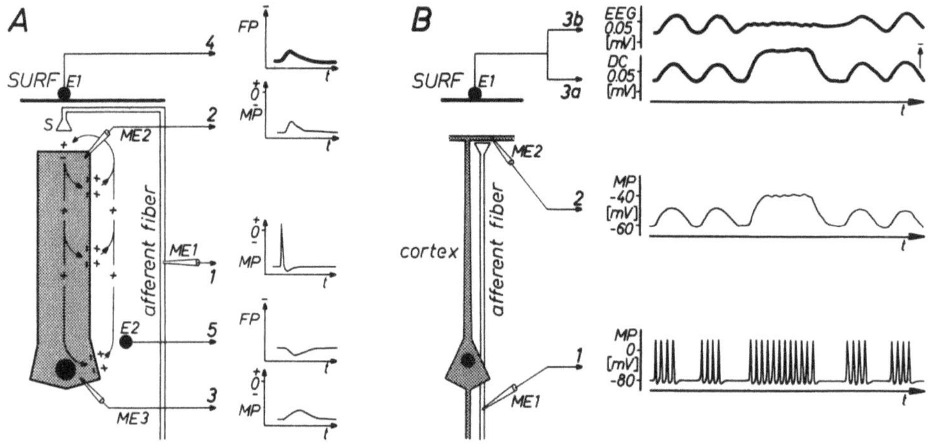

Fig. 1A, B. Elementary mechanisms underlying the generation of field potentials in the central nervous system. (A) Mechanisms responsible for a single potential fluctuation. *1*, Recording the membrane potential (*MP*) of the afferent fiber by means of an intracellular microelectrode (*ME1*). *2*, and *3*, MP of a neuronal element recorded at both its superficial and deep end with microelectrodes ME2 and ME3. *S*, excitatory synapse. *4* and *5*, Field potentials are picked up at the surface (*SURF*) of a central nervous structure with extracellular electrode E1 and in the vicinity of ME3 with electrode E2. (B) Mechanisms responsible for wave-like fluctuations at the cerebral cortex. *1* and *2*, Recording the MP of an afferent fiber and of a superficial dendrite with intracellular microelectrodes ME1 and ME2. *3*, Measurement of the DC potential (*3a*) and the EEG (*3b*) at the cortical surface with the extracellular electrode E1. (From Speckmann 1986; modified after Speckmann et al. 1984 and Speckmann and Elger 1982)

place, similar to those described for the superficial excitatory synapse. Thus, in the model of Fig. 1A, a negative field potential develops at the surface of the cortex when a superficial excitatory or a deep inhibitory synapse is activated. The generation of superficial field potentials of positive polarity can be explained on corresponding principles (Hubbard et al. 1969; Rall 1977; Lopes da Silva and van Rotterdam 1982; Speckmann and Elger 1982; Speckmann et al. 1984).

Single potential fluctuations, the basic mechanisms of which have been described (Fig. 1A), can combine into wave-like fluctuations when the afferent fiber activity is appropriately distributed in time and space (Andersen and Andersson 1968). In the model experiment on cortical units shown in Fig. 1B, grouped discharges occur in the afferent fiber, which are then replaced by a sustained activity (Fig. 1B1). The ascending action potentials evoke excitatory postsynaptic potentials at the upper neuronal element. Corresponding to the discharge pattern of the afferent fiber, the individual postsynaptic potentials summate to long-lasting depolarizations (Fig. 1B2). These depolarizations give rise to field potentials that can be picked up at the cortical surface (Fig. 1B3). The extracellular potentials elicited by grouped and sustained afferent activity are reflected precisely in epicortical DC recordings (Fig. 1B3a), whereas for technical reasons, only faster field potentials appear in conventional EEG recordings (Fig. 1B3b) (Caspers and Speckmann 1969, 1974; Gumnit et al. 1970; Goldring 1974; Speckmann and Caspers 1974, 1979b; Caspers et al. 1979, 1980).

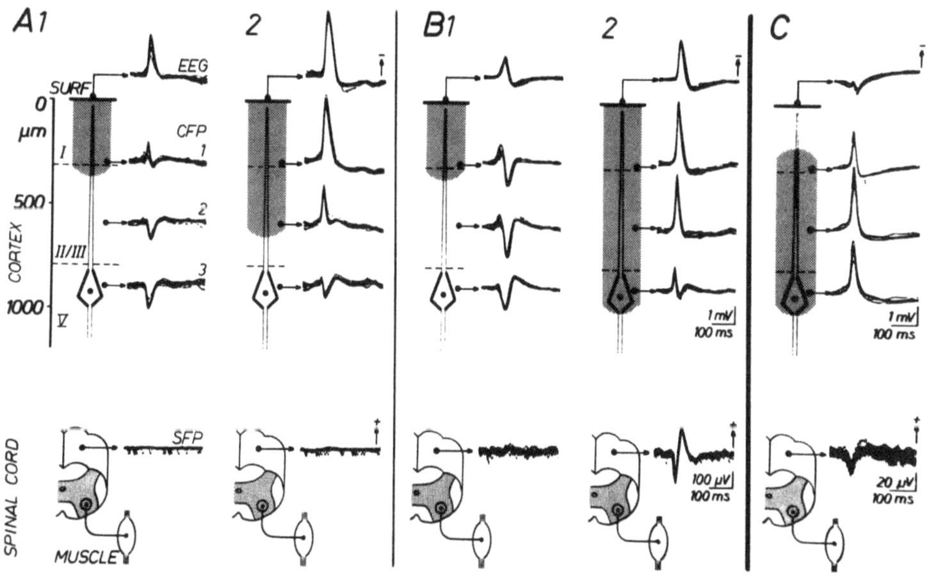

Fig. 2A–C. Cortical field potentials during focal seizure activity. Distribution of field potentials within the motor cortex and descending neuronal activity to the spinal cord. Simultaneous recordings of epicortical (*EEG*), intracortical (*CFP1–3*), and spinal (*SFP*) field potentials. Superimposition of 15 single potentials. *SURF*, cortical surface. *I*, *II/III*, *V*, cortical laminae. Seizure activity was elicited by local application of penicillin to the cortical surface (A, B) and in deeper laminae of the cortex (C). (A1, A2, B1, B2), Recordings 5 min (*1*) and 15 min (*2*) after the epicortical penicillin application. Additional epicortical application of penicillinase before (A) and after (B) the penicillin administration. Vertical extension of seizure activity is indicated by *hatched columns*. (From Speckmann 1986; modified after Elger et al. 1981)

3 Cortical Field Potentials During Focal Seizure Activity

The functional significance, the extension, and the laminar distribution of field potentials can be studied in greater detail using seizure activity as a model. This section deals with focal seizure activity. After the topical application of an epileptogenic agent – e.g., penicillin – to the surface of the cerebral cortex, steep negative field potentials develop. These focal epileptiform potentials are associated with characteristic changes in membrane potential of individual neurons in the area of drug application. The neuronal reaction consists of a steep depolarization, accompanied by a group of high-frequency action potentials, a plateau-like diminution of the membrane potential in the course of which action potentials are blocked, and a more or less steep repolarization. Such membrane potential fluctuations have been found to be characteristic for neuronal epileptiform activity and labeled "paroxysmal depolarization shifts" (Jasper et al. 1969; Purpura et al. 1972; Speckmann et al. 1972, 1978; Elger et al. 1981; Klee et al. 1981; Elger and Speckmann 1983; Speckmann 1986).

Typical experiments in which focal seizure activity was elicited in cortical motor regions for the forelimbs and hindlimbs are presented in Fig. 2. In the experiments shown in Fig. 2A and B, penicillin was applied to the surface of the cortex, and in the experiment shown in Fig. 2C, penicillin was injected into deeper cortical layers. Field potentials were recorded simultaneously from the surface as well as from different cortical laminae, with an interelectrode distance of 300 μm. To get a preliminary idea of the functional significance of seizure potentials recorded at the surface of the motor cortex, neuronal activity descending to the spinal cord was controlled. For this purpose, spinal field potentials were measured from cervical and lumbar segments. Figure 2 shows that negative seizure potentials were recorded 5 min (A1, B1) and 15 min (A2, B2) after epicortical penicillin application. There were stereotyped epileptiform potentials at the cortical surface, but the intracortical potential distribution differed considerably. Thus, the field potential in layer V was mainly positive in A1, A2, and B1, whereas it was mainly negative in B2. As indicated by the hatched columns, only in the latter case did seizure activity reach pyramidal tract cells. Consequently, neuronal activity descending to the spinal cord was restricted to these conditions, as indicated by synchronized spinal field potentials. When penicillin was applied in deeper cortical layers, negative field potentials were generated in the region of application (Fig. 2C). Simultaneously, atypical potential fluctuations appeared at the cortical surface (Elger and Speckmann 1980, 1983; Elger et al. 1981; Petsche et al. 1981; Pockberger et al. 1983, 1984a, b).

From these findings, the conclusions may be drawn that (a) a vertical segment of the cerebral cortex reacts in homogeneously in the given experimental conditions; and (b) the field potentials measured at the cortical surface are not necessarily representative of the field potentials in deeper cortical layers and, under certain conditions, may not even be representative of those in superficial laminae.

4 Cortical Field Potentials During Generalized Tonic-Clonic Seizure Activity

Tonic-clonic seizure activity can be elicited in animal experiments using a variety of techniques. In the experiments reported in this section, convulsive activity was induced by repetitive systemic administrations of pentylenetetrazol. Epileptic seizures elicited in this way are characterized by slow negative displacements of the cortical DC potential with superimposed fast potential fluctuations (Fig. 3). In what follows, the slow DC shifts and the fast DC fluctuations will be considered successively (Caspers and Speckmann 1969; Gumnit 1974).

As a first step, the *shifts* in the cortical DC potential are compared with changes in the membrane potential of pyramidal tract cells. Typical recordings are presented in Fig. 3. Parts B and C of Fig. 3 demonstrate that the negative shift of the epicortical DC potential is associated with a series of paroxysmal depolarization shifts. Also, the DC potential measured from the pyramidal tract cell layer shows a negative shift during tonic-clonic activity (Fig. 3D). Thus, the mean neuronal depolarization is accompanied by DC shifts of the same polarity in epicortical and laminar recordings.

Besides the described correspondence between the shifts in epicortical and laminar DC potentials and the changes of membrane potential in pyramidal tract cells

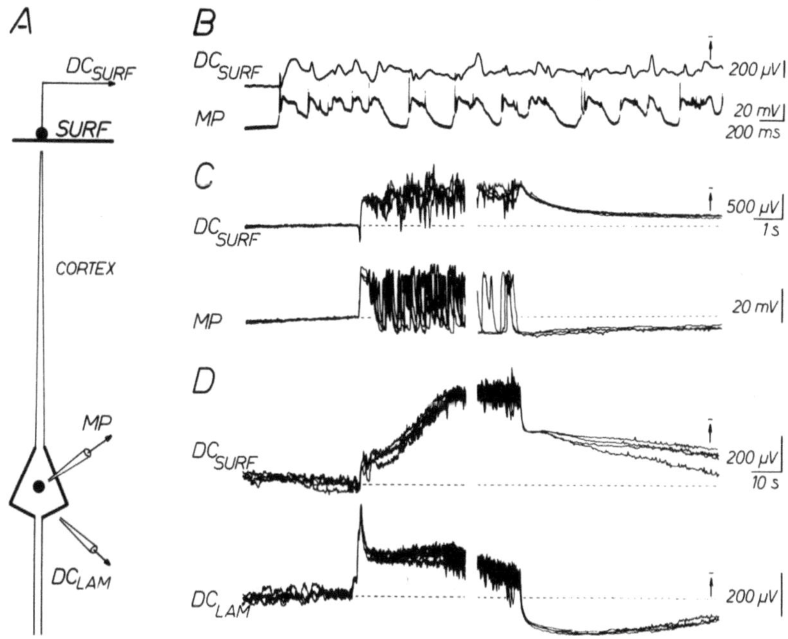

Fig. 3A–D. Cortical field potentials during generalized tonic-clonic seizure activity. Relationship between changes in epicortical and laminar DC potentials and in membrane potential (*MP*) of pyramidal tract cells. Seizures were elicited by systemic administration of pentylenetetrazol. (A) Schematic drawing of the electrode arrangement. *SURF*, cortical surface; DC_{SURF} and DC_{LAM}, epicortical and laminar DC potential. (B) Simultaneous recording of DC_{SURF} and MP of a pyramidal tract cell. (C, D) Graphical superimposition of simultaneous recordings of DC_{SURF} and MP (C) and of DC_{SURF} and DC_{LAM} (D). Interruption of recordings: 30–60 s. (From Speckmann 1986; modified after Speckmann and Caspers 1979b)

during epileptic activity, there are also discrepancies between these bioelectric events, which become apparent especially at the onset and termination of seizures. On one hand (a) the first neuronal paroxysmal depolarization shift coincides with monophasic negative or positive or biphasic positive-negative fluctuations in the epicortical DC recording; and (b) the neuronal hyperpolarization occurring at the end of an attack parallels a slow redecline of the negative DC shift at the cortical surface (Fig. 3C). On the other hand, there is a close correlation between membrane potential changes in pyramidal tract cells and shifts of the laminar DC potential (Fig. 3C and D). From these findings, the conclusion may be drawn that the laminar DC potential in question is predominantly generated by somata of pyramidal tract cells and adjacent neuronal structures and that superficial cortical elements, e.g., apical dendrites, have to be taken into account as generative structures for the epicortical DC potential. The correspondence of all the described cortical bioelectric phenomena during ictal activity can be explained by a simultaneous excitation of superficial and deep generator structures by afferent neuronal activity (Speckmann et al. 1972, 1978; Speckmann and Caspers 1979b).

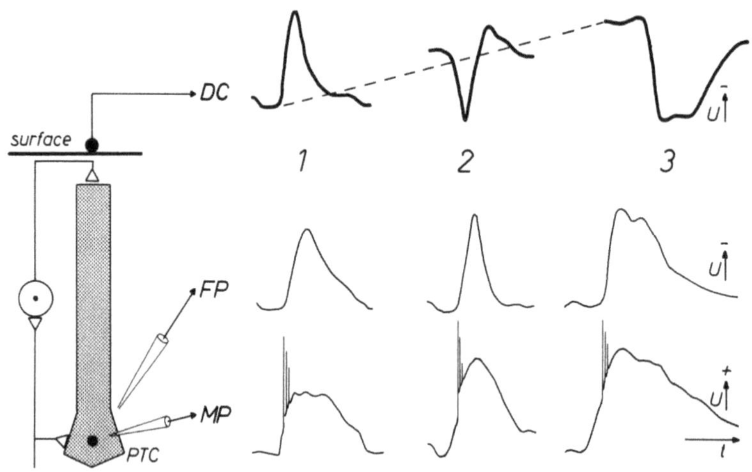

Fig. 4. Comparison of potential fluctuations at the cortical surface *(DC)*, of field potentials *(FP)* in the fifth cortical lamina, and of the membrane potential *(MP)* of a pyramidal tract cell *(PTC)* during generalized tonic-clonic seizures. Epileptic activity was induced by systemic administration of pentylenetetrazol. Motor cortex; cat. The negative DC shift occurring during the seizure is indicated by the *dashed line* in the *upper row*. *U*, voltage. (From Speckmann et al. 1984)

As a second step, the *fast potential changes* superimposed on the shifts of the DC potential are compared with changes in the membrane potential of pyramidal tract cells. A typical experiment is shown in Fig. 4. Field potentials recorded at the cortical surface and in the pyramidal tract cell layer are shown simultaneously with the membrane potential changes in pyramidal tract cells. The recordings show that every paroxysmal depolarization of a pyramidal tract cell is associated with a solitary negative fluctuation of the laminar field potential. These stereotyped intracellular and extracellular potential fluctuations in deep cortical layers are accompanied by field potentials at the cortical surface with either solitary negative or positive configurations or with positive-negative configurations (Creutzfeldt et al. 1966a, b; Speckmann et al. 1972, 1978; Speckmann and Caspers 1979b).

Whereas the laminar fluctuations in the DC potential are obviously generated by the somata of pyramidal tract cells, the generation of the different waves of the superficial EEG may be explained on the basis of the following three observations:

(a) correlating the shape of the seizure potentials in the epicortical EEG with the extent of the negative DC shift, it becomes apparent that surface negative fluctuations tend to be associated with a small DC shift and that surface positive fluctuations appear only if the negative shift at the cortical surface exceeds a critical value (Fig. 4; Speckmann et al. 1972, 1978; Speckmann and Caspers 1979b; cf. also Caspers 1959, 1963);

(b) a close correlation exists between the amplitude of the negative DC shift at the cortical surface and the discharge frequency in afferent systems (Speckmann et al. 1978; Speckmann and Caspers 1979b);

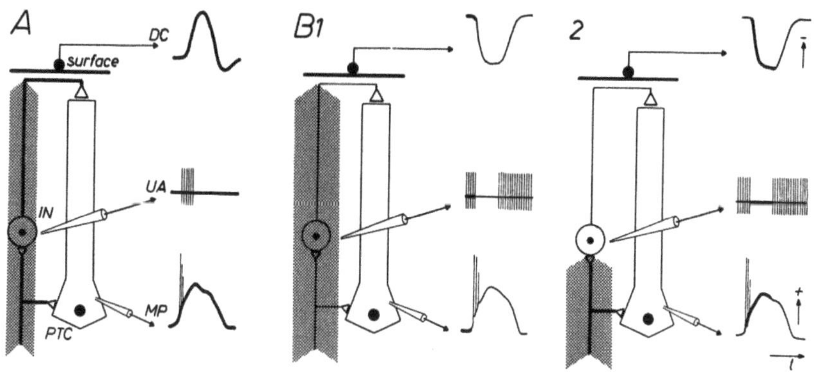

Fig. 5A, B. Hypothetical diagram of an interpretation of fast DC waves of opposite polarity occurring during generalized tonic-clonic seizures. *Hatched arrows,* asynchronous input to the cerebral cortex; *heavy lines,* phasic volleys initiating single convulsive discharges; *PTC,* pyramidal tract cell; *IN,* interneuron; *MP,* membrane potential; *UA,* extracellularly recorded unit activity. (A) With a moderate asynchronous input to the cortex (small hatched arrow), a burst of UA triggers a paroxysmal depolarization shift in a PTC and evokes a depolarization of superficial neuronal structures and thus a negative fluctuation in the DC recording at the cortical surface. (B) With an increased asynchronous input to the cortex (*widened hatched arrow;* B1), a phasic volley reaching the cortex triggers paroxysmal depolarization shifts in the PTC and interrupts the asynchronous input by an inactivation of IN (B2). The latter process results in a disfacilitation of superficial neuronal structures and thus in a positive fluctuation of the epicortical DC potential. (From Speckmann et al. 1984; drawings after original tracings from Speckmann et al. 1978)

(c) the afferent input to superficial cortical structures is enhanced at the beginning of negative fluctuations and reduced during positive ones (Speckmann et al. 1978; Speckmann and Caspers 1979b).

These observations may be combined into a hypothesis which is illustrated in Fig. 5. In this diagram, the afferent input to superficial generator structures, which can be regarded as being predominantly asynchronous, is symbolized by the width of the hatched arrows. As a consequence, the afferent input and the resulting negative DC shift at the cortical surface are smaller in part A than in part B. In the situation shown in part A, a highly synchronized afferent input, symbolized by heavy lines, reaches not only deep cortical structures but also superficial ones, via intercalated neurons. This triggers a paroxysmal depolarization shift in pyramidal tract cells and depolarizes neuronal elements close to the cortical surface. Thus, the paroxysmal depolarization of the soma of pyramidal tract cells parallels a negative potential fluctuation at the surface. In the situation shown in part B, the pre-existing afferent inflow is high (widened hatched arrows), which causes an elevation of interneuronal activity. If a synchronized afferent input then takes place, paroxysmal depolarizations are still triggered in the pyramidal tract cells. The activity of interneurons, however, is transiently interrupted because of inactivation. This process decreases the excitatory input to superficial neuronal elements and thus leads to a positive field potential at the cortical surface, associated with a stereotyped paroxysmal depolarization and a monophasic negative field potential in the depth (cf. also Fig. 4; Speckmann et al. 1978, 1984; Speckmann and Elger 1982; Speckmann 1986).

5 Conclusions

In summary, the results presented demonstrate: (a) neuronal activity within the cerebral cortex does not show a constant relationship to the epicortical field potential; and (b) deep and superficial field potentials are generated at functionally different generator structures. Therefore, epicortical field potentials can be interpreted in different ways as far as neuronal activity in the cortex is concerned.

References

Andersen P, Andersson SA (1968) Physiological basis of the alpha rhythm. Meredith, New York

Caspers H (1959) Über die Beziehungen zwischen Dendritenpotential und Gleichspannung an der Hirnrinde. Pflügers Arch 269:157–181

Caspers H (1963) Relations of steady potential shifts in the cortex to the wakefulness-sleep spectrum. In: Brazier MAB (ed) Brain function. University of California Press, Berkeley-Los Angeles, pp 177–213

Caspers H, Speckmann E-J (1969) DC potential shifts in paroxysmal states. In: Jasper HH, Ward AA Jr, Pope A (eds) Basic mechanisms of the epilepsies. Little Brown, Boston, pp 375–388

Caspers H, Speckmann E-J (1970) Postsynaptische Potentiale einzelner Neurone und ihre Beziehungen zum EEG. Z EEG EMG 1:55–65

Caspers H, Speckmann E-J (1974) Cortical DC shifts associated with changes of gas tension in blood and tissue. In: Remond A (ed) Handbook of electroencephalography and clinical neurophysiology, vol 10A. Elsevier, Amsterdam, pp 41–65

Caspers H, Speckmann E-J, Lehmenkühler A (1979) Effects of CO_2 on cortical field potentials in relation to neuronal activity. In: Speckmann E-J, Caspers H (eds) Origin of cerebral field potentials. Thieme, Stuttgart, pp 151–163

Caspers H, Speckmann E-J, Lehmenkühler A (1980) Electrogenesis of cortical DC potentials. Prog Brain Res 54:3–15

Caspers H, Speckmann E-J, Lehmenkühler A (1984) Electrogenesis of slow potentials of the brain. In: Rockstroh B, Lutzenberger W, Birbaumer N (eds) Self-regulation of the brain and behavior. Springer, Berlin Heidelberg New York Tokyo, pp 26–41

Creutzfeldt O, Houchin J (1974) Neuronal basis of EEG waves. In: Remond A (ed) Handbook of electroencephalography and clinical neurophysiology, vol 2C. Elsevier, Amsterdam, pp 5–55

Creutzfeldt O, Lux HD, Watanabe S (1966a) Relations between EEG phenomena and potentials of single cortical cells. I. Evoked responses after thalamic and epicortical stimulation. Electroencephalogr Clin Neurophysiol 20:1–18

Creutzfeldt O, Lux HD, Watanabe S (1966b) Relations between EEG phenomena and potentials of single cortical cells. II. Spontaneous and convulsoid activity. Electroencephalogr Clin Neurophysiol 20:19–37

De Robertis EDP, Carrea R (eds) (1965) Biology of neuroglia. Prog Brain Res 15

Eccles JC (1964) The physiology of synapses. Springer, Berlin Göttingen Heidelberg

Elger CE, Speckmann E-J (1980) Focal interictal epileptiform discharges (FIED) in the epicortical EEG and their relations to spinal field potentials in the rat. Electroencephalogr Clin Neurophysiol 48:447–460

Elger CE, Speckmann E-J (1983) Penicillin induced epileptic foci in the motor cortex: vertical inhibition. Electroencephalogr Clin Neurophysiol 56:604–622

Elger CE, Speckmann E-J, Prohaska O, Caspers H (1981) Pattern of intracortical potential distribution during focal interictal epileptiform discharges (FIED) and its relation to spinal field potentials in the rat. Electroencephalogr Clin Neurophysiol 51:393–402

Goldring S (1974) DC shifts released by direct and afferent stimulation. In: Remond A (ed) Handbook of electroencephalography and clinical neurophysiology, vol 10A. Elsevier, Amsterdam pp 12–24

Gumnit R (1974) DC shifts accompanying seizure activity. In: Remond A (ed) Handbook of electroencephalography and clinical neurophysiology, vol 10A. Elsevier, Amsterdam, pp 66–77

Gumnit RJ, Matsumoto H, Vasconetto C (1970) DC activity in the depth of an experimental epileptic focus. Electroencephalogr Clin Neurophysiol 28:333–339

Hubbard JI, Llinas R, Quastel DMJ (1969) Electrophysiological analysis of synaptic transmission. Arnold, London (Monographs of the Physiological Society)

Jasper HH, Ward AA, Pope A (eds) (1969) Basic mechanisms of the epilepsies. Little Brown, Boston

Klee MR, Speckmann E-J, Lux HD (eds) (1981) Physiology and pharmacology of epileptogenic phenomena. Raven, New York

Kuffler SW, Nicholls JG (1966) The physiology of neuroglia cells. Ergeb Physiol 57:1–90

Kuffler SW, Nicholls JG, Orkand RK (1966) Physiological properties of glial cells in the central nervous system of amphibia. J Neurophysiol 29:768–787

Lopes da Silva F, van Rotterdam A (1982) Biophysical aspects of EEG and MEG generation. In: Niedermeyer E, Lopes da Silva F (eds) Electroencephalography. Urban and Schwarzenberg, Munich, pp 15–26

Orkand RK, Nicholls JG, Kuffler SW (1966) Effect of nerve impulses on the membrane potential of glial cells in the central nervous system of amphibia. J Neurophysiol 29:788–806

Palay SL, Chan-Palay V (1977) General morphology of neurons and neuroglia. In: Kandel ER (ed) The nervous system. American Physiological Society, Bethesda, pp 5–37 (Handbook of physiology, vol 1/1)

Petsche H, Pockberger H, Rappelsberger P (1981) Current source density studies of epileptic phenomena and the morphology of the rabbit's striate cortex. In: Klee MR, Lux HD, Speckmann E-J (eds) Physiology and pharmacology of epileptogenic phenomena. Raven, New York, pp 53–63

Pockberger H, Petsche H, Rappelsberger P (1983) Intracortical aspects of penicillin-induced seizure patterns in the rabbit's motor cortex. In: Speckmann E-J, Elger CE (eds) Epilepsy and motor system. Urban and Schwarzenberg, Munich, pp 161–178

Pockberger H, Rappelsberger P, Petsche H (1984a) Penicillin-induced epileptic phenomena in the rabbit's neocortex. I. The development of interictal spikes after epicortical application of penicillin. Brain Res 309:247–260

Pockberger H, Rappelsberger P, Petsche H (1984b) Penicillin-induced epileptic phenomena in the rabbit's neocortex. II. Laminar specific generation of interictal spikes after the application of penicillin to different cortical depths. Brain Res 309:261–269

Purpura DP, Penry JK, Tower DB, Woodbury DM, Walter RD (eds) (1972) Experimental models of epilepsy. Raven, New York

Rall W (1977) Core conductor theory and cable properties of neurons. In: Kandel ER (ed) The nervous system. American Physiological Society, Bethesda, pp 39–97 (Handbook of physiology, vol 1/1)

Shepherd GM (1974) The synaptic organization of the brain. Oxford University Press, London

Somjen GG (1973) Electrogenesis of sustained potentials. Prog Neurobiol 1:199–237

Somjen GG (1975) Electrophysiology of neuroglia. Annu Rev Physiol 37:163–190

Somjen GG, Trachtenberg M (1979) Neuroglia as generator of extracellular current. In: Speckmann E-J, Caspers H (eds) Origin of cerebral field potentials. Thieme, Stuttgart, pp 21–32

Speckmann E-J (1986) Experimentelle Epilepsieforschung. Wissenschaftliche Buchgesellschaft, Darmstadt

Speckmann E-J, Caspers H (1974) The effect of O_2- and CO_2-tensions in the nervous tissue on neuronal activity and DC-potentials. In: Remond A (ed) Handbook of electroencephalography and clinical neurophysiology, vol 2C. Elsevier, Amsterdam, pp 71–89

Speckmann E-J, Caspers H (eds) (1979a) Origin of cerebral field potentials. Thieme, Stuttgart

Speckmann E-J, Caspers H (1979b) Cortical field potentials in relation to neuronal activities in seizure conditions. In: Speckmann E-J, Caspers H (eds) Origin of cerebral field potentials. Thieme, Stuttgart, pp 205–213

Speckmann E-J, Elger CE (1982) Neurophysiological basis of the EEG and of DC potentials. In: Niedermeyer E, Lopes da Silva F (eds) Electroencephalography. Basic principles, clinical applications and related fields. Urban and Schwarzenberg, Munich, pp 1–13

Speckmann E-J, Elger CE (1984) The neurophysiological basis of epileptic activity: a condensed overview. In: Degen R, Niedermeyer E (eds) Epilepsy, sleep and sleep deprivation. Elsevier, Amsterdam, pp 23–34

Speckmann E-J, Caspers H, Janzen RWC (1972) Relations between cortical DC shifts and membrane potential changes of cortical neurons associated with seizure activity. In: Petsche H, Brazier MAB (eds) Synchronization of EEG activity in epilepsies. Springer, Wien New York, pp 93–111
Speckmann E-J, Caspers H, Janzen RWC (1978) Laminar distribution of cortical field potentials in relation to neuronal activities during seizure discharges. In: Brazier MAB, Petsche H (eds) Architectonics of the cerebral cortex. Raven, New York, pp 191–209 (IBRO monograph series, vol 3)
Speckmann E-J, Caspers H, Elger CE (1984) Neuronal mechanisms underlying the generation of field potentials. In: Elbert T, Rockstroh B, Lutzenberger W, Birbaumer N (eds) Self-regulation of the brain and behavior. Springer, Berlin Heidelberg New York Tokyo, pp 9–25

IV. Cognitive Potentials

The Many Faces of Neuroreductionism

G. WERNER

1 Introduction: Reductionism

The generally held belief that a person's psychological functions such as perception or cognition are in some way related to processes and events in the nervous system is the essence of neuroreductionism. However, this seemingly straightforward proposition is subject to ambiguities in that the concept of reductionism allows for several interpretations, which in turn have also triggered antireductionist arguments. To escape from these ambiguities, it has become necessary to impose certain constraints on reductionist thinking, with corresponding implications for empirical research. However, while the practicing neuroscientist continues to go unperturbed about his business of applying his methodological repertoire to the systematic examination of processes, events, and the structural organization of the nervous system, the task of integrating his observations into a coherent conceptual context, has largely become relegated to the philosopher of science, the theoretical psychologist and, more recently, the computer scientist and system theoretician. Yet the neuroscientist, too, has a vote in these deliberations.

My intent in this overview is to take a position somewhere between the experimentalist and the pure theoretician and to obtain – in a metaphorical vein – a fusion of the two images in stereoscopic vision. This intent is based on the conviction that observation and theory are inextricably tied together in a recursive relationship, with neither one nor the other being able to run its course profitably for any length of time. This is, of course, simply another way of stating the doctrine of Francis Bacon.

The desired outcome of such an examination with stereoscopic vision is to be able to match empirically testable propositions with logically consistent concepts of reductionism in order to remove the latter from the pedestal of an ontological faith to an empirically contingent and methodologically reputable account. From the pragmatic point of view the experimentalist, resolving ambiguities of reductionism has important implications, for different concepts of reductionism entail different scopes of desirable and attainable goals of experimentation.

In its boldest form as a regulative principle in science, reducibility to physics – and, hence, the unity of science – has been considered for some time the rite of passage for theories in the special sciences. As a result, the latter would disappear to the extent to which this program would succeed. The expectation of the reductionist program in this strong form as "eliminative reductionism," endorsed by Rorty (1970) and Feyerabend (1970), is that each explanation using psychological terms can be rendered in physiological terms without unaccounted residue: where there are

Springer Series in Brain Dynamics 1
Edited by Erol Başar

psychological explanations, there will be neurological explanations. The success of this program is said to be achieved by the uncovering of "bridge laws," which are symmetrical, transitive relations containing predicates of both the reducing and reduced domain. The theory of psychoneural identity is a special case of this radical reductionism insofar as it claims that each "type" of mental event is identical with a type of neurological event, whereby "type" refers to classes of entities, events, and properties. Almost no one subscribes to this view nowadays. Dennett (1978) offered a telling analogy that makes the implausibility of this form of "type – type reductionism" transparent: a clock is the type of thing that tells time; stated in this form, "clock" is described in functional terms. Now, when trying to define "clock" as a type of physical thing, it becomes readily apparent that there is no one physical property, event, etc., by virtue of which all individual ("token") clocks belong to the physical type of clock. Hence, each individual clock is a physical thing, but there is no vocabulary in physics that can meet all the requirements of talking about clocks in terms of the function they subserve.

The token-physicalist concludes that a weaker doctrine than eliminative reductionism must do: reduction must be restricted to instances of individual pairs of events in the reducing and the reduced domain. But Fodor (1976) warns that even if individual (token) psychological events are token neurological events, there is no ontological warrant that all kind predicates of psychology strictly correspond to kind predicates in any other discipline. Rather, as Fodor (1976) put it "it is an institutionalized gamble" that such lawful connections between event predicates in the reducing and the reduced science can, in fact, be found. Putnam's (1981) reasoning supports this skeptical attitude: it is in principle possible to design vast numbers of automata which would satisfy kind predicates of psychology without satisfying any neurological predicates at all; moreover, it is conceivable that identical psychological states can be instantiated at different times by different neurological states. Consequently, functional accounts are, in principle, not uniquely reducible to mechanistic accounts. As Fodor (1968) put it, "psychological states are not available for microanalysis". However, notice that this premise is itself contingent on empirical validation. I will argue later that the considerable evidence generated by the previously mentioned institutionalized gamble weakens the credence one may at first grant the a priori status of this postulate. Nonetheless, the question remains: is there more to reductionism than an institutionalized gamble in Fodor's sense, and is it possible to delineate conceptually sound and pragmatically useful positions that can guide the experimentalist's pursuits?

In a recent monograph, Clark (1980) examined in detail the scope of a "model reductionism" that assures internal consistency of the relationship between reduced and reducing science and avoids unwarranted ontological commitments. In this context, a model comprises a structure of relationships between states and processes, shared by two separate domains of which one (i.e., the reducing domain) is well understood, and the other contains hypothetical terms under empirical investigation. Explanatory reduction is then said to occur if events and processes in the reducing model domain can be assigned to each theoretical term in the reduced domain. Model reduction succeeds if an isomorphism of relationships between reduced and reducing domain can be justified. To illustrate: the behavior of a closed volume of gas is described by certain quantitative relationships between volume, pressure, and temperature. In the first step, certain identity claims are made, e.g., pressure and average kinetic energy of particles impinging on the container's wall are considered equivalent; then, justification for the identity claim is established by quantitative correspondence between temperature on the one hand, and pressure as well as mean kinetic energy, on the other. Hence, applicability of, say Boyle's law corresponds to appropriate equations in statistical mechanics. Thus, model reduction is justified.

In contrast to mere analogy, model reduction imposes a strong constraint. If, for instance, a model is found that can account for a set of psychological phenomena, but requires processes that are known not to be realizable in the nervous system, it is considered disconfirmed. For example, consider a model of psychological functions in vision that would require computations that could demonstrably not be executed by the nervous system. The impossibility would require abandoning this particular psychological model. Accordingly, model building in psychology and neurology are complementary, and both disciplines work together towards defining a unified theory.

In the conventional sense, a theory is a partially interpreted calculus in which the nonlogical terms receive sense and reference through association with observational predicates. Since the theory is common to both domains of empirical discourse, the assignments of observational predicates are admissible in both domains; hence, there is validity in both psychological and neurological discourse running side by side. In contradistinction from eliminative reductionism, two levels of discourse coexist and are joined by a common theory. Explanations in psychological and neurological terms can employ a different vocabulary, and address different observables as long as the relationships between these terms in their respective domains are of the same structure (i.e, stand in the same antecedent–consequent relationship). Accordingly, the exile imposed on mental processes and events by positivism and logical behaviorism is waived, and "mentalese" is once again elevated to the status of a reputable scientific dialect.

In the structuralist tradition, sameness of structure goes beyond the static description of relationships between components; it also requires identical transformation rules, such that a perturbation in one domain is accompanied by a corresponding transformation in the isomorphic domain.

Model reduction seeks to offer an escape hatch from the dead end of an ontologically encapsulated, eliminative reductionism by offering a different concept of "bridge laws:" in place of translation rules for properties in "type physicalism" or mechanistic bridge explications in "token physicalism," model reductionism posits structural isomorphism as the connecting principle, while granting continued autonomy to the connected disciplines. Clark (1980) attempts a valiant refutation of the indictment by the critics of type and token physicalism by adducing some seemingly successful examples of neuroreduction of motivated behavior; but it still remains a matter of debate whether his illustrations fall more into the category of analogies than of genuine model reduction.

The transposition of these general considerations to the central topic of this overview motivates, in the first place, the delineation of some currently prevailing doctrines in psychology as models of psychological processes. It will then become possible to scrutinize the extent to which these psychological models are suited to conceptualize research of brain events. This, in turn, will allow the application of the acid test of neurological realizability of psychological models as the prerequisite for stipulating isomorphism in the sense of model reductionism.

2 Models of Computation and Representation

The sorting out of deficencies of token physicalism in interaction with arguments from machine analogies led to the emergence of the doctrine of functionalism: mental events are to be characterized in terms of their causal relations to input-events, output-events, and other mental events (Fodor 1981). As Dennett (1978) put it, "to say that a particular belief or pain, for instance, is a particular functional state is to say that anything, regardless of its composition, chemistry or other physical features that fulfill the same functional (i.e. causal) role in a functionally equivalent system would be the same belief, pain, etc." A given psychological state is said to have a content in virtue of the causal role it plays in regulating behavior. In the specific form of "machine functionalism," this doctrine provides the umbrella for much of the current work in cognitive psychology, psycholinguistics and artificial intelligence; it is based on the notion that the computing paradigm is the best bet for granting scientific status to psychological theories.

Starting from an observed competence, say for stimulus discrimination, the task is to devise a performance model (often in the form of a computer program) that would display that same competence. A complex operation is decomposed into elementary processes that can be stated in formal, syntactic terms. The next step is to seek support for the claim that the performance model is isomorphic to the "real" situation. Dennett (1978) captured this idea in a delightful pun: "getting the cat skinned at all can be a major accomplishment; getting it skinned in the way people seem to get it skinned is even better." Functionalism is in accord with the research strategy of model reductionism. Consider the competence of learning: getting it skinned at all amounts to devising models that exhibit this competence; finding a model that satisfies neurologically realizable and identifiable process and events is skinning it even better, as it affirms isomorphism between psychological and neurological competence.

As an historical side remark. Freud was cognizant of this functionalist approach. In his posthumously uncovered manuscript "Project for a Scientific Psychology," he took ideas of psychological competences, such as wish fulfillment in dreams, repression, etc., as points of departure and designed the blueprint of a model that would be functionally equivalent; except that we would now say that he skinned it in a way that is different from the way the nervous system does it, at least as we now know (Hobson and McCarley 1977).

A considerable number of current and recent studies adhere to the same paradigm: consider, for instance, the numerous efforts to design associative networks that mimic learning and memory (Sutton and Barto 1981; McClelland and Rummelhart 1981). These attempts consist of either implementing a psychological competence with mechanisms known to be realized in the nervous system or, else, inventing mechanisms that can be searched for in the nervous system to achieve closure on the skinning. Several of these "neurologizing" architectures were recently reviewed by Anderson (1984).

Within the framework of functionalism, psychology in general and cognitive psychology in particular are currently inspired to a large extent by the intuition of a fundamental similarity between computation and cognition. This intuition is nourished by the notion that brains, like computers, are physical systems whose performance can be described in terms of rules operating on symbolic representations. The functionalist views representations as provisions for successful causal interaction with the environment. Although the notions of representation and computation are currently the implicit or explicit reference points for burgeoning fields of study, it has not been possible to muster any stronger endorsement than that they are "the only detailed hypothesis available for exploring how it is possible for a physical system to exhibit regularities that must be explained as rule following or even as governed by goals and beliefs" (Pylyshyn 1980), or "it is the best we have got – and it is overwhelmingly likely that computational psychology is the only one we are going to get" (Fodor 1980).

Even though cognitive psychology appears to flourish on the soil of the computation/representation intuition, there seems to be reason for some caution against grasping at the "only straw floating," as J. Lettvin is reported to have once said (quoted from Dennett 1981).

The appeal to representation in cognitive psychology entails a further specification of the nature of the computational processes as being both formal and symbolic, the latter to the extent to which they involve semantic attributes such as truth, reference, and meaning.

While the concept of computation and representation appears heuristically useful in cognitive psychology, it has in the last analysis led to a disavowal of the complete realization of any form of reductionism. Some cognitivists reason along these lines: conditions for ensuring the internal consistency of a view of mental activity as literal computation include a fundamental distinction between two kinds of explanation of behavior (Pylyshyn 1980, 1984); these are, first, the operation of causal or biological laws as necessary and sufficient conditions and, secondly, the availability of an internal representation with semantic properties. This distinction separates "fixed" mental capacities such as the mind's "functional architecture" from capacities that are context dependent and can vary from situation to situation. Paraphrasing an illustrative example given by Pylyshyn (1980) may help clarify the distinction: while preparing notes for this overview, I have ideas in mind that I selected as the goals and objectives I wish to communicate. I proceed writing notes. Clearly, my brain states cause me to move my hand in particular ways as I lead the pencil over the paper, but these movements are members of a larger equivalent class of brain states that encompass reference to the as yet not accomplished goals of formulating my ideas. This latter relationship to my brain states of writing are not causal in any direct sense; instead, these states are representations of an intention with which the execution of my hand movements are connected by certain rules. The goal state of completing the outline is seen as an independent, autonomous level of representation (a semantic description) for which the brain states of executing the hand movements are the causal-executive channel; it can be accessed by many different semantic representations. This example is to underscore the distinction between semantic-interpretive symbolic codes and functional architecture as a multiple-realization relationship.

When speaking of contents of representations, cognitivists have conceptual content in mind, and not the reference of this content to the real world. As the morning star and the evening star have the same referent in the real world, but different sense à la Frege, they are said to be intensionally different, and this difference is reflected in the different contents of their representations.

The internally consistent elaboration of the general distinction between a fixed, informationally encapsulated form of brain processes and representations with semantic content has run a pessimistic course of resignation that reached its most explicit formulation in Fodor's (1983) monograph entitled "The Modularity of Mind." Fodor cuts the homogeneous perception-cognition cake of the "new look" psychologists into vertical slices of domain-specific, hard-wired, autonomous faculties, which one may in first approximation equate with the functional architecture of the sense modalities; their output is thought to covary with environmental states in the manner of "compiled transducers" (Fodor and Pylyshyn 1981). This is to connote the independence of their function from such situational fractors as a goals, beliefs, or wishes. Their isolation from the organism's background knowledge limits their function to supplying primarily input-data-driven presentations of proximal stimulus configurations to the top layer of the cake after some unspecified degree of computational elaboration by neuroanatomically and genetically specified mechanisms.

In contrast, the top layer is said to extend isotropically over the cake's entire surface, with information "flowing every which way" (Fodor 1985), lacking articulate architecture and with no markings that tell how to divide it up. This top layer is considered the domain of symbolic representations; wholistically encompasses semantic contents and is resistant to analysis of global interdependencies between sets of propositions and beliefs. It would functionally correspond to Quine's "webs of belief" or, perhaps more remotely, to Husserl's Noema (Dreyfus 1982), or to the closed semiotic systems of de Saussure and Peirce.

Fodor (1983) gloomily concludes, "just as the earlier Turing machine models precluded any serious neuropsychology, so does the account of the stipulated isotropic central processor, for – as Dreyfuss has independently and incessantly emphasized – we have no idea how a computational formalism would operate in such a setting." With some irony, we must notice that the most prominent offspring of functionalism has defeated model reductionism, which was seen by some as one of its promising offsprings. *Sic transit gloria mundi!*

I shall return to this question after some detour.

3 Quo Vadis, Neurophysiologist?

The implications for the neuroscientist are far reaching: some cognitivists have carved the brain for him into a territory he has no business entering, lest he take the risk of transgressing alleged conceptual boundaries; his travel permit is restricted to the land of functional architecture. This territory bears the stamp of information-processing machines, running on compiled programs which cognitivists declare out-of-bounds to the world of propositional attitudes such as beliefs, expectations, and goals; a criterion of "cognitive penetrability" has been set up as frontier guard. This criterion has two sides: on one side, it circumscribes processes in the sensory transducers which can fully be accounted for by stable, causal mechanisms; on the other side are processes whose regularities and general features cannot be captured in behavioral or neurological terms, but are thought to require the appeal to semantic, representational, and intentional accounts. By way of illustrating, consider the perceptual process. To the extent to which it requires inference and is subject to the perceiver's belief system, it is cognitively penetrable; but a noninferential contribution of transducers with fixed modes of operation is also part of the process. Accordingly, fluctuations in late components of visually evoked responses relative to a person's expectancies must be considered cognitively penetrable, but the transduction of the visual stimulus in the retina is not. Although in each case physical processes cause the organism's response, the explanation of the entire class of evoked cortical responses requires recourse to generalizations in cognitive (semantic, intentional, symbolic) terms.

As a methodological device, cognitive penetrability purports to divide the entire range of psychology (and by implication, neurology) into two classes: those functions that are and those that are not modified by a persons's goals or beliefs. This ist not to be construed as a departure from materialism, but merely reflects the stance of some cognitivists that generalizations over certain classes of psychological functions call for explanations in cognitive terms, which, as a class, are not reducible to causal-mechanistic accounts. However, at least at first glance, this conclusion conflicts with the extensive, recent evidence for neurobiological mechanisms of "state control," reflected in regular relationships between states of motivation, attention and prior knowledge, and neuronal activity (Mountcastle et al. 1981; Wurtz et al. 1984; Hobson 1984).

This radical position departs in a fundamental way from the more conservative cognitive neuropsychology that treats perceptual-cognitive faculties in a unitary fashion within the framework of information processing; encoding, retrieving, and transforming information are the units of its currency. With this currency, the perceiver is thought to resolve the "poverty-of-the-stimulus argument". The critical issue is whether or not the information contained in the stimulus per se supplies all that is needed for the perception of the object. Granting the importance of the data-driven, "bottom-up" processing of stimulus information does not eliminate the need – so it is argued – for some contribution by stored, prior knowledge. Once this position is adopted, the puzzle becomes: how does the form of these mental contents differ from the propositional format of conscious knowledge (Rock 1985)? When Fodor (1985) proposed that "very much wanting the Mueller-Lyer illusion to go away does not make it disappear," Rock (1985) would presumably answer: "because unconsciously represented knowledge in the form of visual memories overrides consciously apprehended knowledge."

The approach of Palmer (1982) can serve as an illustrative example of the dilemma: a chain of sequentially arranged "analyzers" is thought to partition the output from lower-order analyzers progressively, according to rules of transformational equivalence. The task of higher-order analyzers is to extract relations over the spatial distribution of the energy flux impinging on receptors from the environment, which display transformational invariance under the group of similarity transformations of Euclidean geometry. If one subscribes to a relatively naive realism, one could argue

– as Shepard (1984) implies – that the embodiment of the similarity transformation group of Euclidean geometry evolved under evolutionary pressure: it would have become part of the functional architecture. However, perceived shape is relative to a reference frame, roughly analogous to coordinate systems in analytical geometry. The selection of the reference frame by the perceiver, although multiply determined by properties of the target stimulus, also appears influenced by an "intentional component" as a biasing process, based on expectations (Palmer 1985).

Of course, this – shall we say, moderate – type of cognitive theory is also computational and representational, as any transformation requires some data structure to be operated on by an algorithm of sorts. But note the different place assigned to intentional-semantic functions as the distinguishing feature that separates the "radicals" from the "moderates."

4 How Functional Is Functional Architecture

The concept of functional architecture is noncommittal with regard to the nature of the input signal: are the functionally relevant properties of the impinging stimulus array extracted by algorithms, such as those elaborated in great detail for the optic array (Marr 1982), or is there some form of "direct" pick-up of "invariant properties" from the stimulus array, as Gibson's ecological approach proposes (Gibson 1979)? Some cognitivists, for instance Fodor and Pylyshyn (1981), seem to see possibilities for reconciliation in what, for years, had been a radical dichotomy between these two positions. In their mind, the issue is no longer whether input processing is computational or not, but rather what it is that is being computed, and what are the primary data for computation: is it possible to extract computationally useful concepts from Gibson's genuine insights, even though they may entail consequences that are diametrically opposed to his radical refutation of algorithmic approaches?

Gibson said, "invariants are detected in the optic flow." Recent investigations have, in fact, established principles of computational information extraction from optic flow (Clocksin 1980; Buxton 1983; Riseman and Arbib 1977) and thus began to eliminate one of the objections to Gibson's position, which were based on the alleged failure to specify operationally how to capture the information in optic flow. This circumstance could bring the dialectic between computation and anticomputation to an end. But is this really the central issue? Perhaps not, and the arguments go like this: Gibson's "heresy" is fundamentally related to the objection against separating syntax and semantics; however, this very separation is an essential aspect of formal computation (Haugeland 1981; Hopcroft and Ullman 1969). Once this separation has occurred – so says the ecological psychologist – any subsequent assignment of semantics by interpretive functions is arbitrary. Hence, the computed representation has no structural and functional resemblance to its referent. Any attempt to reconcile the ecological with the representational view is thus doomed on first principles (see Carello et al. 1984).

I will now suggest considerations that may enable the neurophysiologist to contribute to this dialectic. After Hubel and Wiesel's discorvery of the feature-detecting neurons in area 17 of the visual cortex, structurally and functionally fixed line and edge detectors (and, later on, spatial frequency detectors) were taken to be the exclusive raw material for any theory of vision, and – in generalization – for much speculation about any perceptual process; see for instance Werner (1974). The sophisticated algorithms developed by Marr and his associates were designed first to generate and then to operate on the "primal sketch" as the basis of segmentation of the neural response panorama into regions that would correspond to "real" objects. Based on psychophysical data of Stevens (1983), Marr (1982) also explored other kinds of measurements that can be extracted from variations of texture, and how shape could be recovered from shading; the computational problems are considerable (Grimson 1981). Despite their computational sophistication, currently existing artificial intelligence vision systems require generally substantial a priori knowledge of real objects (Mackworth 1976). An alternative approach to figural synthesis seeks to bridge the step from local to global processes differently: visual contours would emerge as invariants under transformations of vector fields (Hoffman 1984; Dodwell 1984).

One of the main difficulties with the implementation of computer vision algorithms in the conventional (von Neuman) architecture of digital machines resides in the temporal constraints: a perceptual process carried out by the brain in, say, 100 ms involving computational elements (i.e., neurons) with a basic speed of about 1 ms per operation, requires in the machine implementation millions of time steps. Therefore, alternatives deserve serious consideration.

Beginning with the input stage: is it conceivable that line and edge detection are but one component of the output made available by the registration devices? If it were possible to identify neurons that could form canonical groupings over the energy flux of their receptors, a radically different situation would emerge; for instance, edges and slants would fall out secondarily as boundaries between regions. The issue is the "forest before the trees" phenomenon (Navon 1977), in the sense that analysis of aggregates could precede the analysis of their components. Riseman and Arbib (1977) and Ullmann (1983), amongst others, have emphasized the considerable economy resulting from a dual approach to segmentation by region-based as well as edge-detecting methods. Here is a specific question to the neurophysiologist: have we, in fact, exhausted the search for different forms of "raw measurements" that may be available at the input stage? Actually, it did not become apparent until recently that neurons in the visual projection pathway signal more complex and also time-varying aspects of the visual scene in anything but the rigidly fixed, static manner of the Hubel-Wiesel neurons (Fischer and Krueger 1974; Regan and Beverley 1979; Zeki 1979; Smith and Marg 1974).

The question is actually of a more general type. Experimental design in neurophysiology is generally guided by a search for the encoding of physical variables that are basic in Newtonian physics or Euclidean geometry, although there is no a priori reason to assume that perceptual systems must necessarily begin by registering elementary variables in physics. Runeson's (1977) startling illustration of the possibility of "smart" perceptual mechanisms that would use shortcuts and register complex variables directly, efficiently, and economically can serve as an alerting signal to the neurophysiologist. The lesson from Runeson is that suitable concatenations of elementary physical variables to fixed units of operation can turn complex into simple computational tasks.

The point of this excursion is that, conceivably, the computational intuition may possess more and still unexplored power than cognitivists and neurophysiologists have granted it: first, at the level of transducers, by considering edge detectors to be not the only format of input. This entails a search for smart neurons that directly register composites of Newtonian variables.

A second limitation of the conventional computational paradigm in cognitive science is attributable to its restriction to the notion of von Neuman machines, which are based on the sequential operation of a central processor on passive data structures. The accumulating evidence from the neurosciences calls the appropriateness of this prototype for computation in the nervous system into serious doubt. Consider, for instance, that the visual pathway, at least in higher mammals, consists of different categories of neurons with distinctive physiological properties and separate regional distribution, each contributing distinctively to vision; see Lennie (1980) for a review. Likewise, consider the degree to which different cortical areas specialize for different details in the visual field (Zeki 1979). The parallel to a system of hardware-connected, interacting processors is inviting (Ballard et al. 1983), but the principles of computation in such a system are, at this stage, far from transparent, both in the theory of computation and in the neurosciences (Fahlman 1981; Feldman 1981). Nevertheless, the rapid rise of interest in computing architectures consisting of multiple processors working together has significant implications for the neurosciences. Computer architectures with massive parallel structure make it possible to replace the time-consuming algorithms for symbolic information processing and the constraints of content-addressable memories in the conventional von Neuman configuration with suitable connectivity between separate processors; the connectivity is equivalent to wired-in semantics (Brown 1984). Thus, the burden of neural computation shifts to the structure of connections in networks of neurons.

Consider the difference: conventional computation is based on serial operations with abstract symbols whose meaning originates with the rules that contain and manipulate them. This symbol-processing approach mimics certain forms of human mental competences remarkably well (Newell and Simon 1972). The parallel architecture is, in some sense, the flip-side of this: it grants the activity in a collection of processing elements the status of meaning-carrying symbols and enables the activity pattern in the collection of processors to causally determine the interaction with other collections of processors. Minsky (1980) sketched an outline for the way in which neural computation may be organized by this principle: the place of the central processor with content-addressable memory of the conventional paradigm is taken by a "society" of processors of simple complexity, and direct access to a limited number of other, local processors is postulated. The agents communicate by passing excitatory or inhibitory signals between them, according to their respective internal states. The patterns of their activity are thought to correspond to particular mental states. Although not yet developed in operational details, this approach seems to offer promising possibilities for avoiding the troublesome problem of computing representations of the external world in real time by assembling its tokens from hardware-connected, ready-made components.

The formal characterization of the logic that governs the behavior of such a society of agents is a problem that has occupied McCulloch in a very substantial way (McCulloch 1965), and it may now be possible to capitalize on the extensive studies of non-Aristotelian logic that Guenther (1964) undertook to find a suitable decision algorithm in contexts of interrelated agents.

History is also coming full circle in another way, though enriched with new conceptual foundations and technical tools. The random nets of elementary computing units with self-organizing capability of the 1960s (Rosenblatt 1962) surface now as fixed connectionist or random interconnected networks (Amari and Arbib 1982; Grossberg 1982; Feldman and Ballard 1982) that can model an action-oriented notion of perception and the transition from a random to a quasi-stable functional organization.

The growing interest in the formal properties of connectionist models is being reciprocated by a corresponding emphasis on coalitions in groups of neurons at the theoretical level (Anderson and Hinton 1981), at the level of neurophysiological instrumentation development (Reitboeck and Werner 1983), and with initial efforts to capture quantitative indices of the activity patterns in clusters of neurons (Reitboeck 1983). The trend is away from the "localist" view that each single neuron represents a particular feature of the environment to a view of coalitions of neurons as dynamically functioning units, in the sense of Hebb's cell assemblies (Jusczyk and Klein 1980).

Based in part on the work of Minsky and Papert, Ullman (Ullman 1984) cautions appropriately against facile generalizations. There is room for both serial and parallel operations, much as there is evidence from psychophysical observations for a division of labour between seeing "where" and seeing "what," the former being attributable to preattentive parallel processes, and the latter requiring serial search by focal attention (Sagi and Julesz 1985). At a more intuitive level, the image of multiple knowledge sources, each with in incomplete view of the world but scheduled to interact for refinement and error correction, presents itself as an attractive paradigm (Kohler 1983). The computational prototype for this is the Hearsay II speech-understanding system (Erman et al. 1980) or a hard-wired or simulated implementation of NETL (Fahlman 1982).

5 The Exorcism of the Homunculus

Of course, computation and representation, naively viewed, can resurrect an ugly ghost. Unless conceptual rigour is observed, the image of homunculus watching some display in the brain is conjured up. Worse still, cortical maps with at least some superficial resemblance to the spatial organization of perception seem to lend some credence to a tangible role of an inner screen. The specter of dualism and infinite regress looms threatingly on the horizon! Koenderink (1984) recently generated one of the most elegant exorcisms in connection with the time-honored problem of "local sign:" how can a place label be assigned to a peripheral nerve fiber carrying impulse responses from excitation of its receptors? The line of thought runs roughly as follows: the only thing the brain has access to are the fluxes of nerve impulses in a matrix of receptive fields; such a record has structure insofar as there are constraints on the possible simultaneous/successive orders in which the matrix elements can become active. Hence, the totality of all possible patterns of simultaneous/successive orders of activity is, de facto, isomorphic with the geometrical structure of some abstract space. In this sense only is it then permissible to say that the matrix of neural elements carries (represents) geometric information. But note that the geometric structure – as Koenderink put it – is "not in the record but in the description of the apparatus." The place of the homunculus watching the picture on the screen is now taken by an algorithm that computes correlations over the record; the geometry is implicit in the activity of the network. As Dennett said, "the more procedural knowledge becomes in a representation, the less homuncular is the problem."

In Koenderink's view, the data structure and the algorithms are available for two purposes: to govern sensory-motor behavior reflexly; and to acquire experientially a "key" for coordinating categorial descriptions of successive/simultaneous orders of neural activity in the record with perceptual "images." For both situations, activity patterns in groups of nerve fibers or receptive fields are the truly relevant neural datum.

We can see in this principle of neural population function the shades of Pitts and McCulloch (1947) and their elaboration by Arbib (1980) to the concept of cooperative computation in somatotopically organized neural networks: a population of neurons governs behavior through their joint activity, except that in Koenderink's view somatotopy is itself a function of population activity. Indeed, it has been proposed that somatotopic organization of sensory projections may be more an evolutionary expedient than have primary functional significance (Werner 1970). McCulloch's principal concern with the "redundancy of potential command" in neuronal populations, to which he sought the solution in various forms of non-Aristotelian logic (Guenther 1979), then reduces to some form of correlation algorithm that feeds on the constraints on simultaneous/successive orders of activity in neural assemblies as the equivalent of geometrical structure in abstract spaces. Similarly, the allocentric and egocentric spatial maps of which Lieblich and Arbib (1982) speak could be construed to reflect differences within a class of related algorithms with distinct spatial isomorphisms.

Liebich and Arbib (1982) endow their maps explicitly with motivational and action-oriented functions. Along with Koenderink's previously stated view, this implies the causal functions of neural representations and implies in addition that neural representations are recursively updated and enriched with each action performed. Shaw and Mingolla (1982) swiftly "ecologized" the Lieblich-Arbib concept of representations by equating allocentric maps (and trajectories for their traverses) with Gibson's (1979) affordance structure for subject-environment transactions.

In the next section, I will follow a path that also highlights an action-oriented and structure-oriented approach. Although this path has been available for some time, it is not much traveled, perhaps because it carries these elements to consequences that violate some deeply rooted habits of thought.

6 Semantics or Structure? This Is the Question

The preceding sections of this overview brought us at several junctions up against semantics as a controversial issue. This was the case with the split amongst the Cognitivists, with the split between cognitivists and the ecological psychologists, and with the advice of some cognitivists that we shall never know how "higher nervous system" functions operate. I alluded to possible and yet inadequately explored expansions of the experimental and theoretical knowledge base that could, perhaps, heal some of these splits within a representational and information-processing framework. But we have not yet considered the possibility of the basic flaw in the representational paradigm.

One basic fact – self-evident to the point of often not receiving adequate attention – is that all that is accessible to the nervous system are the states of activity of its neurons, in turn giving rise to other states of activity (see Koenderink 1984). In this sense, the nervous system is a self-referring system (Maturana 1970). Any change produced by an independent, external source instigates a change in the state of activity of some of the elements in the system. Here is the crucial issue: is this change an "image" of the external event, or is it – as Maturana and Varela (1980; Varela 1979) propose – a response to an external perturbation that is primarily determined by the system's internal organization and structure. In the latter view, a firm distinction is drawn between the organism's response to an external event and the way in which an observer, beholding both the organism and the environment, can conceptualize this response. As far as the organism (and its nervous system) is concerned, the interaction with the environment sets in motion a perturbation of its internal organization and structure which attain recursively a new equilibrium state. For the observer, on the other hand, the interaction can be viewed as an informational transaction between the organism and its environment. In other words, to speak of representation in the nervous system is to obscure the constructive, recursive interdependence between organism and environment, which is a function of the organism's structure. The semantic discourse of the observer has no referent qua nervous system. Without intending a pun, here is a conceptual connection to the connectionist trend of parallel processing: the semantics is wired into the system.

From this position it follows that three (and only these three) problem areas are valid approaches for empirical research in neuropsychology:

(a) to correlate behavior of the organism with observable changes in the environment;

(b) to describe the neurophysiological processes set in motion by the external perturbations as evidence for the nervous system's homeostatic, self-referring structure; and

(c) to treat the organism's responses as ways in which it specifies its environment (rather than its ways of representing the environment) (Maturana et al. 1972).

Here is an example how the story goes. Consider size constancy as the divergence between perceived object size and retinal image size; conventionally, this phenomenon is analyzed in terms of distance and size as features of the external environment. In contrast, Maturana et al. (1972) ask, are there processes internal to the perceiver's organism that would account for size constancy without recourse to features of the environment? Their observationally supported answer is yes; size constancy is a function of accommodation as a manifestation of the nervous system's compensatory response to the environment. Instead of distance being a feature of the environment that needs to be "grasped" by the perceptual system, the sensory event of size constancy is specified by the organism's response to changes in the stimulus input that results in a change of the ciliary muscle innervation. The system's response to a perturbation, which is determined by its internal organization, becomes manifest as a sensory response.

The change of perspective is radical; one might even be inclined to say it is of quasi-Copernican scale. Note the shift in outlook, from the environment as the beholder of physical features of percepts, to the nervous apparatus as a self-contained system which specifies features of the environment by its internal reactions to perturbations. The objects of perception in an observer-included epistemology as "tokens of stable behavior" are for the nervous system equilibrium states in recursions of sensory-motor interactions (von Foerster 1984). Accordingly, perception and perceptual space do not represent features of the environment, but are expressions of the anatomical and functional organization of the nervous system in its interactions. Kant and Leibniz still cast long shadows!

Whether this change in perspective illuminates a successful path around the roadblocks of semantics and intentionality in neuropsychology remains to be seen; but, considering the difficulties with them, it may be worth the gamble. Disentangling the knots of reductionism would be one of the trophies.

7 Summary

Recent trends in cognitive science invite a reexamination of neuroreductionism. The role assigned to semantics and intentionality challenges the scope of legitimate problem formulation in the neurosciences. This overview traces some of the implicit and explicit conceptual strands to their consequences, and indicates approaches that may circumvent apparent inconsistencies and conflicts. Of the several considerations raised, a primary role is given to the replacement of the serial processing paradigm of conventional computing by a parallel architecture of multiple, interconnected processors in models of psychological functions and in the interpretation of neurological data. In the parallel mode, some of the semantics of the information interchange can be attributed to the structure and the strength of connections between processors. The question is: how much semantics is wired into the nervous system?

Acknowledgement. This article was prepared during the author's tenure of a Senior Scientist award from the Humboldt Foundation. Professor Reitboeck, Biophysikalisches Institut of the University of Marburg, and his staff provided gracious hospitality and offered stimulating discussions.

References

Amari S, Arbib MA (1982) Competition and cooperation in neural nets. Springer, Berlin Heidelberg NewYork (Lecture notes in biomathematics, vol 45)
Anderson JR (1984) Cognitive psychology. Artif intell 23:1–11
Anderson JA, Hinton GE (1981) Models of information processing in the brain. In: Hinton GE, Anderson JA (eds) Parallel models of associative memory. Erlbaum, Hillsdale, pp 9–48
Arbib MA (1980) Visuomotor coordination: from neural nets to schema theory. Cognit Brain Theor 4:23–39
Ballard DH, Hinton GE, Sejnowski TJ (1983) Parallel visual computation. Nature 306:21–26
Brown CM (1984) Computer vision and natural constraints. Science 224:1299–1305
Buxton BP, Buxton H (1983) Monocular depth perception from optical flow by space time signal processing. Proc R Soc Lond [Biol] 218:27–47
Carello C, Turvey MT, Kugler PN, Shaw RE (1981) Inadequacies of the computer metaphor. In: Gazzaniga S (ed) Handbook of cognitive neuroscience. Plenum, NewYork, pp 229–248
Clark A (1980) Psychological models and neural mechanisms. Clarendon, Oxford
Clocksin WF (1980) Perception of surface slant and edge labels from optical flow: a computational approach. Perception 9:253–269
Dennett DC (1978) Current issues in the philosophy of mind. Am Philos Q 15:249–261
Dennett DC (1981) A cure for the common code? In: Brainstorms: philosophical essays on mind and philosophy. MIT Press, Cambridge, MA
Dodwell PC (1984) Figural synthesis. In: Dodwell PC, Caelli T (eds) Figural synthesis. Erlbaum, Hillsdale, pp 219–248
Dreyfus HL (1982) Introduction. In: Dreyfus HL (ed) Husserl-intentionality and cognitive science. MIT Press, Cambridge, MA, pp 1–27
Erman L, Hayes-Roth F, Lesser V, Reddy D (1980) The Hearsay-II speech understanding system: integrating knowledge to resolve uncertainty. Comput Surv 12:213–253
Fahlman SE (1981) Representing implicit knowledge. In: Hinton GE, Anderson JA (eds) Parallel models of associative memory. Erlbaum, Hillsdale, pp 145–159
Fahlman SE (1982) NETL: a system for representing and using real-world knowledge. MIT Press, Cambridge, MA
Feldman JA (1981) A connectionist model of visual memory. In: Hinton GE, Anderson JA (eds) Parallels models of associative memory. Erlbaum, Hillsdale, pp 49–81
Feldman JA, Ballard DH (1982) Connectionist models and their properties. Cognit Sci 6:205–254
Feyerabend PK (1970) Materialism and the mind-body problem. In: Borst CV (ed) Materialism and the mind-body problem. McMillan, London
Fischer, B, Krueger J (1974) The shift-effect in the cat's lateral geniculate neurons. Exp Brain Res 21:225–227
Fodor JA (1968) Psychological explanation. Random, NewYork
Fodor JA (1976) The language of thought. Harveston, Hanox
Fodor JA (1980) Methodological solipsism considered as a research strategy in cognitive psychology. Behav Brain Sci 3:63–109
Fodor JA (1981) Representation. MIT Press, Cambridge, MA
Fodor JA (1983) The modularity of mind. MIT Press, Cambridge, MA
Fodor JA (1985) Precis of the modularity of mind. Behav Brain Sci 8:1–42
Fodor JA, Pylyshyn ZW (1981) How direct is visual perception?: some reflections on Gibson's 'ecological approach'. Cognition 9:139–196
Gibson JJ (1979) The ecological approach to visual perception. Houghton Mifflin, Boston
Grimson WEL (1981) From images to surfaces: a computational study of the human early visual system. MIT Press, Cambridge, MA
Grossberg S (1982) Studies of mind and brain: neural principles of learning, perception, development, cognition, and motor control. Reidel, Boston (Boston studies in the philosophy of science, vol 70)
Guenther G (1961) Das Problem einer Formalisierung der transzendental-dialektischen Logik. In: Nicolin F, Poeggeler O (eds) Hegel Studien. Bonn, Bouvier, pp 65–130

Guenther G (1979) Beiträge zur Grundlegung einer operationsfähigen Dialektik. Meiner, Hamburg
Haugeland J (1981) Semantic engines: An introduction to mind design. In: Haugeland J (ed) Mind design. MIT Press, Cambridge, MA, pp 1–34
Hobson JA (1981) How does the cortex know when to do what? A neurobiological theory of state control. In: Edelman GM, Gall WE, Cowan WM (eds) Dynamic aspects of neocortical function. Wiley, New York, pp 219–257
Hobson JA, McCarley RW (1977) The brain as dream state generator: an activation synthesis hypothesis of the dream process. Am J Psychiatry 134:1335–1348
Hoffman WC (1984) Figural synthesis by vectorfields: geometric neuropsychology. In: Dodwell PC, Caelli T (eds) Figural synthesis. Erlbaum, Hillsdale, pp 249–282
Hopcroft JE, Ullman JD (1969) Formal languages and their relation to automata. Addison-Wesley, Reading MA
Jusczyk PW, Klein RM (1980) The nature of thought: essays in honour of D.O. Hebb. Erlbaum Hillsdale
Koenderink JJ (1984) The concept of local sign. In: van Doorn AJ, van de Grind WA, Koenderink JJ (eds) Limits in perception. VNU Science, Utrecht, pp 495–547
Kohler RR (1983) Integrating nonsemantic knowledge into image segmentation processes. PhD thesis, University of Massachusetts
Lennie P (1980) Parallel visual pathways: a review. Vision Res 20:561–594
Lieblich I, Arbib MA (1982) Multiple representations of space underlying behavior. Behav Brain Sci 5:627–659
Mackworth AK (1976) Model driven interpretation in intelligent visual systems. Perception 5:349–370
Marr D (1982) Vision: a computational investigation into the human representation and processing of visual information. Freeman, San Francisco
Maturana HR (1970) Neurophysiology of cognition. In: Garvin PL (ed) Cognition: a multiple view. Spartan, Wahsington, pp 3–23
Maturana HR, Varela FJ (1980) Autopoiesis and cognition. Reidel, Boston (Boston studies in the philosophy of science, vol 42)
Maturana HR, Varela FG, Frenk SG (1972) Size constancy and the problem of perceptual spaces. Cognition 1:97–104
McClelland JL, Rummelhart DE (1981) An interactive activation model of context effects in letter perception. I. An account of basic findings. Psychol Rev 88:375–405
McCulloch WS (1965) Embodiments of mind. MIT Press, Cambridge MA
Minsky M (1980) K-lines: a theory of memory. Cognit Sci 4:117–133
Mountcastle VB, Anderson RA, Motter BC (1981) The influence of attentive fixation upon the excitability of the light-sensitive neurons of the posterior parietal cortex. J Neurosci 1:1218–1235
Navon D (1977) Forest before trees: the precedence of global features in perception. Cognit Psychol 9:353–383
Newell A, Simon HA (1972) Human problem solving. Prentice-Hall, Englewood Cliffs
Palmer SE (1982) Symmetry. Transformation and the structure of perceptual systems. In: Beck J (ed) Organization and representation in perception. Erlbaum, Hillsdale, pp 95–144
Palmer SE (1985) The role of symmetry in shape perception. Acta Psychol (Amst) 59:67–90
Pitts WH, McCulloch WS (1947) How we know universals, the perception of auditory and visual forms. Bull Math Biophys 9:127–147
Putnam H (1981) Reductionism and the nature of psychology. In: Haugeland J (ed) Mind design. MIT Press, Cambridge, MA, pp 205–219
Pylyshyn ZW (1980) Computation and cognition – issues in the foundation of cognitive science. Behav Brain Sci 3:111–169
Pylyshyn ZW (1981) Computation and cognition. MIT Press, Cambridge, MA
Regan D, Beverly KI (1979) Visually guided locomotion: psychophysical evidence for a neural mechanism sensitive to flow patterns. Science 205:311–313
Reitboeck HJ (1983) A 19-channel matrix drive with individually controllable fiber microelectrodes for neurophysiological applications. IEEE Trans Syst Man Cybern 13:676–682
Reitboeck HJ, Werner G (1983) Multi-electrode recording system for the study of spatiotemporal activity patterns of neurons in the central nervous system. Experientia 39:339–341

Riseman EM, Arbib MA (1977) Computational techniques in the visual segmentation of static scenes. Comput Graphics Image Process 6:221–276
Rock I (1985) Perception and knowledge. Acta Psychol (Amst) 59:3–22
Rorty R (1970) Mind-body identity, privacy and categories. In: Borst CV (ed) The mind-brain identity. McMillan, London
Rosenblatt F (1962) Principles of neurodynamics. Spartan, New York
Runeson S (1977) On the possibility of smart perceptual mechanisms. Scand J Psychol 18:172–179
Sagi D, Julesz B (1985) Where and what in vision. Science 228:1217–1219
Shaw RE, Mingolla E (1982) Ecologizing world graphs. Behav Brain Sci 5:648–650
Shepard RN (1984) Ecological constraints on internal representation: resonant kinematics of perceiving, imaging, thinking and dreaming. Psychol Rev 91:417–447
Smith JD, Marg E (1974) Zoom neurons in visual cortex: receptive field enlargement with near fixation in macaques. Experientia 31:323–326
Stevens KA (1983) Slant-tilt: the visual encoding of surface orientation. Biol Cybern 46:183–195
Sutton RS, Barto AG (1981) Toward a modern theory of adaptive networks: expectation and prediction. Psychol Rev 88:135–170
Ullmann JR (1983) Aspects of visual automation. In: Braddick OJ, Sleigh AC (eds) Physical and biological processing of images. Springer, Berlin Heidelberg New York, pp 15–32
Ullmann S (1984) Visual routines. Cognition 18:97–159
Varela FJ (1979) Principles of biological autonomy. North-Holland, Amsterdam
Von Foerster H (1984) Objects: tokens for eigenbehaviors. In: von Foerster H (ed) observing systems. Intersystems, Seaside, CA, pp 274–285 (The systems inquiry series)
Werner G (1970) The topology of the body representation in the somatic afferent pathway. In: Schmitt FO (ed) The neurosciences: second study program. Rockefeller University Press, New York, pp 605–616
Werner G (1974) Neural information processing with stimulus feature detectors. In: Schmitt FO, Worden FG (eds) The neurosciences: third study program. MIT Press, Cambridge, MA, pp 171–183
Wurtz RH, Richmond BJ, Newsome WT (1984) Modulation of cortical visual processing by attention, perception, and movement. In: Edelman GM, Gall WE, Cowan WM (eds) Dynamics aspects of neocortical function. Wiley, New York, pp 195–217
Zeki S (1979) The mapping of visual functions in the cerebral cortex. In: Katsuki Y, Norgren R, Sato M (eds) Brain mechanisms of sensation. Wiley, New York, pp 105–127

The Endogenous Evoked Potentials

T. W. Picton

1 Introduction

This paper reviews some "endogenous" potentials that can be recorded from the human scalp in relation to the processing of sensory stimuli into behavioral responses. Twenty years ago, Sutton et al. (1965) distinguished between the endogenous and exogenous evoked potentials. Exogenous potentials are determined by the physical nature of the evoking stimulus, whereas endogenous potentials are related to the psychological significance of the stimulus to the subject.

2 The N100 and Attention

The human brain is able to attend to selected channels of incoming information and to ignore others. A typical example of selective attention occurs at the cocktail party. The experienced guest at such a party can ignore the monologue of the boring person before him and attend to a more interesting conversation at some other location. In 1973, Hillyard et al. described an evoked potential paradigm based on this cocktail party effect. Two trains of tones were presented to the subject, one train of low-pitch tones to the left ear and one train of high-pitch tones to the right ear. In one condition, the subject attended to the low tones in order to detect an occasional subtle change in frequency, and ignored the high tones in the other ear. In a reciprocal condition, the subject attended to the high tones and ignored the low tones. The attended tones evoked a negative wave at 100 ms (the N100) that was greater in amplitude than when the tones were ignored. More recent experiments have shown that this enhanced N100 is probably caused by a superimposed negative process that begins during the N100 wave and that may outlast it. This "processing negativity" (Näätänen and Michie 1979; Näätänen 1982) or "negative difference wave (Nd)" (Hansen and Hillyard 1980, 1983; Picton et al. 1985) is best demonstrated by subtracting the evoked potentials obtained when the tones are ignored from those obtained when the tones are attended. This is shown in Fig. 1.

The nature of this negative process is not known. It could represent some additional sensory evaluation of the attended stimuli, a fixing of such stimuli in memory for later comparison, or some supervisory processes that activate and direct the sensory evaluation or the memory access. Regardless of the nature of the generator, the Nd wave does indicate a relatively early selective processing of incoming stimuli. As

Springer Series in Brain Dynamics 1
Edited by Erol Başar

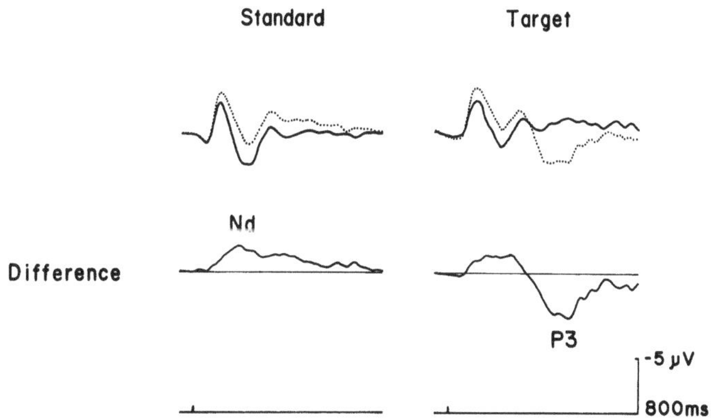

Fig. 1. Selective attention. In this experiment, the subjects were presented with two trains of 50 ms 65 db nHL tones. One of the trains was presented to the left ear and the other to the right ear. One of the trains contained "standard" tones of 500 Hz and occasional ($p = 0.2$) "target" tones of 475 Hz; the other train contained standard tones of 1000 Hz and target tones of 1050 Hz. The subject was required to attend to one ear in order to detect the occasional target tones in that ear and to ignore the tones being presented in the other ear. This figure shows the grand mean evoked potentials from eight subjects. The data have been collapsed over both ear and tonal frequency. Waveforms were recorded from the vertex using a chest reference with negativity at the vertex plotted upward. The *upper waveforms* represent the evoked potentials to the standard and target tones when they were attended *(dotted line)* or ignored *(straight line)*. The *lower waveforms* represent the difference between the attend and ignore conditions. When tones are attended, there is an extra negative wave superimposed upon the response. This shows up in the difference waveform as the Nd wave. The detected target tones are associated with a large positive wave called the P300. This only occurs if the target tones occur in an attended ear

such, it may be helpful in investigating the problems that occur in patients with attentional disorders. McGhie and Chapman (1961) have suggested that one of the main deficits in schizophrenia is an inability to attend selectively to incoming information. The schizophrenic subject therefore finds himself bombarded with far too much information to process appropriately. Baribeau-Braun et al. (1983) demonstrated that schizophrenic subjects show a normal attentional change in the N100 when stimuli were presented at rapid rates but not when stimuli were presented more slowly. The deficit in schizophrenic patients is therefore not an absent filtering mechanism but rather an inability to maintain the setting of the filter over prolonged interstimulus intervals.

3 The P300 and Information Processing

The detection of an improbable "target" stimulus in a train of "standard" stimuli is associated with a large late positive wave in the evoked potential with a peak latency of approximately 300 ms, called the P300. The more improbable the target stimulus, the larger the amplitude of the P300 wave (Duncan-Johnson and Donchin 1977; Fitzgerald and Picton 1981). As the amount of information in a stimulus is inversely proportional to the logarithm of its probability, the P300 appears to reflect some aspect of information processing (Campbell et al. 1979). The P300 is clearly an endogenous wave, as it may be elicited by the absence of a stimulus if such a stimulus omission is both unpredictable and task relevant (Sutton et al. 1967). In simple tasks, the P300 occurs a little later than the reaction time (Goodin and Aminoff 1984). It manifests some cerebral event occurring after the sensory stimulus has been evaluated (McCarthy and Donchin 1981; Magliero et al. 1984).

The peak latency of the P300 wave increases with age. Goodin et al. (1978a) initially reported that the P300 wave increased in latency by 1.8 ms per year between the ages of 15 and 76 years, whereas the earlier N100 wave did not change significantly with age. Similar age-related increases in latency have been found in many different laboratories using many different kinds of stimuli (Pfefferbaum et al. 1984a; Picton et al. 1984). Some papers have reported that the increase in latency is greater the older one gets (Brown et al. 1983), but most papers have shown a fairly linear increase of 1–2 ms per year. Many patients with dementia have P300 latencies that are significantly longer than the normal range for their age (Goodin et al. 1978b; Pfefferbaum et al. 1984b).

In order to determine what the age-related increase in P300 latency reflects, Stuss and I have related this latency to performance on various tests of mental function (Army Individual Test 1944; Shipley 1946; Thurstone and Thurstone 1962). Although we used many different tests, there were basically two kinds. The first kind evaluated verbal memory or vocabulary. In this kind of test, the subject must give or recognize the definition or synonym of a word. The subject either knows the answer or not. The test evaluates the content and accessibility of long-term memory. The second kind of test evaluates reasoning ability. In this kind of test, the subject must go through a sequence of mental operations. For example, in the trail making test (Army Individual Test 1944) the subject has to link up a set of numbers in order on a page. This requires the subject to search for the first number, move the pencil to that number, determine which is the next number, search for that number, move the pencil to that number, and so on. In other tests, the subject determines the next

Table 1. Correlations with P3 latency

Variable	Coefficient of correlation
Age	0.54
Trail making test	0.56
Shipley institute of living scale	
Vocabulary	−0.25
Abstraction	−0.45
Primary mental abilities	
Verbal meaning	−0.36
Numerical facility	−0.47
Reasoning	−0.55
Spatial relations	−0.50

Data obtained from 52 normal subjects between the ages of 21 and 78 years

members of a sequence of numbers or shapes. The subject must first examine the sequence, hypothesize a rule governing the sequence, test the typothesis, if it fails formulate another hypothesis, test the hypothesis, if it works determine the next shape or number in the sequence, and so on. These kinds of tests require the continual updating of a working or scratch-pad memory. The P300 latency is related to performance on such tests much more than to performance on vocabulary tests (Table 1). One possibility is therefore that the P300 reflects the updating of working memory (Donchin 1981). Its latency on a simple task would therefore assess the speed of memory updating.

4 N400 and Language

Kutas and Hillyard (1980) recorded a late negative wave with a peak latency of approximately 400 ms following the presentation of a semantically inappropriate word at the end of a sentence: "He spread the warm bread with socks." This late negative wave did not occur if the final word in the sentence was physically inappropriate rather than semantically incongruous. Stuss et al. (1983) found a similar late negative wave in the response to a picture that the subject must name (Kaplan et al. 1976). The N400 wave therefore seems to be associated with some access to semantic memory. Two recent experiments illustrate that the N400 wave varies with the amount of memory search necessary to obtain the meaning of a stimulus. Fischler et al. (1983) presented sentences wherein the subject was an example and the predicate was a

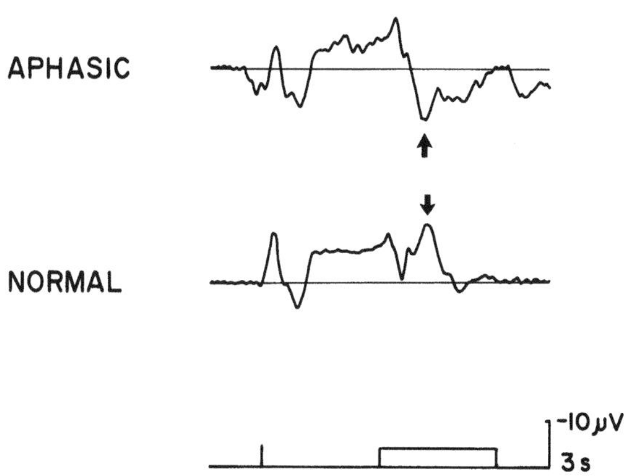

Fig. 2. Evoked potentials during naming. In this experiment, the subjects were presented with pictures from the Boston naming task. These were displayed on a video monitor for 1 s. The visual stimulus was preceded by a warning tone. The subject was required to determine the name of the picture and to repeat this 2 s later. The evoked potentials were recorded from a vertex electrode using a chest reference. Negativity at the vertex was plotted upward. Normal subjects showed a large late negative wave – the N400 – in response to the picture *(arrow)*. The *upper tracing* represents the recording from a mildly aphasic patient. The responses were limited to those in which the patient actually gave the correct name for the picture. This patient shows a totally different waveform from that observed in normal subjects. There is a large late positive wave at about the same latency as the normal N400 *(arrow)*

category. The sentences could be either negative or affirmative and either true or false. A larger N400 wave occurred when the sentence was negative and true ("A robin/is not/a tree") or when the sentence was affirmative and false ("A hammer/is/a building"). Kutas and Hillyard (1984) found that the N400 to the final word of a sentence was inversely related to the predictability of that word occurring in the context set up by the sentence. A larger N400 occurred for "The bill was due at the end of the hour" than for "He mailed the letter without a stamp."

Stuss and I have been investigating the N400, by recording during a naming task in patients with aphasia. Although aphasia varies with the location of the lesion, all aphasic patients have some disorder of naming. It is therefore possible that different morphologies of the evoked potential during naming may relate to different language disorders. The evoked potentials shown in Fig. 2 illustrate one pattern of abnormality that we have seen in aphasic patients. The evoked potentials of the aphasic patient were recorded when the patient named the stimulus correctly. Despite being correct, the actual waveform of the response is quite different from that seen in normal subjects. There is a large positive wave at about the same latency as the normal N400. This indicates that although the aphasic patient is performing correctly, the cerebral mechanisms underlying this performance are quite different from those used by the normal subject.

5 The Contingent Negative Variation and Anxiety

Walter et al. (1964) reported that when one stimulus warned a subject about a second "imperative" stimulus that required a response, a negative baseline shift developed between the two stimuli. This negative baseline shift was called the "contingent negative variation" (CNV). It occurred when a subject realized an important association between two stimuli; i.e., that the second stimulus was contingent upon the first stimulus.

Anxiety can decrease the amplitude of the CNV. There are two possible explanations for this effect. One is that the anxious subject cannot concentrate on the task because he or she is distracted by the thoughts and feelings related to the anxiety (Tecce et al. 1976). The second explanation is that the anxious subject has a higher baseline cortical negativity, and that the CNV can increase only a little above this already high baseline because of a ceiling negativity (Knott and Irwin 1973). The relationship between anxiety and the CNV has been investigated during a simple learning task (Proulx and Picton 1984). Subjects with low levels of anxiety were easily able to learn the association between a pair of tones and a buzzer that they were asked to turn off as quickly as possible. This learning was associated with a contingent negative variation (Fig. 3, top). Subjects with high levels of anxiety showed two patterns of response. One type of subject learned the association and developed a larger CNV than the low-anxiety subjects (Fig. 3, middle). This type of subject experienced high levels of anxiety during the experiments. The other type of high-anxiety subject did not realize the association and did not develop any measurable CNV (Fig. 3, bottom). This type of subject, however, did not experience any high level of anxiety during the task. These results suggest that a subject who is prone to

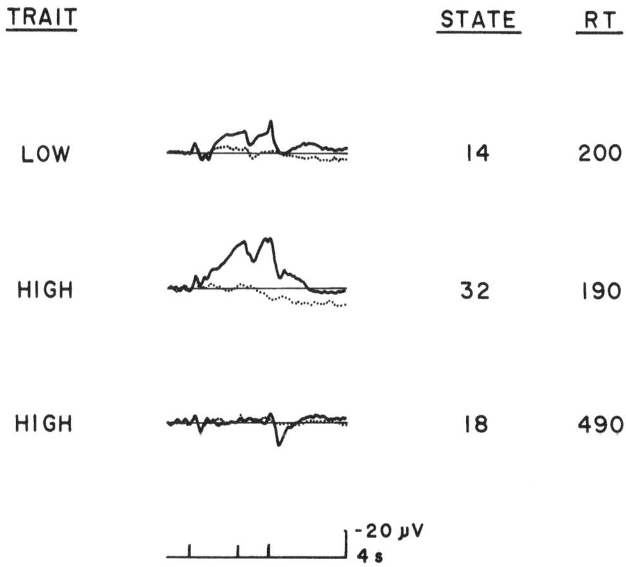

Fig. 3. Anxiety and expectancy. In this task, subjects were presented with pairs of tones separated by 1 s. The tones were either high (2 kHz) or low (1 kHz) in frequency, and there were therefore four different possible pairs. One particular pair was followed by a buzzer. The subjects were instructed to turn off the buzzer as quickly as possible, but they were not informed of the association between the paired tones and the buzzer. If the subject recognized the association between the tones and the buzzer, he could react more quickly to the buzzer. The evoked potential waveforms plotted with the continuous lines show the responses to the high-high pair of tones when they were associated with the buzzer. The *dotted line* tracing shows the event-related potentials recorded in association with the low-low pair of tones. The *upper responses* are from a subject who scored low on measures of trait anxiety. A negative wave occurs in association with the high-high pair of tones and the buzzer. The subject shows little state anxiety and turns off the buzzer quite quickly. The *middle tracings* are taken from a subject who scored high on measures of trait anxiety. A large contingent negative variation develops in association with the high-high pair of tones. The subject shows high levels of state anxiety and performs relatively well. The *lower tracings* are from another subject who scored high on measures of trait anxiety. No contingent negative variation develops. The subject turns off the buzzer very slowly, but does not develop much state anxiety during the taks

anxiety can choose either to become involved in a task or not. If the subject becomes involved, there is more cortical activation than normal and the subject experiences increased stress. If the subject does not become involved, there is little stress but the performance is poor.

6 Conclusion

The endogenous evoked potentials recorded from the human scalp can suggest some of the cerebral mechanisms underlying attention, information processing, language perception, and anxiety. At present, they are helpful in delineating possible cerebral mechanisms in normal subjects and in suggesting the abnormal processes that may occur in patients with disorders of higher nervous function. Because of the wide range of normality, the endogenous evoked potentials are not sufficiently precise to provide diagnostic information in individual subjects.

Acknowledgements. This paper presents unpublished data obtained in association with several colleagues - Anna Cerri, Sandra Champagne, Anita Maiste, Guy Proulx, and Don Stuss. I appreciate their kindness in allowing me to review these data. The Medical Research Council and the Canadian Geriatrics Research Society provided financial support. Janice O'Farrell prepared the manuscript.

References

Army Individual Test (1944) Manual of directions and scoring. War Department, Adjutant General's Office, Washington

Baribeau-Braun J, Picton TW, Gosselin JY (1983) Schizophrenia: a neurophysiological evaluation of abnormal information processing. Science 219:874–876

Brown WS, Marsh JT, LaRue A (1983) Exponential electrophysiological aging: P3 latency. Electroencephalogr Clin Neurophysiol 55:277–285

Campbell KB, Courchesne E, Picton TW, Squires KC (1979) Evoked potential correlates of human information processing. Biol Psychol 8:45–68

Donchin E (1981) Surprise!... Surprise? Psychophysiology 18:493–513

Duncan-Johnson CC, Donchin E (1977) On quantifying surprise: the variation of event-related potentials with subjective probability. Psychophysiology 14:456–467

Fischler I, Bloom PA, Childers DG, Roucos SE, Perry NW (1983) Brain potentials related to stages of sentence verification. Psychophysiology 20:400–409

Fitzgerald PG, Picton TW (1981) Temporal and sequential probability in evoked potential studies. Can J Psychol 35:188–200

Goodin DS, Aminoff MJ (1984) The relationship between the evoked potential and brain events in sensory discrimination and motor response. Brain 107:241–252

Goodin DS, Squires K, Henderson B, Starr A (1978a) Age related variations in evoked potentials to auditory stimuli in normal human subjects. Electroencephalogr Clin Neurophysiol 44:447–458

Goodin DS, Squires KC, Starr A (1978b) Long latency event-related components of the auditory evoked potential in dementia. Brain 101:635–648

Hansen JC, Hillyard SA (1980) Endogenous brain potentials associated with selective auditory attention. Electroencephalogr Clin Neurophysiol 49:277–290

Hansen JC, Hillyard SA (1983) Selective attention to multidimensional auditory stimuli. J Exp Psychol [Hum Percept] 9:1–19

Hillyard SA, Hink RF, Schwent VL, Picton TW (1973) Electrical signs of selective attention in the human brain. Science 182:177–180

Kaplan EF, Goodglass H, Weintraub S (1976) Boston naming test. Experimental edition. Boston VA Hospital, Boston

Knott JR, Irwin DA (1973) Anxiety, stress and the contingent negative variation. Arch Gen Psychiatry 29:538–541

Kutas M, Hillyard SA (1980) Reading senseless sentences: Brain potentials reflect semantic incongruity. Science 207:203–205

Kutas M, Hillyard SA (1984) Brain potentials during reading reflect word expectancy and semantic association. Nature 307:161–163

Magliero A, Bashore TR, Coles MGH, Donchin E (1984) On the dependence of P300 latency on stimulus evaluation processes. Psychophysiology 21:171–186

McCarthy G, Donchin E (1981) A metric for thought: a comparison of P300 latency and reaction time. Science 211:77–80

McGhie A, Chapman S (1961) Disorders of attention and perception in early schizophrenia. Br J Med Psychol 34:103–116

Näätänen R (1982) Processing negativity: an evoked potential reflection of selective attention. Psychol Bull 92:605–640
Näätänen R, Michie PT (1979) Early selective-attention effects on the evoked potential: a critical review and reinterpretation. Biol Psychol 8:81–136
Pfefferbaum A, Ford JM, Wenegrat BG, Roth WT, Kopell BS (1984a) Clinical application of the P3 component of event-related potentials. I. Normal aging. Electroencephalogr Clin Neurophysiol 59:85–103
Pfefferbaum A, Wenegrat GB, Ford JM, Roth WT, Kopell BS (1984b) Clinical application of the P3 component of event-related potentials. II. Dementia, depression and schizophrenia. Electroencephalogr Clin Neurophysiol 59:104–124
Picton TW, Stuss DT, Champagne SC, Nelson RF (1984) The effects of age on human event-related potentials. Psychophysiology 21:312–325
Picton TW, Rodriguez RT, Linden RD, Maiste AC (1985) The neurophysiology of human hearing. Hum Commun Canada 9:127–136
Proulx GB, Picton TW (1984) The effects of anxiety and expectancy on the CNV. Ann NY Acad Sci 425:617–622
Shipley WC (1946) Institute of living scale. Western Psychological Services, Los Angeles
Stuss DT, Sarazin FF, Leech EE, Picton TW (1983) Event-related potentials during naming and mental rotation. Electroencephalogr Clin Neurophysiol 56:133–146
Sutton S, Braren M, Zubin J (1965) Evoked potential correlates of stimulus uncertainty. Science 150:1187–1188
Sutton S, Tueting P, Zubin J, John ER (1967) Information delivery and the sensory evoked potential. Science 155:1436–1439
Tecce JJ, Savignano-Bowman J, Meinbresse D (1976) Contingent negative variation and the distraction-arousal hypothesis. Electroencephalogr Clin Neurophysiol 41:277–286
Thurstone LL, Thurstone TG (1962) Primary mental abilities. Science Research Associates, Chicago
Walter WG, Cooper R, Aldridge VJ, McCallum WC, Winter AL (1964) Contingent negative variation: an electric sign of sensori-motor association and expectancy in the human brain. Nature 203:380–384

Cognitive Processing in the EEG

H. Pockberger, P. Rappelsberger, and H. Petsche

1 Introduction

The usefulness of the EEG for the detection of brain diseases, particularly epilepsies, has been shown convincingly over the years. Its clinical application has been based mainly on a visual description of more-or-less regular potential fluctuations, in terms of frequency bands and voltages and of several specific grapho-elements such as spikes, spike-wave complexes, and others. Thus, different focal or generalized EEG patterns could be empirically assigned to pathological or normal activities of the brain.

With the introduction of computer-aided EEG analysis, a new challenge emerged and has led, among other things, to a deeper examination of the "normal" EEG and its possible alterations by mental processes. Such analyses, however, require the choice of appropriate parameters to be extracted from EEGs recorded during the performance of different cognitive or decisive acts.

The selection of such parameters ultimately determines the choice of the experimental design. For instance, studies of event-related potentials put the greatest emphasis on amplitude, waveshape, and latency of certain grapho-elements occurring before, during, and after the stimulus. Since these responses are usually small, they have to be extracted from background noise by averaging a large number of trials. Therefore, such parameter extractions can be done only with tasks of short duration, which puts narrow limitations on the experimental design.

Contrary to this, studies of the ongoing EEG recorded during different mental tasks have the advantage of greater versatility of experimental design, so that it is possible to study complex mental activities, such as reading, mental arithmetic, and others.

The application of spectral analytical methods makes possible the quantification of the EEG in terms of power and coherence, yielding information about activities recorded at different electrode sites on the scalp and also giving insight into the dynamic interactions between different neocortical areas.

The aim of our studies is to show task-dependent changes in the ongoing EEG during mental activities, using these analytical methods.

2 Methods

EEGs were recorded from 19 electrodes placed according to the 10–20 system. Linked earlobe electrodes served as a common reference. Healthy volunteers were

Springer Series in Brain Dynamics 1
Edited by Erol Başar

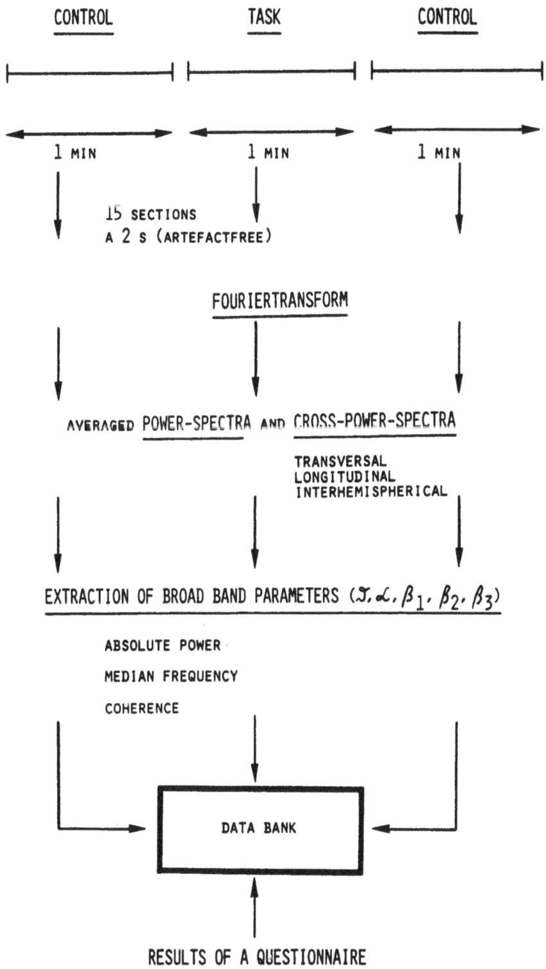

Fig. 1. Flowchart of parameter extraction from different EEG recordings (control – task – control) for further processing by statistical means. See text for further details

seated in a comfortable chair and instructed about the different tasks (reading, listening to a story, mental arithmetic, contemplating a picture, and listening to music). Each task lasted 1 min and was preceded and followed by a control period of 1 min during which subjects were asked to relax and either to keep their eyes closed or to look at a black spot about 2 m in front of them. These two control periods served as a comparison for the EEG recorded during the task.

Figures 1 and 2 outline the general timetable of such an experiment and the subsequent analysis in a digital computer. The 19 EEG signals were stored on analogue tape and digitized at 256 per second. From every minute of recording, 15 2-s epochs were selected for spectral analysis after visual inspection for artefacts. Then power spectra and cross power spectra were computed between adjacent electrodes in the transversal and longitudinal direction and between electrodes on homologous areas on both hemispheres (interhemispheric coherences). From these spectra, broadband parameters were selected for the theta (4–7.5 Hz), alpha (8–12.5 Hz), $beta_1$ (13–18 Hz), $beta_2$ (18.5–24 Hz) and $beta_3$ (24.5–31.5 Hz) band. Finally, coherence

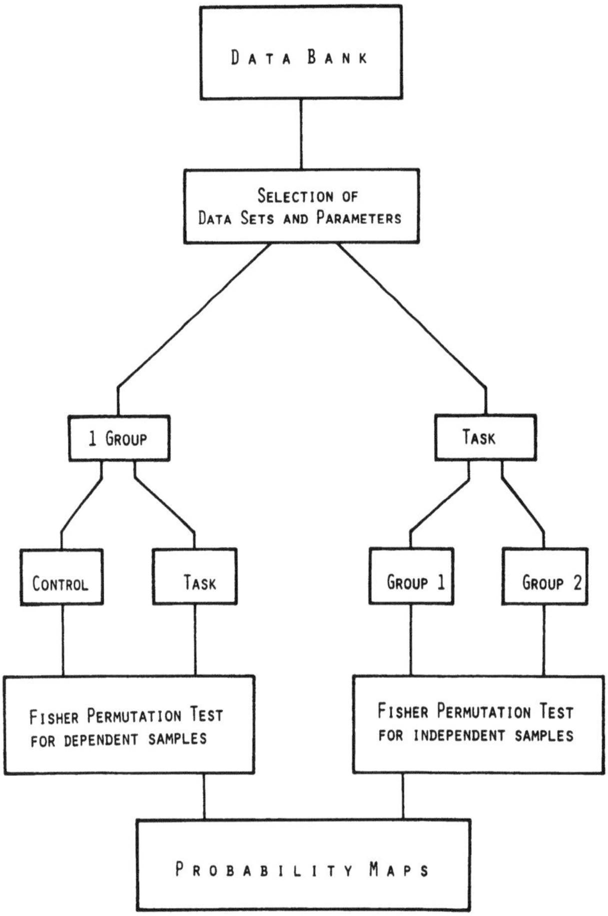

Fig. 2. Flowchart of the two main pathways for the computation of probability maps. Either the EEGs of one group are compared (control – task) with statistical tests or the EEGs of different groups recorded under the same situation are compared

spectra were computed. These broadband power and coherence parameters were entered, along with results of a questionaire (right-left hander, male-female, musician, mathematician, etc.), into a data bank that served as a basis for further statistical evaluation of EEG changes. Significant changes of power and coherence parameters were presented in topographical maps. Thus, significant EEG differences caused by mental activity, or other differences between two groups of subjects (e.g., right-handers vs. left-handers) could be plotted in so-called probability maps.

3 Results

For demonstration, examples of a study with healthy volunteers are given. Figure 3 shows the probability of changes in EEG power and coherence in connection with

READING $N = 16$ (KV-LE)

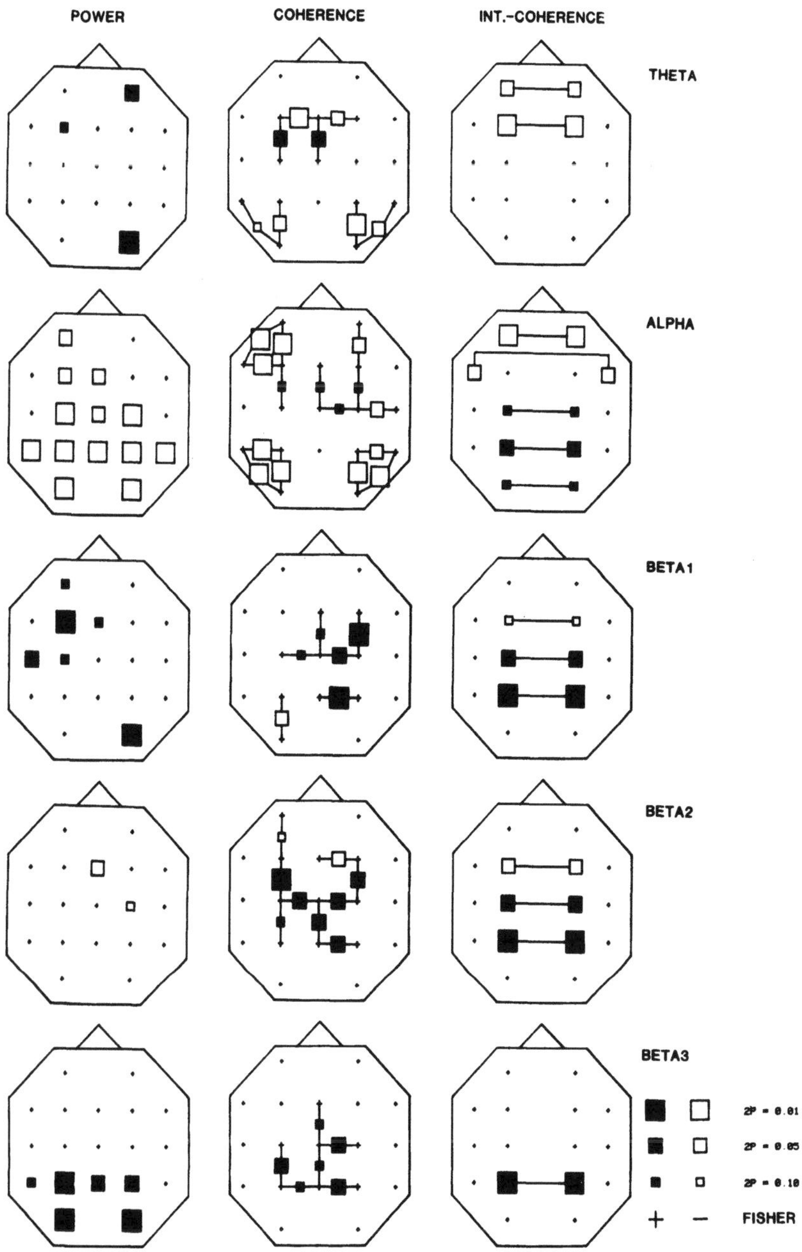

Fig. 3. Probability maps for power and coherence changes during reading [Fisher permutation test (Edgington 1980) for a group of 16 right-handed males] as compared with a resting condition immediately before. *Squares* indicate either an increase *(black)* or decrease *(open)*. Their size corresponds to an error probability of 0.90, 0.95, and 0.99. The *vertical bars* between the squares in the *middle* and the *right* columns indicate the pairs of electrodes for which coherence was calculated (*middle column*, transversal and longitudinal coherence; *right column*, interhemispheric coherence). Whereas alpha power decreases, beta power increases in different areas. Coherence increases primarily in the beta bands and decreases in the theta and alpha bands

MENTAL ARITHMETIC (1+2+3+...) N=16 (KV-KR)

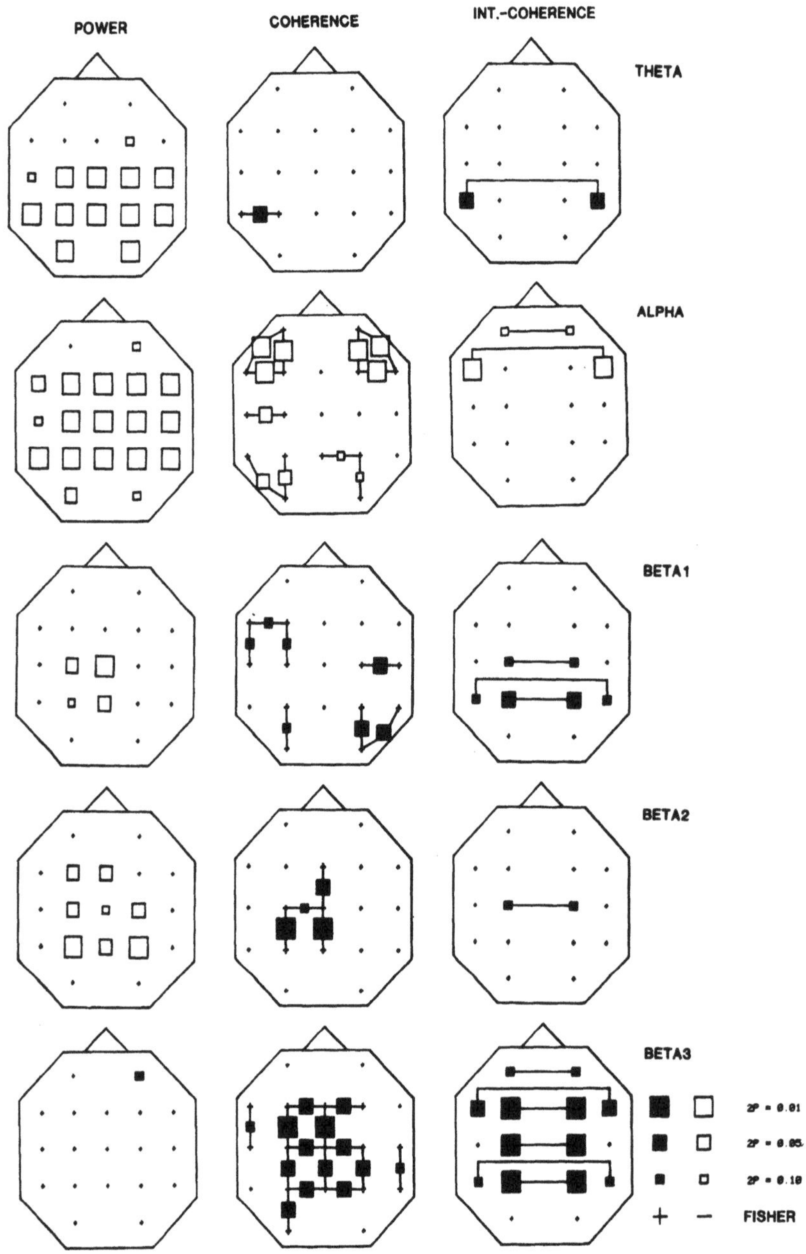

Fig. 4. Probability maps for EEG changes during mental arithmetic. Note different decreases in power in the five frequency bands and localized increases and decreases in coherence. Further details are given in the text. Symbols as for Fig. 3

silent reading as compared with a control period prior to this task (relaxed and looking at a black spot at a distance of 2 m) for a group of 16 right-handed male students. Regarding power, one observers a decrease in alpha power, primarily in the caudal part of the scalp. In contrast to this, theta, $beta_1$ and $beta_3$ power increase in different locations (see left column of Fig. 3).

Coherence changes are different. First, increases and decreases occur in a single frequency band. Secondly, the changes are more circumscribed. The probability maps indicate that theta and alpha coherences decrease, whereas coherences increase in the beta bands. Furthermore, the decrease in alpha coherence is strictly confined to the left and right frontal and left temporooccipital regions. The increase in beta coherence occurs instead in the central and parietal areas. Interhemispheric coherences show a decrease in the frontal regions in the theta and alpha band and an increase in the paramedian regions (see left column of Fig. 3).

A different pattern of power and coherence changes is observed during *mental arithmetic* (adding the natural numbers silently beginning with one, with eyes closed) when compared with the EEG when relaxed with eyes closed (see Fig. 4). The probability maps show that power decreases significantly in all five frequency bands, particularly in the theta and alpha band and to a lesser degree also in the $beta_1$ and $beta_2$ band. No clear lateralization to either hemisphere can be seen.

Again, coherence changes are more localized than during the reading task. Alpha coherence decreases in the right and left frontal and also in the left temporooccipital regions. Contrary to this, an increase in beta coherences is found, particularly in the $beta_3$ band, which is more confined to the central and parietal regions (see middle column of Fig. 4).

The interhemispheric coherences (left column of Fig. 4) reflect an increase and decrease in the alpha and beta bands, similar to the transversal and longitudinal coherences. Particularly the paramedian regions show an increased beta coherence during mental arithmetic.

As a third example, power and coherence changes during *listening to music* with eyes closed are shown in Fig. 5. A period of being relaxed with eyes closed was recorded prior to this task and served as a comparison. Significant power decreases are observed in the theta, $beta_1$, and alpha bands, the alpha decreases being more confined to the left hemisphere.

Coherences change in both directions, but are less significant than during reading and mental arithmetic. However, interhemispheric coherences show a characteristic change in that a decrease in the theta and alpha band in contrast to an increase in the beta bands in the frontal and temporal regions.

Evidently these three examples, derived from the same group of male students, show remarkable differences and therefore support the long-held belief that information about mental activity can be drawn from the EEG. In comparison with the EEG changes during mental arithmetic and during listening to music, in which power is decreased in all frequency bands, reading shows an increase and decrease of power in different frequency bands. The cause for this difference might be that during reading the eyes were open, whereas the other two tasks were performed with eyes closed.

One also observes different changes in coherence during these three different tasks. The smallest changes in coherence are observed during listening to music, whereas mental arithmetic is accompanied by the most significant changes, particu-

LISTENING TO MUSIC (MOZART KV458) $N = 16$ (KV-MU)

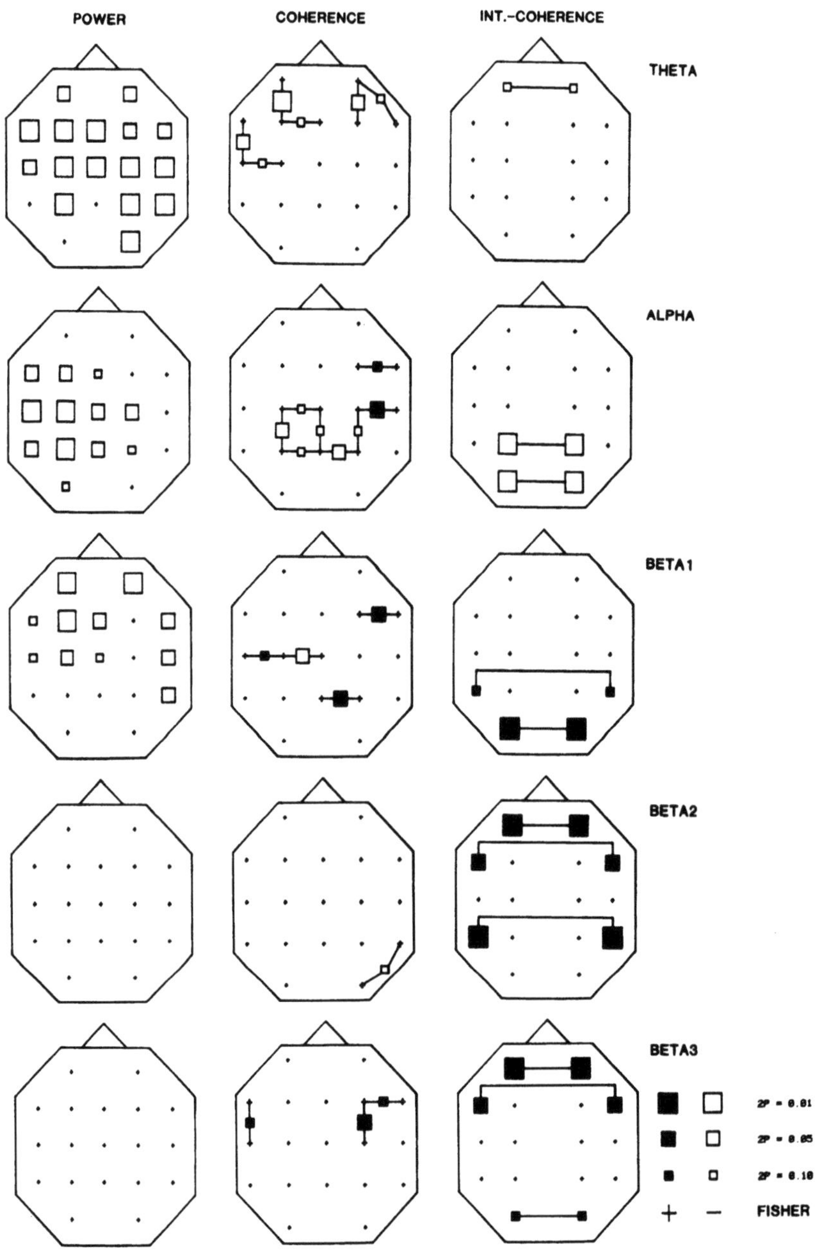

Fig. 5. Probability maps for EEG changes during listening to music. Power as well as coherence changes are less significant than during reading or mental arithmetic (compare Figs. 3 and 4). For further details, see text. Symbols as for Figs. 3 and 4

larly in the $beta_3$ band. This might be caused by the different degrees of difficulty of the three tasks and the different degrees of personal engagement in every task. Moreover, personal talents and interests were not taken into account.

4 Discussion

The development of computer-aided EEG analysis has brought into being an entirely new source of information. Several groups of investigators have developed new methods of EEG mapping, but have obtained results that unfortunately differ to some extent (Buchsbaum et al. 1982; Duffy et al. 1981; Giannitrapani 1985; Gevins et al. 1979). The major cause of the diversity of these results seems to lie in the investigators' selection of different parameters, methods, and paradigms. It is important to select the parameters properly and to try any meaningful interpretation.

With regard to power, the traditional view of the alpha as a mere "idling" rhythm has to be discussed anew. The same holds true for activities in the beta bands. Coherences additionally add to the problem of interpretation. According to our experience, changes in power and coherence do not necessarily occur at the same place and in the same direction, so that these two parameters have to be viewed as complementary. The coherence function describes the linear correlation of two signals in the frequency domain and is independent of the amplitude of the signals. It has been argued that the basic level of coherence depends in part on the fiber connections between two areas (Busk and Galbraith 1975; Beaumont et al. 1978). Shaw and Ongley (1972) suggested that a decrease in coherence may indicate the involvement of an area in processing information. Our findings prove that such a generalized statement does not conform with reality, as coherence may increase in one frequency band but decrease in another. For example, the reading and mental arithmetic tasks showed primarily a decrease in coherence in lower frequency bands and an increase in the higher ones. Additionally, we observed that the theta and alpha coherence decreases in quite different regions, whereas the beta coherence increases, the latter being largely confined to the central and parietal regions (compare Figs. 3 and 4).

Two further comments should be made. The first deals with the concept of "lateralization". Most of the phenomena hitherto observed by us during mental tasks proved to involve both hemispheres, though usually at different locations and to different extents. These findings underline the point of view held by LeDoux (1983), who criticized the misconception regarding hemispheric dominance often found in psychophysiological publications. Similarily, Geschwind (1984) regarded hemispheric dominance as an often misunderstood concept. According to his point of view, this misunderstanding originated in wrong interpretations of histological observations related to speech disturbances, which led to the concept of a left-sided dominant hemisphere and a right hemisphere of minor importance in right-handed people.

The final point to be discussed regards the problem of how the brain may adapt to different situations. In an earlier study, we found that EEG changes are different if a control period before listening and after listening to music is compared with listening to music (Petsche et al. 1985). Another study showed different EEG changes for depressive patients and healthy persons during verbal tasks, although both groups performed the tasks equally well (Pockberger et al. 1985).

All these observations led us to the conclusion that the experimental design – i.e., the sequence of tasks presented to a person – might influence the results derived from a comparison of EEG records during different situations. It seems that any change of neocortical activity caused by a certain task situation depends heavily on the prior state of activity. This implies that one is confronted with a dynamic system, which not only reacts by switching on or off certain subsystems, but also modulates the interaction of different specialized subsystems. Therefore one has to expect many EEG changes that are not necessarily localized to a certain area or even lateralized to either hemisphere during cognitive processing.

Acknowledgements. This research was supported in part by the Karajan Foundation of the Gesellschaft der Musikfreunde, Vienna. The authors wish to thank Mrs. E. Genner and Mrs. G. Luger for experimental assistance and typing the manuscript.

References

Beaumont HG, Mayer AR, Rugg MD (1978) Asymmetry in EEG alpha coherence and power: effects of task and sex. Electroencephalogr Clin Neurophysiol 45:393–401

Buchsbaum MS, Rigal F, Coppola R, Cappeletti JC, King C, Johnson J (1982) A new system for grey-level surface distribution maps of electrical activity. Electroencephalogr Clin Neurophysiol 53:237–242

Busk J, Galbraith GC (1975) EEG correlates of visual motor practice in man. Electroencephalogr Clin Neurophysiol 38:415–422

Duffy FH, Bartels PH, Burchfield JL (1981) Significant probability mapping: an aid in the topographic analysis of brain electrical activity. Electroencephalogr Clin Neurophysiol 51:455–462

Edgington ES (1980) Randomization tests. Dekker, New York

Geschwind N (1984) Historical introduction. In: Geschwind N, Galaburda AM (eds) Cerebral dominance – the biological foundations. Harvard University Press, Cambridge MA, pp 1–18

Gevins AS, Zeitlin GM, Yingling CD, Doyle JC, Dedon MF, Schaffer RE, Roumasst JT, Yeager CL (1979) EEG patterns during "cognitive" tasks. I. Methodology and analysis of complex behaviors. Electroencephalogr Clin Neurophysiol 47:693–703

Giannitrapani D (1985) The electrophysiology of intellectual functions. Karger, Basel

LeDoux JE (1983) Cerebral asymmetry and the integrated function of the brain. In: Young AW (ed) Functions of the right hemisphere. Academic, New York, pp 203–216

Petsche H, Pockberger H, Rappelsberger P (1985) Musikrezeption, EEG und musikalische Vorbildung. EEG-EMG 16:183–190

Pockberger H, Petsche H, Rappelsberger P, Zidek B (1985) On-going EEG in depression: a spectral analytical pilot study. Electroencephalogr Clin Neurophysiol 61:342–348

Shaw JC, Ongley C (1972) The measurement of synchronization. In: Petsche H, Brazier MAB (eds) Synchronization of EEG activities in epilepsy. Springer, Vienna New York, pp 204–214

An Analysis of Preparation and Response Activity in P300 Experiments in Humans

H. G. STAMPFER and E. BAŞAR

1 Introduction

The limitations of the averaging technique in studies of sensory evoked potentials (EPs) and "endogenous" event-related potentials (ERPs) have been discussed by various authors (Sayers 1974; Başar et al. 1975, 1976a, b; Squires and Donchin 1976; Van der Tweel et al. 1980). These authors have questioned the assumptions underlying the averaging technique (identical stimuli, identical responses, and random stationary background EEG activity that is not correlated with the EP and ERP response) and have suggested that EP and ERP slow potentials may be manifestations of stimulus-evoked and response-evoked synchronization, frequency stabilization, frequency selective enhancement and damping, and phase reordering of spectral components of the spontaneous EEG activity already present. That is to say, these slow potentials signify a dynamic change in the spontaneous EEG activity associated with an "event" of signal or information processing in the brain. From this perspective, the "background" EEG activity should not be considered as "noise" that has to be averaged out, nor should the EP and ERP signal be interpreted in terms of some "additive component" to the spontaneous EEG activity.

Evidence from previous studies (Başar 1983; Başar et al. 1979a, b) has suggested that prestimulus EEG activity also bears an important functional relationship to poststimulus response characteristics. We have used the expression "preparation changes" to refer to alterations in prestimulus EEG activity during ERP experiments and have suggested that a study of this activity is important in understanding ERP response characteristics.

In view of the above considerations, frequency domain analysis may hold greater potential for understanding the genesis of these slow potentials, especially if attempts are made to relate poststimulus changes to prestimulus EEG activity. Such an approach offers greater opportunities for understanding the brain's state-dependent system dynamics, since it is not possible to measure response variation in averaged data, nor is it possible to investigate the functional relationship between pre- and poststimulus EEG activity.

The findings presented here were obtained by applying a combined analysis procedure to ERPs elicited by auditory stimuli under different experimental conditions. The combined analysis procedure includes single-sweeps analysis, selective averaging, response adaptive filtering of pre- and poststimulus EEG epochs, and frequency domain analysis.

The studies were undertaken specifically to explore the frequency perspective and underlying brain dynamics of the P300 change in humans, which had hitherto been

Springer Series in Brain Dynamics 1
Edited by Erol Başar

described almost exclusively in terms of averaged latency/amplitude measures. Apart from studying the P300 response characteristics in the frequency domain, we were particularly interested in further testing our general hypothesis that prestimulus activity also changes under different experimental conditions, and that these changes in turn have an important influence on ERP response characteristics.

2 Experimental Design

Our studies consisted of three different experiments. The first was a conventional auditory "oddball" experiment, in which subjects were asked to count random and infrequently occurring target tones. In the second experiment, we presented a series of alternating target and nontarget tones, with constant interstimulus interval. Subjects were again asked to count the number of target tones. In the third experiment, we presented the same regular tone series as in the second experiment, but, differently from the first and second experiments, we also informed subjects that target/nontarget tones would occur in a regular, alternating sequence. Subjects were again asked to ignore the nontarget tones and mentally count the target tones, but were additionally asked to mentally anticipate the occurrence of target tones.

2.1 Baseline Data

Background EEG and auditory sensory EP data were recorded for each subject prior to the above-mentioned experiments to provide graduated "baselines" for the comparison of ERP changes.

2.2 Subjects

Studies were carried out on 12 healthy, drug-free, volunteer subjects, whose ages ranged from 25 to 45 years. A monetary reward was promised if subjects counted the correct number of target tones.

2.3 Experimental Setup

EEG activity was recorded at the vertex Cz position of the international 10–20 system for electrode placement, with the reference electrode placed on the left ear. The filter band-pass of the EEG amplifier was set to 0.1–70 Hz. EEG activity was recorded continously on a paper trace during all data collection. Facilities were available for silent behavioral observation of subjects via closed circuit television.

Subjects were seated in a quiet, but not fully soundproof room, with their eyes comfortably closed during every experiment and period of background data collection. Auditory stimuli consisted of 1500 Hz nontarget tones, which subjects were asked to ignore, and 1550 Hz target tones, which they were asked to count mentally. These stimuli were step-function tones with 500 μs rise time and 800 ms duration and were delivered at 70 db above hearing threshold. They were presented through a loudspeaker mounted 2 m directly in front of the subject.

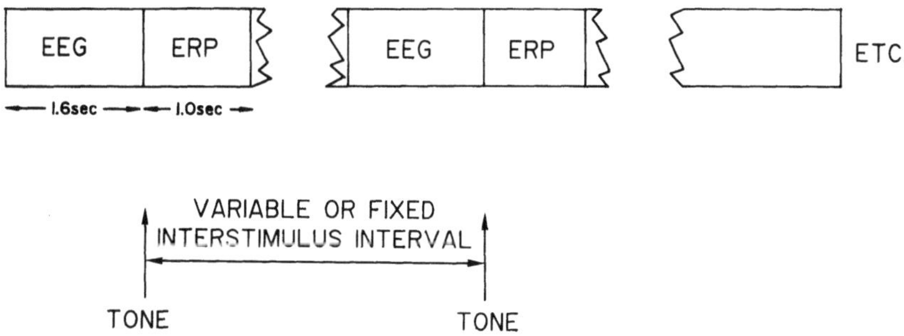

Fig. 1. EEG-EPograms and EEG-ERP epochs

In the oddball experiment, target tones accounted for about 18% of the stimuli presented. The interstimulus interval in the oddball experiment varied randomly between 4.3 and 10.6 s. In the experiments with regular stimuli, the constant interstimulus interval was 2.65 s. A total of 80 tones was presented in all experiments.

2.4 Data Collection

With every stimulus presented, 1.6 s of EEG activity preceding the stimulus, and 1 s of EEG following the stimulus were sampled at a rate of 800 μs. Digitized data were labeled and stored on computer disk. Paper-trace EEG records were also kept. This sequence of EEG-EPograms, and, in terms of this study, also EEG-ERP epochs, is shown in Fig. 1. Background EEG activity used for comparison of EP and ERP changes was obtained by recording 80 single sweeps of the EEG activity, with the same timing and data collection methods as if subjects had received tone stimuli.

2.5 Data Analysis

The application of the combined analysis procedure to pre- and poststimulus EP and ERP epochs, and the relevant theoretical considerations have been discussed by Başar in previous publications (Başar et al. 1975, 1976a, b, 1979a, b). For present purposes it is necessary only to mention the main aspects of this analysis procedure.

2.5.1 Averaging and Selective Averaging

By not averaging our data progressively during the study, we can evaluate various "selective averages" at any time after the study. All the necessary raw data is always available on disk or magnetic tape. This approach also permits more efficient artefact rejection than various direct, on-line methods.

2.5.2 Evaluation of the Amplitude Frequency Characteristics of EPs and ERPs

The selectively averaged EP and ERP data are transformed to the frequency domain by means of the fast Fourier transform (FFT) to obtain the amplitude frequency characteristics which reflect the quantitative properties of the studied system in terms of its frequency components (see Fig. 3 below).

2.5.3 Response-Adaptive Filtering

The amplitude maxima of the amplitude frequency curves are used to set the band limits of digital filters used to analyze various data. These filters do not create any phase shift and their band limits depend upon the response characteristics for each subject in different experiments. That is to say, these filters are adapted to the time-locked frequency selectivities detected in the averaged sensory EPs and ERPs of the subject.

3 Results

Figure 2 shows typical averaged data from three of our 12 subjects. The top half of Fig. 2 shows the sensory EP for each of the three subjects. The bottom half shows the ERPs from each of the same three subjects in the oddball experiment. The dashed curves show the response to nontarget tones, and the solid curves show the response to the target tones. It can be seen that the sensory EP and the ERP to nontarget tones show a similar configuration in all three subjects. By contrast, the averaged response to target tones shows a large positive deflection at around 380 ms, which represents the late positive component or P300 peak.

Figure 3 shows the amplitude frequency characteristics from one of the subjects whose averaged EP and ERPs are shown in Fig. 2. These curves were used to define the band limits of the digital filters used to analyze the EP, ERPs, and single sweep epochs from this subject. In this case, band limits were selected as 1–3.5, 3.5–8, and 8–13 Hz. These filter limits were based on the amplitude maxima peaks shown in the curves.

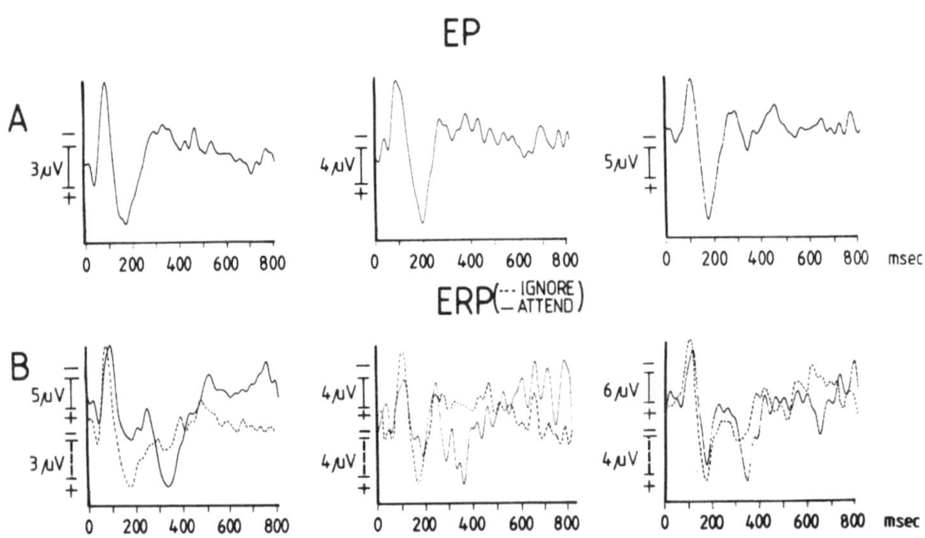

Fig. 2. (A) Averaged sensory EPs for each of three subjects. (B) ERPs for the same three subjects in an oddball experiment

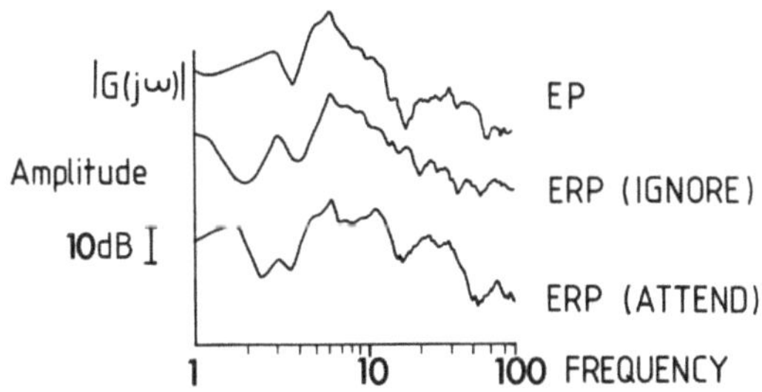

Fig. 3. Amplitude frequency characteristics from one subject

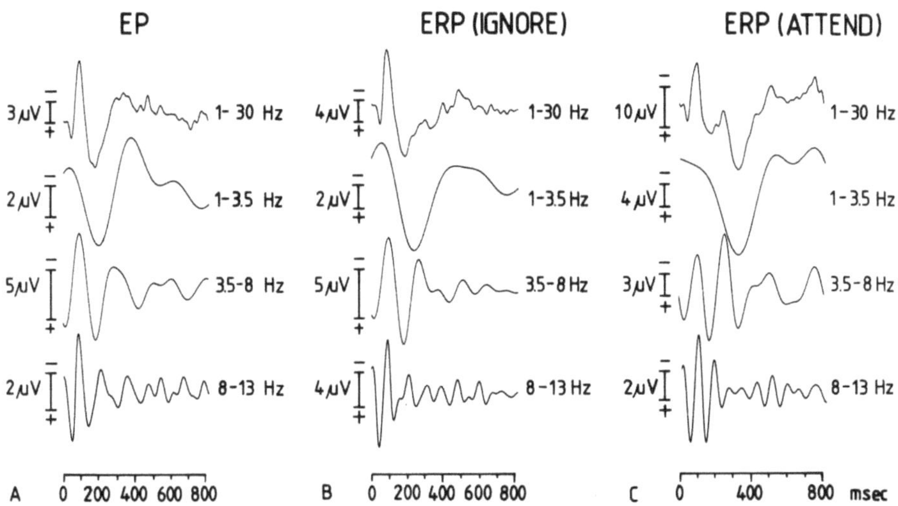

Fig. 4. Averaged response and filtered components of the averaged EP and ERP

3.1 Filtered Components of Averaged Responses

Figure 4 shows the averaged response and the filtered components of the averaged EP and ERPs from the subject whose frequency amplitude curves are shown in Fig. 3. The top traces are the averaged responses (EP, nontarget ERP, target ERP). Below each trace are the various filtered components, with their respective band widths indicated on the right side of each trace (1.0–3.5, 3.5–8.0, 8.0–13.0 Hz). Assuming that the averaged peaks shown in the top traces are formed by the superposition of different frequency components, it can be seen that in the sensory EP, for example, the N100 peak is formed mainly by theta and alpha activity (3.5–8.0 Hz and 8.0–13.0 Hz). The P200 peak is formed predominantly by the first positive deflection of theta activity, although the first positive deflection of delta activity also contributes.

Figure 5 shows the same filtered frequency components arranged to provide a comparison, within each of the three frequency bands, of the particular frequency activity in the EP, ERP to nontarget tones, and ERP to target tones for the same subject.

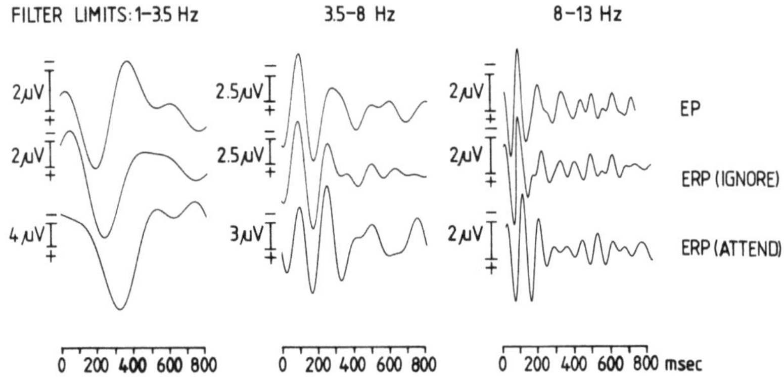

Fig. 5. The same filtered frequency components as in Fig. 4, arranged to provide a comparison of the particular frequency activity in the EP, ERP to nontarget tones, and ERP to target tones

4 Comparative Analysis of Poststimulus EP, Nontarget ERP and Target ERP Frequency Changes and Their Contribution to Different Latency Peaks

4.1 1.0–3.5 Hz Delta Activity

Dealing first with delta activity, it can be seen that the configuration of activity in this frequency band is grossly similar in the EP, nontarget ERP, and target ERP, except for a progressive latency shift to the right of the major positive deflection, which is maximally delayed in the response to target tones. This maximum positive deflection in the target ERP was often followed by a negative deflection peaking at around 600 ms. Our data suggest that this N600 deflection should be considered as a continuation of the positive deflection at around 380 ms and as part of the mechanism giving rise to changes in the delta bandwidth.

4.2 3.5–8.0 Hz Theta Activity

Activity in the theta range showed the greatest and most consistent task-related changes, i.e., in the response to oddball target tones. In the EP and nontarget ERP, the maximum amplitude, or poststimulus enhancement, is usually seen in the first poststimulus wave, whereas in the target ERP response, the maximum amplitude is usually seen in the second and sometimes in the third wave. More generally, there is prolonged theta oscillation in target ERP responses, such that this frequency band often makes a significant contribution to the P300 deflection and N200 peak, which will be discussed below. The above-mentioned theta changes were observed in all 12 subjects after filtering the averaged responses.

4.3 8.0–13.0 Hz Alpha Activity

Stimulus-elicited oscillations in the alpha band are usually maximally desynchronized at around 380 ms in the response to attended target tones. However, the alpha activity of the ERP response to target tones, as seen in Fig. 5, shows a frequently observed finding that the second major oscillation is greater for target than for EP and nontarget ERP responses. This second oscillation contributes to the N200 peak. Our findings suggest that whereas alpha activity does not appear to make any direct contribution to the P300 peak, it does show task-related ERP changes.

4.4 N200 Peak

A negative peak at around 200 ms was a conspicuous feature of the averaged ERP to nontarget and especially to target tones. This N200 peak can be seen readily in Fig. 2b. Frequency analysis of the averaged response revealed that this peak results mainly from the delayed theta amplification, or enhancement, and, to a lesser extent, from the more prominent second oscillation in the alpha band. This peak cannot be observed in the sensory EP, and it is suggested that it should be regarded as part of the changes involved in the oddball P300 response.

4.5 P165 Peak

In the sensory EP, a positive peak at around 200 ms is formed by the superposition of delta, theta, and more variably, alpha time-locked oscillations (see Fig. 4). It can be seen that in the task-related ERPs, a similar peak occurs earlier, at around 165 ms. According to our analysis, this P165 peak is formed by time-locked theta and alpha activity, which have become more prominent because of the latency shift to the right of the major positive deflection in the delta frequency range.

5 Functional Interdependence and Independence of 1–3.5, 3.5–8.0 and 8.0–13.0 Hz Activity

The results presented to this point have given evidence that a subject's averaged EP, nontarget ERP, and target ERP show poststimulus differences within delta, theta, and alpha frequency bands. We have drawn attention to specific changes within these frequency bands in the averaged ERP response to target tones; the main ones being a delayed positive delta deflection, prolonged theta oscillation and delayed maximum enhancement, as well as poststimulus oscillation changes and desynchronization at around 380 ms in the alpha band. However, analysis of single sweep epochs from the same subject in the oddball experiment revealed that these frequency changes can occur together or independently during the course of the experiment. Figure 6 shows four single sweeps, from the same subject in the oddball experiment, filtered within delta, theta, and alpha frequency bands. It can be seen that in sweep C, there is no evidence of a positive delta deflection or latency shift; there is very little poststimulus alpha oscillation, but there is prolonged theta oscillation as well as delayed maximum enhancement. Sweep B, by contrast, shows characteristic delta (delayed positive deflection), theta (prolonged oscillation and delayed maximum enhancement), and alpha (prominent positive deflection around 165 ms and desynchronization at around 380 ms) changes as mentioned above. These findings suggest that different frequency components may act together or independently. This sug-

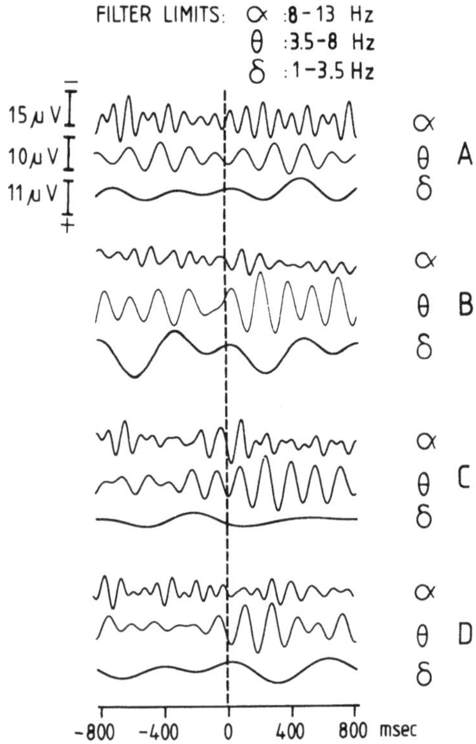

Fig. 6. Four single sweeps filtered within delta, theta, and alpha frequency bands

gests further that different psychophysiological mechanisms may come into operation to a different extent following stimulation in the oddball experiment.

6 Pre- and Poststimulus Variation During the Course of Any Single Experiment and the Relationship Between Pre- and Poststimulus Activity

Figures 7, 8, 9, and 10 show single-sweep EEG-ERP epochs from one subject in different experimental conditions. These epochs have been filtered in different band widths derived from the frequency amplitude curves, to show examples of pre- and poststimulus variation during the course of different experiments within these frequency bands.

Figure 7 shows 0.5–3.5 Hz (delta) activity of target-tone epoch numbers 43–61 in the third experiment, in which subjects were informed that target/nontarget tones would be presented in a regular alternating sequence. It can be seen that the general configuration of activity is very regular in both pre- and poststimulus epochs. Furthermore, there is evidence of a preferred phase angle at the point of stimulation. In the examples shown, there is evidence of a tendency to maximum negativity at the time of stimulation.

By contrast, Fig. 8 shows baseline EEG epochs filtered in the same band width (0.5–3.5 Hz), and it can readily be seen that there is no evidence of pre- or post-

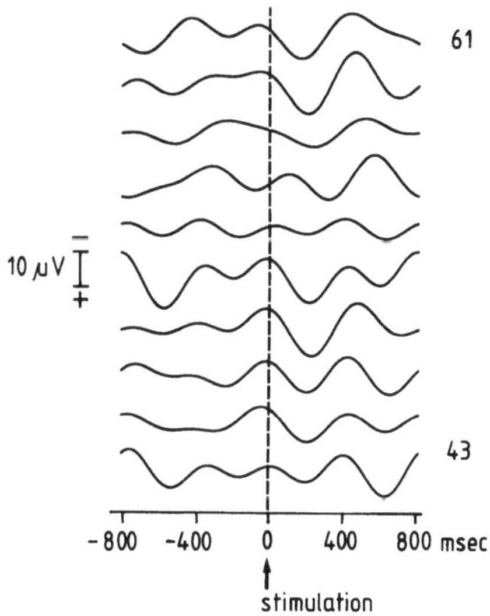

Fig. 7

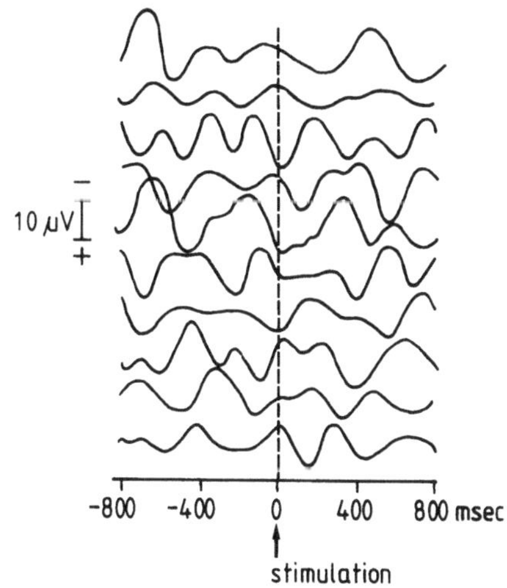

Fig. 8

Fig. 7. Delta activity of target-tone epoch numbers 43–61 in the third experiment

Fig. 8. Baseline EEG epochs filtered in the same band width as in Fig. 7

stimulus regularity. It should be emphasized that the percentage of single sweeps showing phase alignment at the point of stimulation can vary from 35% to 85% between subjects. The subject's cooperation and/or capacity to focus attention may be important in determining this percentage.

Figure 9 shows single sweep epochs, filtered in the alpha range (8–13 Hz), from the same subject in the second experiment, in which subjects were not informed about the regular alternation of target/nontarget tones. Only target responses are shown. It can be seen that the magnitude of prestimulus alpha activity increased progressively during the experiment and was associated with increasing phase alignment at the point of stimulation. These changes were evident in most subjects by about the 20th target tone, although there was variation between subjects. The onset of these changes occurred earlier when subjects were informed about the regular occurrence of target tones and was observed to occur by about the tenth target tone.

Figure 10 shows filtered theta (3.5–8.0 Hz) activity of single sweep target-tone epochs, from the same subject, in the auditory oddball experiment. Prolonged, regular oscillations and delayed maximum enhancement can be seen after sweep No. 49, i.e., the tenth target tone, which also marks a reduction of prestimulus theta amplitude. The onset of these changes varied between subjects and also under different experimental conditions. Further quantification and study is required to understand these variations. For present purposes, we wish merely to report the more obviously observed changes in pre- and poststimulus activity during the course of various ERP experiments.

We have used the expression "preparation changes" for the more regular or synchronized activity and phase alignment, within different frequency bands, that is evident in the prestimulus EEG activity during the course of an experiment and also be-

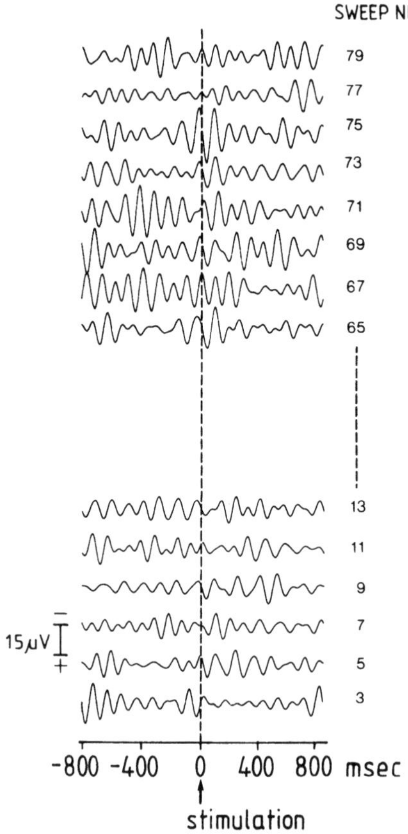

Fig. 9

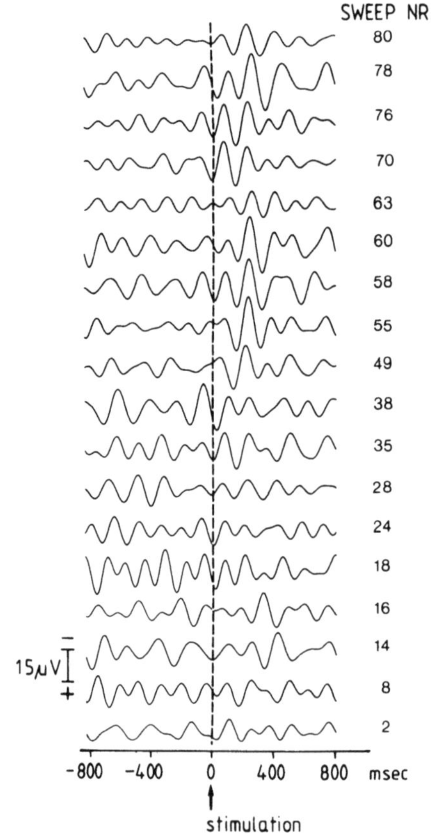

Fig. 10

Fig. 9. Single sweep epochs filtered in the alpha range

Fig. 10. Filtered theta activity of single sweep target-tone epochs

tween different experiments. It would appear that the expectancy and experience of regular target stimuli leads to physiological prestimulus synchronization, which in turn influences the poststimulus response characteristics, such as, for example, the prolonged theta oscillations and delayed maximum enhancement that we have related to different averaged peaks in our oddball experiment.

7 Changes Between the Average of Early and Late Sweeps During an Experiment

One would expect that if the poststimulus frequency characteristics change during the course of an experiment, the change should also be reflected in the average of early and late sweeps. Figure 11 shows averaged data from the oddball experiment

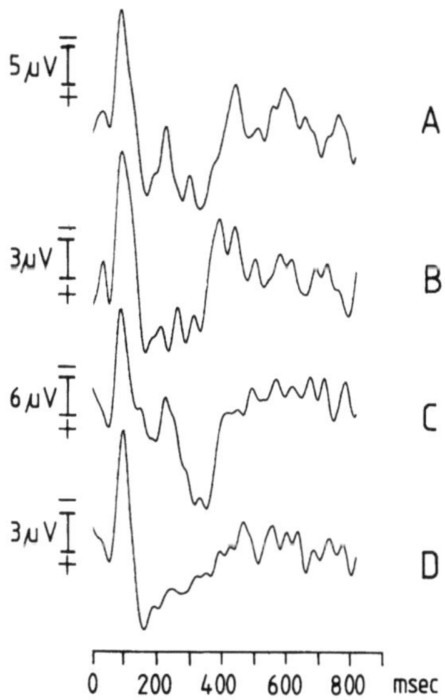

Fig. 11A–D. Averaged data from the oddball experiment. (A) The average of the first eight responses to target tones. (B) The average of the first ten responses to nontarget tones. (C) The average of the last eight responses to target tones. (D) The average of the last ten responses to nontarget tones

from the same subject. It can be readily seen that whereas there is not a definite P300 peak at around 380 ms in the average of the first ten nontarget responses, there is a clear P300-like change that lessens considerably in the average of the last ten nontarget responses.

The changes in the response to nontarget tones during the experiment suggest that the subject's evaluative attention to target and nontarget tones was more equal early in the experiment, and that later the subject was able to discriminate and ignore nontarget tones more confidently.

The same trend was evident in the second and third experiments, and, as might be expected, the least difference between early and late nontarget tones was found in the third experiment, where subjects were informed that the target and nontarget tones would alternate regularly.

8 Summary

In summary, we have described frequency perspectives of sensory EPs and ERPs elicited by auditory stimuli under different experimental conditions. We have described task-related changes within conventionally defined delta, theta, and alpha frequency bands. There is no reason to suppose that other frequencies may not be found to contribute in important and various ways to the development of ERP changes. It may be best to conceive of a family of task-related frequency changes, the identification of which may provide useful measures and understanding of EP and ERP phenomena.

Our finding of interdependent and independent task-related changes in different frequency bands as evidenced by single sweep analysis was thought to be particularly interesting and deserving of further studies. We have found evidence of what we have termed "preparation changes" in prestimulus EEG activity under different experimental conditions, and have suggested that prestimulus activity bears an important functional relationship to poststimulus response characteristics.

We recognize that quantification and further studies are required to substantiate our findings and hypothesis. However, we are of the opinion that the frequency perspective as well as the study of pre- and poststimulus activity may help to further understand sensory EPs and ERPs.

References

Başar E (1983) Toward a physical approach to integrative physiology. I. Brain dynamics and physical causality. Am J Physiol 245: R510–R533

Başar E, Gönder A, Özesmi C, Ungan P (1975) Dynamics of brain rhythmic and evoked potentials. I. Some computational methods for the analysis of electrical signals from the brain. Biol Cybern 20: 137–143

Başar E, Gönder A, Ungan P (1976a) Important relation between EEG and brain evoked potentials. I. Resonance phenomena in subdural structures of the cat brain. Biol Cybern 25: 27–40

Başar E, Gönder A, Ungan P (1976b) Important relation between EEG and brain evoked potentials. II. A system analysis of electrical signals from the human brain. Biol Cybern 25: 41–48

Başar E, Demir N, Gönder A, Ungan P (1979a) Combined dynamics of EEG and evoked potentials. I. Studies of simultaneously recorded EEG-EPograms in the auditory pathway, reticular formation and hippocampus of the cat brain during the waking stage. Biol Cybern 24: 1–19

Başar E, Durusan R, Gönder A, Ungan P (1979b) Combined dynamics of EEG and evoked potentials. II. Studies of simultaneously recorded EEG-EPograms in the auditory pathway, reticular formation and hippocampus of the cat brain during sleep. Biol Cybern 34: 21–30

Sayers B (1974) The mechanism of auditory evoked EEG responses. Nature 247: 481–483

Squires KC, Donchin E (1976) Beyond averaging; the use of discriminant functions to recognize event-related potentials elicited by single auditory stimuli. Electroencephalogr Clin Neurophysiol 41: 449–459

Van der Tweel LH, Estevez O, Strackee J (1980) Measurement of evoked potentials. In: Barber C (ed) Evoked potentials. MTP Press, Lancaster

The CNV Potentials and Adaptive Preparatory Behaviour

D. Papakostopoulos

1 Introduction

The contingent negative variation (CNV) has been described by its discoverers (Walter et al. 1964; McCallum and Walter 1968) and others (Low et al. 1966; Loveless 1973) as the sustained slow negative potential change which develops between two stimuli, separated by one or a few seconds, when the first of these stimuli (S1) acts as a warning for an action or decision to be taken with the occurrence of the second (S2). This commonly adopted description of "the" CNV implicitly stresses the singularity of "the" or "a" CNV wave and waveform. Data will be reviewed in this paper that form the basis of opposing such a notion. Some of the implications of these data, in CNV research and clinical application, will be discussed. The term "CNV potentials" will be used in the present paper, except when referring to the literature.

The discovery of the CNV influenced the experimental outlook of many neuroscientists. Amongst the reasons which determined this influence, some need particular consideration. The CNV offered the first evidence that not only stimulus-related but also subject-related "mental state" activities have a neurophysiological basis in man. This development had occurred in the final period of an era during which mental concepts like attention were regarded by many as unnecessary. At the same time, psychologists were holding themselves "aloof from neurophysiology, claiming that the investigation of behaviour can proceed independently of the investigation of what goes on in the brain" (Berlyne 1969). Those investigators who objected to one or both of these attitudes saw in the CNV the beginning of a new era. The closely following discoveries of the P300 and of the *Bereitschaftspotential* (Sutton et al. 1965; Kornhuber and Deecke 1965) reinforced such views.

Another fundamental contribution of the CNV to the field of cognitive neurosciences is that it focused discussion on the nature of the relationship between particular cognitive processes and particular brain electrical phenomena and consequently on even more fundamental levels of neurobiological operations. The CNV discovery was also instrumental in obliging people to consider the importance of sustained or dc electrical changes in the brain. Finally, as an extension of research on the CNV, Walter et al. (1964) proposed "the application of the new methods to the investigation of neuropsychiatric disorders."

It could be considered as indicative of future developments that the early major paper which confirmed the brain origins of the CNV criticised as inaccurate the term "contingent" and proposed its replacement by the term "conative" because "the exis-

Springer Series in Brain Dynamics 1
Edited by Erol Başar

tence of the shift does not depend upon the occurrence of either S1 or S2" (Low et al. 1966). The use of the CNV to investigate neuropsychiatric disorders was also questioned on theoretical grounds developed on the interpretation of experimental evidence. Weinberg (1975) argued that it is difficult to see how the CNV can be used to investigate psychiatric diseases if every waveform variation can also be recreated in normal subjects. Even the inclusion of the CNV amongst the brain electrical activities related to cognition has been strongly disputed by Rohrbaugh and Gaillard (1983) who remarked that "it is with some irony that 'their' chapter on the CNV appears amidst a collection of papers dealing with cognitive aspects of event-related potentials".

In view of this criticism, it seems justified to re-examine certain fundamental issues of the CNV's neurophysiological nature and psychological relevance.

2 Neurophysiological and Psychological Issues

The neuronal generators of the potential changes constituting the CNV are, like those of all the brain macropotentials, unknown. This indisputable fact led many neurophysiologists to dismiss scalp-recorded field potentials as serious contenders for providing information about brain function. Others, however, observed that, as "in so many other fields of science, if we wait until we understand some of the elementary processes, we find that we don't ever get to the more complicated ones" (Jasper 1969) and advocated that these different approaches "have to be made in parallel". However, "in parallel" should not be interpreted to mean in isolation from one another. Speculations on the nature of brain macropotentials should take into account the established facts obtained with microelectrodes directly from the animal brain. Nevertheless, because of serious species-specific differences, uncritical utilisation of such data could deteriorate into an elaborate but fruitless exercise. Clearly an intermediate step is necessary. This step can be provided from direct recordings with microelectrodes from the human brain during various behavioral sequences. Scarce such data may be, but it is upon them as a basis that the neurophysiological nature of the scalp-recorded CNV potentials should be discussed. A corollary of adopting such an attitude is the suggestion of developing techniques to reveal the characteristics of the cortical and depth data in scalp-recorded activities and not the assumption that the brain behaves according to elaborate reformulations of badly distorted information. The brain will not oblige and frustration could be the not inexpensive penalty.

The relevance of the CNV potentials to psychological theory cannot be separated from their neurophysiological characteristics or be seen in isolation from the multiplicity of meaning attributed to any psychological construct by different authors. To break this vicious circle, the development of new concepts on the basis of neurophysiological data was proposed (Papakostopoulos 1978). These new concepts could prove to be an inadequate beginning, but because their definition will be based on quantifiable criteria, any mistakes could be traced back and necessary corrections could be applied.

Evidence for locally generated CNV potentials has been accumulated from direct cortical and subcortical recordings as well as from scalp electrodes. The method, rationale, and results of cortical and subcortical recordings have been previously discussed. It was shown (Papakostopoulos et al. 1980) that CNV potentials from cortical ares 4 mm apart can be different in waveform or can have similar shapes but with amplitudes varying independently in the two areas. It was also shown that frontal and temporal areas also develop CNV potentials. The CNVs of the temporal areas start considerably later than those recorded simultaneously from the centroparietal cortex. CNV potentials from the frontal areas start as early or earlier than the centroparietal potentials; their amplitude, however, remains small throughout the foreperiod in comparison with other areas. CNV potentials were not recordable from electrodes judged to be located in orbital cortical areas. Recordings from the upper brain stem and rostral to the thalamus (McCallum et al. 1973, 1976) revealed CNV potentials with negative and positive polarity respectively and a difference in latency between those potentials and the cortical potentials.

The indications from the implanted electrodes for area-specific CNV potentials were pursued and verified with scalp electrodes, using specially designed methodological and experimental procedures. With a pattern recognition method, it has been shown that the waveforms of the simultaneously recorded CNV potentials from frontal and central areas are significantly different (Weinberg and Papakostopoulos 1975). It was also shown that the frontal and central potentials correlated differently with motor performance (Papakostopoulos and Fenelon 1975). The area specificity of the CNV potentials was upheld by the results of Papakostopoulos and Jones (1980). Recent results by McKay et al. (1985), based on average source-density profiles, further confirmed the area specificity hypothesis of the CNV potentials. The data shown in Fig. 1 provide additional evidence. These data were recorded during an experiment (Papakostopoulos and Jones 1980) in which the subject was required to press a button using a prescribed level of force after a stimulus (S1) and to release it quickly after another stimulus (S2) appearing 2 s later. In between S1 and S2, a high-frequency or low-frequency tone could appear (interposed stimulus, Si), indicating an increase or maintenance of the applied force, respectively. It is apparent from Fig. 1A that the CNV potentials in frontal and parietal areas behave in different ways after the Si. This area-specific behaviour of the CNV potentials is not easily apparent in all experimental paradigms. Data collected from the same subjects and same locations as in Fig. 1A are shown in Fig. 1B. It can be seen that the CNV amplitude in the parietal location is well maintained. The task, however, was somewhat different. The subjects were required to press a button after S2 and to ignore anything else.

The neuronal organisation underlying the CNV potentials in each particular area seems to be specific and different from that necessary to generate other types of potentials recordable from the same area. Cortical areas with different types of intrinsic rhythmical activities generate CNV potentials, and within the same area, changes in CNV amplitude and abundance in various frequency bands proceed independently. Potentials closely associated with CNV onset and cutoff vary independently from the sustained potentials. Figure 2 shows consistent CNVs associated with variable N100 to S2 and vice versa. The independence of the CNV potentials and the P300 potentials has been shown with cortical recordings (Papakostopoulos and Crow

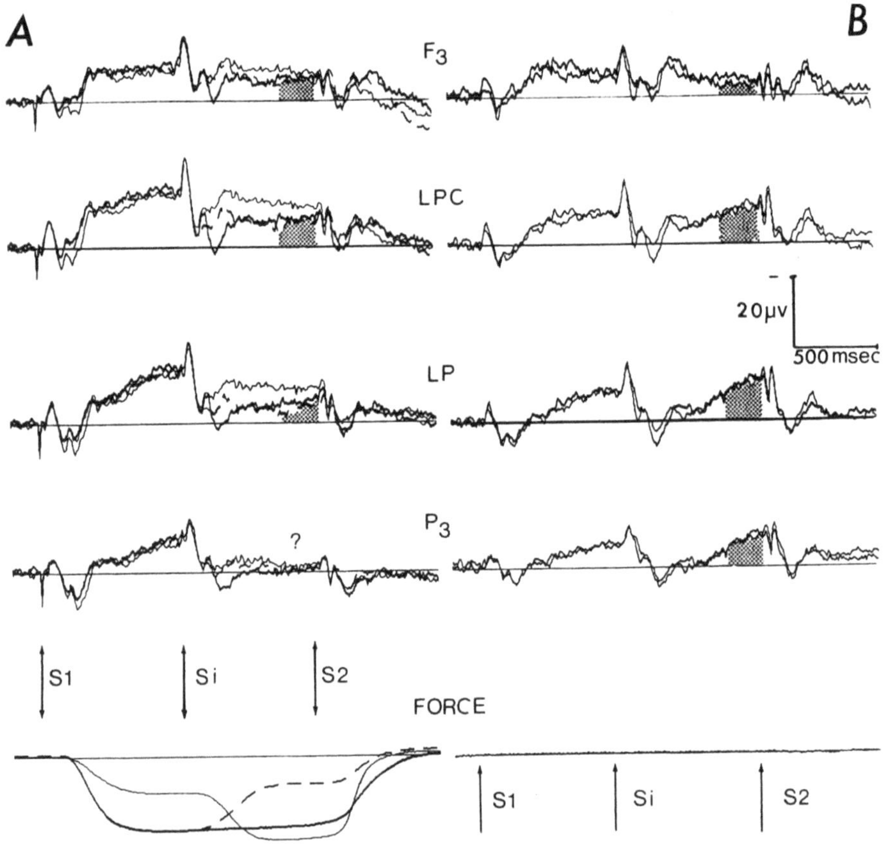

Fig. 1A, B. Superimposed CNV averages from four subjects and four left hemisphere locations. F3 and F4 according to the 10/20 system. LPC and LP about 2 cm anterior and posterior of C3 respectively. (A) *Thick line*, maintenance; *thin line*, increase; *interrupted line*, decrease of original force with the occurrence of S1 (a high- or low-frequency tone); *S1*, click; *S2*, flash. (B) Recordings from the same subjects and locations, with instructions to press a button after S2 and ignore S1. Note the suppression of the CNV potentials at P3 during the S1–S2 interval in all conditions in A(?) but not in B. Note also the different waveform of the CNV potentials during S1 S2 in A and B conditions

1976) and scalp recordings (Donchin and Heffley 1979). In view of this independence, it can be argued that the CNV potentials develop in a particular brain area only when a particular type of information processing takes place. Otherwise they have area-dependent functional specificity.

The overall brain pattern of this "mosaic"-like development of sustained potentials could be different during different types of preparation. Because of the spatial averaging characteristics of the scalp, it is to be expected that the skin-recorded CNV could change in waveform if the experimental situation changed, as indeed happened in the previously maintained experiment of Weinberg (1975). On this basis, the sustained potentials that develop after an unexpected S1, or prior to self-paced movements (*Bereitschaftspotential*), or during prolonged S1–S2 intervals, also have different overall brain patterns of development and thus different waveforms. To segre-

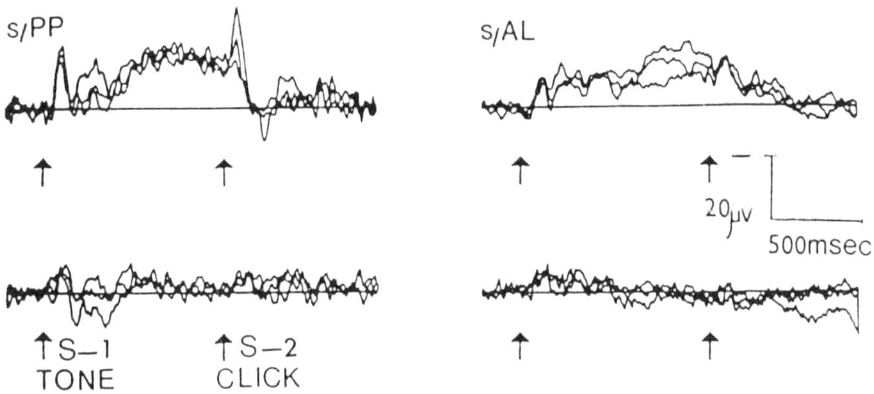

Fig. 2. Superimposed CNV potentials ($N = 6$, Cz referred to linked mastoid electrodes) and the simultaneously recorded eye movement activity at FPz from two subjects (*PP* and *AL*). Note the variability of N100 after S2 in subject PP and of the CNV potentials in subject AL, and the stability of the eye monitor traces and the N100 after S1 in both subjects

gate certain aspects of scalp-recorded waveforms under the titles of "early", "late" or "true" CNV could lead to the same confusions that were generated by the "singularity"-biased description of the CNV.

Faced with the plurality of the CNV potentials and the inherited transformation of their scalp-recorded waveform according to experimental conditions, it was proposed (Papakostopoulos 1978) that "diversity" or "rigidity in experimental paradigms" should alternate according to the specific question under investigation. In a study of head-injured patients, Curry (1983) adopted a rigid recording procedure incorporating diverse experimental conditions and the results seem both diagnostically and therapeutically interesting.

In parallel with the CNV potentials, a constellation of other electrobiological changes takes place during the foreperiod (Papakostopoulos and Cooper 1973, 1976). The excitability of the spinal monosynaptic reflexes increases, the respiration pattern can change, the heart rate decreases, and pupil diameter and blink reflexes also change, as does finger blood volume. By alteration of experimental procedures, it can be shown that these widespread changes can occur independently of each other. This independence suggests that they are controlled and integrated in a certain pattern by different brain systems acting in parallel. A manifestation of simultaneous activity of many brain systems is the area-dependent functional specificity of the CNV potentials. It is proposed that the type of each particular pattern of integration and the adequate realisation of this pattern, during the preparatory period, determines the characteristics of the forthcoming overt behaviour. It is hypothesised that a number of trials are necessary before the establishment of the most appropriate pattern of integration for a particular overt behaviour. Cooper et al. (1979) referred to this stage of nervous activity as the "scopeutic mode". With repetition of a particular behaviour, the integration of the various subsystems is adjusted and finally stabilised. We have referred to this stage as the "categoric mode". It was assumed that each system is more active during the scopeutic mode, whereas less activity is necessary during the categoric mode, to achieve similar results in terms of performance.

3 Support for a Dual Dynamic Concept

This dual dynamic concept receives support from neurophysiological and behavioural data. In Fig. 3, data recorded from four subjects are illustrated. Averages of 11 trials, according to the mode reaction time, are shown from the beginning, the middle and the end of a long experiment. It can be seen that although the reaction time of each subject is the same throughout, the CNV amplitude declines to half of its initial value by the end of the experiment. These results have been confirmed in another series of recordings (McCallum 1978), during which it was also observed that unexpected change in some experimental parameter result edina reversal of the declining CNV trend. In yet another recording (D. Papakostopoulos, in preparation), two experimental conditions were interlaced. In one of them, a click was always followed by a series of clicks to which the subject had to respond with a button press

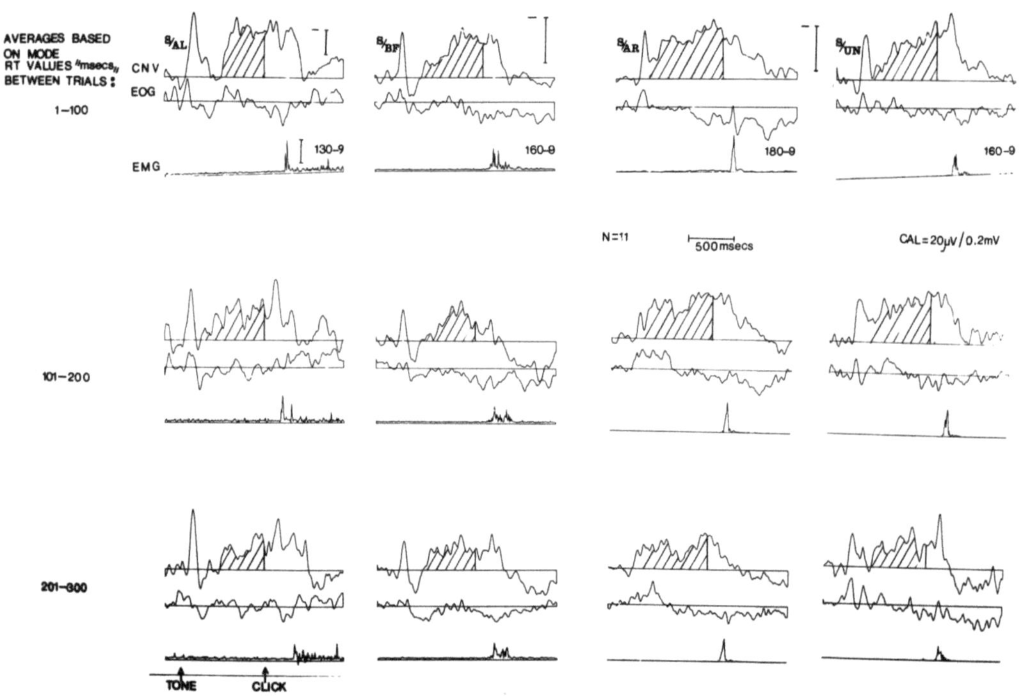

Fig. 3. Three sets of averaged ($N = 11$) CNV potentials. Electro-oculographic activity (EOG) and the electromyograms (EMG) selected according to the mode of reaction time (shown above the EMG of the upper set) from four subjects (*S*/) during the first (*upper set*), the second (*middle set*) and the last (*lower set*) 100 trials of a prolonged experiment. Note the progressive decline in CNV amplitude while the corresponding EOGs remain unchanged

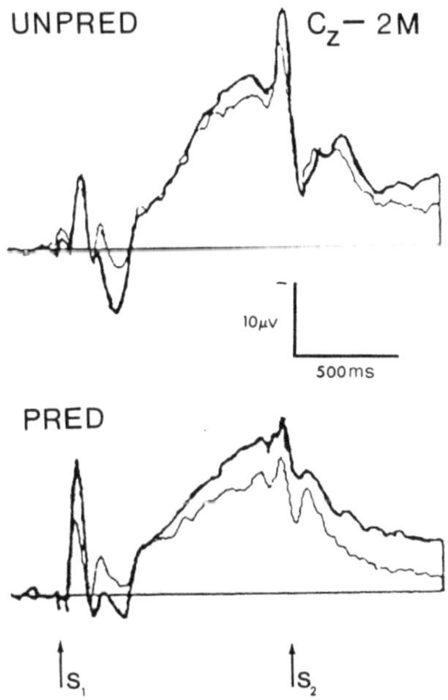

Fig. 4. Superimposed CNV averages from 12 subjects when S1 did not permit prediction and when it indicated (*PRED*) the type of action required after S2 (*UNPRED*). *Thick trace*, the first 32 trials; *thin trace*, the last 32 trials of a total of 160 in each condition. Note the reduction of CNV and N100 amplitude in the predictable condition

(predictable condition). In the other, a flash was followed by either a high-frequency or low-frequency tone but a button press was required with only the high-frequency tone (unpredictable condition). Twelve subjects were tested. The CNV drop in amplitude between the beginning and the end of the experiment was not significant (two-tailed paired *t* test) in the "unpredictable" condition ($T = 1.294$, $P = 0.2$) and was significant ($T = 2.340$, $P = 0.02$) in the condition that allowed consistent prediction. There was no significant difference in reaction times between the beginning and the end of the experiment in either condition. These data suggest that the experimental conditions that prevent the establishment of the categoric mode also prevent a drop in amplitude of the CNV potentials (Fig. 4). The differential drop of CNV amplitude in the two interlaced conditions, without any detrimental effect on performance, strongly suggests that specifically preparatory functions are mediated by the CNV potentials and not some kind of unspecific change in the arousal level during these long-lasting experiments. It also suggests that these preparatory functions are situation specific and adaptable to situation change. It seems that during certain behavioural sequences, a process of adaptive preparation occurs and that sustained potentials underlie this process. Adaptive preparation presupposes continuous sensorimotor, perceptual and cognitive interaction between the subject and his environment. If the CNV potentials reflect the neurophysiological basis of certain aspects of this interaction, their relevance in understanding mental functions cannot be overemphasised.

It could be asked, in view of the above results, whether the proper integration of the preparatory functions could be selectively prevented. Data relevant to this question were collected during the following experiment. While the subject was performing a usual 2-s foreperiod CNV experiment, irrelevant somatosensory stimuli were

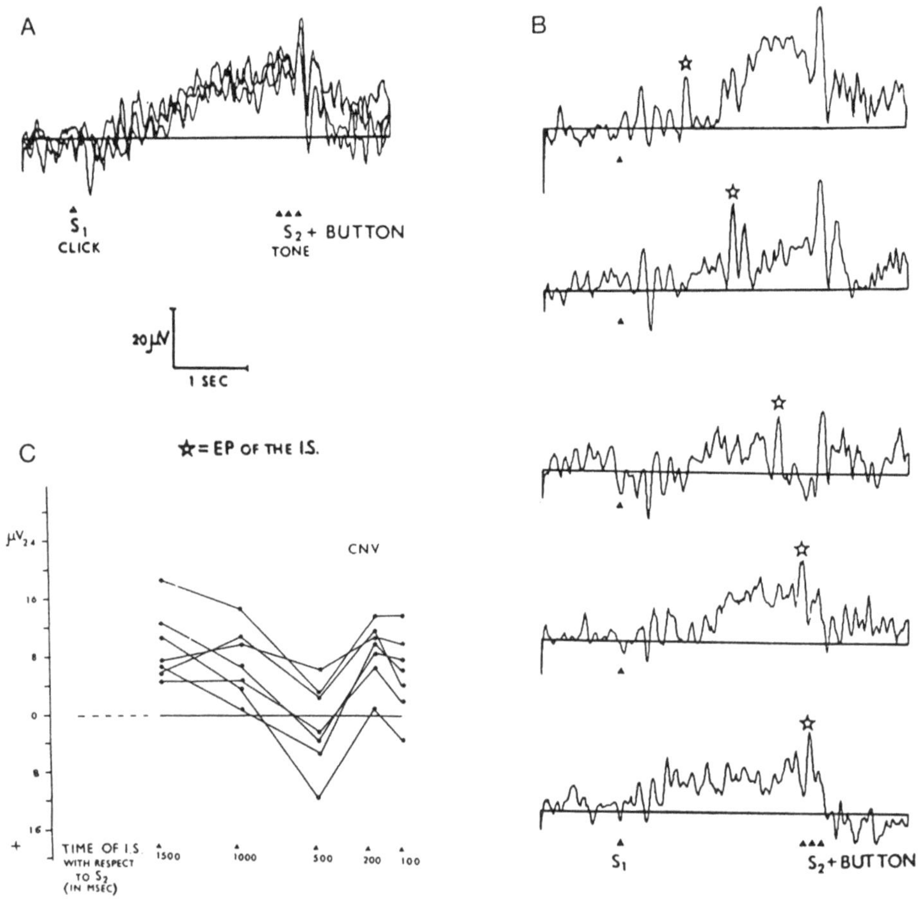

Fig. 5A–C. CNV potentials ($N = 6$) with an irrelevant somatosensory stimulus appearing randomly at various times outside the S1 S2 period (A) and during this period (B) from one subject. In (C) are shown the CNV values measured for 100 ms before S2 in the condition illustrated in (B) from another seven subjects. Note the CNV suppression when the irrelevant stimulus occurred 500 ms before S2

presented randomly but at five specified times during the foreperiod and five outside the S1–S2 interval. Separate averages of the CNV potentials ($N = 6$) for each of the ten experimental conditions were collected from eight subjects. CNV data from one subject and CNV values from the remaining seven subjects are shown in Fig. 5. It can be seen that the CNV development in all but one instance proceeds in a similar way independently of the presence or absence of the irrelevant stimulus. However, when the stimulus occurs 500 ms before S2, the CNV amplitude drastically diminishes. It seems that at that instant, the integration of preparatory functions has been completed but is amenable to disturbance. Some milliseconds later, the whole process is so rigid that it can proceed unaffected by irrelevant environmental interference. However, even at this late stage, 100 ms before S2, the present action can be aborted and the reaction time to the forthcoming imperative signal inhibited if the interposed stimulus conveys such an instruction (Papakostopoulos and McCallum 1973). This suggests that specific preparatory processes are affected and not processes related to stimulus selection.

4 Conclusions

It can be concluded that the term CNV includes a variety of sustained electrical potentials recordable from many cortical and subcortical brain areas. These potentials have area-dependent functional specificity and develop during those behavioural conditions in which flexible and adaptive spatiotemporal integration of many bodily systems is necessary for the preparation and accomplishment of purposeful behaviour. The waveform and amplitude of the local potentials and their integrated pattern are situation specific and adaptable to situation change. It is suggested that the CNV potentials, together with other configurations of sustained potentials, reflect processes underlying adaptive preparation and that they offer a distinct possibility of studying neurophysiologically the organisation of human behaviour with noninvasive techniques in numerous situations and different environmental conditions.

References

Berlyne DE (1969) The development of the concept of attention in psychology. In: Evans CR, Mulholland TB (eds) Attention in neurophysiology. Butterworth, London, pp 1–26

Cooper R, McCallum WC, Papakostopoulos D (1979) A bimodal slow potential theory of cerebral processing. In: Desmedt JE (ed) Cognitive components in cerebral event-related potentials and selective attention. Karger, Basel (Progress in clinical neurophysiology, vol 6)

Curry SH (1983) Event related potentials as tools for the assessment of functional disability and recovery following closed head injury. Thesis, University of Bristol

Donchin E, Heffley EF (1979) The independence of the P300 and the CNV reviewed: a reply to Wastell. Biol Psychol 9:177–188

Jasper HH (1969) Opening remarks. In: Evans CR, Mulholland TB (eds) Attention in neurophysiology. Butterworth, Stoneham, p XXIII

Kornhüber HH, Deecke L (1965) Hirnpotentialänderungen bei Willkürbewegungen und passiven Bewegungen des Menschen: Bereitschaftspotential und reafferente Potentiale. Pflugers Arch Gen Physiol 284:1–17

Loveless NE (1973) The contingent negative variation related to preparatory set in a reaction time situation with variable foreperiod. Electroencephalogr Clin Neurophysiol 35:369–374

Low MD, Borda RP, Frost JD Jr, Kellaway P (1966) Surface negative, slow potential shift associated with conditioning in man. Neurology (Minneap) 16:771–782

McCallum WC (1978) Some anomalies found in relationships between event related potentials and performance. EEG [Suppl] 34:211–223

McCallum WC, Walter WG (1968) The effects of attention and distraction on the contingent negative variation in normal and neurotic subjects. Electroencephalogr Clin Neurophysiol 25:319–329

McCallum WC, Papakostopoulos D, Gombi R, Winter AL, Cooper R, Griffith HB (1973) Event related slow potential changes in human brain stem. Nature 242:465–467

McCallum WC, Papakostopoulos D, Griffith HB (1976) Distribution of the CNV and other slow potential changes in human brainstem structures. In: McCallum WC, Knott JR (eds) The responsive brain. Wright, Bristol, pp 205–210

MacKay DM, MacKay V, Roulon MJ (1985) Local CNVs for different phases of a cognitive task. J Physiol (Lond) 367, 24P

Papakostopoulos D (1978) Integrative models: Macropotentials as a source for brain models. In: Otto D (ed) Multidisciplinary perspectives in event-related brain potential research. US Govt Printing Office, Washington, pp 635–639

Papakostopoulos D, Cooper R (1973) The contingent negative variation and the excitability of the spinal monosynaptic reflex. J Neurol Neurosurg Psychiatry 36:1003–1010

Papakostopoulos D, Cooper R (1976) Brain, spinal cord and autonomic changes before, during and after a planned motor action in man. In: McCallum WC, Knott JR (eds) The responsive brain. Wright, Bristol, pp 114–119

Papakostopoulos D, Crow HJ (1976) Electrocorticographic studies of the contingent negative variation and "P300" in man. In: McCallum WC, Knott JR (eds) The responsive brain. Wright, Bristol, pp 201–204

Papakostopoulos D, Fenelon B (1975) Spatial distribution of the contingent negative variation (CNV) and the relationship between CNV and reaction time. Psychophysiology 12:74–78

Papakostopoulos D, Jones JG (1980) The impact of different levels of muscular force on the contingent negative variation (CNV). Prog Brain Res 54:195–202

Papakostopoulos D, McCallum WC (1973) The CNV and autonomic change in situations of increasing complexity. Electroencephalogr Clin Neurophysiol [Suppl] 33:287–293

Papakostopoulos D, Crow HJ, Newton P (1980) Spatiotemporal characteristics of intrinsic, evoked and event related potentials in the human cortex. In: Pfurtscheller G et al. (eds) Rhythmic EEG activities and cortical functioning. Elsevier/North-Holland, Amsterdam, pp 179–200

Rohrbaugh JW, Gaillard AWK (1983) Sensory and motor aspects of the contingent negative variation. In: Gaillard AWK, Rigger W (eds) Tutorials in event related potential research: endogenous components. North-Holland, Amsterdam, pp 269–310

Sutton S, Brazen M, Zubin J, John ER (1965) Evoked potential correlates of stimulus uncertainty. Science 150:1187–1188

Walter WG, Cooper R, Aldridge VJ, McCallum WC, Winter AL (1964) Contingent negative variation: an electrical sign of sensorimotor association and expectancy in the human brain. Nature 203:380–384

Weinberg H (1975) The contingent negative variation: its clinical past and future. Am J Electroencephalogr Technol 15:51–67

Weinberg H, Papakostopoulos D (1975) The frontal CNV: its dissimilarity to CNVs recorded from other sites. Electroencephalogr Clin Neurophysiol 39:21–28

V. Magnetic Fields of the Brain

Magnetic Fields from the Human Auditory Cortex

K. Saermark, J. Lebech, and C. K. Bak

1 Introduction

Since the first observation by Reite et al. (1978) of auditory evoked magnetic fields (AEF) from the human cortex, several research groups have been actively engaged in such studies (Reite et al. 1978; Elberling et al. 1981, 1982a, b; Hari et al. 1980, 1982; Arlinger et al. 1982; Romani et al. 1982; Zimmerman et al. 1983; Tuomisto et al. 1983; Pellizone et al. 1985; Bak et al. 1985). It appears well established by now that the experimentally measured AEF perpendicular to the skull surface can be accounted for in terms of a model consisting of an equivalent current dipole (ECD) embedded in a volume conductor (for ECD, see for example Williamson and Kaufman 1981), with a source strength, the dipole moment measured in units of amperes times meters, determined by the intracellular axial currents in the neural tissue in question (Plonsey 1981). However, for a dipole oriented in an arbitrary direction, only the tangential component generates the field normal to the skull. The measured magnetic fields are, of course, not caused by a single active neuron, but rather to an array of neurons, which may be modelled as a distribution of elementary current dipoles. If these can be regarded as independent, noninteracting units, excited synchronously, the field may be calculated as if its origin were that of a single equivalent current dipole. The dipole moment, $P(t)$, will in general be time dependent and the magnetic field is given by $B(\bar{r}, t) = P(t)\, G(\bar{r}, \bar{r}_s)$, where G is a geometrical factor that depends on the position of the point of measurement $\bar{r}$, the position of the source $\bar{r}_s$, and on instrument properties, but not on the time t. It follows that spatial mapping studies allow for a determination of both the source location and the source strength P; we note that it also follows that the latency τ, for an ECDE model, is independent of the position of the point of measurement.

In our experiments, the magnetic field is measured by means of a first-order SQUID gradiometer system (SHE); see Bak et al. (1985) for experimental details. Here we mention only that the analog signal from the magnetometer is low-pass filtered from dc to 80 Hz (18 db per octave), time epochs of 800 ms are sampled (256 points), and averaging is based on 60 sweeps employing a fixed noise-rejection level. Off line, the recordings are digitally low-pass filtered from dc to 18 Hz. The stimuli used were tone bursts (TB) and, in some experiments (Arlinger et al. 1982), a frequency-glide stimulus (FG).

Springer Series in Brain Dynamics 1
Edited by Erol Başar

2 Results and Discussion

Most AEF work has concentrated on examinations of the 100-ms signal, i.e., the magnetic equivalent of the electric N100 component, although results for a 50-ms signal have also been presented (Romani et al. 1982; Zimmerman et al. 1983; Pellizone et al. 1984). However, we first comment on the problem of source location for the 100-ms signal.

With respect to source location, the results of our group (summarized in Elberling et al. 1982b) indicate a location near or within the primary auditory cortex at the sylvian fissure in the superior surface of the temporal lope, with the orientation of the dipole being nearly perpendicular to the sylvian fissure and the "100-ms" peak current directed downwards. Further, a significant hemispheric difference with respect to the posterior-anterior source location was pointed out, as were also interhemispheric and aural differences with respect to the source strength (Zimmerman et al. 1983). This source location seems to be in general agreement with the results obtained by the New York group (Pellizone et al. 1984) and the Helsinki group (Tuomisto et al. 1983). Bearing in mind that the AEF is insensitive to a normally oriented current dipole, the location is also in good agreement with a recent reevaluation (Scherg and von Cramon 1985) of mapping studies of auditory evoked (electric) potentials. On the other hand, the evidence presented by Elberling et al. (1982a) for a tonotopical organization has been questioned both by Pellizone (1984) and by Tuomisto et al. (1983). Whereas Elberling et al., using monaural stimulation in the contralateral configuration RH/LE and random ISI (range 1.5–4.5 s), found an anterior shift of the source location for the 100-ms signal with decreasing frequency of the TB, no such shift was found by Pellizone et al. and only for one out of three subjects by Hari et al. However, Pellizone et al. used *binaural* stimulation and random ISI (range 3–4 s) and Hari et al. used monaural stimulation and *constant* ISI (= 4 s). It is therefore not obvious that the three sets of experiments can be compared offhand, and the question concerning a tonotopical organization of the source for the 100-ms signal is still open. We add, however, that based on quite a different approach, namely positron emission tomography, Lauter et al. (1985) presented results in good agreement with those of Elberling et al. (1982a).

However, Pellizone et al. (1984) did find a tonotopical organization for the source of a steady-state signal obtained by applying a sinusoidally amplitude-modulated pure tone (modulation frequency 32 Hz, modulation depth $\lesssim 100\%$) as stimulus (see Romani et al. 1982) and, further, for TB stimuli they found (for one subject) that the source of an observed 180-ms signal, equivalent to the electric P200 signal, was shifted anteriorly (~ 1 cm) relatively to the 100-ms signal source; they further argued that their results were consistent with those of Tuomisto et al. (1983) in suggesting different neural origins for the 100-ms and 180-ms signals. Tuomisto et al. (1983) base this conclusion (see Mari 1982) upon a comparison of the form of the 100–200 ms complex with the form of the (electric) N100–P200 complex for increasing values of ISI, the two types of measurements being performed simultaneously (six subjects only). Especially they note that the magnetic signal observed at a latency of 180–200 ms is poorly developed for increasing values of ISI (1–16 s) in contrast to the (electric) P200 signal. We comment on this problem below.

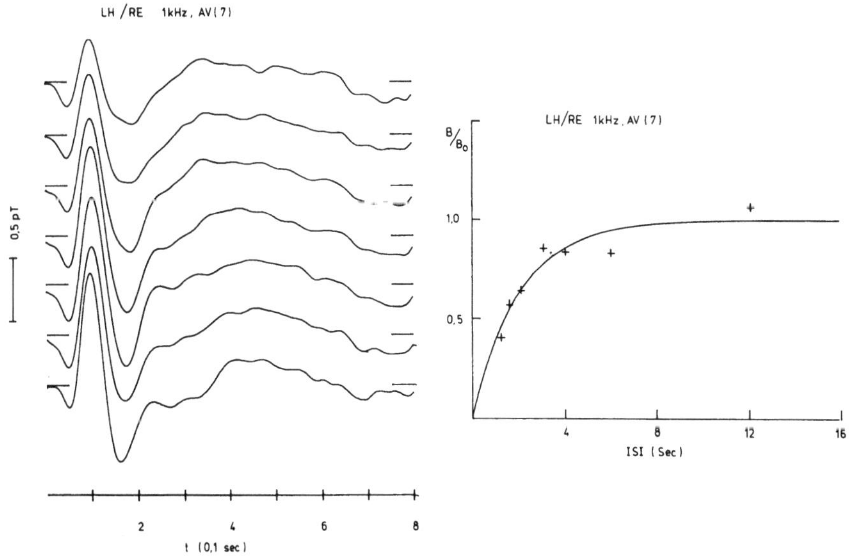

Fig. 1. *Left*, recordings for the ISI values: 1, 1.5, 2, 3, 4, 6, and 12 s *(top to bottom)*. *Right*, amplitude fit for 100-ms signal

In Fig. 1, we show a set of average recordings (seven different sessions scattered over several months) illustrating the development of the 100–200-ms complex for increasing values of ISI. In Fig. 1 we have further fitted the amplitude of the 100-ms signal to a "recovery function" of the type $B = B_0[1 - \exp(t/t_0)]$, where t = ISI and t_0 is a time constant. Similar recordings have been obtained for the same subject in the RH/LE configuration (two sessions) and for three other subjects in the LH/RE configuration (five sessions). From this we find that the latency of the 100-ms signal is practically constant, but the latency of the 180–200-ms signal shows a slight decrease in value with increasing ISI; the amplitude of the 100-ms signal and – in contrast to Hari et al. (1982) and Tuomisto et al. 1983 – of the 180–200-ms signal increases gradually with increasing ISI and follows reasonably well a recovery function of the type given above. Quantitatively, one finds from Fig. 1 a time constant t_0 = 2.1 s for the 100-ms signal; in a similar way, one finds for the 180–200-ms signal a time constant of t_0 = 1.3 s. The difference between the two time constants may substantiate the claim (Hari et al. 1982; Tuomisto et al. 1983) that the two signals are of different neural origins. For the same subject one finds for the other contralateral configuration, RH/LE, a rather longer time constant t_0 = 4.4 s (two sessions) for the 100-ms signal. To illustrate the intersubject variability, we give the median value and range of t_0 for the 100-ms signal in the LH/RE configuration (population: one and three subjects, see above): t_0 (median) = 2.4 s, range 1.8–3.5 s. Finally we remark that in Fig. 1, in the ISI region 4–6 s, there may exist a "bump" structure; this behavior has been observed for nearly all measurements performed until now and could further substantiate a claim for more than one neural source affecting, however, both the 180–200-ms signal and the 100-ms signal.

The intensity dependence of the 100-ms signal was recently discussed in some detail (Bach et al. 1985) and we here summarize only the main findings. We measured

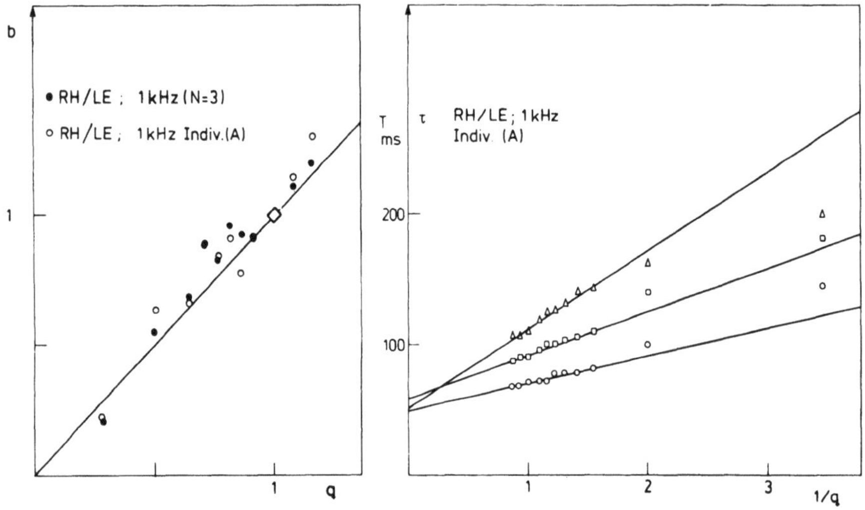

Fig. 2. *Left*, reduced amplitude b vs. reduced intensity q. *Right*, latency *(squares)* and times for maximal slope *(triangles* and *circles)* vs. $1/q$; 100-ms signal. (Reproduced from Bak et al. 1985)

the amplitude, latency, and time positions for the maximal slopes (positive/negative) of the 100-ms signal for the RH/LE configuration ($N = 3$; TB 1 kHz; five sessions per subject) and for the LH/RE configuration ($N = 5$; TB 0.5, 1, 2, and 4 kHz; two sessions per subject) as a function of the applied intensity I_i ($i = 1 \dots n$), the intensities being measured relative to the psychoacoustic threshold. The stimuli were applied with a random ISI in the range 1.5–4.5 s. To analyze the data, we first select among the I_i a reference intensity, say I_o, and next define two dimensionless variables: a reduced field $b = B(I)/B(I_o) = P(I)/P(I_o)$ – the latter form follows as the measurements are performed at a fixed space point – and a reduced intensity q, defined by $q^2 = (I/I_o)$. In Fig. 2, we show a plot of b versus q, both for an individual and for the corresponding population. Analogous results were obtained for the second of the above-mentioned groups at the frequencies cited. These results suggest the validity of the relation $b = q$, which from the definitions of b and q leads to the relation $P^2(I) = P_1^2[10\log_{10}(w/w_{th})]$, where P_1 is a constant and w is the physical sound intensity of the stimulus (th = threshold). As the dipole moment P can be written as $P = L\,i_{av}$, where L is the length of the current dipole and i_{av} is the average axial intracellular current of the current dipole, one also has $i_{av}^2 = i_{av,1}^2\,[10\log_{10}(w/w_{th})]$, where $i_{av,1}$ is a constant. Assuming $L \sim 100\,\mu m$ and using $P(I) \sim 2 \cdot 10^{-8}$ Am (Elberling et al. 1982b), one finds $i_{av,1} \sim 3 \cdot 10^{-5}$ A. These estimates relate, of course, to the equivalent current dipole and not directly to neuronal circuits. In Fig. 2, we also show the characteristic times (latency and maximal slopes) versus $1/q$ for the same individual. One notes the linear behavior and the common intersection at $T_o \sim 60$ ms. Physically, T_o is a transit time from the onset of the stimulus to the activation of the current dipole. Figure 2 further implies a constant relative linewidth with a Q value $1/Q \simeq 1.3$ for the 100-ms "resonance." The population results for the two groups show the same behavior as the individual results in Fig. 2. We remark that similar results (including $T_o \sim 60$ ms) are obtained for FG stimuli with a suitable definition of the parameter q (Saermark 1983).

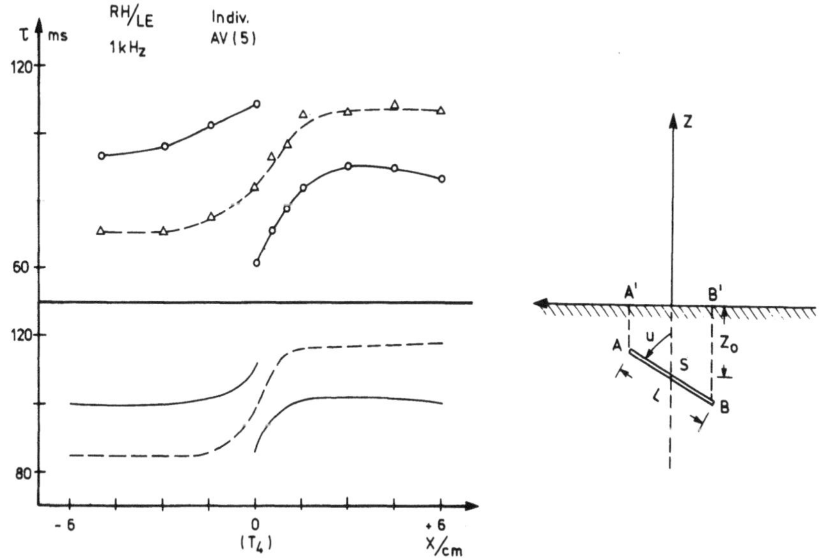

Fig. 3. *Left top*, experimental variation of latency (*circles*) and time of maximal positive slope; 100-ms signal. *Left bottom*, calculated variation for the array of elementary current dipoles shown schematically to the *right*

Finally we discuss a spatial variation of the latency for the 100-ms signal observed in all of our mapping recordings (performed since 1979) for contralateral and ipsilateral configurations, the measurements being performed along tracks parallel with the $T_4(T_3)$-nasion line. The variation consists of a small systematic latency change that becomes rather abrupt close to the point of polarity reversal of the 100-ms signal. We illustrate this in Fig. 3 (individual, five sessions), where we also show the variation of the time position for the maximal (positive) slope of the 100-ms signal. This variation could be caused by the presence of another signal with a separate source (Elberling et al. 1982a) but can, however, also be explained by a different mechanism (Saermark 1983) involving a dynamic aspect. Thus, assume that the elementary current dipoles of the neuronal array giving rise to the 100-ms signal are not excited synchronously, but that a time delay exists across the array. Then the signal parameters (latency, amplitude, line shape, half-width, etc.) depend on the position of the measuring point. A simple computer simulation reveals that a latency variation similar to that experimentally observed is easy to obtain, as shown in Fig. 3; here we have used a monophasic, Gaussian-bell form for $P(t)$ with a halfwidth of 20 ms and a time delay of 30 ms. The array configuration is shown in Fig. 3 ($L = 1$ cm, $u = 45°$, $z_o = 3$ cm, 100 elementary current dipoles uniformly distributed). Qualitatively, the calculated and observed latency variations are very similar; however, it will be next to impossible to deduce the various parameters from the observed variation, except for noting that the projection ($A'B'$) of the array (AB) on the experimental track roughly covers the region where the abrupt change in latency takes place. It should be emphasized, however, that if the concept of a time delay is correct and if $P(t)$ has a form more complicated than a simple Gaussian bell, e.g., a bi- or triphasic form giving rise to a 180–200-ms signal, then the spatial variation of the

180–200-ms signal (amplitude, latency, shape) will in general not parallel that of the 100-ms signal, but there will still be only one generator or source for the signal complex. In addition, there may be other signals (e.g., brain waves) present in the recordings; these problems will be discussed in greater detail elsewhere.

Acknowledgements. Financial support from The Danish Natural Science and Medical Research Councils is gratefully acknowledged. We also thank Dr. C. Elberling for extensive computer program support.

References

Arlinger S, Elberling C, Bak C, Kofoed B, Lebech J, Saermark K (1982) Cortical magnetic fields evoked by frequency glides of a continous tone. Electroencephalogr Clin Neurophysiol 54: 642–653

Bak C, Lebech J, Saermark K (1985) Dependence of the auditory evoked magnetic field (100 msec signal) of the human brain on the intensity of the stimulus. Electroencephalogr Clin Neurophysiol 61: 141–149

Elberling C, Bak C, Kofoed B, Lebech J, Saermark K (1981) Auditory magnetic fields from the human cortex. Influence of stimulus intensity. Scand Audiol 10: 203–207

Elberling C, Bak C, Kofoed B, Lebech J, Saermark K (1982a) Auditory magnetic fields. Source location and "tonotopical organization" in the right hemisphere of the human brain. Scand Audiol 11: 61–65

Elberling C, Bak C, Kofoed B, Lebech J, Saermark K (1982b) Auditory magnetic fields from the human cerebral cortex: location and strength of an equivalent current dipole. Acta Neurol Scand 6: 553–569

Formby C, Zeiger HE, Lauter JL, Raichle ME, Herscovitch P (1982) Evidence of tonotopic representation in the cerebral cortex of man. ASHA Meeting, 21st Nov, Toronto

Hari R, Aittoniemi K, Järvinen ML, Katila T, Varpula T (1980) Auditory evoked transient and sustained magnetic fields of the human brain. Localisation of neural generators. Exp Brain Res 40: 237–240

Hari R, Kaila K, Katila T, Tuomisto T, Varpula T (1982) Interstimulus-interval dependence of the auditory vertex response and its magnetic counterpart: implications for their neural generation. Electroencephalogr Clin Neurophysiol 54: 561–569

Lauter JL, Herscovitsch P, Formby C, Raichle ME (1985) Tonotopic organization in human auditory cortex revealed by positron emission tomography. Hear Res 20: 199–205

Pellizone M, Williamson SJ, Kaufman L (1985) Evidence for multiple areas in the human auditory cortex. In: Weinberg H, Stroink G, Katila T (eds) 5th world conference on biomagnetism, Vancouver. Pergamon, New York, pp 326–330

Plonsey R (1981) Magnetic field resulting from action currents on cylindrical fibers. Med Biol Eng Comput 19: 311–315

Reite M, Edrich J, Zimmerman JT, Zimmerman JE (1978) Human magnetic auditory evoked fields. Electroencephalogr Clin Neurophysiol 45: 114–117

Romani GL, Williamson SJ, Kaufman L, Brenner D (1982) Characterization of the human auditory cortex by the neuromagnetic method. Exp Brain Res 47: 381–393

Saermark K (1983) Some tentative model considerations based on experimental neuromagnetic data. Nuovo Cimento 2D: 438–459

Scherg M, von Cramon D (1985) Two bilateral sources of the late AEP as identified by a spatio-temporal dipole model. Electroencephalogr Clin Neurophysiol 62: 32–44

Tuomisto T, Hari R, Katila T, Poutanen T, Varpula T (1983) Studies of auditory evoked magnetic and electric responses: Modality specificity and modelling. Nuovo Cimento 2D: 471–483

Williamson SJ, Kaufman L (1981) Biomagnetism. J Magn Magn Mat 22: 129–202

Zimmerman JT, Reite M, Zimmerman JE, Edrich J (1983) Auditory evoked magnetic fields: a replication with comments on the magnetic P50 analog. Nuovo Cimento 2D: 460–470

Interpretation of Cerebral Magnetic Fields Elicited by Somatosensory Stimuli

R. HARI

1 Introduction

Magnetoencephalography (MEG), the recording of the weak magnetic fields produced by the brain, has developed rapidly during the last decade. Electroencephalography (EEG) detects cerebral currents regardless of their orientation with respect to the skull, whereas MEG gets contributions only from tangential currents. In the cortex, the main current flow is perpendicular to the surface. Therefore, in most fissures, the currents are tangential to the skull and thus detectable with MEG. In man, about half of the cortex is theoretically accessible to MEG. Because MEG is superior to EEG in locating current sources, it essentially increases the possibilities of studying the functional organization of the human cortex.

The interpretation of field patterns of magnetic evoked responses – i.e., locating the underlying neuronal activity – is based on several hypotheses about the physiological and anatomical properties of the structures studied. The validity of these hypotheses can be tested with simulations and measurements of activity whose origin is known. For example, neuromagnetic studies of the somatosensory projection areas SI and SII, whose locations and internal organizations are rather well known on the basis of stimulation studies in man and recordings in the monkey, illustrate clearly the usefulness of the method.

In the following, I describe the basic procedures of the analysis of magnetic field data and their application to the study of somatosensory cortices. The significance of MEG in locating current sources underlying evoked potentials is also briefly discussed.

2 Neuromagnetic Signals and the Inverse Problem

It is generally assumed that the magnetic field recorded outside the head is generated by intracortical currents associated with postsynaptic potentials. In order to locate the source currents of the evoked magnetic responses, the measurements are made at several points over the scalp. Field maps at different latencies are then constructed by computer. The patterns change rapidly with time: clearly dipolar patterns appear at certain latencies, usually during the peaks of the responses, whereas at other latencies the patterns are more complicated.

The inverse problem – i.e., the determination of source currents within a sphere on the basis of the magnetic field outside the sphere – does not have a unique solu-

Springer Series in Brain Dynamics 1
Edited by Erol Başar

tion: several current distributions can generate identical field patterns. To restrict the solutions to those that are possible regarding the brain, both anatomical and physiological knowledge is needed.

3 Source Models

Despite the fact that current configurations are complex at the neuronal level, simple models help to organize existing knowledge and to test hypotheses about brain func-

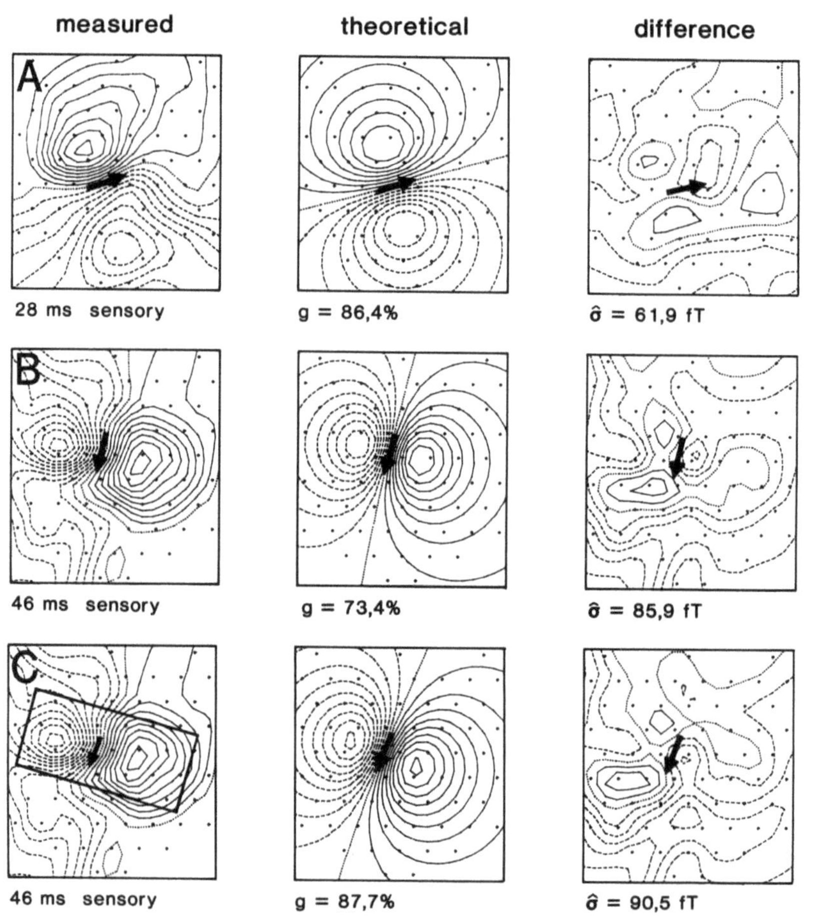

Fig. 1A–C. Examples of dipole fitting. The field patterns were obtained from data gathered over the right central sulcus at latencies of 28 ms (A) and 46 ms (B, C) after stimulation of the sensory branches of the left median nerve. *Left*, measured field patterns; *middle*, theoretical patterns generated by the equivalent dipole in the middle; *right*, the residual fields (measured minus theoretical). The goodness of fit values *(g)* and the residual noise values *(σ̂)* are given. Comparison of (B) and (C) shows that the area included in the fitting affects both *g* and *σ̂*. *Crosses*, measurement locations, which are separated by 16 mm. The isofield lines are separated by 35 fT. *Continuous lines*, flux out of the skull. *Dotted lines*, flux into the skull. (Adapted from Kaukoranta et al. 1986)

tion. It is reasonable to begin modeling with the simplest model – i.e., the current dipole model. The equivalent dipole, that best explains the measured field distribution is usually found by a least-squares fit. The local radius of curvature of the measurement surface is a good estimate for the radius of a spherical model. The goodness of fit can be expressed in a way analogous to that used in linear regression analysis (Kaukoranta et al. 1986). This value must, however, be judged critically because it depends strongly on the choice of the measurement points, as shown in Fig. 1.

If the model is adequate, the "residual" – i.e., the measured minus the theoretical field (Fig. 1) – should not show any systematic features and should not differ from noise. The sources and magnitude of noise must therefore be estimated. The standard deviation of an averaged response gives an estimate for the noise produced by the variability of the signal, by the cerebral "background" activity, and by the instrumental noise. When 150 responses to median nerve stimulation are averaged with a passband of 100 Hz, this type of noise is in our laboratory of the order of 60 fT. An important additional source of noise is that caused by the inaccuracy in the positioning of the probe in successive measurements. This noise depends on local field gradients. Assuming a typical dipolar source 3 cm beneath the scalp and a positioning inaccuracy of ± 3 mm, the noise is of the order of 15 fT (Kaukoranta et al. 1986). The estimate of the total noise in the experiment illustrated in Fig. 1 is then about $\sqrt{(60^2 + 15^2)} = 62$ fT. The residual noise at the latency of 28 ms (Fig. 1A) does not exceed this value and thus the dipole model can be considered adequate. At the latency of 46 ms, however, the conclusion is that either a single dipole does not explain the pattern satisfactorily or our noise estimate is too small.

The physiological interpretation of the data in Fig. 1 is that the hand area of SI in the posterior wall of the central sulcus is activated at the latency of 28 ms. Other similar data suggest that the activity continues at SI for at least 150 ms. The current source detected after stimulation of the thumb is lateral to the source activated by stimulation of the little finger (Fig. 2), and the stimulation of the peroneal nerve at

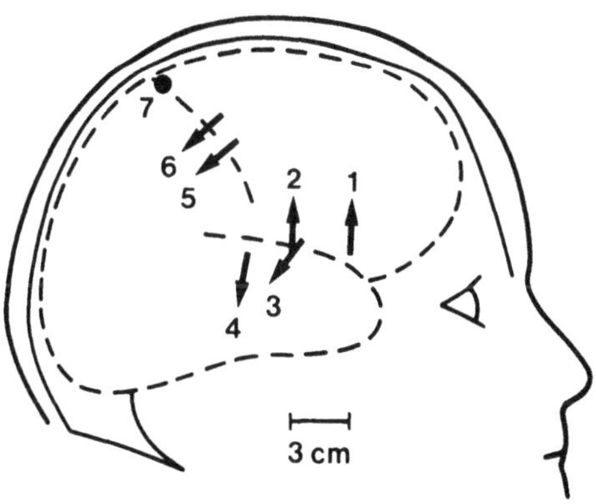

Fig. 2. A summary of sources activated by different types of stimuli in one subject. *Arrows*, projections of the equivalent dipoles on the skull during the main peak of the responses. *1*, Painful electrical stimulation of the tooth pulp; *2*, electrical stimulation of the left peroneal nerve (activation of SII); *3*, painful carbon dioxide stimulation of nasal mucosa; *4*, auditory stimulation with short tones; *5*, electrical stimulation of the left thumb; *6*, electrical stimulation of the left little finger; *7*, electrical stimulation of the left peroneal nerve (activation of SI; the dipole is pointing towards the reader)

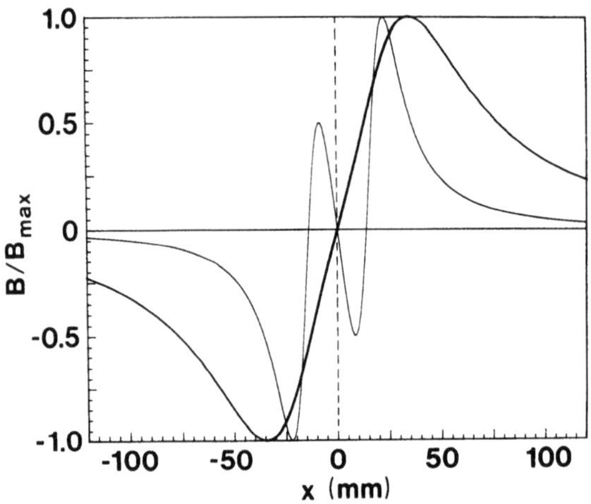

Fig. 3. The dependence of the radial component of the magnetic field on the distance between the source and the detector. Two current dipoles are situated in a homogeneous sphere (radius = 100 mm), 10 mm beneath the surface, 30 mm from each other, symmetrically in respect to the origin. The field was calculated on an arc perpendicular to the orientation of the dipoles *(x-axis)*. On the surface of the sphere (*thin line*) the pattern is complicated, whereas, 20 mm outside the sphere (*thick line*), the higher spatial frequencies have faded away and the dipolar term dominates. The amplitudes have been normalized according to the maximal value; the maximum field is about seven times stronger on the surface than 20 mm above it

the ankle evokes activity at the mesial surface of the hemisphere. These findings agree with with the known somatotopy of SI. Further, it is possible with MEG recordings to locate activity originating in SII (Hari and Kaukoranta 1985).

The success of dipole models in neuromagnetism has clear physical reasons. Any current distribution can be described by a current multipole expansion whose lowest-order term is a current dipole. The higher-order terms (quadrupolar, octupolar, etc.) decrease rapidly with distance. Therefore, even a complex source looks dipolar when observed from far away (see Fig. 3), and from the typical measurement distance we cannot detect the actual complexity of the source. Of course, good agreement of the data with a current dipole model does not mean that the real source is a dipole.

Interpretations of field patterns are generally based on the assumption that the head is spherical with concentric electrical inhomogeneities. However, real heads also contain nonconcentric inhomogeneities, which alter volume currents and thus magnetic field patterns as well; such inhomogeneities are for example the falx and the skull beneath the frontal lobe. Therefore even a dipolar source may generate a nondipolar field pattern in unfavorable conditions.

4 Comparison with Electrical Data

Simultaneous electrical and magnetic recordings give more information about the underlying neural events than either method alone. Magnetic fields are generated by tangential currents or tangential components of tilted currents. We can resolve

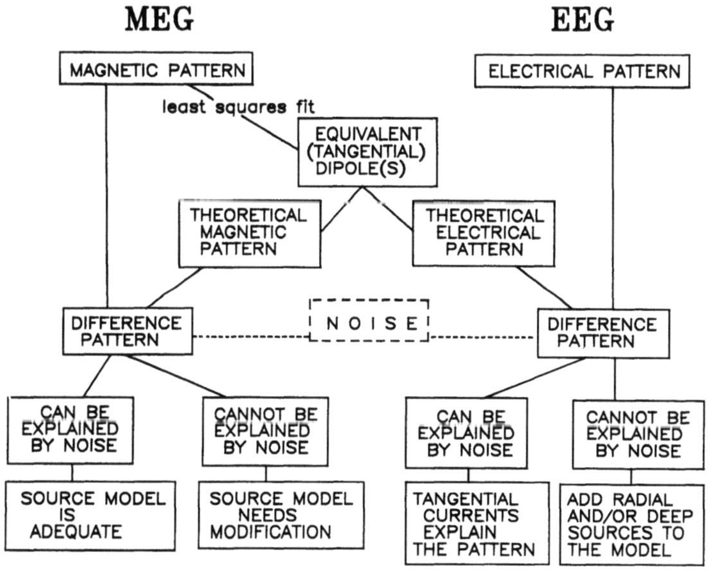

Fig. 4. A flowchart of procedures when the sources of electrical potentials are studied with combined electrical and magnetic recordings

tangential currents from magnetic recordings and then compute forward the potential produced by these sources (Fig. 4). If the theoretical potential distribution agrees with the measured one within noise, tangential currents can be considered the only source of both electrical and magnetic records. In the opposite case, radial and/or deep sources must be taken into consideration. This kind of approach was used by Wood et al. (1985), who studied the origin of the early responses to median nerve stimulation with combined electrical and magnetic recordings.

Electrical potential distributions are usually very complex when recorded directly from the cortex. This fact has provoked criticism directed against dipole models. As discussed above, the details of the current configuration cannot be detected with MEG because the measurement distance is long. This limitation is also true of the electrical scalp recordings; both types of records give a simplified picture of the source. Therefore magnetic recordings are more comparable to electrical scalp recordings than to electrocorticography.

5 Conclusion

MEG is a new noninvasive means to study mass actions of the human nervous system. MEG studies can be used to confirm in the human brain findings that have been made in electrophysiological studies of experimental animals or to reveal new principles of cerebral function. Multichannel recordings will, in the near future, considerably increase the spatial resolution of the technique. Although simple source models have been successful in the analysis of different types of neuromagnetic data, one important task in the future is to improve the interpretation of complex – i.e., non-dipolar – field patterns.

Acknowledgements. This work has been supported by the Academy of Finland. I thank M. Hämäläinen, J. Huttunen, E. Kaukoranta, J. Knuutila, L. Leinonen, O. V. Lounasmaa and J. Sarvas for constructive criticism and collaboration.

References

Hari R, Kaukoranta E (1985) Neuromagnetic studies of somatosensory system: Principles and examples. Prog Neurosci 24:233–256

Kaukoranta E, Hämäläinen M, Sarvas J, Hari R (1986) Mixed and sensory nerve stimulations activate different cytoarchitectonic areas in the human somatosensory cortex SI. Neuromagnetic recordings and statistical considerations. Exp Brain Res 63:60–66

Wood CC, Cohen D, Cuffin BN, Yarita M, Allison T (1985) Electrical sources in human somatosensory cortex: identification by combined magnetic and potential recordings. Science 227: 1051–1053

Auditory Evoked Magnetic Fields

M. Hoke

1 Localizing the Origin of Evoked Potentials

Evoked potentials are remotely detectable traces of neural excitation processes. They are commonly used as scalar measures to monitor the functional state of a sensory system. The time domain waveform, however, if recorded in the far field of the generators, only very poorly reveals time-variant aspects (with respect to time elapsing since stimulus onset) of the underlying processes. Potentials from differently excited and differently located generators superimpose, and the anisotropic body tissue distorts considerably the distribution of the propagating electric field. The attribution of a voltage, measured over the scalp at a specified instant, to a specified process occurring in neural generators at a specified location is at least ambiguous, if not impossible. Even if the field distribution over the scalp is determined with multichannel recordings, the solution of the inverse problem still remains equivocal. Consequently, the only reliable measures of far-field recorded evoked potentials are peak latencies and, with less reliability, amplitudes – i.e., in a certain sense "static" measures – obtained at specified instants, while the remaining portion of the time domain waveform – i.e., the "dynamic" aspect (though I am aware of the fact that I use the term "dynamic" in a different sense) – is disregarded.

The introduction of superconducting quantum interference device (SQUID) magnetic field sensors has opened a new dimension of noninvasive investigation of neural processes, both spontaneous and evoked. The SQUID, which makes possible the measurement of extremely weak magnetic fields, provides a more reliable means for localizing the source of neural activity (under certain limiting assumptions), because the propagation of magnetic fields within the human body is virtually undistorted. Therefore SQUID measurements seem to be the most appropriate tool not only for studying the time-dependent aspects of evoked neural activity, but also for deciding between contradictory hypotheses, which cannot be done on the basis of electric potentials. There has been a controversy for almost two decades over whether scalp-recorded cortical evoked potentials originate from the specific primary cortical projection area or from "nonspecific cortex." The latter locus of origin has been favored by quite a few authors because, independently of the modality of the stimulus, the largest potential amplitudes are recordable over the vertex region [a finding that led to the often-used term "vertex potentials" (Davis 1976)]. On the other hand, Keidel and his group (Keidel 1976) inferred as early as 20 years ago (David et al. 1966) from animal experiments that, except for painful stimuli, in which a general excitation all over the cortex can be observed, auditory and vibratory stimuli exclusively evoke excitation over the specific primary projection areas.

Springer Series in Brain Dynamics 1
Edited by Erol Başar

In 1978, the transient auditory evoked magnetic field (AEF) elicited by brief clicks was first detected (Reite et al. 1978). [For a comprehensive history of the investigation of AEF, see the excellent review by Okada (1983).] Two years later, Farrell et al. (1980) gave convincing evidence for the polarity reversal of the field direction of the 50-ms click-evoked transient component over both parietal regions. This pattern pointed to the existence of two symmetrical current dipole sources located in the parietal region and oriented in the vertical direction. Hari et al. (1980) demonstrated the similarity of the waveforms of both the auditory evoked potential and the simultaneously recorded AEF. Precise mapping experiments (Elberling et al. 1980, 1982; Hari et al. 1980) clearly showed maxima of the outward-going and inward-going flux at the ends of the Sylvian fissure, which led to the conclusion that the scalp distribution was generated from two bilateral current dipole sources located at the primary auditory cortices (near the superior surface of the temporal lobes approximately 20 mm below the surface of the skull). The dipoles are oriented in the superior-inferior direction, almost perpendicular to the Sylvian fissure. Three reproducible hemisphere/ear differences have been established: (a) the dipole is located approximately 14 mm posterior in the left hemisphere as compared to the right hemisphere; (b) the strength of the dipole generally is stronger over the left hemisphere than over the right hemisphere; and (c) contralateral stimulation produces stronger fields than does ipsilateral stimulation. In addition to these amplitude differences, latency differences can also be established for contralateral versus ipsilateral stimulation. The 100-ms field component of the transient response occurs approximately 9 ms earlier with contralateral stimulation as compared to ipsilateral stimulation (Elberling et al. 1980). A similar measure (7.5 ms) was derived from the phase of the steady-state response by Romani et al. (1982).

Tonotopic representation was recently investigated by Elberling et al. (1982) using tone bursts of different carrier frequency and by Romani et al. (1982) using amplitude-modulated tones. Location in the tangential x–y plane and depth of the equivalent current dipole source show an almost logarithmic dependence on the carrier frequency of the stimulus.

Therefore we can state that AEF measurements have clearly answered the question about the origin of evoked potentials. It is now unequivocal that evoked potentials (and corresponding evoked magnetic fields) originate from the specific primary cortical projection area.

Though AEFs allow sufficient localization of its source (the locus of the "equivalent current dipole"), localization calculations have been commonly restricted to one or two specified instants after stimulus onset (the time of maximum of the 100-ms and 180-ms component of the transient response; see Hari et al. 1980; Elberling et al. 1980, 1982). Therefore, the calculated location is true only for that very moment. The experimental paradigm employed in the study of Romani et al. (1982) is not appropriate for attributing the calculated location to a specified instant. A modulated tone with a modulation frequency (or a click with a repetition rate) of 32 Hz elicits the steady-state response, as was first described by Galambos et al. (1981). This implies that, during stimulus presentation, the entire auditory system is permanently excited so that magnetic fields from multiple sources in different parts of the auditory system must necessarily superimpose. The recorded signal is a temporal and spatial average of auditory excitation. Consequently, the concept of one single current dipole source is not applicable for a modulated tone. That there are at least four different cortical *and* subcortical sources for the steady-state response was recently experimentally proven by Yoshida et al. (1984).

2 Dynamic Behavior of the Locus of Excitation

Since the dynamic behavior of the locus of excitation – i.e., its dependence on time after stimulus onset – has not been considered so far, we have focused our research on that very point – i.e., on whether the equivalent current dipole source of the tone burst-evoked transient response is stationary or whether its location changes during excitation. The results presented here are based on data collected from five individuals with normal hearing, of whom one is left-handed (Lehnertz 1985; Spittka 1985). Technical details of stimulation, recording, and data processing must be omitted here; they are being published elsewhere (Lehnertz 1985; Pantev et al. 1986a–c).

Isofield contour maps of one individual, obtained from data recorded over the right hemisphere with contralateral stimulation, are reproduced in Fig. 1. Though the style of this set of maps, calculated for instants at every 8th ms, is certainly inadequate for revealing all of the hidden information, it gives at least a certain impression of the underlying process. It vividly demonstrates the emergence from noise and vanishing back into noise of the 100-ms and 160-ms components, as well as the polarity reversal around 110 ms after stimulus onset. The variation of the dipole parameters (location and orientation) is, in that particular case, only gradual, except for the reversal of dipole orientation.

It will be substantiated later in this paper that comparisons of dipole parameters obtained at specified instants under different conditions as to side of recording and side of stimulation may be misleading in some cases. The particular case demonstrated in Fig. 1 is one exception in which a comparison is possible, because the dipole location remains largely constant around the extrema of the dipole moment. Only cases like this one clearly exhibit the typical hemisphere/ear differences described in the literature. Figure 2 allows a comparison of the dipole location in the subject of Fig. 1 for both hemispheres with both contralateral and ipsilateral stimulation, at those instants when the dipole moment assumes an extremum (88 ms and 160 ms after stimulus onset). It concerns a case which has never been published before: a left-handed individual whose results differ from all published data insofar as the dipole is located approximately 15 mm posteriorly on the right hemisphere rather than on the left hemisphere. This is the only case where the dipole was found to be located posteriorly on that hemisphere, while in all remaining (right-handed) individuals the dipole was found to be located posteriorly on the left hemisphere. Corresponding to the side reversal of dipole location in the horizontal direction, we also found a side reversal in dipole strength for the left-handed subject: the dipole momentum is stronger on the right hemisphere rather than on the left hemisphere. A consistent finding with the data published in the literature is that contralateral stimulation generally produces stronger dipoles than does ipsilateral stimulation, independently of right- or left-handedness. Though it would be premature to draw

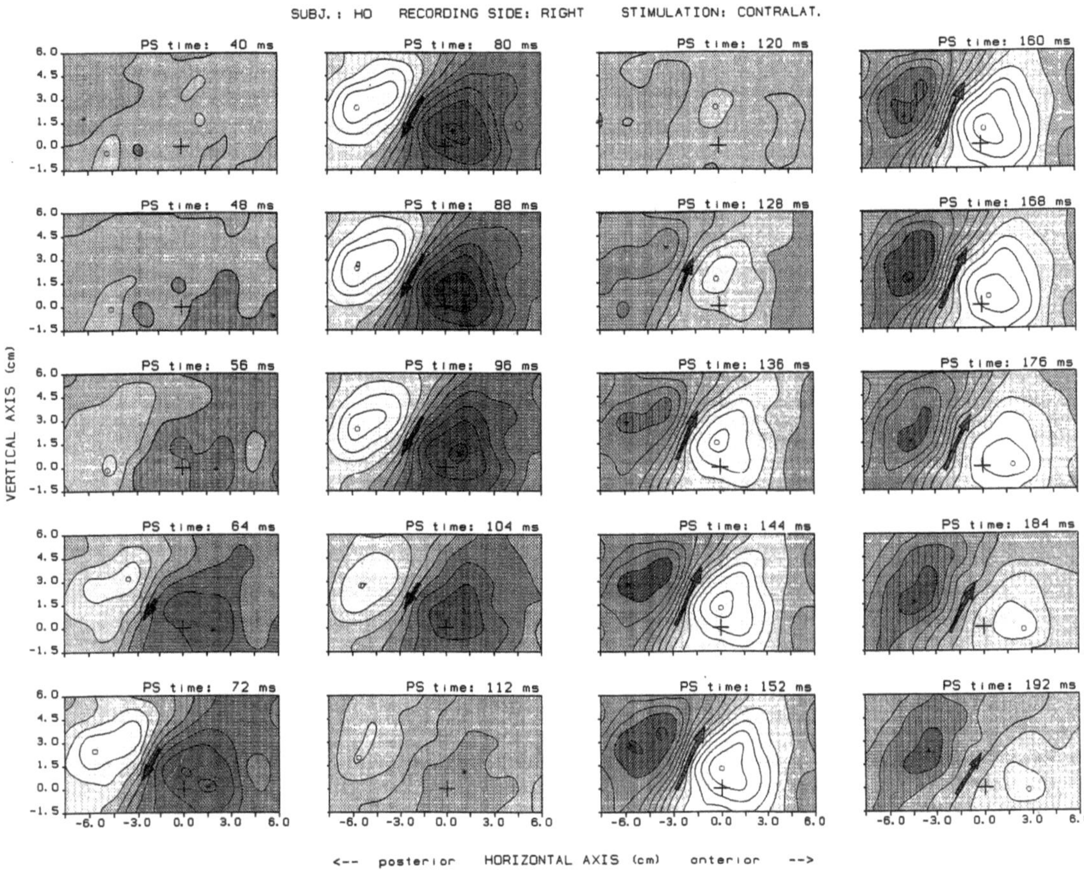

Fig. 1. Isofield contour maps of auditory evoked magnetic fields (transient response to 1000-Hz tone bursts, monaurally presented with an intensity of 60 db sound pressure level), calculated for time instants every 8th ms, beginning 40 ms after stimulus onset (inserted at the *top* of each map), from a left-handed subject. Recording side: left hemisphere; contralateral stimulation. The *space* between two adjacent isocontour lines represents 60 fT. Areas between isocontour lines are shaded according to the strength of the recorded field (*dark*; outward-going flux; *light*; inward-going flux). *Heavy line:* zero field strength; *crosses*, origin of the coordinate system (corresponding to T3/T4); *dark arrow*, location and orientation of the equivalent current dipole in the tangential plane, with length proportional to the dipole moment

final conclusions from findings with just one subject, they at least strongly suggest a dependence of the relative horizontal location and the strength of the dipole moment on right- or left-handedness. The vertical position – if we continue to regard only the instants determined by the extrema of the dipole momentum – between the left and right hemispheres of the same individual differs only slightly: the dipole is generally located approximately 3 mm superior on the left hemisphere; there is no obvious dependence on handedness and side of stimulation. If, however, we compare maps from different individuals, obtained under identical experimental parameters, we

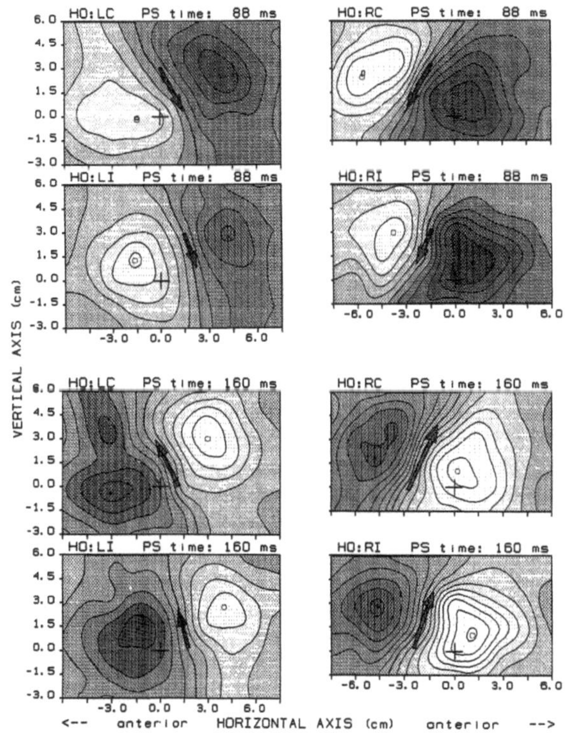

Fig. 2. Isofield contour maps obtained from the same subject as in Fig. 1, computed for two time instants (88 ms and 160 ms after stimulus onset). Maps for both hemispheres with both contralateral and ipsilateral stimulation are shown for comparison. For full explanation, refer to the caption of Fig. 1

find considerable interindividual differences in the absolute position in the tangential plane (as much as 2 cm in the horizontal direction and 2.5 cm in the vertical direction) as well as in depth and orientation of the dipole.

That a comparison like this is often inadmissible can easily be inferred from the study of the time dependence of the dipole parameters (i.e., changes of the dipole's location in the tangential plane, depth, direction, and moment with time elapsed since stimulus onset). The results we obtained showed as great a variability as did the interindividual comparison. Figure 3 gives examples for the time dependence of dipole location, calculated for two subjects. In case A (the left-handed individual of Figs. 1 and 2), location in the tangential plane, depth, and direction remain constant for a relatively long time (70–110 ms and 130–170 ms), especially around the maximum of the first (100 ms) component. The moment exhibits the advantage of contralateral stimulation already described with respect to shorter latency and greater strength. In case B, however, the dipole parameters exhibit quite a different behavior. Only the depth is relatively constant around the maximum of the 100-ms component. Its angle, however, as well as angle and depth of the 180-ms component, shows a dramatic but monotonic change with time. The movement of the location in the tangential plane is much more complex; it does not even assume a monotonic shape around the extrema of the dipole moment.

Systematic trends of the parameter changes have not yet been possible to infer. There certainly exist several possible sources for the enormous time dependence,

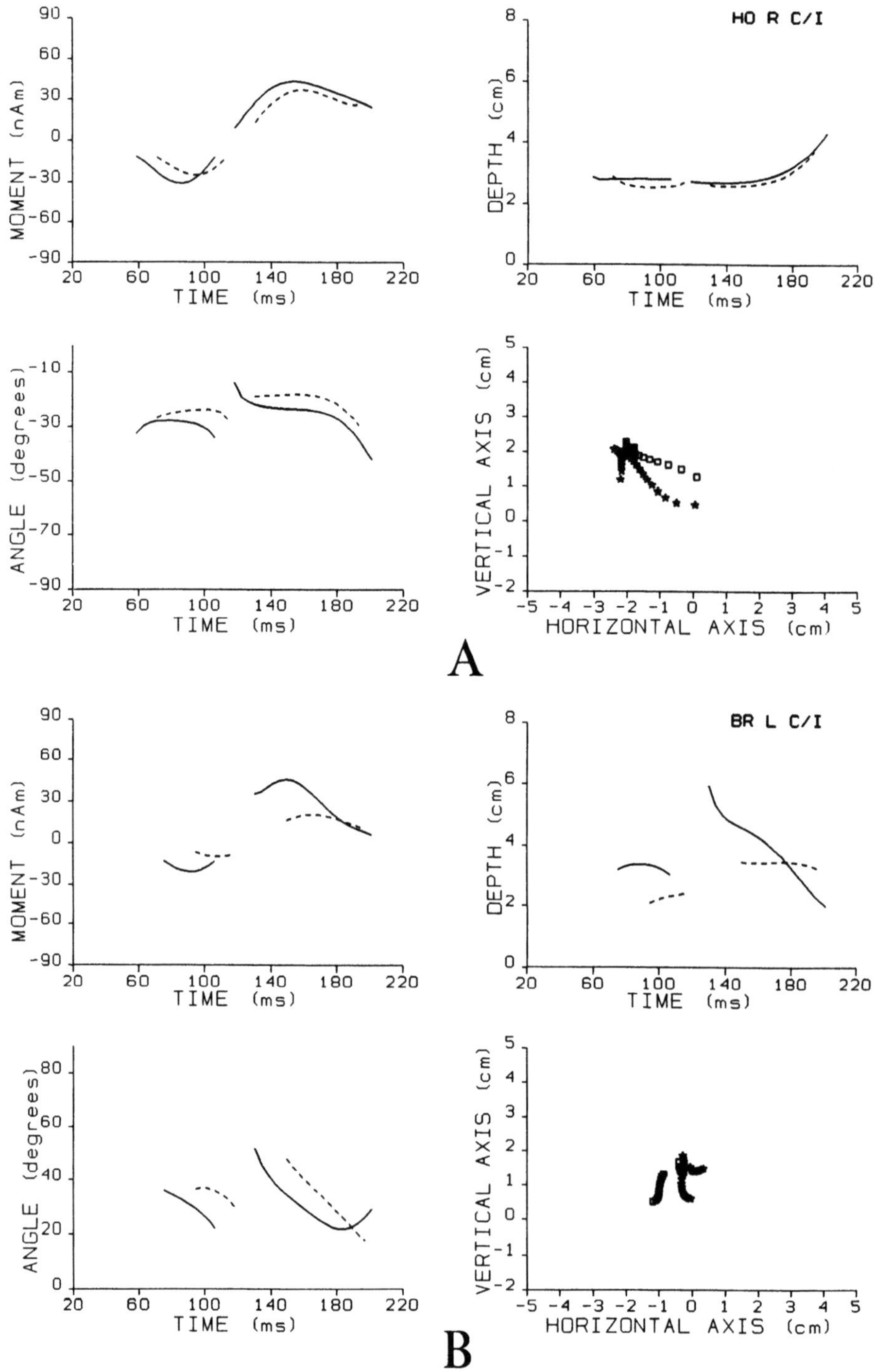

Fig. 3A, B. Dipole moment, orientation, depth, and location in the tangential plane as a function of time after stimulus onset, calculated for every 4th ms, from data obtained from two different individuals. Recording side: (A) (left handed) right hemisphere, (B) (right-handed) left hemisphere. *Solid line and open squares*, contralateral stimulation; *dashed line and asterisks*, ipsilateral stimulation. The space between symbols represents 4 ms

especially of the dipole location. First, one should bear in mind that we measure exclusively the tangential component of the dipole vector; i.e., we might sometimes miss the major vector component. This, however, cannot account for the entire extent of the time dependence. If we adhere to the hypothesis that the locus of origin of evoked potentials and evoked fields is the specific primary cortical projection area, we then have to assume that there must be an anatomical variability of comparable extent. However, the correlation with precise morphological data still has to be done. In view of the excellent signal-to-noise ratio, as expressed in the continuous course of the isocontour lines and the time functions of the dipole parameters, technical or computational reasons seem to be improbable. In our opinion, the results indicate rather that a single equivalent current dipole model, instead of a multipole expansion, does not represent the appropriate approach. This hypothesis is also favored by field distributions from other individuals (not shown in this paper) that definitely cannot be generated by single current dipole sources.

We have been able to demonstrate that the study of the time dependence – the dynamic aspect – of the dipole parameters of evoked responses in order to obtain information on the underlying cortical excitation processes, is much superior to simply studying one or two specified time instants. The results strongly suggest the need for comparative anatomical data as well as for the development of more sophisticated mathematical models. The model that Saermark presents elsewhere in this volume, which assumes a delayed progression of cortical excitation, seems to represent an important and promising approach towards a better understanding of the time dependence of cortical excitation processes.

Acknowledgements. Dr. B. Lütkenhöner, Dr. C. Pantev, Mr. K. Lehnertz, and Mr. J. Spittka made essential contributions to the results presented.

This work has been supported by the Deutsche Forschungsgemeinschaft and the Alexander-von-Humboldt-Stiftung.

References

David E, Finkenzeller P, Spreng M (1966) Komplexe akustisch evozierte Potentiale der wachen Katze (implantierte Elektroden-AI). Pflügers Arch Gesamte Physiol 291:45

Davis H (1976) Electrical response audiometry, with special reference to the vertex potentials. In: Keidel WD, Neff WD (eds) Auditory system. Part 3: Clinical and special topics. Springer, Berlin Heidelberg New York, pp 85–104 (Handbook of sensory physiology, vol 5)

Elberling C, Bak C, Kofoed B, Lebech J, Saermark K (1980) Magnetic auditory responses from the human brain. Scand Audiol 9:185–190

Elberling C, Bak C, Kofoed B, Lebech J, Saermark K (1982) Auditory magnetic fields from the human cerebral cortex: Location and strength of an equivalent current dipole. Acta Neurol Scand 65:553–569

Farrell DE, Tripp JH, Norgren R, Teyler TJ (1980) A study of the auditory evoked magnetic field in the human brain. Electroencephalogr Clin Neurophysiol 49:31–37

Galambos R, Makeig S, Talmachoff P (1981) A 40-Hz auditory potential recorded from the human scalp. Proc Natl Acad Sci USA 78:2643–2647

Hari R, Aittoniemi K, Järvinen ML, Katila T, Varpula T (1980) Auditory evoked transient and sustained magnetic fields of the human brain. Localization of neural generators. Exp Brain Res 40:237–440

Keidel WD (1976) The physiological background of the electric response audiometry. In: Keidel WD, Neff WD (eds) Auditory system. Part 3: Clinical and special topics. Springer, Berlin Heidelberg New York, pp 85–104 (Handbook of sensory physiology, vol 5)

Lehnertz K (1985) Measurement of evoked biomagnetic fields using a squid system (in German). Diploma thesis, University of Münster

Okada K (1983) Auditory evoked field. In: Williamson SJ, Romani GL, Kaufman L, Modena I (eds) Biomagnetism. An interdisciplinary approach. Plenum, New York London, pp 433–442

Pantev C, Lütkenhöner B, Hoke M, Lehnertz K (1986a) Comparison between simultaneously recorded auditory evoked magnetic fields and potentials elicited by ipsilateral, contralateral and binaural tone-burst stimulation. Audiology 25:54–61

Pantev C, Hoke M, Lehnertz K (1986b) Randomized data acquisition paradigm for the measurement of auditory evoked magnetic fields. Acta Otolaryngol Suppl 432:21–25

Pantev C, Hoke M, Lütkenhöner B, Lehnertz K, Spittka J (1986c) Causes of differences in the input-output characteristics of simultaneously recorded auditory evoked magnetic fields and potentials. Audiology 25:263–276

Reite M, Edrich J, Zimmerman JT, Zimmerman JE (1978) Human magnetic auditory evoked fields. Electroencephalogr Clin Neurophysiol 45:114–117

Romani GL, Williamson SJ, Kaufman L (1982) Characterization of the human auditory cortex by the neuromagnetic method. Exp Brain Res 47:381–393

Spittka J (1985) Acoustically evoked magnetic fields. Time dependence of the parameters of the equivalent current dipole source (in German). Diploma thesis, University of Münster

Yoshida M, Lowry LD, Liu JJC, Kaga K (1984) Auditory 40 Hz responses in the guinea pig. Am J Otolaryngol 5:404–410

VI. Clinical Applications

Iterative Estimation of Single-Trial Evoked Potentials

H. J. Heinze, H. Künkel, and M. Scholz

1 Introduction

Evoked potentials offer an excellent opportunity to study the stages of cerebral information processing in a clearly defined experimental setting. This applies especially to the so-called endogenous potentials (see Picton and Stuss 1980 for a definition), for which a relation to various cognitive functions is widely accepted. The task-dependent changes of these potentials permit an experimental test of models of perceptual functions developed by cognitive psychology.

Adequate decoding of the information contained in these electrophysiological phenomena is a crucial prerequisite of this research. A key problem is the extraction of the evoked potentials from the spontanous EEG. In evoked potential research, signal extraction is commonly achieved with a very simple filter, by so-called averaging. This synchronous stimulus summation of a number of EEG segments is based on the hypothesis that the spontanous EEG can be viewed as a stochastic process with an expected value of zero, whereas the evoked potentials can be regarded as being invariant over all segments. A number of considerations can be held against this assumption. First, habituation and learning processes might occur over an experimental period, which might lead to an uncontrolled change of the cognitive set and consequently to changes of the components of the potentials. Secondly, a relationship and interference between the evoked potential and the spontanous EEG has been shown by Başar (1980, 1983, 1985) and others. Conventional averaging will – in the best case – yield only the time-invariant portions of the potentials; the question of interference with the ongoing EEG cannot be answered. The extraction of single-trial potentials is therefore highly desirable for the detailed analysis of evoked potential data. However, the separation of spontaneous EEG and evoked potentials constitutes a difficult problem, for reasons discussed below. Up to now, most methods for single-trial analysis have employed a reference signal (template) to which the single segments were compared, using the least-squares principle. Generally, the averaged evoked potential is used as such a template. Pattern recognition techniques have been used as an alternative to the least-squares approach.

In the present paper, a method for single-trial potential evaluation is presented, using a model of time-series analysis for the construction of an adaptive digital filter. The selectivity of this filter is improved by an iterative procedure. The quality of this estimator is controlled exclusively by criteria derived from the mathematical model. No reference signal is needed and no filter criteria have to be specified a priori from the averaging procedure.

Springer Series in Brain Dynamics 1
Edited by Erol Başar

2 Selective Filtering of Single-Trial Response from the EEG

2.1 Definition of the Problem

The core problem of single-trial potential analysis is the lack of a priori criteria that would allow a reliable separation of spontanous EEG from evoked potential in a single EEG segment. Many of the proposed methods, specifically those using least-squares estimators, define separation criteria using the average evoked potential. In doing so, one is faced with two substantial drawbacks. First, the variance of the single potential can be considered only relatively to the average evoked potential. The second problem is with the validity of the filtered data, since it can be shown that some least-squares methods will extract a "potential" from an EEG segment which by construction does not contain a signal.

A powerful alternative approach uses internal criteria; i.e., criteria derived from the structure of the filter constructed for the extraction of single-trial potentials and their validation without any external specifications, in particular without the need of a reference signal. In such an approach, a basic distinction is made between the spontanous EEG as a (largely) stochastic signal and the evoked potential as a (largely) deterministic signal. This distinction is based on the experimental definition of the evoked potential as being a triggered signal. If no additional information is at hand for single-trial analysis, a signal can be considered deterministic only with a certain probability if it occurs repeatedly; i.e., at least twice. This means that at least two successive EEG segments are needed for this method. In these circumstances, an "evoked potential" is defined as being the periodic signal contained in both segments. No additional assumptions are made about the shape of the potentials. A method for single-trial analysis based on these considerations can be constructed by adapting a mathematical model to the signal configuration in question – a periodic signal masked by colored noise – and then deriving an algorithm for the extraction of the periodic portion.

2.2 Basis

A suitable mathematical model, describing the so-called autoregressive moving average (ARMA) processes, is found in the field of time-series analysis. An ARMA process ist the stochastic equivalent of higher-order differential equations and characterizes a linear stochastic process, which is additively mixed from a periodic and a stochastic portion. Above a certain signal-to-noise ratio (SNR), the spectrum of such an ARMA process permits the distinction between periodic and stochastic signal portions and is therefore suitable for the construction of a filter for certain periodic frequency portions. By now, powerful algorithms for the calculation of stable ARMA spectra are available (Cadzow 1980; Kay 1980; Heinze et al. 1984), permitting an identification of periodic signal portions even at a relatively low SNR.

Figure 1 gives the basis for the use of the ARMA spectrum as an adaptive filter by a computer simulation. Figure 1a shows a periodic signal, which is additively composed from three sine wave frequencies; in Fig. 1b this signal has superimposed white noise with a SNR of 0.3. The ARMA spectrum (Fig. 1c) that is calculated from this noisy signal identifies clearly the three frequencies of the original signal. If this

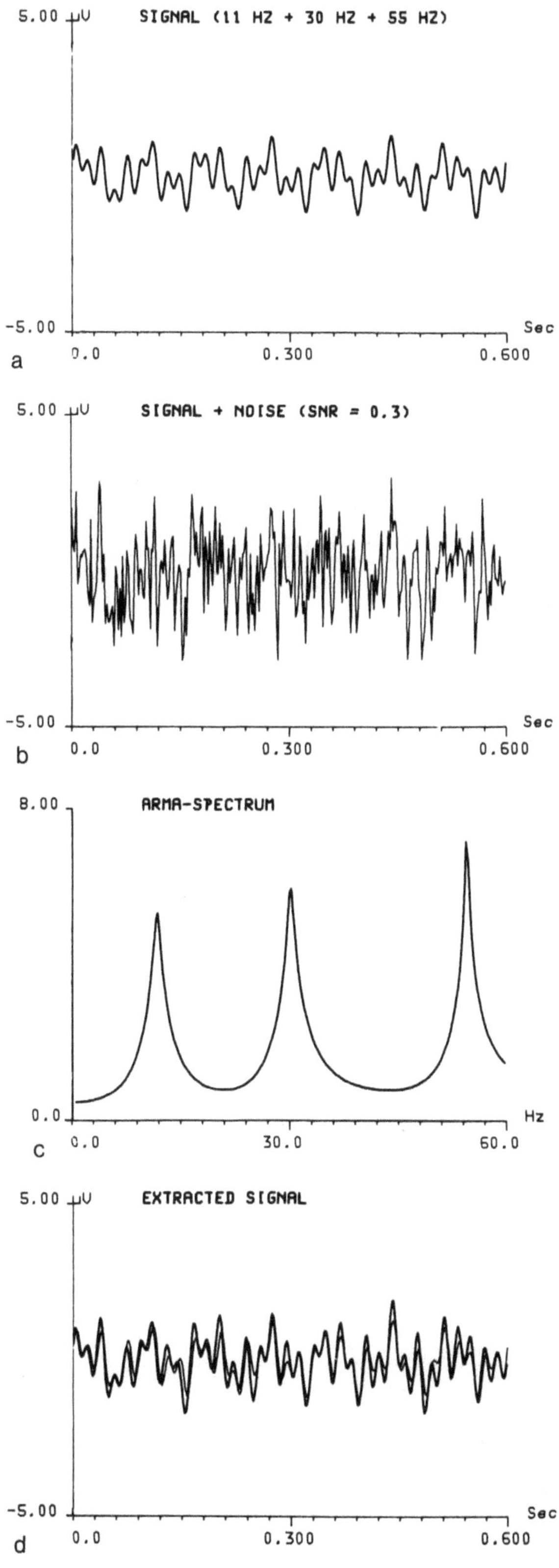

Fig. 1a–d. Basis of the ARMA filtering procedure. (a) Periodic signal additively composed of three sine wave frequencies (11 Hz + 30 Hz + 55 Hz). (b) Signal additively mixed with noise (1 Hz – 250 Hz) in a signal-to-noise ratio (SNR) of 0.3. (c) ARMA spectrum of the noisy signal. The three frequencies of the original signal are clearly identified. (d) ARMA filter of the noisy signal. The original signal is recovered with a negligible amplitude difference

spectrum is used as the transfer function of a nonrecursive digital filter and the noisy signal is subjected to this filter, the original signal is recovered (with a neglectable amplitude difference (Fig. 1d). It should be emphasized that the algorithm for the calculation of the ARMA spectrum is not constrained to white noise conditions, but works just as efficiently if the target signal has superimposed colored noise. Note also that no previous knowledge about the extracted signal is required with this method.

2.3 Development of the Algorithm

As described in the previous two sections, the principle of the desired algorithm is to extract the periodic portion from two consecutive EEG segments and to test whether

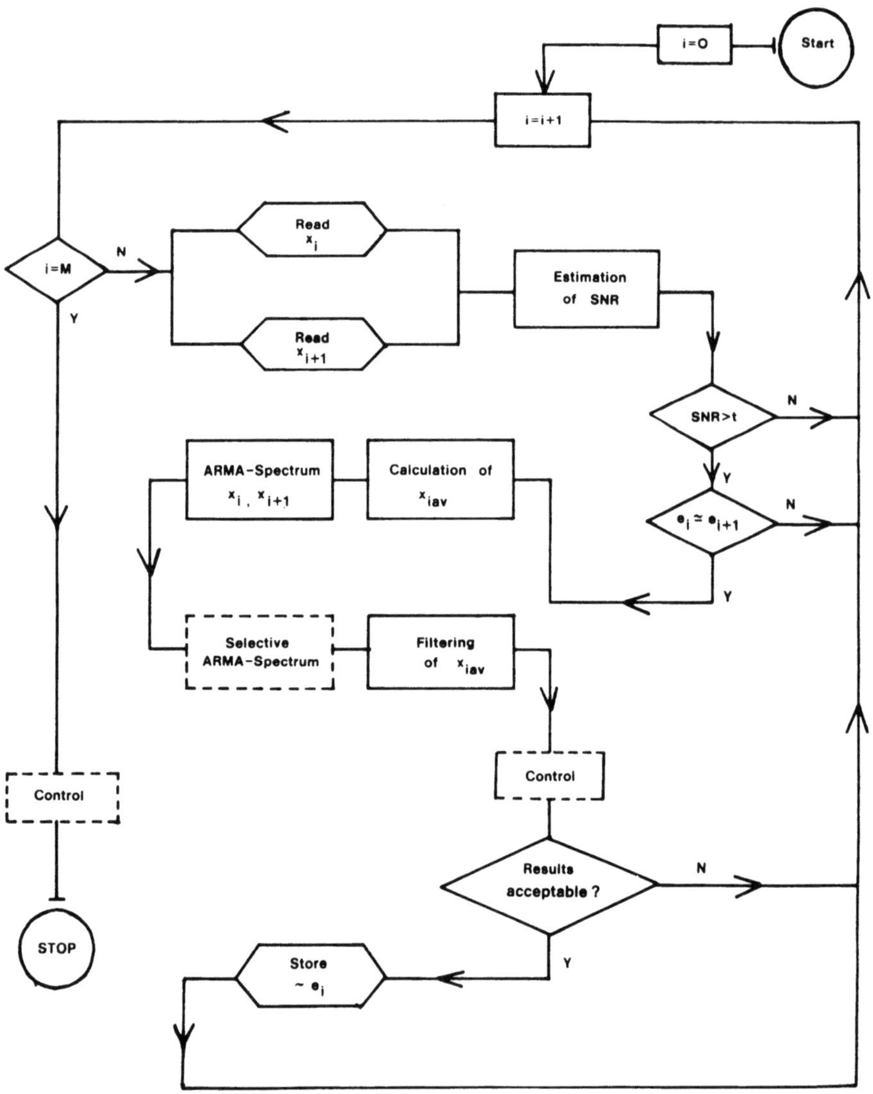

Fig. 2. Flowchart of the filter algorithm

the extracted signal meets the criteria for an evoked potential. The algorithm can be subdivided into several steps (Fig. 2). The first step involves the estimation of the SNR in two consecutive EEG segments, from which two criteria for the use of the ARMA filter are derived (see below). If these criteria are fulfilled, the target signal is extracted by the ARMA filter. The filtered signal is finally subjected to certain control criteria. If these criteria are met, the selectivity of the filter and thus the quality of the estimator is optimized iteratively. This stepwise procedure is described in the following section. Mathematical derivation and proofs cannot be considered here because of space limitations. A detailed account and program listings are being published elsewhere (Heinze et al., in preparation).

2.3.1 Estimation of SNR

The estimation of the SNR in two consecutive EEG segments is aimed at determining the share of the total variance accounted for by the periodic signal component. If this share exceeds a certain threshold, the periodic frequencies can be identified in the ARMA spectrum. If the SNR is defined as the ratio $\frac{\text{power (signal)}}{\text{power (noise)}}$, a formula can be derived; under the assumption that

$$e_1 = ce_2 \quad \text{and} \quad c = 1$$

the SNR can be estimated as

$$\mathrm{SNR} = \frac{\mathrm{Kov}(x_1,x_2)}{\frac{1}{2}(\mathrm{Var}(x_1)+\mathrm{Var}(x_2)) - \mathrm{Kov}(x_1,x_2) + \frac{1}{4}[E(x_1)-E(x_2)]^2} \tag{1}$$

where x_i, with $i = 1,2$, denote the vectors of the data points of both EEG segments, e_i the data values of the two evoked potentials, and n_i the data values of the spontanous EEG, c denotes a scalar factor; Kov stands for covariance, and E stands for expected value.

Equation (1) is based on the implicit assumption of identical potentials in both segments. Since potentials of the same shape, albeit different amplitude, can simulate an acceptable SNR, it is necessary to estimate the ratio of both amplitudes – i.e., the factor c – explicitly. Under the assumption of equal variances of the spontanous EEG in segments x_1 and x_2, the following equation holds:

$$\mathrm{Var}(x_1) - \mathrm{Var}(x_2) = \mathrm{Kov}(x_1,x_2)(1/c - c) \tag{2}$$

In the simulation studies, a SNR value of greater than or equal to 0.3 and an interval of 0.5–1.5 for the factor c were determined as necessary prerequisites for the use of ARMA filtering.

2.3.2 Filter Algorithm

The algorithm for the calculation of the ARMA filter has been described by Cadzow (1980) and Heinze et al. (1984) and will not be considered here in detail. The principle goal is to determine the parameters of the transfer function of the stochastic process. Of great importance is the definition of the model order. Unfortunately, until

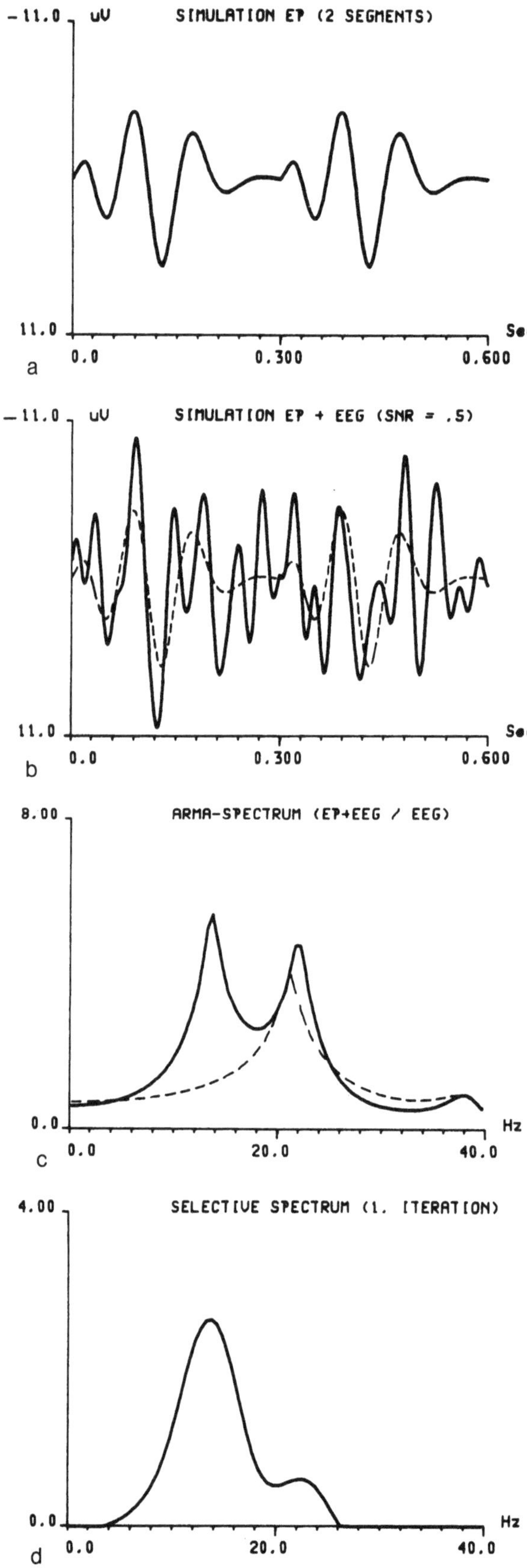

Fig. 3a–f. Simulation study I. (a) Visual evoked potentials in two consecutive segments (spectrum between 1 Hz and 15 Hz). (b) These potentials have a superimposed EEG (spectrum between 18 Hz and 26 Hz) with a SNR of 0.5. (c) Overall ARMA spectrum of both segments *(solid line)*. The spectra of the evoked potential and of the EEG are identified by two peaks. The spectrum of the difference of the segments is shown by a *dotted line*. (d) Selectively improved spectrum after one iteration. The spectrum of the evoked potential is significantly enhanced. (e) Signal extracted from both segments by the selective filter *(solid line)*, compared with the original evoked potential *(dotted line)*. (f) Signal extracted from both segments by simple bandpass filtering (1 Hz–40 Hz) of the average of these segments *(solid line)*, compared with the original signal *(dotted line)*

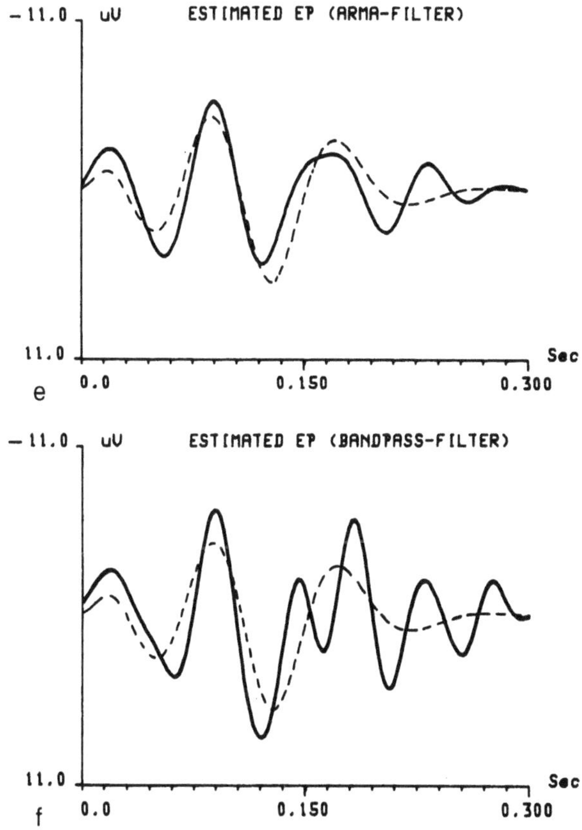

Fig. 3e, f

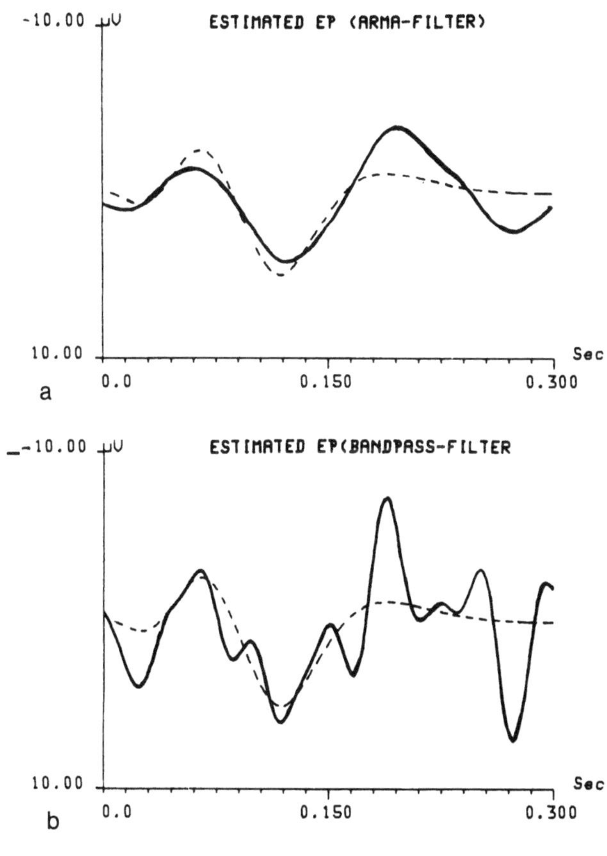

Fig. 4a, b. Simulation study II. (a) Signal extracted from both segments by the selective filter *(solid line)*, compared with the original evoked potential *(dotted line)*. (b) Signal extracted from both segments by simple bandpass filtering (1 Hz–40 Hz) of the average of these segments *(solid line)*, compared with the original signal *(dotted line)*

now no sufficient method for determination of the ARMA model order has been described. However, in the case considered here, characterization of the process is secondary to a high resolution of the spectrum. Therefore, following the suggestions of Chan and Landford (1982) we calculated the so-called pseudo-inverse, encountered in the determination of the AR coefficients a_i, by an orthonormalization procedure. The resulting increase in stability of the estimation of the ARMA spectrum permits choosing the filter order according to the desired frequency resolution independently of possible instabilities of the spectrum. The digital filter used is a modification of the algorithm of Stearns (1979).

2.3.3 Selective Spectrum and Iterative Filtering

In principle, the ARMA spectrum will yield peaks for the periodic signals according to the underlying frequencies (see Figs. 1c, 3c). As the spontanous EEG is colored rather than white noise, it is to be expected that the spontanous EEG portions in two consecutive segments will lead to peaks as well, which, in the worst case, might have the same spectral distribution as the evoked potential. It is therefore necessary to calculate a selective spectrum in order to reduce maximally the portion of the spontanous EEG. A solution to this problem can be found in the following way. If the evoked potentials in two consecutive segments are assumed to be highly identical (as follows from Eqs. 1 and 2), the ARMA spectrum of the spontanous EEG can be estimated from the difference of the two segments. This estimation is based on the area properties of the integrated spectrum and uses empirically determined correction factors (Heinze et al., in preparation).

After subtraction of the spectrum of the difference signal from the spectrum of both segments, the portion of the spontanous EEG in the calculated new spectrum is markedly reduced. The new spectrum is again used as a filter. If the control criteria for the extracted signal are met, this procedure is repeated iteratively until no further optimization can be obtained.

2.4 Convergence Conditions and Control Criteria

The iteration is based on several criteria, which at the same time represent a validity control for the filtered signal:

1. The SNR (Eq. 1) is improved by a factor of at least 1.5
2. The factor c falls into the interval 0.5–1.5 after filtering (Eq. 2)
3. Certain frequency and time characteristics must be congruent. For details see Heinze et al. (in preparation).

Two examples, simulations of visual (pattern) evoked potentials, are given to clarify the procedure (Figs. 3 and 4). The signals were constructed with the use of spline functions and digital filtering of white noise. It is apparent in Fig. 3 that the main components of the single-trial potentials are sufficiently well estimated. Figure 4 demonstrates a borderline case for this method with a SNR of 0.3. Again the single potential is clearly identified. In these cases, an improvement of the SNR by a factor of 10–20 was obtained after the first iteration. The consistency of the estimator was tested in simulation studies, for which details are given in Heinze et al. (in preparation).

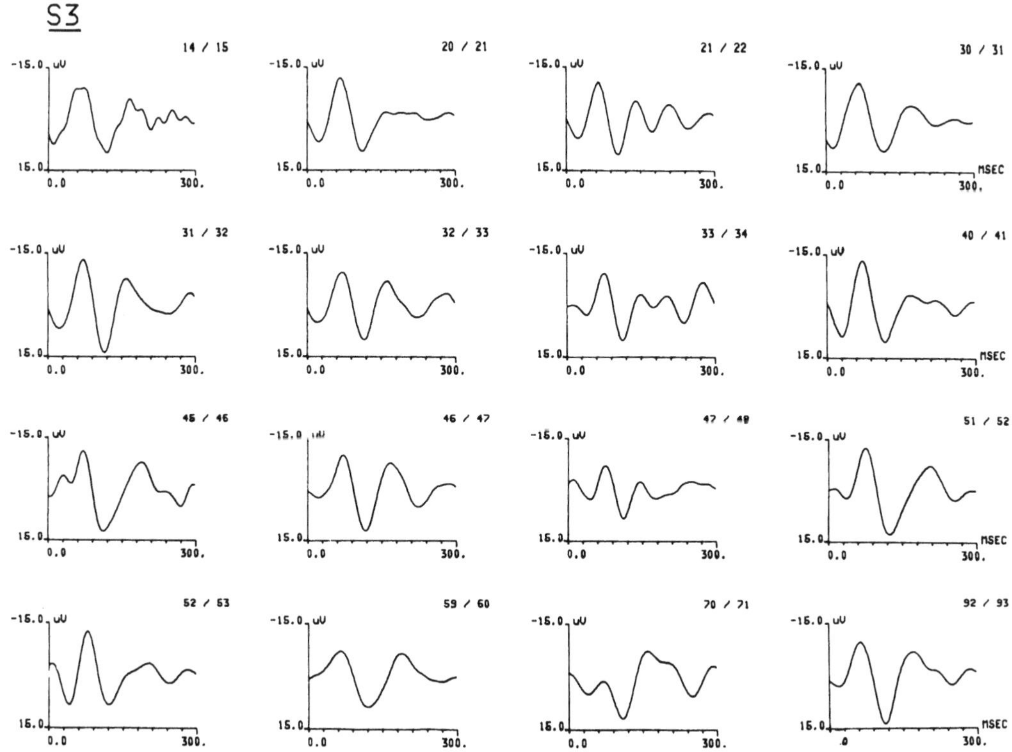

Fig. 5. Examples of single visual evoked potentials from one trial with a normal subject

3 Application to Visual Evoked Potentials

Investigations by Coppola et al. (1978) showed the SNR for single-trial visual evoked potentials and spontanous EEG to be between 0.4 and 0.6. As our simulation studies yielded a threshold SNR of 0.3, the visual evoked potential in certain ways constitutes a borderline case for this method. It is of some interest whether the single-trial potentials of this exogenous brainwave phenomenon show variability across an experimental period and whether the single-trial potentials are altered by pathological conditions. For further control, the averages of the selected segments were compared before and after filtering. Apart from small amplitude differences, identical averages were obtained in all cases.

Figure 5 shows visual pattern-reversal evoked potentials recorded from a healthy male subject during a single run. A considerable variation in the shape of the single-trial potentials is apparent, although the so-called P1 component is identifiable in all cases. An interesting repetition of certain patterns is observed, for example in segments 20/21, 30/31, and 40/41, but no specific conclusions about the nature of this phenomenon can be drawn at the present stage of investigations. P1 latencies of the single-trial potentials of four healthy subjects are given in Fig. 6. As expected, the latencies are distributed around mean values between 100 and 120 ms, as they are commonly found for the average visual evoked potential. Markedly shortened as

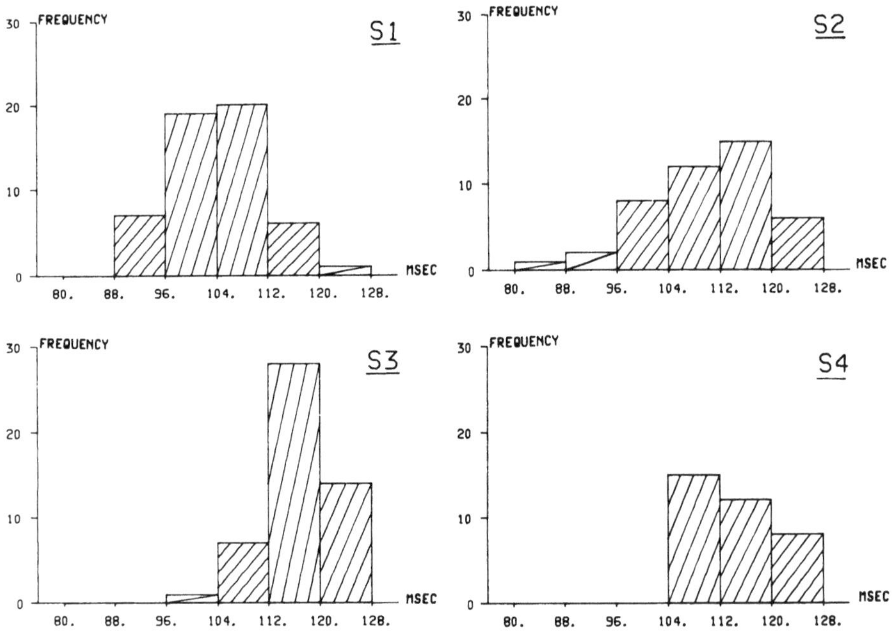

Fig. 6. P1 latencies of single visual evoked potentials of four healthy subjects

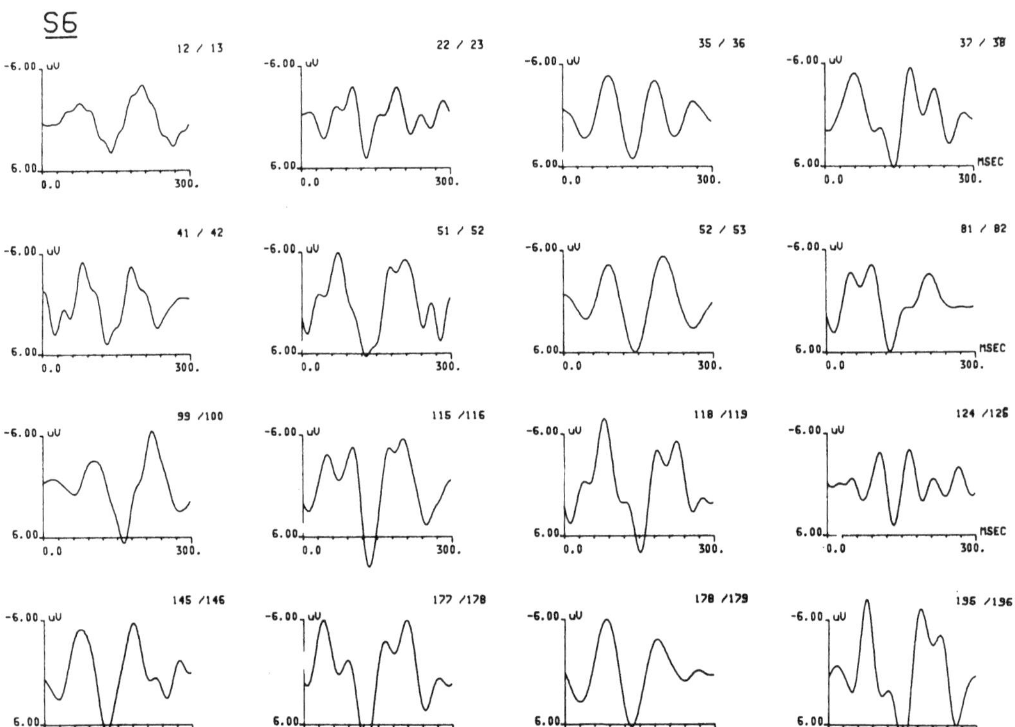

Fig. 7. Examples of single visual evoked potentials from one trial with a patient suffering from multiple sclerosis

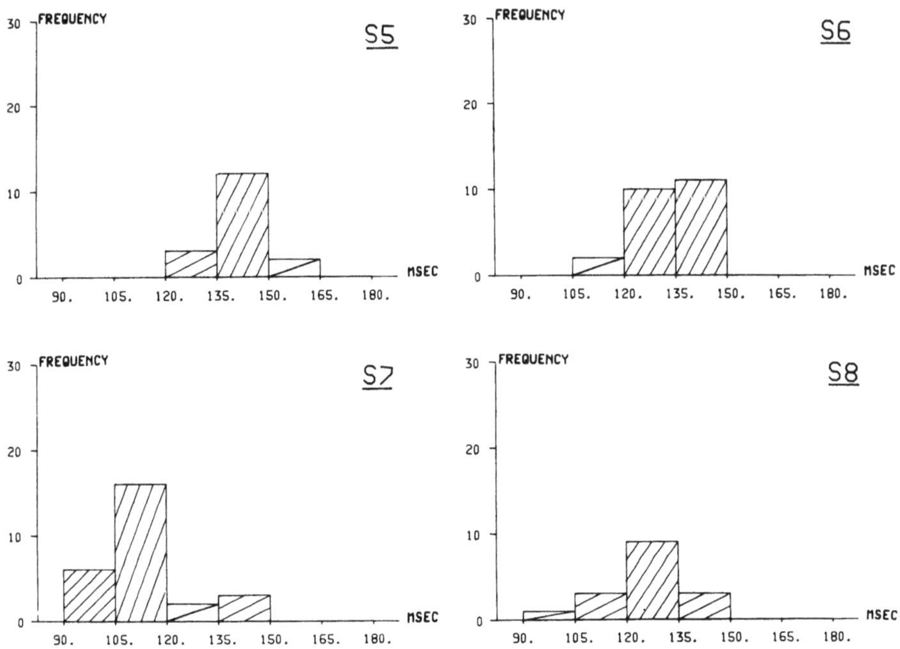

Fig. 8. P1 latencies of the single visual evoked potentials of four patients suffering from multiple sclerosis

well as lengthened latencies are also obtained, of which the latter, considering the empirically determined confidence intervals, would have to be regarded as significantly pathological. Results for the P1 amplitude show a similar pattern.

Analogous pictures can be seen with patients suffering from multiple sclerosis (Figs. 7, 8). P1 latencies are on average lengthened far beyond the usual 120 ms, although potentials with P1 latencies in the normal range are observed in most of the subjects.

4 Discussion

The results presented show that visual evoked potentials cannot be regarded as stationary events. This is remarkable, because visual evoked potentials are generally viewed as exogenous, stimulus-dependent potentials. Their relatively precisely defined neuroanatomical substrates lets them seem to be functions of simple stimulus-registration processes; an interference with higher-order cognitive functions seems unlikely. As yet, little is known about the variability of the exogenous potentials. Gasser et al. (1983) and Möcks et al. (1984) developed a method for estimating the variability of visual evoked potentials. Barber and Galloway (1979) investigated adaptation processes of these potentials with the aid of blockwise averaging. Of some interest in this respect are also the observations of Zerlin and Davis (1967), who recorded auditory evoked potentials from a patient with exceptionally large N1/

P2 amplitudes, permitting analysis without prior averaging. They found an amplitude variability with a generally normal distribution and observed no systematic trend effect.

Similar results are seen in the present study. Latencies and amplitudes of the visual evoked potentials are randomly distributed and no systematic effects have yet been seen. This fact stands against the assumption of a simple linear relationship between stimulus and evoked response in the EEG. Coppola (1979), based on the Wiener model for nonlinear systems, could show that the visual evoked response is strongly influenced by nonlinear effects. In this respect, Başar's model (1980) of a synchronization of spontanous EEG and the evoked potential, in which the evoked response is proposed to depend largely on endogenous synchronization processes, is important. The study of single-trial potentials will – among other effects – ease the investigation of the possible interactions of spontanous EEG and evoked potentials. Application of the method elaborated in this paper to endogenous potentials will prove valuable in the study of short-term fluctuations of cognitive functions. Specifically, single-trial potential analysis might be used for the prediction of the results of cognitive tasks executed by subjects.

References

Barber C, Galloway NR (1979) Adaptation effects in the transient evoked potential. In: Lehmann D, Gallaway E (eds) Human evoked potentials. Plenum, New York

Başar E (1980) EEG-brain dynamics. Relation between EEG and brain evoked potentials. Elsevier/North-Holland, Amsterdam

Başar E (1983) Synergetics of neuronal populations. In: Başar E, Flohr H, Haken H, Mandell AJ (eds) Synergetics of the brain. Springer, Berlin Heidelberg New York, p 183

Başar E (1985) Evozierte Potentiale and EEG-Dynamik. In: Bente D, Coper H, Kanowski S (eds) Hirnorganische Psychosyndrome im Alter II. Springer, Berlin Heidelberg New York Tokyo, p 83

Cadzow JA (1980) High performance spectral estimation – a new ARMA method. IEEE Trans Acoust Speech Signal Proc ASSP-28: 524–529

Chan YT, Landford RT (1982) Spectral estimation via the high-order Yule-Walker equations. IEEE Trans Acoust Speech Signal Proc 30: 689–698

Coppola R (1979) A system transfer function for visual evoked potentials. In: Lehmann D, Gallaway E (eds) Human evoked potentials. Plenum, New York

Coppola R, Tabor R, Buchsbaum MS (1978) Signal to noise ratio and response variability measurements in single trial evoked potentials. Electroencephalogr Clin Neurophysiol 44: 214–222

Gasser T, Möcks J, Verleger RS (1983) A method to deal with trial-to-trial variability of evoked responses. Electroencephalogr Clin Neurophysiol 55: 493–504

Heinze HJ, Künkel H, Massing W (1984) Detection of periodic signals in noise – an iterative procedure. Comput Programs Biomed 19: 55–59

Kay SM (1980) A new ARMA spectral estimator. IEEE Trans Acoust Speech Signal Proc 28: 585–588

Möcks J, Gasser T, Tuan PD (1984) Variability of single visual evoked potentials evaluated by two new statistical tests. Electroencephalogr Clin Neurophysiol 57: 571–580

Picton TW, Stuss DT (1980) The component structure of the human event-related potentials. In: Kornhuber HH, Deecke L (eds) Motivation, motor and sensory processes of the brain: electrical potentials, behaviour and clinical use. Elsevier/North-Holland, Amsterdam, p 17

Stearns SD (1979) Digitale Verarbeitung analoger Signale, Oldenburg, München

Zerlin S, Davis H (1967) The variability of single evoked vertex potentials in man. Electroencephalogr Clin Neurophysiol 23: 468–473

Auditory Evoked Potentials: Topodiagnostic Value, Spatiotemporal Aspect, and Phase Problem (Rarefaction Versus Condensation)

K. Maurer

1 Introduction

Of the many waves elicited during acoustic stimulation, two of the most important with regard to new trends in human electrophysiology are waves with short latency, the so-called auditory brain stem response (ABR), and one wave with long latency, the so-called P300. The ABR has been applied in diagnostis for about 10 years and has contributed to a better understanding of auditory function and to a more precise diagnostic procedure in brain stem lesions (Starr and Achor 1975; Stockard and Rossiter 1977; Maurer et al. 1979). P300 has been known since Sutton et al. (1965) described evoked potential correlates of stimulus uncertainty.

After an acoustic stimulation, approximately 15 brainwaves appear, which can be divided according to their latency into components with early, middle, and late latency. The earlier such a wave occurs, the more exact is its association with well-defined structures in the CNS. This applies in particular to the ABR, which originates in structures such as auditory nerve and brain stem. In longer latency conditions, the influence of psychological factors increases.

An important technical factor influencing latencies and amplitudes of the ABR is the phase of the stimulus. Responses to condensation (C) stimuli (which push forward on the tympanic membrane) differ from responses to rarefaction (R) stimuli (which pull backward on the tympanic membrane). The phase problem has been known since Kiang et al. (1965), Goblick and Pfeiffer (1969), Salomon and Elberling (1971), Maurer et al. (1980) and Maurer (1985) described earlier peaks to R stimuli. The difference between responses to C and R stimuli is crucial, because threshold estimation and topodiagnosis often vary, depending on the phase of the stimulus.

The topodiagnostic and spatiotemporal aspects and the significance of taking into account acoustic polarity (R vs. C) will be demonstrated in normal persons and patients with neuro-otological and psychiatric diseases.

2 Topodiagnostic Aspect

Wave I of the ABR originates at the beginning of the auditory nerve. This is certain for wave N_1 has long been used in electrocochleography (ECochG) for audiological purposes. Wave I is a volume-conducted farfield potential of potential N_1.

In experiments with animals, it was possible to show the origin of wave I in the distal part of the auditory nerve (Maurer and Mika 1983). The rabbit exhibits the

Springer Series in Brain Dynamics 1
Edited by Erol Başar

Table 1. Topodiagnostic categories distinguished by means of the ABR

Peripheral hearing disturbance	External ear
	Middle ear
	Cochlea
Neural hearing disturbance	Cerebellopontine angle
	Auditory nerve
Central hearing disturbance	Medulla
	Pons
	Midbrain

same anatomical characteristics as the human concerning the cerebello-pontine angle; this animal is therefore suitable for topodiagnostic examination. As in the human, one detects five waves, however with shorter latencies. If impulse propagation is disrupted by means of a pressure lesion upon the auditory nerve, wave I, which originates distally to the injury, can still be recorded. Waves III and IV originate according to our best information in brain stem structures such as the pons and midbrain, but the origin of waves II and V is still a matter of discussion.

This association of waves with structures makes possible a clinical classification of hearing disturbances as peripheral, neural, and central, which can be used in topodiagnosis (Table 1). The ABR provides a topodiagnostic information mainly in the longitudinal plane. The fact that both the left and the right auditory pathways produce independent potentials makes it possible to evaluate not only the longitudinal but also the transverse extension of a lesion. If both auditory tracts are involved, one may find lesions on the same as well as at different levels. Besides longitudinal and transverse aspects, one must also consider the aspect of lateralization; that is, that left-sided and right-sided lesions can be distinguished. The origin of waves I and II on the side of stimulation is certain and one may also assume an origin ipsilateral to the stimulated ear for waves III to V. This is proven by the effects of unilateral brain stem lesions, which lead to unilateral decay of waves on the same side.

3 Spatiotemporal Aspect

Whereas the ABR is caused by circumscribed activity of nuclei in the brain stem and helps in identifying the site of a lesion, the long-latency auditory evoked potentials (AEP) originate in complex cortical structures and exhibit a topographic distribution over the skull and an evolution over time. A newly developed computerized topographic technique called "brain mapping" enables the visualization of spectral, spatial, and temporal characteristics of evoked potential data (Duffy et al. 1979; Duffy 1981; Morihisa and McAnulty 1985; Maurer and Dierks 1987). One wave, the so-called P300, is of special interest with regard to the dynamics of sensory and cogni-

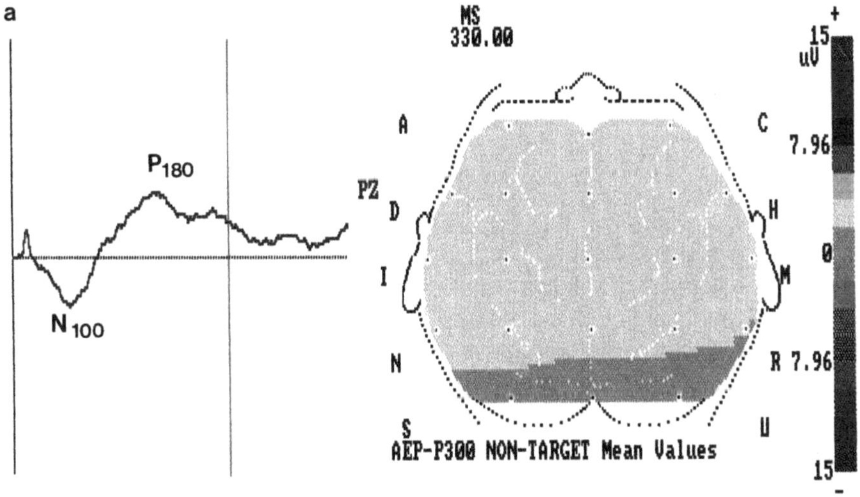

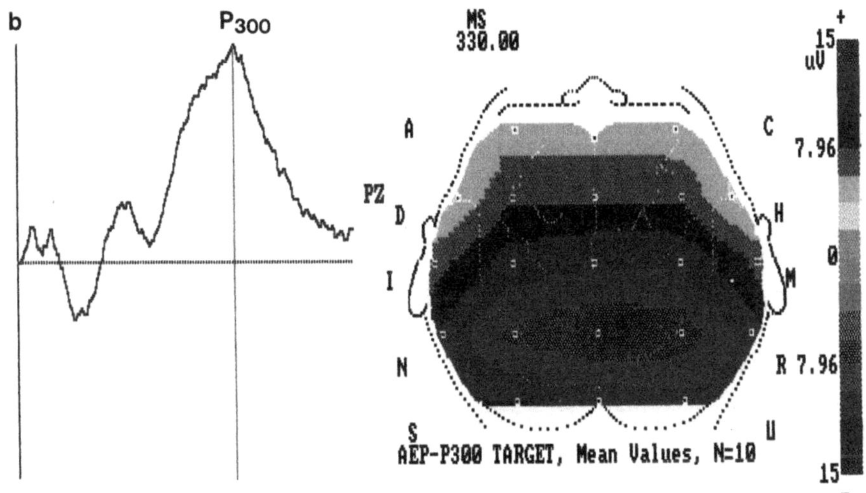

Fig. 1a, b. AEP and topography of the auditory evoked P300. (a) *Left:* Display of late AEP with peaks N100 and P200. *Right:* Topography of AEP at a time interval of 330 ms and after stimulation with a frequent 1000 Hz tone. A wave P300 is not present. (b) *Left:* Display of late AEP and a P300 complex at 330 ms. *Right:* Topography of P300 with a maximum of positivity at points P3, PZ and P4 and a decrease towards the frontal and temporal areas. The colored calibration bar indicates positive and negative amplitude values in the range of +15 to −15 μV

tive processing in the brain. The P300 is a late component of evoked potentials and is elicited by change in attended stimuli in any modality. Because they provide new information about the brain's response, the P300 and its spatiotemporal aspect are of special interest in psychophysiological conditions.

Figure 1 shows the topogrpahic distribution of P300 waves elicited by acoustic stimulation in a normal individual. Whereas presenting a frequent tone with a frequency of 1000 Hz causes clear definable waves N_1 (~100 ms) and P_2 (~180 ms), adding an infrequent 2000 Hz tone causes a late positive peak at about 300 ms. This wave has a maximal distribution in the centroparietal area (9 μV at point C_z).

Results in schizophrenic patients support the view that a major factor in the disorder may be the inability of the brain to focus on relevant stimuli. According to Morstyn et al. (1983), and consistent with several of our own observations (Maurer et al. 1987), a deficiency in activity in frontal and temporal areas may be assumed. Although preliminary, the brain mapping method provides new insight into late evoked potential behavior in endogenous psychoses and dementia.

4 Aspects of Stimulus Phase

4.1 Phase-Dependent Alterations in Normal Persons

Whereas waves I, II, and III were stable in shape, waves IV and V had variable configurations, as shown in Fig. 2, in which groups a–f were selected to characterize the different responses. The wave variations did not occur systematically; with the same stimulus phase, the two ears of a given subject could give rise to identical or to different patterns. A total of 100 ears of 50 subjects were evaluated. Figure 3 shows the re-

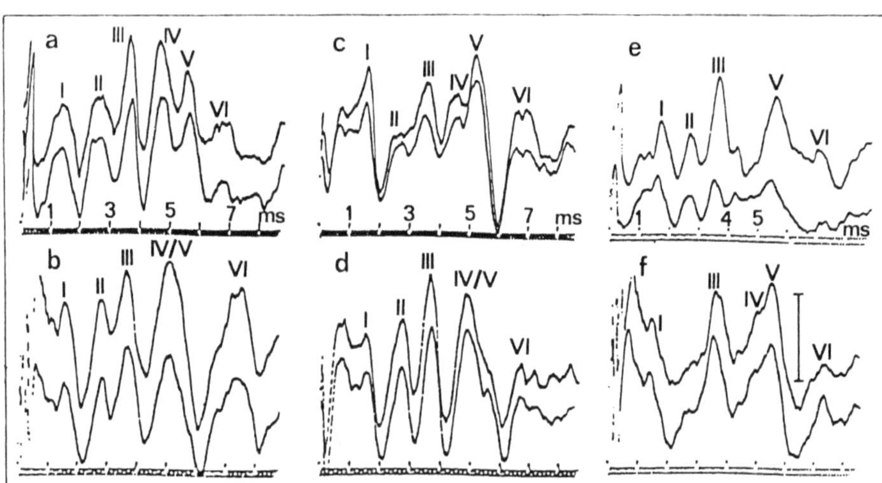

Fig. 2a–f. Alterations of waves IV and V in normal hearing subjects. (a) Separated waves IV and V; (b) single IV/V complex; (c) wave IV as shoulder of wave V; (d) wave V as shoulder of wave IV; (e) absence of clear definable wave IV; (f) absence of clear definable wave II. Calibration 200 nV

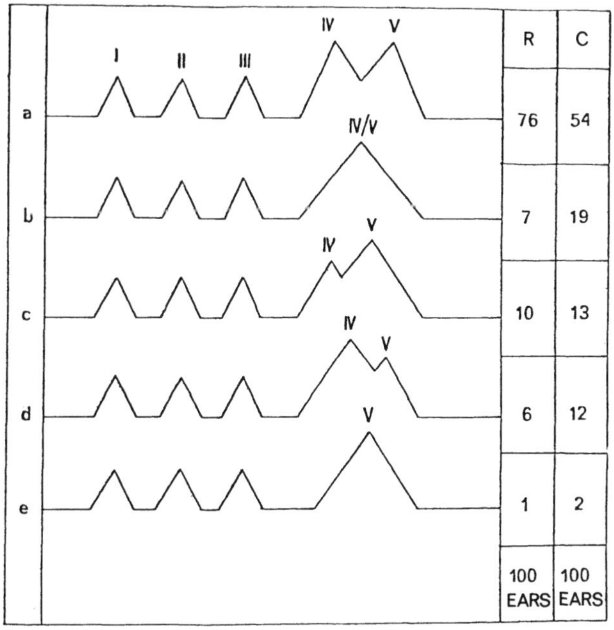

Fig. 3. Schematic drawing of wave alterations and percentage occurrence in the different phases. *R*, rarefaction; *C*, condensation

sults. R stimuli produced a high percentage of separated waves IV and V (about 80%). C stimuli tended to elicit either a IV/V complex (group b) or a pattern with wave IV as a shoulder of wave V (group c) or wave V as a shoulder of wave IV (group d).

4.2. Latencies with Respect to Stimulus Phase

Statistical calculations to find phase-dependent latency differences are problematic because wave patterns are not systematic. Although about 80% of ears gave rise to waves of group a after R stimuli, patterns b–e occured in about 20% of the ears.

When the Wilcoxon test for the peak latencies of waves I–V was applied in 50 subjects, using separate R and C stimuli, waves I, II, and IV appeared earlier after R stimuli, whereas wave V appeared earlier after C stimuli (Table 2).

A statistical method that considers the whole wave pattern along a time axis seemed to be more promising. For each ear and each polarity, the cross-correlation function (R_{xy}) was determined. The cross-correlation function is a measure of the similarity of a function $x(t)$ to another function $y(t)$, if $y(t)$ moves along a time axis with respect to $x(t)$. τ is the shifting time. The greater the similarity, the larger $R_{xy}(\tau)$. Therefore $R_{xy}(\tau)$ reaches a maximum if the ABR after rarefaction is similar to the response after condensation. A time shift of R_{xy} towards positive values of τ signifies shorter latencies to stimuli. To avoid differences between the right and left ear, the statistical calculations were done separately for each ear. Table 3 shows latencies of the maxima of R_{xy} for 12 normal subjects. Figure 4 shows the ABR from

Table 2. Latency change correlations with rarefaction and condensation stimuli

Wave	Stimulus conditions	Probability of an earlier peak
I	R–C	5%
II	R–C	1 per 1000
III	No difference	–
IV	R–C	1 per 1000
V	C–R	1%

R–C, rarefaction before condensation
C–R, condensation before rarefaction

Table 3. Latencies of the maxima of the cross-correlation function from the ABR in 12 subjects with normal hearing

No.	Right ear	Left ear
1	0.15	0.10
2	−0.02	0.07
3	0.17	0.04
4	0.01	0.01
5	0.13	0.03
6	0.09	0.09
7	0.11	0.09
8	0.07	0.09
9	0.05	0.03
10	−0.09	−0.04
11	0.07	0.05
12	0.05	0.20

−, Condensation before rarefaction; otherwise rarefaction before condensation

a 22-year-old woman in response to R and C stimuli and the corresponding cross-correlation-function. The Wilcoxon test showed a significant difference ($\alpha < 1\%$) between responses to R and C stimuli in such a way that latencies were shorter in the R mode.

4.3 Phase-Dependent Alterations in Audiology

Threshold determinations with R and C stimuli are time consuming. Near threshold, at least two trials have to be done for each polarity to guarantee reproducibility. Ten patients suffering from sensorineural hearing loss combined with loudness recruitment were investigated. Four patients exhibited reproducible components near threshold only in the R mode. In six patients, the threshold could be determined more precisely in the R mode (Fig. 5).

4.4 Phase-Dependent Alterations in Otology

A total of 28 patients with surgically identified acoustic neuromas were tested. In 17 cases, the wave pattern was identical in the two phases. In 11 cases, however, R and C gave rise to different wave patterns, resulting in a different topodiagnosis.

4.5 Phase-Dependent Alterations in Neurology

Phase-dependent differences often occur in diseases that affect the brain stem, and they obscure the diagnostic decision as to the lesion level. A total of 31 patients with

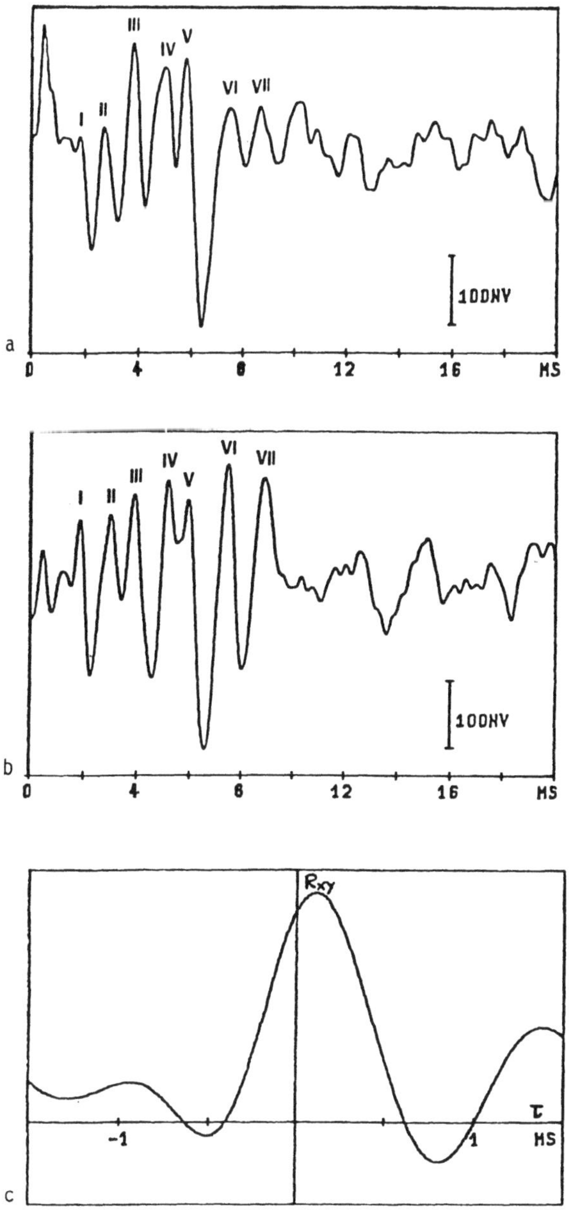

Fig. 4a–c. ABR in response to rarefaction (a) and condensation (b) stimuli. (c) Cross-correlation function (R_{xy}). τ, shifting time

"definite" multiple sclerosis (MS) were tested. Eleven cases had normal waves I–V; in 20 cases, altered waves were indicative of a brain stem lesion. In the 20 cases with altered waves, the responses of 40 ears could be evaluated. Nineteen ears had identical patterns in the two phases, whereas 21 of the ears gave rise to different wave patterns, depending on the phase of the stimulus (Fig. 6). More than 50% of the MS patients exhibited different patterns after R and C stimuli, resulting in a different topodiagnosis.

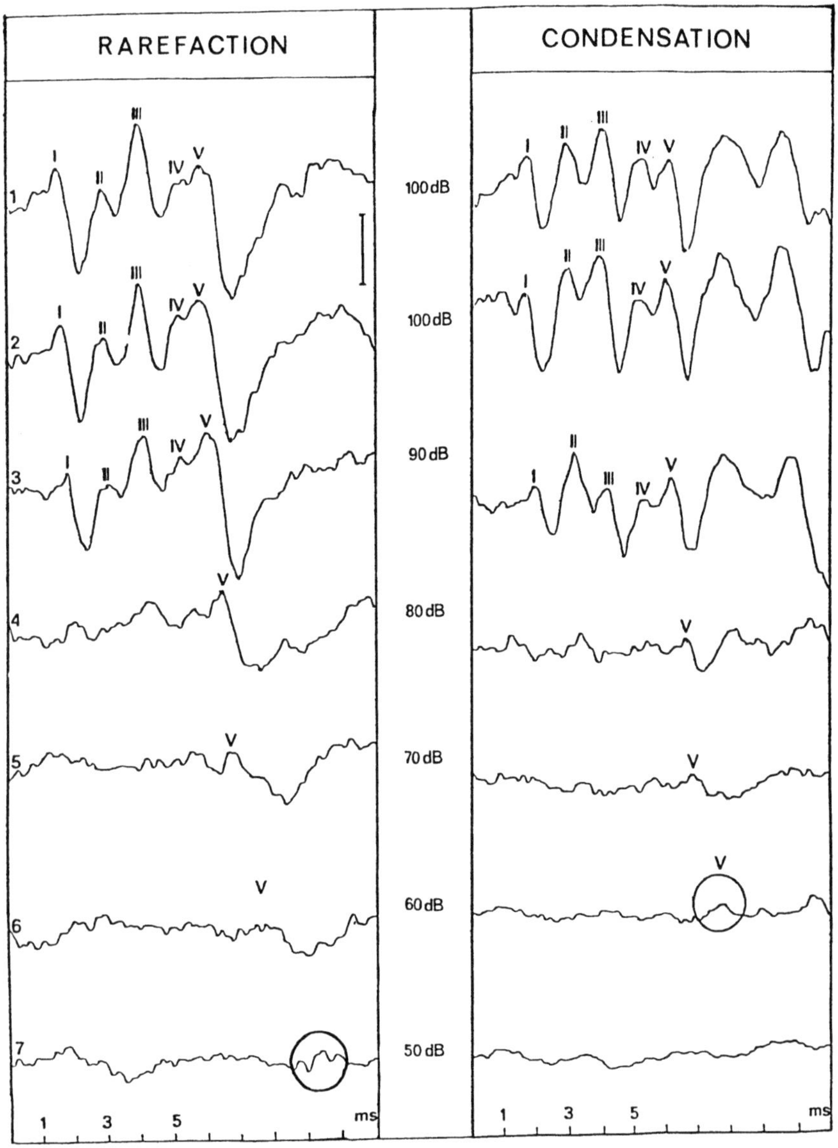

Fig. 5. Threshold determination in the rarefaction and condensation mode. Thresholds: R = 50 dB; C = 60 dB

5 Conclusions

In earlier times, technical limitation permitted the detection only of the late AEP. These late waves, however, did not contribute to topodiagnosis in neurology. The topodiagnostic aspect became clinically relevant only about a decade ago, when Sohmer et al. (1974) and Starr and Achor (1975) found early-latency AEP (1–10 ms);

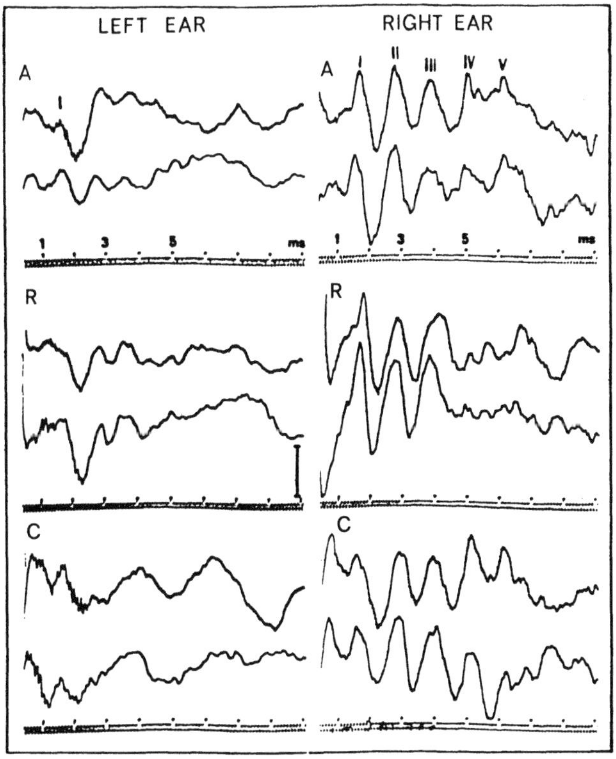

Fig. 6. BAEP in MS. *Upper part*, alternating mode *(A)*; *middle part*, rarefaction mode *(R)* with waves I–III *(right ear)*; *lower part*, condensation mode *(C)* with waves I–V *(right ear)*. Calibration 300 nV

these waves originate in well-defined brain stem structures, such as the auditory nerve, medulla, pons, and midbrain.

Whereas the origin of waves I, III, and IV is well established, the origin of waves II and V is still not fully understood. Their extracranial origin was assumed by Møller et al. (1981). This question can most likely be clarified by studying wave alterations in well-defined neurological lesions, such as vascular syndromes; e.g., the Wallenberg syndrome. In this disease, among other neural lesions, those of the cochlear nucleus eliminate AEP components starting with wave II. As for wave V, findings in conditions with increased intracranial pressure point to a mesencephalic origin.

As long as the late-latency AEPs were recorded by a limited number of electrodes (mainly by a vertex electrode), their diagnostic value was restricted to threshold estimations. This limitation was overcome when Duffy et al. (1979) introduced a spatiotemporal technique by attaching multiple electrodes (16–20) at the surface of the skull. In this way it was possible to evaluate the topographic distribution and spread of electrical activity over time. Evoked potentials with long latency became attractive again, because lateralization of electrical activity in the brain could be examined in psychiatric patients. Although there have been only a few studies in patients with endogenous psychoses, the results show that brain mapping is able to de-

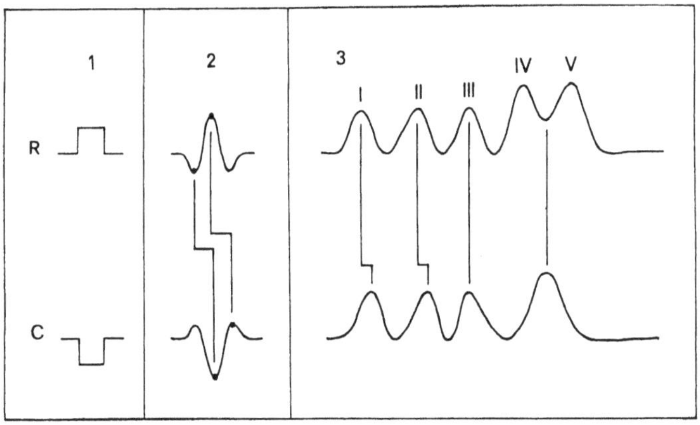

Fig. 7. Schematic drawing of the stimulation modality in the rarefaction and condensation mode. *1*, electrical waveform; *2*, acoustic waveform; *3*, *upper part*, waveform pattern due to rarefaction; *3*, *lower part*, waveform pattern caused by condensation with delay of latencies of waves I and II

fine regional differences in spectral content between schizophrenic patients and a control group (Maurer and Dierks 1987). Figure 1 showed the chronologic resolution of the P300 wave. It may be possible to do further studies of the cognitive dynamics of the brain's electrical activity.

The phase-dependent latency shift and change of pattern of AEP is generally accepted. However, the reason why R and C stimuli create different responses in subjects with normal hearing and in patients with neuro-otological diseases is not yet fully understood. Salomon and Elberling (1971) discussed the possibility that N_1 (wave I) firing is phase locked to the outward movement of the eardrum (R); they assumed a latency shift of half a cycle caused by reversal of the stimulus (Fig. 7). This hypothesis, however, does not explain the shorter latencies after C stimuli found in about 20%–30% of subjects with normal hearing.

The assumption of an excitation exclusively linked to the R mode could not be confirmed by our experiments. One reason may be the incalculable behavior of sound in the external auditory canal; only the electrical waveform can be determined exactly. Experiments similar to those done by Salomon and Elberling (1971) and Maurer et al. (1984) should therefore be carried out with probe measurements of sound pressure in the ear canal after R and C tone pips.

The phase problem can be studied only in part by electrically producing half-sine waves of different durations. If one assumes a nearly biphasic acoustic pattern and an initial upward deflection (R) responsible for excitation, a similar upward deflection should occur in the opposite phase.

According to probe measurements in the auditory canal in man (Maurer et al. 1984), this latency shift is of the order of 0.1–0.2 ms. Salomon and Elberling (1971) described an increase in latency of N_1 in the same range as the phase shift of sound pressure in the ear canal when reversing stimulus polarity.

The phase-dependent differences complicate threshold determination in audiology and obscure topodiagnosis in neurology. What hearing loss is to be diagnosed if R and C result in different hearing levels (Fig. 5)? What diagnosis is to be made if R indicates a medullary brain stem lesion whereas C points to mesencephalic damage (Fig. 6)?

In cases with different patterns and marked latency shifts, the alternating phase mode has to avoided, as the summating process may obscure the peaks specific to each phase; false positives may then be generated, including even pathological wave forms.

References

Duffy FH (1981) Brain electrical activity mapping (BEAM): Computerized access to complex brain function. Int J Neurosci 13:55–65

Duffy FH, Burchfield JL, Lombroso CT (1979) Brain electrical activity mapping (BEAM): a method for extending the clinical utility of EEG and evoked potential data. Ann Neurol 5: 309–321

Goblick J, Pfeiffer RR (1969) Time-domain measurements of cochlear nonlinearities using combination click stimuli. J Acoust Soc Am 46:924–938

Kiang NY, Watanabe T, Thomas EC, Clark LF (1965) Discharge pattern of single fibers in the cat's auditory nerve. MIT Press, Cambridge (Research monographs, vol 35)

Maurer K (1985) Uncertainties of topodiagnosis of auditory nerve and brain-stem auditory evoked potentials due to rarefaction and condensation stimuli. Electroencephalogr Clin Neurophysiol 62:135–140

Maurer K, Dierks T (1987) Brain Mapping – topographische Darstellung des EEG und der evozierten Potentiale in Psychiatrie und Neurologie. Z EEG-EMG 18:4–12

Maurer K, Mika H (1983) Early auditory evoked potentials (EAEPs) in the rabbit. Normative data and effects of lesions in the cerebellopontine angle. Electroencephalogr Clin Neurophysiol 55:586–593

Maurer K, Leitner H, Schäfer E, Hopf HC (1979) Frühe akustisch evozierte Potentiale, ausgelöst durch einen sinusförmigen Reiz. Dtsch Med Wochenschr 104:546–550

Maurer K, Schäfer E, Leitner H (1980) The effect of varying stimulus polarity (rarefaction vs condensation) on early auditory evoked potentials (EAEPs). Electroencephalogr Clin Neurophysiol 50:332–334

Maurer K, Schröder K, Schäfer E (1984) Programmable auditory stimulus generator and electroacoustic transducers – measurements of sound pressure in an artifical ear and human ear canal. Electroencephalogr Clin Neurophysiol 58:77–82

Møller AR, Janetta P, Bennett M, Møller MB (1981) Intracranially recorded responses from the human auditory nerve: new insights into the origin of brain stem evoked potentials (BAEPs). Electroencephalogr Clin Neurophysiol 51:18–27

Morihisa JM, McAnulty GB (1985) Structure and function: brain electrical activity mapping and computed tomography in schizophrenia. Biol Psychiatry 20:3–19

Morstyn R, Duffy FH, McCarley RW (1983) Altered P300 topography in schizophrenia. Arch Gen Psychiatry 40:729–734

Salomon G, Elberling C (1971) Cochlear nerve potentials recorded from the ear canal in man. Acta Otolaryngol (Stockh) 71:319–325

Sohmer H, Feinmesser M, Szabo G (1974) Sources of electrocochleographic responses as studied in patients with brain damage. Electroencephalogr Clin Neurophysiol 37:663–669

Starr A, Achor LJ (1975) Auditory brainstem responses in neurological disease. Arch Neurol 32: 761–768

Stockard JJ, Rossiter VS (1977) Clinical and pathologic correlates of brainstem auditory response abnormalities. Neurology (Minneap) 27:316–325

Sutton S, Braren M, Zübin J, John ER (1965) Evoked potential correlates of stimulus uncertainty. Science 150:1187–1188

VII. Workshop: How Brains May Work

Co-Chairmen: E. Başar and T. Melnechuk

Panelists: W. R. Adey, E. Başar, T. H. Bullock, W. J. Freeman, R. Galambos, E. R. John, and W. Leinfellner

Introduction

T. Melnechuk

We now begin the final session of this conference with remarks by seven speakers on "How brains may work." The title is not "How *the brain* may work" because, as Dr. Bullock has pointed out, different brains may work somewhat differently, and the title is not "*How* brains work" because we still do not know.

The purpose of the session is to facilitate the explication of new models of brain function. Therefore each of the seven panelists has been asked to do three things – to give the essence of the old way of thinking that needs modification, to give the essence of his new alternative view, and to make its main implications explicit – and to do these three things in only 10 min.

The last of these short talks will be given by Dr. Başar, who, like the six speakers before him, will talk about his own ideas, instead of improvizing a synthesis of the different conceptions of neural dynamics that we will have heard. However, I expect that he will enrich the proceedings volume with an epilogue that will place in a broad perspective the ideas we are about to hear.

After the seven short talks on how brains may work, Dr. Freeman will conclude the session with his second talk of the afternoon, which will review relevant methodology, both technical and conceptual. I expect him to argue for, among other things, the importance of interpreting one's own findings and the findings of others in terms of explicit models. Neuroscience has been described as rich in data and poor in theory. I for one have an intellectual hunger for neuroscientific synthesis, as expressed in the following light yet serious poem by the mythical Dr. Orpheus.

Each question Nature answers, like a seed
Of grain, yields seven new ones at its gleaning,
Whose answers fill the journals that we read
With kernels not yet cerealed to meaning.

As good empiricists pile up new facts
Upon the heaps whose bulk already wearies,
I bless each integrator who contracts
The mounds of data to a few clear theories.

Painstakingly, as in a film by Kubrick
That cumulates the details of a dooming,
The theorist, beneath his roomy rubric,
Connects phenomena in their subsuming.

I honor bits, though they are focal and minute,
And think the mind is not a thing to coddle,
Yet oh, my bliss when facts are not just brute
But evidence for some synthetic model!

In this spirit of welcoming new theoretical ideas, I commend our seven speakers in advance, and also call to your attention the interesting synthetic concepts of other participants in this conference – including, for example, the theory of Schild (1984) on the coordination of neuronal signals as structures in state space, and the theory of Harth and Unnikrishnan (1985) of perception and imagery as a process involving cyclic interaction between the brain stem and higher sensory relays.

Now we turn to the first of our scheduled talks.

References

Harth E, Unnikrishnan KP (1985) Brainstem control of sensory information: a mechanism for perception. Int J Psychophysiol 3: 101–119

Schild D (1984) Coordination of neuronal signals as structures in state space. Int J Neurosci 22: 283–298

Melnechuk: Our first speaker, in honor of Leibniz, Kant, and Hegel, is a philosopher, both in the current disciplinary sense and also in the older sense of "natural philosopher," since he was a working scientist in an earlier phase of his career. He is Dr. Werner Leinfellner, of the University of Nebraska and of the Technical University of Vienna.

The Brain-Wave Model as a Protosemantic Model

W. Leinfellner

For the last 40 years, the human brain has generally been regarded as an analog of the digital electronic computer, as a vast switching network of 100 billion neurons wired up as in a computer or a telephone net and communicating with one another by means of impulses transmitted electrically along fibers and electrically or chemically across synaptic connections.

This computer model of the brain has not gone unchallenged. At the philosophical level, it is objectionable to some for reinstating the dualism of mind and body, by considering the relationship of mentation to the brain only in terms of the relationship of computer programs to hardware.

At the empirical level, both psychological and biological evidence has suggested the inadequacy of the computer model of the brain, as I will now show.

To begin with, the psychological objections to the computer model: the scope and character of human intelligence are very different from those of a computer. In recent years, Dennett, Searle, Simon, Fodor, and I (Leinfellner 1984, 1985, 1986) have pointed out several limitations of the computer model.

First, a computer functions only like a formal logical system. The results of Weizenbaum with ELIZA in 1965, of Pribram in 1971, of Winograd with SHRDLU in 1973, of Schank and Abelson in 1977, and of Hofstadter in 1980 can all be summarized as establishing that a computer is a Turing simulation of formal thinking. As a result, it can understand only a small part of natural language. Dennett in his "spaceship" example and Searle in his "Chinese learning" example deny vigorously that computers will ever understand semantics or the full use of language. Secondly, formal manipulations of symbols and programs do not capture meaning; a computer knows a symbol but not what it symbolizes. Thirdly, a computer's program is in most cases syntactic but not semantic.

In brief, computers generally are realizations of only the logical and computable functions of intelligence. Thus they perform only a part of what we call "intelligence," which also includes cognitive and semantic functions, the maximization of survival and replication.

Complete intelligence, whether biological or artificial, requires the effective

Springer Series in Brain Dynamics 1
Edited by Erol Başar

synergetic cooperation of several subsystems: (a) a sensorium for receiving external stimuli; (b) a representational and memory subsystem; (c) a subsystem for making evaluations in terms of pleasant and unpleasant; (d) a subsystem for drawing inferences (here is the place for the computer); (e) an effector device, a motor subsystem.

The biological purpose of such an intelligent system is to solve problems under uncertainty and risk and under the constraint of maximizing survival. The question arises: could computers maximize survival? Hofstadter thinks that in the future, once computers become able to replicate, they will have feelings as mammals do and consequently will maximize their survival. At this time, that idea belongs to science fiction; contemporary computers do not have feelings, do not evaluate their chances for survival, and cannot respond to the huge variety of external stimuli.

Clearly, electronic computers and human brains are two fundamentally different embodiments of intelligence. The question is not what the brain can learn from the computer, but what the computer can learn from the brain, since the computer is inadequate as an analogy for a brain that provides the intelligence that maximizes survival and uses natural language.

How could one expect a brain modeled after a computer, with its circuits able to perform only logical and recursive operations, to underlie the intelligent and cognitive semantic functions of our language? Even a model like the advanced "associative computers" of Hopfield, which are neural-net simulations constructed "in the spirit of biology," cannot explain how we perceive gestalts which integrate thousands of bits of incoming information.

Happily, there is a brain model that is inherently compatible with representation, cognition, and semantics, and thus with intelligent behavior that maximizes survival, so it can explain how we evolved language and how we use language.

The model I speak of was made necessary by neurophysiological discoveries that weakened the case for the computer model. For example, neurons were discovered to function in a much more continuous and complicated way than do yes/no switches.

However, the most profound challenge to the computer model of the brain has been made by such neuroscientists as Adey, John, and Başar, who, for more than 20 years, have been finding evidence for faster transneural communication than can be performed by action potentials, and for quantitative correlations between brain wave patterns and stimuli – in short, evidence that led them to propose versions of an alternative model of the brain that de-emphasizes neuronal connections and emphasizes neural fields.

As is well known, the brain is pervaded by a complex electromagnetic field. The waves in it, which range in frequency from less than ten to several hundred hertz, can be picked up by electrodes on the scalp and displayed and measured on oscilloscope screens.

In the new model, perceptions, memories, thoughts, and computations are encoded in specific patterns of brain waves, which can resonate and build up superimpositions. Such communication by brain wave is certainly one way, and may be the only way, in which millions or billions of neurons can cooperate almost simultaneously, as there is evidence they do. I will call this communication via electromagnetic field "the silent internal language of the brain."

Perhaps the most important feature of the new brain wave model is that it provides an empirically based mechanism that is appropriate for realizing the cognitive, semantic, and linguistic functions of thought and language. The brain wave model can be called "protosemantic," because the electromagnetic field activity of the 100 billion neurons in the human brain is perfectly suited to instantiate an operative-operational model of semantics I developed several years ago (Leinfellner and Leinfellner 1978) and have since explained in more detail in recent publications (Leinfellner 1984, 1985, 1986).

My semantic model is a speculative linguistic and philosophical model that employs the concept of a unified cognitive semantic intelligence. Therefore it is able to explain what the digital-computer analogy for the brain could not explain – namely, the causal connections of thinking and language, of unconscious and conscious mental processes. Since it is based on the empirical brain wave activity of cerebral neurons, one can hope that ultimately it will be tested rigorously by neurophysiological methods and thus subjected to correction or refutation.

The main neurophysiological assumption of this cognitive semantic model is that cerebral neurons are able to represent invariant sensory input, including dynamic light wave patterns (visual input), sound wave patterns (auditory input), mechanical patterns (tactile input), and chemical patterns (olfactory and gustatory input) by mapping them, in the form of frequency patterns, onto quasi-isomorphic or homomorphic brain wave patterns (evoked encephalograms), which are available and understandable to all of the brain's neurons.

In terms of such specific wave patterns generated by neuronal activity, the underlying empirical basis of the semantic functions of thought and language can now be explained for the first time.

Concept formation can be explained by the well-known neurophysiological phenomenon of the superimposition of brain wave patterns to form a new, complex pattern. This is the process that underlies abstraction, as when, for example, we comprehend all the individuals who smoke tobacco as the class of smokers.

Memory can also be easily explained. Brain wave patterns are stored dynamically and can be reactivated (for instance, by resonance with an environmentally stimulated current instance of the same pattern), for, besides underlying representations, they are also the neurophysical basis of remembering.

The brain wave model also explains the cognitive semantic function of our language in a way that the computer analogy cannot. With regard to the process of naming and the reference relationship between thing and name, the brain wave model assumes that activity of the 100 billion neurons creates, among other semantic activities, a cognitive linguistic representation of the sensory input coming in from an external object. It does so by encoding the input in an invariant evoked brain wave pattern, which, about 3 ms later, via Wernicke's and Broca's areas, causes the speech muscles to utter an invariant linguistic expression, a name, which also comprises a a sound wave pattern of our language. Thus, when we see a rose, the word that we utter is invariably "rose." I call this process the "translation" of the silent brain wave language into the spoken language.

Ross Adey's experiment, which found that different evoked brain wave patterns were associated with the utterance of the two phonologically identical words "rose" and "rows," demonstrated the semantic sensitivity of brain waves. Because the meanings of the two words are different, the two patterns that represent them are different – that is, they do not resonate, and so the linguistic expressions triggered by the patterns are different too. Conversely, if the brain wave patterns that represent two words do resonate, as say in the case of "body" and "extended object," then the linguistic expressions will be found to be synonymous.

The translation of the silent brain wave language into speech means that we have two windows into the brain, one direct and one indirect. The brain wave patterns that we observe on screens are the direct window and the linguistic expressions of our spoken language are the indirect window.

Wittgenstein, in his later philosophy, assumed that thinking, whatever it may be in itself, is objectified as language, and he defined analytic philosophy as the analysis of the semantic functions of language. Wittgenstein's reduction hypothesis of thinking to language is actually a philosophical anticipation of this neurophysiological translation hypothesis.

I will conclude with the main philosophical consequences of the new model.

1. The brain wave model revolutionized the theory of knowledge and permits constructing an evolutionary epistemology. It explains cognitive processes even on the level of mammals that are intelligent but incapable of language, as I have shown in detail (Leinfellner 1984, 1985, 1986).
2. It explains the survival mechanisms of intelligent animals that function even without the help of language. If invariant representation is guided by survival maximization, intelligence will be manifested both in how the environment is perceived and in how problems and conflicts are solved.
3. It regards spoken language as the external communication of intelligent brains. Başar (1980) pointed out that all mammalian brains have similar brain wave characteristics. Chimpanzees, lacking vocal cords like ours, cannot speak, but their brains very quickly adopted the American sign language of the deaf to "translate" their silent internal mammalian brain language into external signs. The female chimpanzee Washoe even taught the sign language to her baby. Therefore the protosemantic model can explain the phylogenetic and ontogenetic origin of spoken language from the silent internal brain wave language of the neurons. This translation hypothesis, which was first proposed by Jaynes, could well be called after Pallas Athena, who sprang full-grown from Zeus's head.
4. Since the model is based on dynamic storage of brain wave patterns, it is consistent with current dynamic theories of semantic memory – for example, those of Tulving (1983) and Schank. Moreover, the superimposition of brain waves posited by the model would permit the brain to possess a far greater storage capacity than the 10^{13} bits permitted by the computer model.
5. Since the demonstration (Leinfellner and Leinfellner 1978) that logic can be reduced to operative-operational semantics, the brain wave model includes the logic-oriented computer model as a special case.
6. It throws light on Wittgenstein's hypothesis that human consciousness is the objectification of the silent internal brain language as spoken language, moreover, the overflow of individual memory into spoken language creates a cultural memory, an interindividual linguistic storage system, comparable to the "third world" of Popper and Eccles (1977).

7. If spoken language originates from and reveals the silent internal language of individual brains, which it translates into language, then language is the inter-individual communication of individual brains that makes possible the inter-individual cooperation we call "culture."

As an example of that cooperation, I cite the welcome help and comments I received during discussions with Dr. Erol Başar and other members of the Institute of Physiology on the occasion of my visit in Lübeck, and from Dr. W. Ross Adey and Dr. E. Roy John, which it is a pleasure to acknowledge. I also acknowledge with gratitude the help and support of the Research Council of the University of Nebraska and of Dean Meisels, who made this work possible.

References

Başar E (1980) EEG brain dynamics. Elsevier, London

Leinfellner W (1984) Evolutionary causality, theory of games and evolution of intelligence. In: Wuketits FM (ed) Concepts and approaches in evolutionary epistemology. Reidel, Boston, pp 223–276

Leinfellner W (1985) Intentionality, representation, and the brain language. In: Chisholm R, et al (eds) Philosphy of mind-philosophy of psychology. Reidel, Boston, pp 44–55

Leinfellner W (1986) Vom Apriori zum semantischen Prius. In: Lutterfelds W (ed) Transzendentale oder evolutionäre Erkenntnistheorie? Wissenschaftliche Buchgesellschaft, Darmstadt (to be published)

Leinfellner W (1986) Mach's Sinnesphysiologie. In: Haller R, Stadler F (eds) Die Philosophie Ernst Machs. Hölder-Pichler-Tempski, Wien (to be published)

Leinfellner W (1987) Evolutionäre Erkenntnistheorie und Spieltheorie. In: Riedl R, Wuketits F (eds) Die Evolutionäre Erkenntnistheorie. Parey, Berlin, pp 195–216

Leinfellner W, Leinfellner E (1978) Ontologie, Systemtheorie und Semantik. Duncker und Humblot, Berlin

Popper K, Eccles J (1977) The self and its brain. Springer, Berlin Heidelberg New York

Tulving E (1983) Elements of episodic memory. Oxford University Press, Oxford

Melnechuk: Our second speaker is Dr. Theodore H. Bullock, Professor of Neurosciences at the University of California, San Diego. Dr. Bullock is an expert on the nervous systems of all the marine creatures from jellyfish and bony fish to cetaceans and surfers. He has long been arguing for the conceptual importance of both the similarities and the differences between nervous systems.

How May Brains Work? A View from Comparative Physiology

T. H. BULLOCK

The first assertion that must be made about the title question is that for the present and a long time to come we expect divergent answers from experts who do not disagree but have different expectations of explanation or understanding. It may help to state my personal view of what constitutes understanding in the context of how the brain works. In the broadest terms, to explain or understand the mechanism of a phenomenon is basically to describe it in the language of any lower integrative level, even the very next level. There are always levels below us (nervous system, subsystem, circuit, cluster, cell, organelle, molecule, atom, particle, etc.); thus any explanation is just pushing the mystery down to phenomenology at a lower level. This is relevant because the extreme diversity of points of view about understanding the brain, many of them resulting from outstanding successes in cellular, membrane and molecular approaches, underlines a diversity in the conceptual goals of various integrative levels of inquiry (Gerard 1940).

What Do I Mean by "Understanding" a System like the CNS?

Focusing on the brain as a system, I will illustrate one view with the somewhat analogous problem "How does a university work?" When the discourse is aimed at the system level, the chemistry of typewriter ribbons and the physics of telephones, even the composition of sample individuals are not the most insightful initial approaches, although they are relevant and even important, especially in explaining malfunction. Findings such as "it works by shuffling material called paper," or "it works by temperospatial configurations of sets of units called committees" are also not satisfying, although they may represent quite significant advances if they are based on evidence that excludes some plausible alternative.

Somewhat closer to an initial understanding of the system would be such claims as the following. The university is found to work by (a) interactions among partially equivalent but nonredundant individuals; (b) each with rich but fragmentary and

Springer Series in Brain Dynamics 1
Edited by Erol Başar

filtered inputs; (c) making decisions at widely different levels of consequence; (d) based on those inputs but integrated with endogenous tendencies; (e) taking actions partly in concert, partly quite out of phase with others, (f) every individual unique but none indispensable; (g) the system adjustable by reason of a network of connectivity and shared competences; and (h) though normally the individuals operate with distinct responsibilities.

Now, in the sense that these statements provide an explanation, though quick and superficial, of how universities work, I see our chairman's challenge about the brain as calling for these kinds of statements as soon as we have even tentative evidence about the working brain. At least this may account for some of my biases and pet peeves.

Three Ps in the Brain: Partners, Properties and Processes

On this view, the working brain is not a soup of chemical messengers or a skull full of sparks or circuit boards. It might be understood by simultaneous consideration of the three Ps: specified *partners*, built-in but malleable *properties,* and parallel as well as serial *processes* – just as we might understand the university by simultaneous consideration of three Cs: intra-individual *capacities*; inter-individual *codes*; and departmental, senate, and regental (trustee's) *committees*.

The concept embodied in the three Ps is compatible with but goes beyond the widely held imagery of circuits of specified neurons in extended arrays. Even this is not to be taken as agreed, without qualification – as Roy John will tell us. If we want agreement, we may have to go back to the neuron doctrine – still seminal, still the great simplifying principle which has made successive generations of research possible and which has held up amazingly well. There are significant revisions, but not upsetting revisions – for instance, that there are signals of many kinds, both chemical and electrical, that signals are not only all or none but also graded, that signals are both received and emitted by dendrites, by somata, and by axonal terminals, that there are microcircuits and local circuits in the angstrom to micrometer range and macro- and field effects in the multimicron to ten micrometer range.

Specified Circuits Account for a Lot

The idea that large arrays of more or less well-specified neurons are connected in functioning circuits may not be universally agreed upon for higher levels in mammalian brains but it has been the guiding principle for a large and astonishingly successful enterprise in "circuit-breaking," both in the invertebrate and in the vertebrate nervous system. We now know major parts of some two dozen or so circuits, such as that for locomotion in a leech, feeding in a gastropod, and prey recognition and capture in a toad. Some are based on individually identified neurons. Others are based on defined classes of neurons, especially in the vertebrates, where the example of a piece of normal behavior best understood in terms of known neuron classes, connections, and properties is a social response in electric fish which we can follow from the first order afferents to the motor neurons through at least 14 orders of neurons, with a great deal of detail in both structure and dynamics. Whether or not some other principles come into play with higher cognitive functions in mammals, this principle of well-specified circuits certainly obtains and we cannot as yet place limitations on what it can explain.

As a concept it satisfies many people in the sense that they believe if we only knew the circuitry we would be able to account for most if not all of behavior. This is a large extrapolation from the known circuits. It has historical roots as an article of faith engendered by the extrapolation of Warren McCulloch and Walter Pitts from their theoretical neuron model more than 40 years ago. It is not upset in principle by the generally accepted addition of a more or less large stochastic component that is thought to improve reliability and signal discrimination in noise by population averaging. In vertebrates the population is generally believed to be a largely redundant number of cells; in invertebrates it is thought to be at the level of terminals and synaptic contacts.

Limitations of the Circuit Concept

There are two major limitations of this connectivity concept. The first I have touched upon: we cannot safely extrapolate to all higher nervous functions the belief that circuitry suffices. Certainly it exists, even in higher levels of the brain, as evidenced by neurons in the monkey superior and inferior temporal gyri that respond best to faces (Rolls 1984; Desimone et al. 1984). Specified connectivity is widespread in the cortex where complex, responsible decisions are made. But we do not know how far circuitry goes in explaining total behavior. We also lack a well-formulated, testable alternative principle, though vague appeals to large assemblies of cells in spatiotemporal configurations of activity, statistically characteristic for given states or cognitive events, are apparently satisfying to many authors and regarded as basically different from the circuitry concept. To me, the latter is not self-evident before the model is better specified.

The other main limitation is that "circuitry" as a term and image does not take account of much that we know about the operating system. It is a necessary and valid term and image but inadequate today. For example, it does not ordinarily or by reasonable stretch of the usage include specification for each part of each neuron of its particular specialization among the many kinds of electrical potentials that act as both responses and signals (e.g., long hyperpolarizations with decreased conductance, plateau potentials, and others), or its chemical messengers and their receptors and lytic enzymes, or its dependence on the three-dimensional geometry of the axon terminal ramifications and dendritic arbors. We now believe there are common field potentials whose extracellular currents can be influential over many micrometers; transmitters and modulators can act over some tens of micrometers. Beyond all these, there is a long list of personality traits of neuronal loci (e.g., facilitation or its opposite, spontaneity, irregularity and type of interval distribution, bursting tendency, after-effects or rebound, etc.) which are decisive integrative properties that must be spelled out for each locus of each neuron, the spike-initiating locus, the pacemaker locus, the point near the terminal where the spike fails, each axonal and dendritic branch point.

We Need a New Concept to Fit the Known Data

We are ready for and we desperately need a more adequate language and working model, to include and go beyond the "local circuits" (Rakic 1975) and the "parasynaptic processes" (Schmitt 1984) – heuristic and indeed revolutionary ideas but neither one nor both embracing all the foregoing, known integrative variables (Bullock 1959, 1967, 1968, 1976, 1977, 1979, 1980, 1984). I maintain that we are significantly handicapped in our evaluations of proposals about how arrays of cells work together by this lack.

My strong belief is that even when we have such a language and have incorporated its imagery into our everyday thinking, overcoming this handicap, that we will need to invoke or discover and recognize properties of organized systems of neurons that can only be called "emergents" – properties and processes which we could not have anticipated from knowing a lot about the elements. Very likely we are dealing with some of them now, without realizing their emergent character – aspects of the EEG and evoked and event-related potentials, for example.

Finally, I believe that it is crucial to think of the stages of evolution, starting with the simpler nervous systems of jellyfish, worms, and insects, which have all the basic properties we are familiar with – the fundamental mechanisms most of the literature is concerned with but which do not distinguish us from lower invertebrates. A meeting like this underlines that the human species is different in cognitive achievement. I think we have to strive harder to go beyond the commonalities and find the differences in partners, properties, and processes relevant to our humanity.

References

Bullock TH (1959) The neuron doctrine and electrophysiology. Science 129:997–1002

Bullock TH (1967) Signals and neuronal coding. In: Quarton GC, Melnechuk T, Schmitt FO (eds) The neurosciences: a study program. Rockefeller University Press, New York, pp 347–353

Bullock TH (1968) Representation of information in neurons and sites for molecular participation. Proc Nat Acad Sci USA 60:1058–1068

Bullock TH (1976) In search of principles in neural integration. In: Fentress JC (ed) Simpler networks and behavior. Sinauer, Sunderland, pp 52–60

Bullock TH (1977) Introduction to nervous systems. Freeman, New York

Bullock TH (1979) Evolving concepts of local integrative operations in neurons. In: Schmitt FO (ed) The neurosciences: fourth study program. MIT Press, Boston, pp 43–49

Bullock TH (1980) Reassessment of neural connectivity and its specification. In: Pinsker HM, Willis WD (eds) Information processing in the nervous system. Raven, New York, pp 199–220

Bullock TH (1984) A framework for considering basic levels of neural integration. In: Reinoso-Suarez R, Ajmone-Marsan C (eds) Cortical integration: basic, archicortical and association levels of neural integration. Raven, New York, pp 27–36

Desimone R, Albright TD, Gross CG, Bruce C (1984) Stimulus-selective properties of inferior temporal neurons in the macaque. J Neurosci 4:2051–2062
Gerard RW (1940) Higher levels of integration. Biol Symp 8:67–88
Rakic P (1975) Local circuit neurons. Neurosci Res Prog Bull 13:290–446
Rolls ET (1984) Neurons in the cortex of the temporal lobe and in the amygdala of the monkey with responses selective for faces. Hum Neurobiol 3:209–222
Schmitt FO (1984) Molecular regulators of brain function: a new view. Neuroscience 13:991–1001

Melnechuk: Our third speaker is Dr. Robert Galambos, Emeritus Professor of Neurosciences at the University of California, San Diego. Dr. Galambos has made many important discoveries in auditory neurophysiology and continues to make them in his laboratory at the Children's Hospital Research Center in San Diego. He is also the person who, 25 years ago, pointed out that since 50% of the brain's volume and 90% of the brain's cells were neuroglia, it was thinkable that glia played a role in neural information handling (Galambos 1961). Dr. Galambos may therefore be interested in a report that the ratio of astrocytes and oligodendroglia to neurons was unusually high in sections of the right and left superior frontal gyrus and inferior parietal lobes of the brain of Albert Einstein (Diamond et al. 1985).

Thoughts on "How Brains May Work": The Truism, the Guess, and the Prediction

R. Galambos

Glia are much more respectable than they were in the old days. Now I will try to make more respectable still another often-slighted aspect of brain function, by identifying what I think is an excellent way to search for answers to one of the most important questions about how brains work.

Back home, as I prepared for this moment, I found myself returning repeatedly to the supposed contrast between innate and learned behavior, and this is what I want to talk about. My interest in this problem began about 30 years ago, when Clifford T. Morgan and I were writing a chapter on the neural basis of learning (Galambos and Morgan 1960). Morgan kept asking whether we could point out neurological differences between innate and learned behaviors. In the end, we decided we could not.

One fact was obvious even then: the fly behaves like a fly, and the dog behaves like a dog, because genes build different behaviors into the structures of their nervous systems. Since then, the huge variety and complexity of the behavioral responses that genes create has never failed to impress me. There are millions of animal species, each with its unique arrangement of neurons – few or many – that unfailingly delivers exactly the behavior its owner must display in order to succeed in its world.

I marvel at how effectively the genes of each species have been selected to equip each organism with a body able to function under the physical conditions – temperature, barometric pressure, available nutrients, etc. – of the ecological niche that it will occupy, and including, as part of that body, a brain so constructed as to limit its owner's behavioral responses to those exactly suited for survival in that niche.

I would rather be able to list the general rules by which genes build species-specific behavior into brains than be able to recite all there is to know about what goes on

Springer Series in Brain Dynamics 1
Edited by Erol Başar

within an auditory system. Generalizations that one can list about species-specific behavior include that it is:

1. Elaborate, appropriate, exact.
2. Inevitable: simple and complex reflexes make it impossible for an animal to avoid making certain responses to specified physical changes in the environment.
3. Highly predictable: animals "know" how to find their food and mates; they build the correct nests for raising their young; they indeed do all the remarkable things Darwin described.

My thoughts can be summarized with a truism, a guess, and a prediction.

The Truism. An animal's phylogenetic structural heritage and its phylogenetic behavioral heritage are both read out by the genes passed on to it from its parents.

The Guess. The mechanism that reads out the one heritage does not differ in principle from the mechanism that reads out the other. Consider the bird: the way its genes construct the wing cannot be all that different from the way they organize neuronal aggregates so that the wing will be properly used in flying.

The Prediction. Progress in solving one of these problems will move in parallel with progress in solving the other.

Since the 1950s I have followed, amateurishly, the increasingly exact descriptions that molecular geneticists have given of how genes establish eye color, hormone levels, the presence or absence of enzymes, and so forth. The number and comprehensiveness of these successes has been growing at an accelerating rate. One needs only a superficial grasp of the recent discoveries about the development-determining genes called "homeotic" (Gehring 1985) to realize that the important details of the way drosophila genes read out the body of the fruit fly can reasonably be expected soon. Can we reasonably expect someone soon to describe exactly how drosophila genes construct the fruit fly's brain so that it produces the exact behavior the fruit fly needs for survival?

According to at least one expert, the answer must be negative. I remember reading his assertion that the number of available genes is too small to specify the development of any complex nervous system. Some time ago, when I mentioned this claim to a geneticist friend, she said, "That may be so, but the fact is we know the function of only about 5% of the DNA in a eukaryotic cell like the neuron – the 5% or so that encodes the suite of proteins typically produced by that cell." When I asked for more details about the remaining 95%, she said there were no more details. She added that geneticists called that 95% "excess DNA" or even "junk DNA" because no one could specify what it did.

Since then, functions have been found for various portions of the junk DNA (Lewin 1986). For example, it includes quasi-autonomous sequences called "selfish DNA," it includes noncoding sequences called "introns" that are contained within genes and that seem to facilitate mutation, and it includes novel control elements such as those called "enhancers." Can it be that in sorting through the remaining "junk," someone will find a fraction that automatically delivers, during ontogeny, exactly one package of species-specific behavioral responses?

Whatever the fate of that speculation, I urge you to keep an eye on neurogenetics, for I expect workers in that field of neuroscience soon to give us news that will lead to an understanding of the brain mechanisms of innate behavior, and thus to a far greater understanding of how brains work.

References

Diamond MC, Scheibel AB, Murphy GM, Harvey T (1985) On the brain of a scientist: Albert Einstein. Exp Neurol 88: 198–204

Galambos R (1961) A glia-neural theory of brain function. Proc Nat Acad Sci USA 47: 129–136

Galambos R, Morgan CT (1960) The neural basis of learning. In: Field J (ed) Handbook of physiology, sect 1, vol 3. American Physiological Society, Washington, pp 1471–1499

Gehring WJ (1985) The homeo box: a key to the understanding of development? Cell 40: 3–5

Lewin R (1986) "Computer genome" is full of junk DNA. Science 232: 577–578

Melnechuk: Our fourth speaker is Dr. W. Ross Adey of the Jerry L. Pettis Memorial Veterans' Hospital and Loma Linda University in Southern California. Besides his well-known role as a pioneer investigator both of the EEG and of weak local fields, Dr. Adey was the neurophysiologist who sent an instrumented monkey into space and also analyzed the EEG of human astronauts. He brought the first big modern computer to UCLA for brain wave analysis and he is also an amateur radio astronomer, so he ranges all the way down from quasars to quarks – or at least to calcium ions.

Do EEG-like Processes Influence Brain Function at a Physiological Level?

W. R. Adey

I have been asked to speak on (a) the essence of the old way of thinking; (b) the essence of the new way of thinking; and (c) what this implies about the basic functions of neural tissue. In a very brief survey, primarily from an historical perspective, the old thinking about brain mechanisms was very confining in its heavy emphasis on the so-called synaptic connectionist brain.

It was not so many years ago that Professor Sir John Eccles delivered the 1953 Waynflete Lectures in Hilary term at Magdalen College, Oxford with the title "The Neurophysiological Basis of Mind," in which he equated the physiology of the spinal motor neuron with mental processes (Eccles 1953b). I will say of my fellow Australian that nothing could be more inappropriate than the use of that approach, but he for his part dismissed the EEG as an epiphenomenon. It was considered at that time to be nothing more than the noise of the brain's motor.

In that era of the 1950s, it was necessary to take something of a defensive stance, because he for one defined wave processes in the brain in terms of postsynaptic potentials and little loops of neurons within the cortex that had appropriate time constants by reason of synaptic delays. His viewpoint was explicitly developed in a paper (Eccles 1953a).

So, from my somewhat defensive castle in the sky, I recognized the importance of seeking correlations of the EEG with cognitive processes on a quantitative basis. In other words, it was necessary to perform studies that might reveal a constancy of brain wave response in relationship to particular task performances. If successful, one might then proceed to look for some aspect of causality; in other words, can an EEG-like process influence brain function at a physiological level?

Our work at that time involved, for example, studies with chimpanzees playing tick-tack-toe (Fig. 1). They could do this with avidity, and, with some objectivity, they would often give their candy reward for winning to the observer, rather than eating it themselves when they lost the game. EEG analysis was based on spectral techniques and the first cross-spectral measurements, including coherence measures

Springer Series in Brain Dynamics 1
Edited by Erol Başar

Fig. 1. Chimpanzee playing tick-tack-toe during ongoing recording of EEG for spectral analysis that established positive correlations between brain wave responses and particular task performances. (From Hanley et al. 1968)

(Fig. 2). By simple pattern recognition procedures like discriminant analysis it was possible, using these spectral parameters, to arrive at a quite high level of correctness after the fact as to whether or not the animal was going to make a correct decision in playing a game like tick-tack-toe (Hanley et al. 1968).

I emphasize how cumbersome these methods were. It was necessary to do a very elaborate computer analysis based on data that involved many superimposed segments in order to achieve any level of reliability. The application of the same techniques to human EEGs allowed us to detect lying and other aspects of distorted higher nervous functions.

Passing over the interim, which I have addressed in part in my paper, it has become clear that certain aspects of causality do attach to EEG-like processes when they are imposed as artificial manipulanda on the brain.

And so I come to the new thinking, and list several emergent needs. The first is that we should determine far more accurately than so far possible the sign of the local process at the cell and molecular level, in terms of transaction, storage, and retrieval of information. The following remarks are not offered in any polemic sense, but I was somewhat surprised and even a little disappointed to see that the epiphenomena

a Case 2 Sample size: 26

Parameter in order of choice	Location	Band	Direction with Performance
1. Sum of spectra	L Hipp	Delta	Value
2. Mean frequency	R F-T	Alpha	Slowed
3. Sum of spectra	L Amyg	Theta	Value
4. Bandwidth	L Hipp	Alpha	Narrowed
5. Mean frequency	R CM	Alpha	Increased

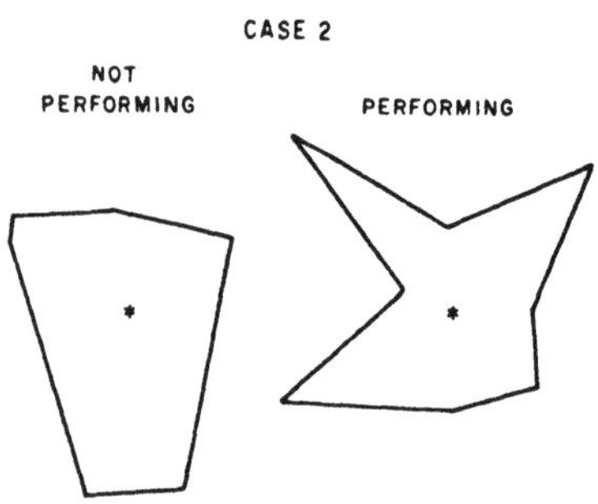

b Case 6 After

Parameter in order of choice	Location		Band	Comparison with correct and incorrect decision
Mean frequency	Left ventral anterior thalamic nucleus		Delta	Increased in correct decision
Coherence	Paracentral nucleus	LVA	Delta	Decreased in correct decision
Mean frequency	Paracentral nucleus		Beta-1	Decreased in correct decision
Coherence	LVA	L Hipp	Alpha	Decreased in correct decision
Coherence	L Hipp	L Amyg	Beta-1	Decreased in correct decision
Coherence	LVA	L Caudate	Beta-1	Decreased in correct decision

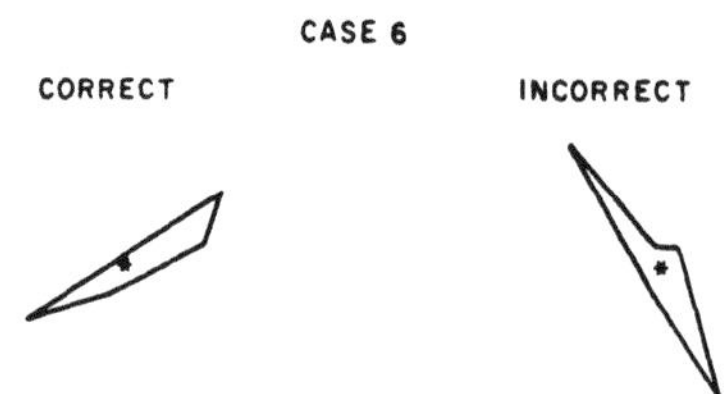

Fig. 2. (a) Pattern recognition techniques applied to EEG data from chimpanzee performing tick-tack-toe task (see text). The *boundaries* enclose all samples of the particular situation; the *asterisks* indicate the position of the group means. Five steps were required to separate completely 26 samples of the two situations, not performing and performing. However, the first selection correctly classified them with more than 90% accuracy. All samples were obtained the same day, and alle were correct. (b) EEG patterns accompanying correct and incorrect decisions in the performing chimpanzee. Each set of two parameters classified the decision with 100% accuracy; the sets are in order of increasing success. (From Hanley et al. 1968)

of brain electrical activity still take pride of place in much of our thinking when we talk about the N300, the P300, the P400 in relation to contingency, to current memory stores, and to long-term memory storage. I do not decry the need to develop correlates of that kind, but it is to me a chimera on reality, a fugue on reality that says very little about what is indeed going on in a brain if we are forced to limit our thinking to that level. It seems to me to be like trying to walk on water, and none of us are very successful at that. Perhaps we demean ourselves a little in failing to look beyond those recurring processes to the events that are the sign of local process at the cell and molecular level and that must surely underlie these phenomena observed at the cortical level.

In the same context, I think there are dangers in simplistic concepts of the local domain – for example, as a column of cortex, expressed simply in terms of numbers of cells, or in dimensional terms. The concept of a local domain so expressed says nothing about the effects of humoral influences, or the global effects of the EEG itself, on the behavior of that dynamic, and the domain must indeed be a dynamic – as I view it, under no circumstances will it be fixed in dimension or in function.

The question has been addressed to me as to whether the EEG does influence the neuronal wave. My answer, as I said at a 1973 symposium at Pebble Beach (Adey 1975), is that there must indeed be a feedback process, in which that portion of the wave which is distributed in the extracellular space comes to have some influence on a broader domain of which that cell is a part.

The second need that we may address concerns signs of the local process in terms of signal space, channel capacity, and channel occupancy. I find it surprising that we have heard virtually nothing here about the communication sciences in that context. I was forcefully reminded of it yesterday, when Roy John showed that very beautiful picture of the EEG of a schizophrenic patient who, while concentrating on some visual task, received a flash of light, of which virtually no sign at all appeared in the occipital leads. It bespeaks our need to recognize that we do not live in a tachistoscopic world. The imposition of punctate stimuli probably does little more than reflect, at the physiological level, the resetting of a brain which is in operation as a continuum in so many ways.

I am mindful that 30 years ago this month, Robert Galambos and his colleagues published their findings on habituation of evoked responses to click stimulation in the cat and their dishabituation by a brief electric shock to the skin of the chest (Galambos et al. 1956). It was also in the summer of 1955 that Raoul Hernandez-Peon visited the UCLA Brain Research Institute and, with Harald Scherrer, noted habituation of responses to brief tone bursts in the cat's cochlear nucleus, with dishabituation if the tone's frequency was changed. They also noted that if the cat's attention was sustained, as in watching two mice, even a novel tone burst did not cause an habituated evoked response to return (Hernandez-Peon and Scherrer 1955). Therefore one wonders about the validity of an approach that preoccupies itself with punctate stimuli.

Returning to the question of signal space and channel capacity, it's important to mention that the Russians have done very important work on what they call "information overload neurosis." These studies by Bechtereva's group have looked at the Russian fighter pilot's ability to handle a variety of contingencies in very stressful circumstances, with most interesting results.

Our third need is to develop concepts for the understanding of the nonlinear electrodynamics of transductive coupling. This is a matter I have addressed in detail in my earlier paper at this conference, but I ask again: how can we know the signal space with our tunnel-vision approach to the electromagnetic spectrum of biological phenomena? We have chosen to preoccupy ourselves essentially with things that happen at frequencies below 100 Hz.

I turn to some work with imposed electromagnetic fields, which had no place in what I said earlier. Figure 3 shows the use by Grundler and Keilmann (1983) of a far infra-red electromagnetic field, with which it was possible to produce directly resonant interactions with tissue macromolecules. As the frequency of this imposed field was varied over a range of a couple of hundred megahertz at 41 GHz, Grundler and Keilmann noted emergent periodicities in the growth of yeast cells.

Figure 4 shows the cross-correlogram between two separate experiments, in which the periodicity is about 8 MHz. The peaks and troughs of the correlograms follow the growth rates of yeast cell cultures. At the same time, by extrapolation, Hubert Fröhlich and others have pointed out that there is good reason to think that, arising from imposition of very high-frequency fields, there will be low-frequency phenomena at frequencies below 100 Hz (Adey and Bawin 1977). In view of this relationship, it is inadmissible to limit our awareness of the brain to what we see

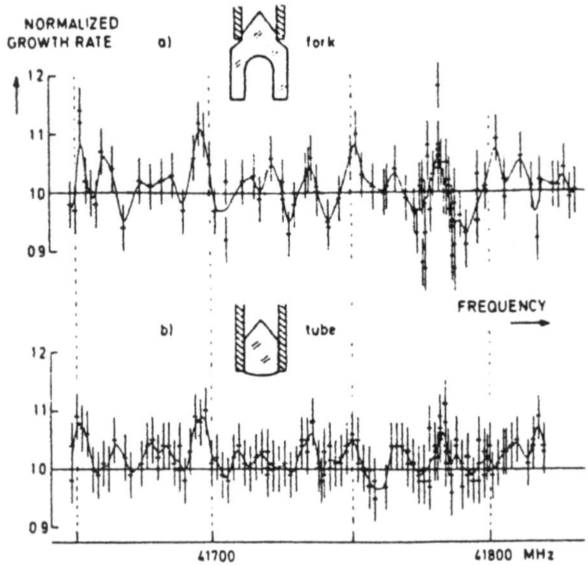

Fig. 3a, b. Microwave effect on yeast growth vs. frequency with use of either (a) fork antenna or (b) tube antenna. (From Grundler and Keilmann 1983)

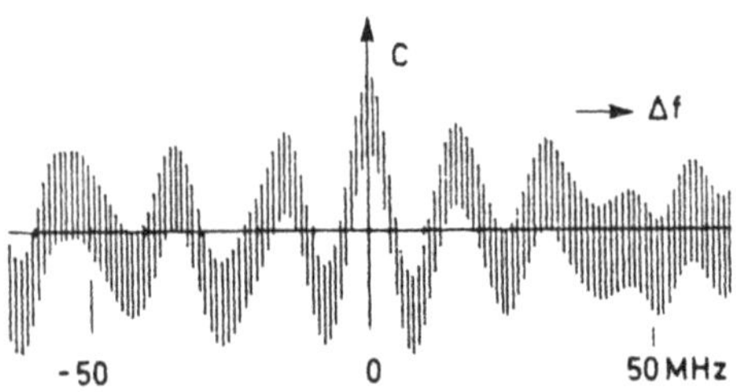

Fig. 4. Central part of cross-correlation of yeast response spectra Fig. 3, with $C(\Delta f) \sim \int \bar{\mu}_a(f)\bar{\mu}_b(f-\Delta f)\,df$. The significant maximum at $\Delta f = 0 \pm 1$ MHz proves together with the mirror symmetry around this point that both spectra *a* and *b* agree and reproduce the resonance positions to within ± 1 MHz, while the width of the resonances at half-height is 8 MHz. (From Grundler and Keilmann 1983)

through the narrow channel of 100 Hz or so at the low-frequency end of the spectrum. I am reminded of the remark of my mentor at Oxford, LeGros Clark, who said that if the brain had been placed upside-down in the skull, so that the hippocampal system was exposed, how much more we would know about the limbic system than about the sensorimotor cortex.

My last remark is a quotation (Trullinger 1978) about nonlinear electrodynamics, referring to solitons:

> Once the uninitiated can bring himself to revolt from his "linear" upbringing (via the harmonic oscillator and Schrödinger's equation), and accept the fact that large-amplitude localized objects can exist which *do not* spread in time, then he stands on the threshold of developing that wonderfully simple and beautiful pattern of thought that emphasizes the remarkable stability of such objects, their very natural use as elementary excitations or as fundamental objects present in the ground state, their coexistence in many cases with extended linearized solutions, their striking particle-like behavior, and perhaps most important of all, their essential role in describing so many physical phenomena that can not be explained in any other way.

References

Adey WR (1975) Evidence for cooperative mechanisms in the susceptibility of cerebral tissue to environmental and intrinsic electric fields. In: Schmitt FO, Crothers DM, Schneider DM (eds) Functional linkage in biomolecular systems. Raven, New York, pp 325–342

Adey WR, Bawin SM (1977) Brain interactions with weak electric and magnetic fields. Neurosci Res Prog Bull 15:1–129

Eccles JC (1953a) Interpretation of action potentials evoked in the cerebral cortex. Electroencephalogr Clin Neurophysiol 3:449–464

Eccles JC (1953b) The neurophysiological basis of mind. Oxford University Press, London

Galambos R, Sheats G, Vernier VG (1956) Electrophysiological correlates of a conditioned response in cats. Science 123:376–377

Grundler W, Keilmann F (1983) Sharp resonances in yeast growth prove nonthermal sensitivity to microwaves. Phys Rev Lett 51:1214–1216

Hanley J, Walter DO, Rhodes JM, Adey WR (1968) Chimpanzee performance: computer analysis of electroencephalograms. Nature 220:879–881

Trullinger SE (1978) Where do solitons go from here? In: Bishop AR, Schneider T (eds) Solitons and condensed matter physics. Springer, Berlin Heidelberg New York, pp 338–340

Melnechuk: Our fifth speaker is Dr. E. Roy John, Director of the Brain Research Laboratories at New York University Medical Center, where he studies endogenous and exogenous brain waves, their distribution in the brain, and their application in psychiatric diagnosis. Ever since he reported findings inconsistent with the prevalent switchboard theory of brain function and proposed a statistical theory, connectionists have grudgingly respected him as a maverick; now, as last, his time seems to have come.

Resonating Fields in the Brain and the Hyperneuron

E. R. John

Classical theory proposes that learned responses, from which it is largely derived, are mediated by activity in discrete circuits linked by experience. These circuits are essentially labelled lines reserved for the mediation of those experiences, although they may be in profuse parallel. This is essentially a computer analogy, with pieces of information residing in places.

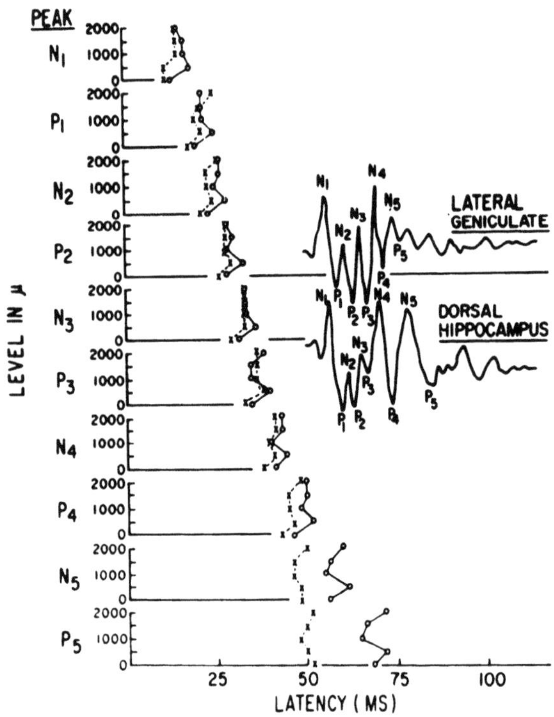

Fig. 1. Analysis showing similar latencies of acoustic evoked response (AER) components recorded from the lateral geniculate *(crosses)* and the dorsal hippocampus *(circles)*. Latency of component is plotted along the *abscissa* versus depth of penetration along the *ordinate*. Successively later components are depicted by *graphs* from *top* (N_1) to *bottom* (P_5). Each point is based on an average response to 500 stimulus presentations in multiple behavioral trials. (Data from John and Morgades 1969b)

Springer Series in Brain Dynamics 1
Edited by Erol Başar

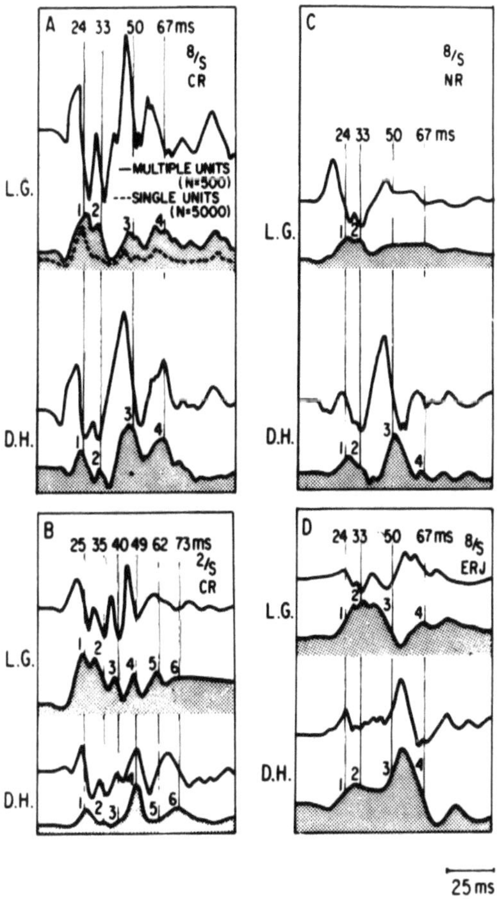

Fig. 2. (A) AERs *(solid curves)* and PSHs *(shaded areas)* simultaneously recorded from microelectrodes in the lateral geniculate body on the left side *(L.G.)* and the dorsal hippocampus on the right side *(D.H.)* during correct performance *(CR)* to the 8-Hz stimulus by cat 2. *Numbered vertical lines* indicate components considered to correspond with respect to relative latency. These and all other responses illustrated in this figure computed from 500 stimulus presentations, except for the PSH derived from a single unit in LG, shown as a *dotted line* (N = 5000). Note the correspondence between the curve describing the probability of firing of this single neuron observed over a long period of time and the PSH for the neural ensemble observed for one-tenth that time. (B) AERs and PSHs simultaneously recorded from LG and DH during correct performance *(CR)* to the differential 2-Hz conditioned stimulus. (C) AERs and PSHs simultaneously recorded from LG and DH during presentations of the 8-Hz conditioned stimulus which resulted in no behavioral performance *(NR)*. (D) AERs and PSHs simultaneously recorded from LG and DH during presentation of a novel stimulus illuminated by the 8-Hz flicker *(ERJ)*. (Data from John and Morgades 1969b)

Consciousness is ignored in such analogies, even though the brain certainly performs more than conditioned reflexes. I do not think our task is to explain only conditioned reflexes. I am interested in human experience and how it is mediated by the brain.

Abundant lesion studies demonstrate the extreme difficulty of deleting a specific memory engram. Classes of responses are affected, rather than specific responses; and retraining with extreme savings is usually possible. There are examples, that I do not have time to cite, showing that even severe interference with sensory input and motor output has little effect; as when Bob Galambos reported that we can take out 98% of the visual tracts bilaterally and preserve patterned vision. Such findings tend to be ignored by those who prefer connectionistic approaches.

I should tell you that I was trained at the University of Chicago by connectionists who had spent a lot of time contributing to classical neurophysiology. Warren McCulloch was one of my mentors. I began my research career not as a maverick but as someone who believed in the orthodox position with complete conviction; I went out into the laboratory expecting to find that things were as I had been told. In fact, hardly anything was as I had been told.

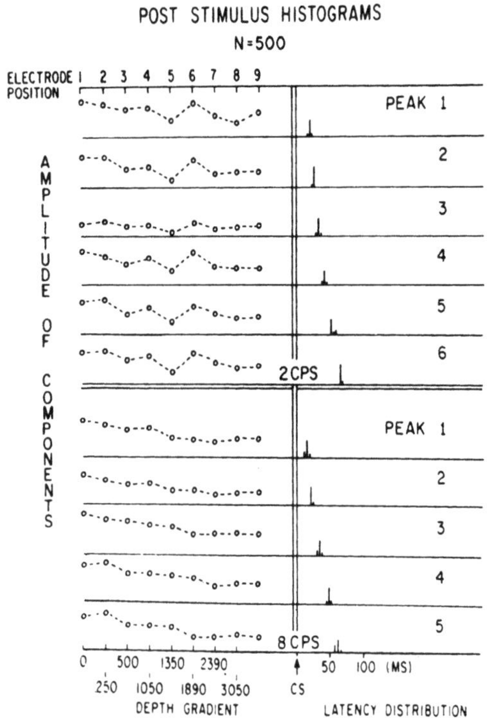

Fig. 3. Graphs on the *left side* illustrate the amplitude gradients of PSH peaks computed for a 2-Hz positive and 8-Hz negative flicker cue. Successively later components are depicted from *top* to *bottom*. In each graph, amplitude of the component is plotted as vertical displacement, while depth of electrode penetration is plotted as horizontal displacement. Each *graph* on the *right* shows the latency distribution of the PSH peak whose amplitude gradients are found in the graph at the same level on the *left side* of the figure. Component latency is plotted along the *abscissa*, while number of positions showing a peak at that latency is represented as the *ordinate*. (Data from John and Morgades 1969b)

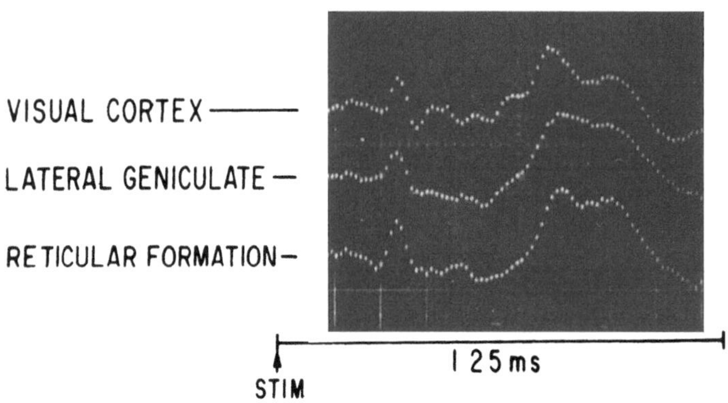

Fig. 4. Difference waveshapes obtained by subtracting average responses computed during three trials resulting in no performance (NR) from average responses computed during five trials resulting in correct performance of the conditioned avoidance response (CAR). All recordings were bipolar, and 75 evoked potentials were used in each of the constituent averages. Note the correspondence in latency and waveshape of the difference process in these various regions. (Data from John 1967)

Our data suggest a type of information processing which is fundamentally different from the connectionist type. To remind you of the basis for these assertions, I am going to recapitulate some points, most of which I presented in my lecture.

In trained animals processing learned *but not novel* stimuli, similar evoked potential (EP) waveshapes and temporal fluctuations in firing density appear in distributed

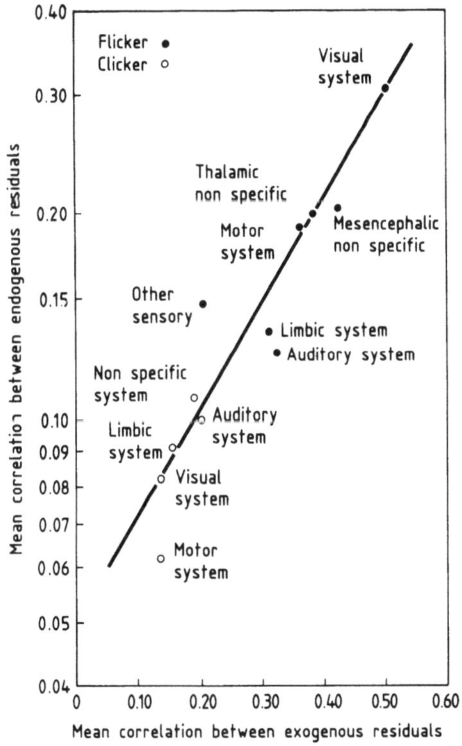

Fig. 5. Plot of mean correlation coefficients between exogenous residuals vs. endogenous residuals for different neural systems and for different cue modalities. *Closed circles*, flicker frequencies as stimuli. Auditory system: $N=305$; auditory cortex (16 cats); medial geniculate (16); brachium colliculi inferioris (1). Limbic system: $N=303$; hippocampus (16); dentate (5); cingulate (5); septum (5); prepyriform (6); medial forebrain bundle (6); mammilary bodies (5); hypothalamus (7). Mesencephalic nonspecific: $N=158$; reticular formation (18); central gray (1); central tegmental tract (1). Motor system: $N=146$; motor cortex (4); substantia nigra (10); nucleus Ruber (4); nucleus ventralis anterior (9); subthalamus (5). Other sensory: $N=54$; sensorimotor cortex (4); nucleus lateralis posterior (1); nucleus ventralis postero lateralis (5); nucleus ventralis postero medialis (1). Thalamic nonspecific: $N=139$; nucleus centralis lateralis (13); nucleus reticularis (6), nucleus reuniens (1); medialis dorsalis (5); pulvinar (1). Visual system: $N=394$; visual cortex (18); lateral geniculate (18); brachium colliculi superioris (2). *Open circles,* click frequencies as stimuli. Auditory system: $N=48$; auditory cortex (5); medial geniculate (5). Limbic system: $N=69$; hippocampus (5); dentate (3); cingulate (3); septum (3); prepyriform (2); medial forebrain bundle (3); mammilary bodies (3); hypothalamus (2). Motor system: $N=37$; motor cortex (1); substantia nigra (4); nucleus ruber (1); nucleus ventralis anterior (5); subthalamus (2). Nonspecific system: $N=50$; mesencephalic reticular formation (6); central gray (1); central tegmental tract (1); nucleus centralis lateralis (3); nucleus reticularis (3). Visual system: $N=55$; visual cortex (6); lateral geniculate (6); brachium colliculi superioris (1). N denotes the number of independent measurements within the designated system. Data from monopolar and bipolar derivations were combined. Replications varied across cats and structures. (Data Bartlett et al. 1975)

neuronal populations. An EP recorded from one location can be strikingly like another recorded some distance away. In Figs. 1 and 2A, the population in the immediate vicinity of one electrode in the lateral geniculate shows an EP and a post-stimulus histogram that has peaks at about the same time as does a population in the

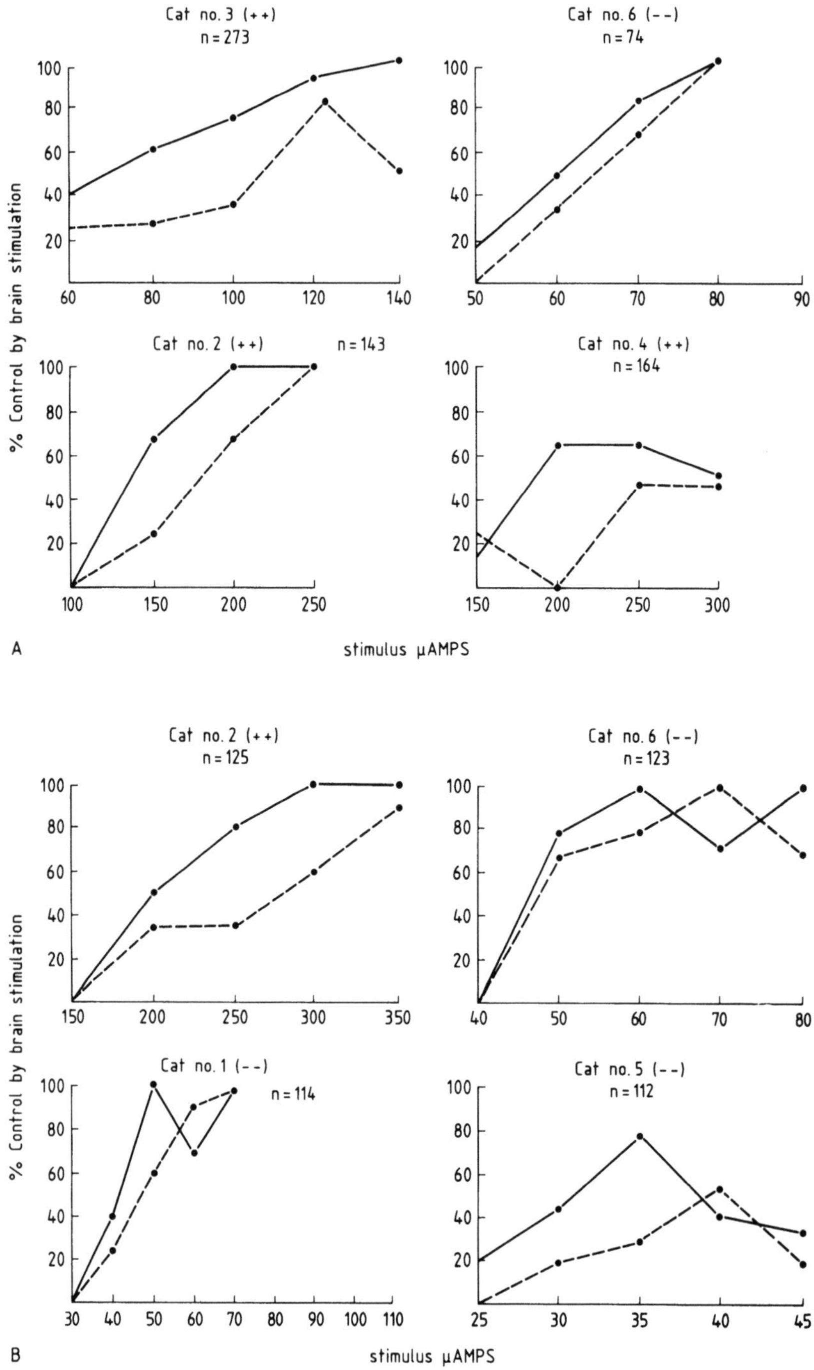
Cat no. 3 (++)
n = 273
Cat no. 6 (--)
n = 74
Cat no. 2 (++)
n = 143
Cat no. 4 (++)
n = 164
% Control by brain stimulation
A
stimulus μAMPS
Cat no. 2 (++)
n = 125
Cat no. 6 (--)
n = 123
Cat no. 1 (--)
n = 114
Cat no. 5 (--)
n = 112
% Control by brain stimulation
B
stimulus μAMPS

dorsal hippocampus on the other side. When the animal fails to recognize the stimulus, this coherent neuronal organization collapses (Fig. 2C).

Note that the histograms have a trough at about 40 ms. If you listen to the firing of these cells, they go brrRRR, pshshsh, BRRrr; that is, there is an input period, then inhibition, and then a second burst, which is the beginning of what we call the endogenous process.

Potential gradients, constructed by chronically implanted moving microelectrodes, are flat across large distances (Fig. 3). This fact excludes volume conduction as an explanation for the coherent electrical phenomenon, as does the fact that geometrically intermediate structures need not belong to the coherent system.

Some of the early activity shown in Fig. 2 is afferent input, or exogenous; but after about 40 ms in the cat, after that inhibitory period mentioned before, a later temporal pattern is released that is endogeneous. It appears to be highly synchronized between some remote regions, far too near to being simultaneous to reflect synaptic transmission between those regions. We have measured the difference from absolute simultaneity down to 200 μs, without being able to show a clear difference in the latency of the released phenomenon in a number of well-separated places (see Fig. 4). However, intermediate positions can be found where these synchronized processes are absent, reduced, or markedly delayed.

The quantification of exogenous and endogenous activity shows that both are extensively distributed throughout the brain, with a signal-to-noise ratio that differs greatly from region to region. With visual stimuli, the highest signal-to-noise ratio is in structures of the visual system (closed circles in Fig. 5). With auditory stimuli, shown by open circles in Fig. 5, the highest signal-to-noise ratio is in the auditory system. The slope of the relationship between exogenous and endogenous shown on the graph is the same for auditory and for visual information. However, the relative position of sensory and nonsensory specific regions on this graph changes according to the stimulus modality. This suggests that what we call "sensory-specific regions" are merely the regions with the highest signal-to-noise ratio for a preferred modality, but the information is distributed throughout many brain regions.

Direct brain stimulation with patterned trains of 200-μs pulses in many different brain regions of trained animals can produce the selective release of differential behaviors and can contradict sensory input signals as a function of the amount of current and timing pattern in the pulse train (Fig. 6 and 7). The volume of the neuronal

Fig. 6A, B. Contradiction of auditory (A) and visual (B) signals by brain stimuli. Each *graph* in the figure shows the effectiveness with which stimulation of the mesencephalic reticular formation at either of two frequencies (RF_1 and RF_2) contradicted simultaneously presented visual stimuli (V_2 and V_1, *left*) or auditory stimuli (A_2 and A_1, *right*), plotted as a function of increasing current intensity. For cats 1, 3, and 6, frequency 1 was 4 per second and frequency 2 was 2 per second. For cats 2, 4, and 5, frequency 1 was 5 per second and frequency 2 was 1.8 per second. *Solid lines* show the outcomes when peripheral stimulation at the higher frequency (V_1 or A_1) was pitted against RF stimulation at the lower frequency (RF_2), while the *dotted lines* show the outcomes when the higher-frequency stimulus was delivered to the RF. Cats 1, 5, and 6 were trained to perform avoidance-avoidance discrimination (− −), while cats 2, 3, and 4 were trained to perform approach-approach discrimination (+ +). *N* refers to the total number of conflict trials carried out in each cat, accumulated in three sessions for cats 2, 5, and 6 and four sessions for cat 1 (visual RF conflict), and in three sessions for cat 2, four for cat 6, five for cat 4, and seven for cat 3 (auditory RF conflict). (Data from Kleinman and John 1975)

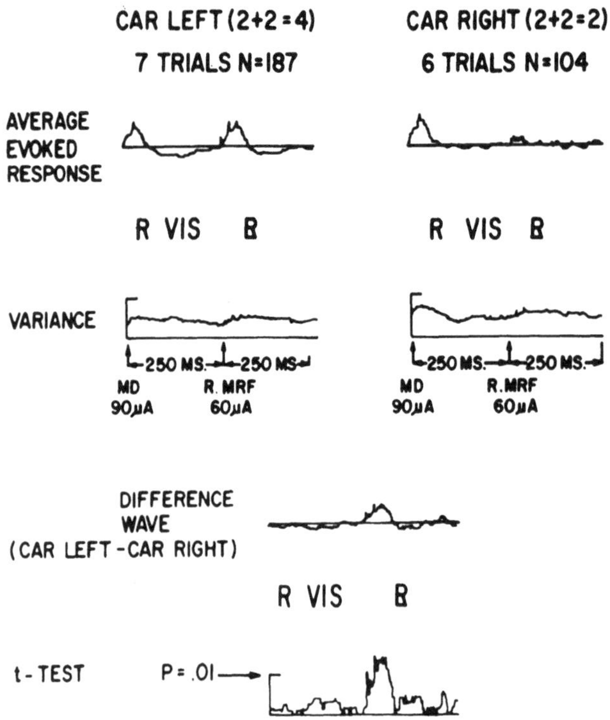

Fig. 7. Averaged evoked responses *(top)* recorded from the visual cortex to two-per-second out-of-phase stimulation delivered to the medialis dorsalis and mesencephalic reticular formation. Evoked response waveshapes (*left* and *right*) are different depending on behavioral outcome. Difference wave and *t* tests between evoked responses on *top*, *left*, and *right at bottom.* (Data from unpublished observations by D. Kleinman and E. R. John 1975)

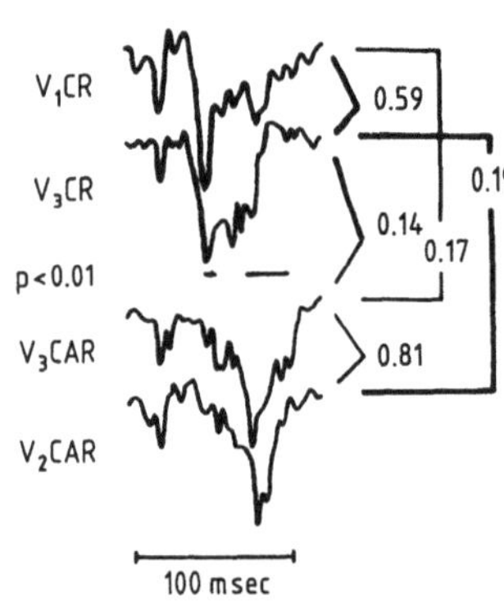

Fig. 8. Response waveshapes with averages based upon sequences of evoked potentials selected by the experimenter from the last 4 s of multiple behavioral trials. Average sample size, 15. See text for explanation. (Data from John et al. 1969)

ensemble that is mobilized by the stimulus is proportional to the amount of current. The more improbable the coherence or nonrandomness of the population (higher current), the more the animal is convinced that internal reality is more valid than external reality. The same result can be obtained when the bursts in a train are *sequentially introduced at different electrodes.*

These effects begin at levels of stimulation that are as low as 10 μamp for 200 μs. Currents of this magnitude apparently flow in the brain during natural stimulation,

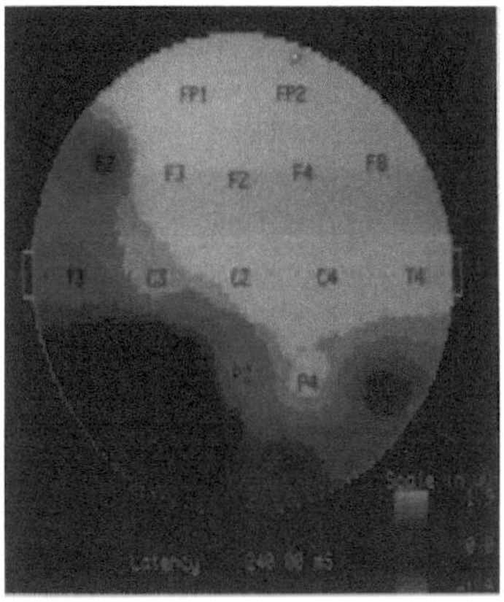

Fig. 9. Topographic map of event-related potential, 240 ms after a click, in a patient aphasic after a stroke

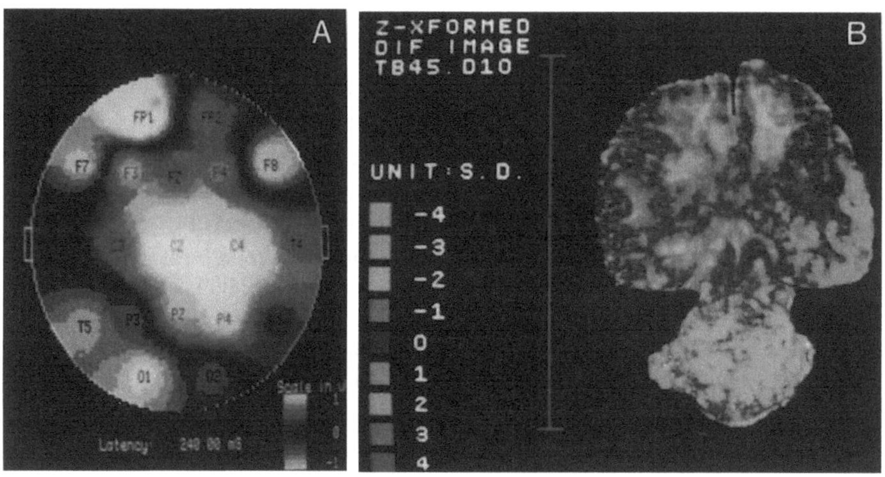

Fig. 10. (A) Topographic map of event-related potential, 240 ms after a click, in the same patient shown in Fig. 9, now pointing at a printed word that is the correct name for a picture. (B) Autoradiograph of a brain section from a split-brain cat, with a visual memory activated on the right but not on the left side. For details, see John elsewhere in this volume

based on the MEG measurements discussed during my earlier presentation at this conference.

These findings argue against connectionism. It is extremely implausible to contend that electrodes inserted into the brain at arbitrary sites, into which one pumps bursts of 200-μs, 10-μamp pulses, are selectively activating labelled lines, let alone that when sequential bursts are distributed among different populations, each burst excites a different labelled line which mediates the same memory and converges to the right place at the right time. The probability of that coincidence in many electrode positions in each of several cats seems rather low.

The same neuronal systems involved in these processes can release various modes of oscillations to the same input, and these modes of oscillations are facsimiles of evoked responses to absent events. We call this "differential generalization." In Fig. 8, V3 is an ambiguous visual stimulus. When V3 is mixed in a train of V1s, for which the appropriate behavior is to press one lever on a panel, as well as in a train of V2s, for which the appropriate behavior is to press another lever, this neutral event is responded to sometimes as though it were V1, and sometimes as though it were V2.

Operationally, the interpretation of V3 is identified by which lever is pressed. When V1's lever is pressed, the response to V3 looks like the EP to V1. When V2's lever is pressed, the response to V3 looks like that to V2, yet V3 is V3 in both cases. Clearly, the shape of the oscillation does not depend on the input. It is a mental product; and that mental product appears in many different places, at about the same time.

Figure 9 shows a topographic map of the scalp distribution of the event-related potential of an aphasic stroke patient at rest, 240 ms after presentation of a click.

Figure 10A shows the map elicited by the same click at the same latency, while that patient points to the name of the picture she sees, which of course activates a memory. To me, the difference between those two maps reflects the activation of millions of cells. In Fig. 10B, we see one slice from the brain of a cat remembering a visual form. That also involves millions of cells. A similar inference can be drawn from the map recorded from the surface of the head in the human and the map extracted from the radioautograph of a slice through the middle of the head of the cat. Let me tell you what they suggest to me.

I propose that afferent input via classical sensory pathways comes to a variety of primary receiving ensembles in multiple regions, via discrete pathways. These inputs activate a significant proportion of cells in many ensembles. These cells recruit other cells by local current flow comparable to or greater than that which we impose in our direct brain stimulation. A resonance is rapidly established between these ensembles, which are oscillating in a common mode. As the density of involved ensembles reaches some critical level, a state change occurs, such that these oscillations become directly coupled by a field that has a characteristic temporal pattern.

The discrete cells that are initially activated are analogous to the particles of dualistic physics, while the resulting coupled interactions are analogous to the waves of that physics. The coupled resonating fields constitute the system which represents the memory. Only that field has the capacity to integrate all of its constituent elements. None of the single neuronal elements can be informed about the whole field. Therefore the subjective awareness of the experience, or consciousness, must be a property of the emergence of a sufficiently organized field of energy within a circumscribed volume. I see no localizable alternative. I suggest the name of "hyperneuron" for this overall field, to indicate explicitly that it transcends the neurons from which it emerges. The individual neuron is important only insofar as it contributes negative entropy to the region.

Melnechuk: Before turning to our next speaker, I wish to mention, as possibly relevant to fast transcerebral electrical effects, a finding by Dr. Mark Ellisman, a neuroanatomist at San Diego, that the microskeletons of adjacent neurons are connected across the synaptic gap by macromolecular filaments, visualizable only because of new developments in electron microscope technology (Ellisman et al. 1985). If this finding is sustained, Golgi and the other reticularists may have been correct after all; and the brain may be a syncitium, if not at Cajal's cell level, at least at the molecular level, with interesting possibilities as a substrate for dynamic mechanisms.

Reference

Ellisman MH, Fields RD, Anderson KL, Deerinck TJ (1985) Transcellular filaments at synapses: a structural continuum linking synaptic membranes and cytoskeletons. Soc Neurosci Abstr 11(1): 646

Melnechuk: Dr. John emphasized gestalts in space. Our sixth speaker, Dr. Walter J. Freeman, emphasizes gestalts in time. Together, they honor the two categories of Immanuel Kant, who might not have been surprised by hemispheric specialization for sequences and for space. Dr. Freeman is Professor of Physiology at the University of California, Berkeley. He has been one of the leaders in replacing the stimulus – response view of Pavlov and Skinner with a dynamic view of the brain as interactive, predictive, and projective.

A Watershed in the Study of Nonlinear Neural Dynamics

W. J. FREEMAN

I also received, as Ross Adey did, a missive from Ted Melnechuk asking for the essence of the old, the new, and what the new implies. I am reminded of an anecdote about the American preacher-philosopher, Jonathan Edwards, who was asked how long it took him to prepare one of the speeches for which he was famous. He said, well, if it was a 10-min speech, it took him a couple of days; if it was an hour's talk, it took him a couple of hours, but if there was no time limit, he could start right now.

Having had a couple of days to think about this, I have decided that it is possible to say what I want to say within 10 min, and to make a contrast between the old and the new. I think it is appropriate to do so, because we are in fact, all of us, standing on a watershed, with one foot in the old, and another in the new. The essence of this difference, as I see it, is expressed in the comparison between the time-honored approach to the nervous system of the past century or more as a ballistic process, rather than what we are now coming to see as an interactive process.

When you look through the literature, particularly the models that people present of what they think their data mean, you see the representation of receptors with the same or differing modalities. The receptors feed into the nervous system with varying degrees of divergence or, alternatively, convergence, onto other sets of neurons. These in turn feed into the associational-integrative systems and eventually we have the motor output of effectors, the glands and the muscles.

There can be parallel input, processing, and output, if you like, but always there is this cascading forward flow, so that a stimulus put in undergoes a sequence of transformations but, nonetheless, always has a forward motion coming through to the output. The phenomena we observe we analyze similarly as sequences of impulses or as sequences of peaks in the evoked potentials, which we label as N1, N2, N3, P1, P2, etc. – a sequence of bongs, beep-beep-boop, like those you hear from a ball going through a pinball game. We do our behavioral modeling similarly, in terms of a forward flow from stimulus to response, in the classical manner that we have taught now for the last century or so, despite repeated criticisms from people like

Springer Series in Brain Dynamics 1
Edited by Erol Başar

John Dewey, Merleau-Ponty, J.J. Gibson, and others. However, we do have a side branch, which Roy John has just alluded to, which is a type of eidetic memory that can hold past stimuli. In appropriate circumstances, some present stimulus can cause a release from memory of the stored trace of the past stimulus. It comes out and initiates the response, but still the action is all in the forward direction.

It has been obvious for well over a century, in conjunction with the observations of the reticularists, led by Golgi, that this approach has serious deficiencies. Certainly, we do have the layers of neurons and successive laminar arrangements, but these cells have not merely their forward connections, they have dense interconnections, and thereby establish feedback relationships between themselves. Overwhelmingly in neocortex and paleocortex, though not in cerebellar cortex, these connections go in large numbers to other cells and back again. So when we ask, what are the implications of this new view, it is essentially that, as Roy John has already said, we cannot reduce the brain to single cells.

I do strongly disagree with one hypothesis that both Roy John and Ross Adey have put before us, which is that field potentials provide the the causal mechanism for the establishment of coherences among large populations. As I see it, that leads to confusion on our part from a failure to see the significance of synaptic interactions now viewed as themselves supporting a mass action property. In other words, as I see it, the electrical fields do have a role in trophic, regulatory, and stabilizing functions, but not in information transactions, and to emphasize these EEG currents is to allow the overlooking of the more fundamental synaptic actions seen as a mass property. Each one of these neurons transmits to thousands of others, those thousands then transmit back as well as to others, and then back again, so what results is the emergence of an interactive assembly.

We talk loosely about populations, but it is important to note that there is a distinct and highly important difference between the interactive assembly and the noninteractive collection. The noninteractive collection simply performs as, let us say, an average neuron, but the interactions provide for the emergent properties of the whole. The classical example that Aharon Katzir-Katchalsky proposed is the distinction between a collection of water molecules in steam and those crystallized in a block of ice, where properties exist in the ice, such as tensile strength, thermal conductance, and so forth, that cannot be defined at the level of the single molecule. The first implication of interaction is negative feedback; it supports the existence of these macroscopic entities in the macroscopic structure of the nervous system.

The second implication has to do with our analyses of the multiple events that we record. You have seen ample numbers, in terms of the brain stem auditory response, N2, P3, etc. Typically, they are analyzed in terms of a series of peaks. What we should be doing is considering these in terms of the properties of interaction among a network of neurons, such that the types of activity we can observe let us say from an excitatory set feeding into an inhibitory set and thereby receiving inhibitory input, can give rise to sequences of responses. When looked at in terms of the more simple-minded approach of one event after another, we have essentially a sequence of peaks, but what actually happens is something more in the nature of a sustained action or a damped oscillation.

This brings me to a major implication regarding education in the analysis of feedback systems. It is my view that neuroscientists as a class suffer from a distressing illiteracy bordering on agnosia about the nature of feedback analysis. Linear systems analysis in particular is one of the major intellectual achievements of this century. It is up there in a class with the double-helix model of DNA and the development of nuclear energy. In large part, it underlies the explosive development of our technology in the last 30–40 years.

It is also exceedingly useful in neurophysiology, in order to be able to go from PSPs, impulse responses, and open-loop responses among noninteractive cells, to the closed-loop condition where one can predict sustained long-term responses, oscillations, and what the frequencies will be, and then from measurements on those frequencies and decay rates infer back to what the strengths of synaptic interaction are. These are the results that should come from these oscillatory events. They seldom do. The implication is that there is a great need for extensive education in the nature of feedback systems, and how to go from a network of neurons, for which we know the time and space constants, to prediction of what it will do under appropriate circumstances, so that we can in fact verify that what we see is interpretable and not the effect of artifacts in our data reduction process.

Most of the responses that we see in evoked potential research have relatively high amplitude near the beginning and then taper off. Why is this so? Because that is what they must do in order for us to perceive them. We must have a system that is in a certain stable state. When we perturb it, it deviates from that state, and then goes back to that state. Only when it is back in that state can we perturb it again.

We require that the system go back to its original state as the basis for our repeated measurements, but we know that these responses can in fact last much longer than our repetition rate will allow, particularly when you train the animal; Roy John has given evidence for this, as have I. We can extrapolate to say that under some appropriate conditions, our evoked response should in fact last indefinitely; it should blow up and take the system to a new stable state.

With that we are getting into a domain which says that the nervous system, once put into a certain state, can generate its own activity over long periods of time. Beyond that, we can say that the nervous system creates its own activity patterns. This means that our old stimulus-response paradigm is as dead as it should be. What we have now is the emergence from the nervous system, or the creation by the nervous system, of its own internal frames of reference, its own internal dynamics.

The real object of behavioral analysis is to understand how our stimuli influence the interactions among these large populations of neurons such that their internally generated behavior is modified and, by virtue of their survival value, those modifications feed back out into the environment to change it in a way that is advantageous to the continuation of the animal.

Well, that is our watershed between the old and the new. We do not have stimuli going like billiard balls through the various relays of the nervous system. We have an interactive neural system that generates and sustains its own activity patterns. Stimuli act to modulate and modify the system, causing it to evolve from one state to another. The proper metaphor for the old view is the pinball machine. The proper metaphor for the new view is the dialog. Each experimental animal has a brain within it that has basically the same capabilities that our brains do, and we, when in our laboratories, are in an interactive dialog with each of our experimental animals.

Melnechuk: The seventh speaker is our chairman and host, Dr. Erol Başar, of the Institute for Physiology of the Lübeck Medical University. Dr. Başar has been a pioneer in applying concepts from modern physics to the search for integrative understanding of brain dynamics. Since he is so much at home in space – in phase space and frequency space – his right hemisphere must be well developed, but so too must be his left hemisphere, for he will address us in the fifth of his languages, which include Turkish, French, mathematics, German, and English.

Thoughts on Brain's Internal Codes

E. Başar

To be the seventh speaker will be very difficult, if what I say has to resonate with what was said by the previous speakers and be a complex reverberation. I hope you will permit me to postpone that chairman's task until I write the introduction and epilogue to the published proceedings of this conference, and instead let me tell you some of my ideas about how brains may work.

However, before I do that, I would like to make a few comments about this session. Some of you have asked me why these particular people were brought together under this title. I had always intended to have a concluding session called something like "Integrative Aspects," but Ted Melnechuk suggested calling it "How Brains May Work," and Bob Galambos suggested that its speakers include some of neuroscience's longstanding "mavericks" – that American expression for nonconformist thinkers that I learned from him last June in Sand Diego.

Now we have heard Bob Galambos say that if he were young, he would investigate neurogenetics, in order to learn how the brain of each species "wires in" its vital species-specific information.

To Ted Bullock, the brain in its entirety seemed to be like a university, with a number of strategies. In his pluralistic way, he compared several different species, advocated using several different windows – for example, single-unit recording of spikes, multi-unit recording, field potentials, EEG, and evoked potentials – and expressed dissatisfaction with mere network models. He seems to be working toward a global theory of the brain, as indicated in an important paper he published last year on quiet revolutions in neuroscience (Bullock 1984).

From Werner Leinfellner, we have heard that the field model for neural communication proposed by Ross Adey and Roy John, which Dr. Leinfellner calls "the internal language of the brain," is consistent with philosophical concepts of perception, cognition, and language.

Ross Adey, in discussing his important early work, told us of the obstacles he met in using the EEG as a building block for this new type of brain model.

Springer Series in Brain Dynamics 1
Edited by Erol Başar

Roy John put an emphasis that I like very much on the complementary dualism of particle and wave models in fundamental physics as an analog of the contention between synaptic and field models in neuroscience. As you know, the first quarter of this century saw a very important new development in physics, quantum theory, which described some aspects of fundamental phenomena in terms of particles and other aspects in terms of waves. Both concepts have been successful, and the marriage of both concepts has been even more successful.

Walter Freeman is more inclined to the synaptic "particle" than to the "wave" model of communication in the brain, although he does consider field potentials to be significant in other roles. He made the important point that the usual form of the synaptic model seriously needs revision, from its old "ballistic" view of unidirectional information flow to a new "interactive" model of recursive information flow generating mass action.

I now repeat what I expressed in my introductory talk and symbolized with an illustration by Escher called "Relativity," that *all* of these windows on brain dynamics are important and necessary. The more windows we look through, the more mechanisms we will discover. Maybe, as I remarked to Roy John yesterday, this session will encourage us to hold future workshops for comparing the views of brain function that he has likened to the particle and wave concepts, because a thorough comparison of the data that support the two views could lead to a better understanding of the brain, as similar comparisons have benefitted physics.

Now for a few personal ideas about how brains seem to work. Bob Galambos has been a physiologist for 40 years, but I have been one for only 20 years, so, unlike him, I cannot yet say what other thing I would do if I had it all to do over again. I can only present ideas from the field I am in, the field of resonances. These ideas are a response to the question asked me last winter by several visitors to my laboratory, "how might the brain use these various resonance phenomena?"

You may recall that in my lecture, I showed that when the brain is excited with various stimuli, it usually responds with waves in the frequency ranges of 4 Hz, 10 Hz, 20 Hz, and 40 Hz. I have called these responses "invariant resonances" or "invariant resonance modes." Such resonance responses also follow the application of various modalities of sensory stimulation and the use of the paradigm of event-related potentials. From these resonance phenomena, I have been able to derive excitability rules, by using the excitability concept of several other physiologists, especially Dr. Sato, who first explained brain excitability.

Now let us consider three different brain structures or networks, A, B, and C, which have excitabilities in the four frequency ranges of 4 Hz, 10 Hz, 20 Hz, and 40 Hz, which I will call alpha, beta, gamma, and theta. Let us assume that structure A can be excited and can excite structure B, which in turn can excite structure C. When structure A excites structure B, there will be a transmission of signals in one of these frequency channels from structure A to structure B; and structure B, if not already excited in a different frequency, will then excite structure C in the same frequency range (see Fig. 1).

Let us now give another name to the specific resonant modes, calling them "passletters," because we have seen that a brain structure into which activity in the 10 Hz frequency range is entering will, by an excitability rule, resonante in that range. In this case the passletter would be the excitability at 10 Hz. A second pass-

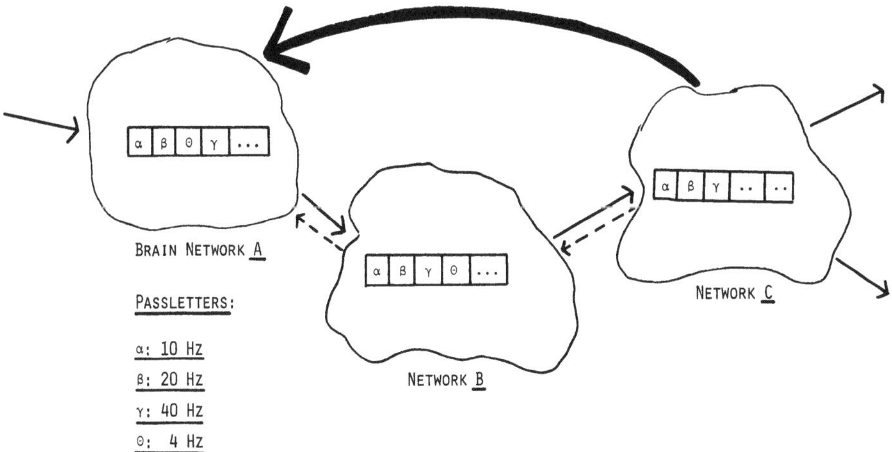

Fig. 1. Schematic explanation for passletters

PASSLETTERS
α, β, γ, Θ
PRESENT
INVARIANT MODES

HOWEVER

COMBINATION OF PASSLETTERS DURING TRANSFER OF MESSAGES THROUGH VARIOUS BRAIN NETWORKS MAY CREATE A GREAT NUMBER OF ELEMENTARY (AT LOWER LEVEL) OR COMPLEX (HIGHER LEVEL) PASSWORDS:

α
α β
α β γ
β Θ γ
.
.
α β Θ γ δ ... ETC.

IN TURN, THIS FACT MAY CREATE AN ENSEMBLE OF PASSWORDS FOR HETEREGENOUS MESSAGES (OR FOR COMPLEX SENSORY/COGNITIVE INPUTS) BY USING REVERBERATION OF PASSWORDS, POSSIBLY VIA AUTOEXCITATION.

Fig. 2. From passletters to passwords

letter is at 20 Hz and a third passletter is at 40 Hz. The term "password" or "pass characteristic" has already been used by Barlow, who in 1961 tried to develop a theory of passwords and pass characteristics in neural networks. My theory is somewhat different from his but I use his expressions.

According to empirical results, alpha, beta, and gamma resonances do occur in the brain. Therefore we can suppose that *the passleters represent resonant modes that help to optimize signal transfer between various brain structures.* This idea of a way in which brains may work was suggested by the excitability rules that I derived from experiments (see the paper in this volume).

Now I will go further and speculate that the passwords alpha, beta, gamma, and theta represent resonant modes that are *invariant.* However, passletters may perhaps combine during the transfer of messages through various brain networks to create a greater number of less elementary or complex combination which we may call passwords. For example, there can be only alpha transmission in all three of the structures A, B, and C, or there can be a combination of alpha and beta (passwords α, β). A still more complicated password would be alpha-beta-gamme (α, β, γ), and so on (see Fig. 2).

As Ross Adey said, the information capacity of a structure can be important, especially when it is busy, for if the information capacity of structure B is limited, or if structure B is busy at the time, the transmission in one or another of these special channels will not be optimal. This fact in turn may lead to the creation of an ensemble of passwords for complex sensory-cognitive inputs and other heterogeneous messages, by using reverberations of passwords, possibly via auto-excitation. How might such auto-excitation occur? One possibility is that structure A could excite structure B again, as in limbic system networks, for example. In such a case, *it is possible that as a consequence of a preliminary input signal, the ensemble of neural structures A, B, and C could start to reverberate.* Such reverberation would be possible only if all the structures had the ability to resonate in the same frequency channels.

Finally, one can speculate further about why it would be advantageous for a brain to transmit by means of resonance phenomena. I have thought of one possibility. The simplest invariant transfer functions are probably represented in the brain by resonant networks, but this *fixed hardware may create a much richter array of useful software.* That is, it could be an economical principle of the brain to achieve good internal communication first by using a small number of similarly structured network as channels for a small number of frequencies, and then to combine the elementary passletters into a large number of compound passwords that could transmit a wide variety of complex patterns. This idea, to echo Ross Adey's phrase for his early belief that the EEG is important in information processing, is my castle in the sky.

References

Barlow HB (1961) Possible principles underlying the transformations of sensory messages. In: Rosenblith WA (ed) Sensory communication. MIT Press, Cambridge, pp 217–234

Bullock TH (1984) Comparative neuroscience holds promise for quiet revolutions. Science 225: 473–478

Hofstadter (1979) Gödel, Escher, Bach: An eternal golden braid. Vintage, New York

Sato K (1963) On the linear model of the brain activity in electroencephalographic potentials. Folia Psychiatr Neurol Jpn 17: 156–166

Melnechuk: For our final scheduled talk we will hear again from Dr. Walter Freeman, who has accepted Dr. Başar's invitation to give us a brief methodological overview covering the various techniques that were or should be mentioned at this conference.

Overview of Methodology

W. J. Freeman

I was asked at the last minute, so to speak, and since, as I have already told you, a 10-min talk takes a couple of days to prepare, I cannot guarantee to be that brief.

Let us consider our present group as an ensemble that is in the process of evolution, so we would like to look at what Waddington referred to as the "epigenetic landscape," to get some kind of overview as to where we are and where we think we are going or would like to be.

We are talking about *brain dynamics.* I take the term "dynamics" very seriously, as a statement of process, of an ongoing flow of energy, of a changing form of structure based on the fluxes of energy. In the classical language of physics and chemistry, what we need to do is to describe these dynamics. In order to do so in a meaningful way, our first reliance must be on the study of behavior. We have to know what the animal is doing during the times that we are making our measurements. For this, we can use, and do use, the techniques of *behaviorism.*

As an explanation, as a philosophical basis for understanding brain function, behaviorism is bankrupt. It has no value for understanding either behavior or the brain, but its residue of functional techniques for conditioning are invaluable as tools for us to use, to put animals – and people, for that matter – into states that will give us stable, reproducible kinds of brain activity that we can then hope to measure and eventually to understand.

Similarly, *computers* were born as a mistaken metaphor or belief about how the nervous system operated. Von Neuman took his language from Warren McCulloch at a time when Warren had a mistaken idea of how nerve cells worked. This has now become the basis for artificial intelligence. Again, that approach, of rule-based machine intelligence – as Gerhard Werner described it for us so eloquently – is bankrupt as a way of trying to understand the brain. But from it we have an incredibly rich residue of hardware and software that we can use to our advantage for the analysis and reduction of our data.

The third element that we need is an understanding of *brain structure*, including *gross anatomy*. Alan Gevins had a beautiful illustration of gross anatomy in relation to function in the NMR sections of one of his subjects. He can now get an accurate placement or a correlation, as it were, between the location of his recording sites and

Springer Series in Brain Dynamics 1
Edited by Erol Başar

the underlying structures that he believes are responsible for the electric fields. Obviously we have to know the *hodology* – that is, the connectivity: which neurons are connected with which others, which centers are connected with which others, and in what manner. Also, as Bob Galambos has repeatedly emphasized, we have to know something about the *chemical anatomy* – the aminergic systems, the cholinergic systems, the ways in which the brain states are expressed in terms of concentrations and distributions of transmitter and modulator chemicals.

Now we come to what occupied a major portion of our deliberations in the course of the meeting – namely, making *observations*. What we do in this procedure is essentially to look at the *energy gradients* that exist during normal function of the brain. The brain is a structure undergoing the expenditure of energy, and in the course of that expenditure it generates *electric* and *magnetic* fields.

Most of the information we have relates to electrical signals, from the passage of current that nerve cells generate to communicate from one part of a nerve cell to another, as an active electromotive force releases metabolic energy. Of course there are both dendritic potentials and action potentials. There is an active membrane site of generation and a passive site of action where the membrane current again crosses the membrane at some other part. Since current always flows in closed loops, it flows back through the extracellular space. With *electrical recording*, we can pick up the potential differences that are created by this flow of current; with *magnetic recording* we can pick up the colinear flows of current by virtue of their tight, high-density magnetic fields – relatively high, because they are indeed miniscule. These are the two main approaches for observables that we can in fact detect in studies of the active brain.

There are also other approaches. One is to measure the heat, negative or positive. When the action potential occurs, there are conductance changes, and ions flow down passive membrane gradients, in what is in essence a flow of ions down their concentration gradient, as in the expansion of a gas. Therefore, with an action potential, one has a cooling off, followed by heating caused by a process of metabolic restoration. As far as I know, the only measurement of the negative heat of nerve was made on isolated axons. We have little else to say about this, because the amount of heat that is produced is not generally a measure of brain function; neither is entropy. In physical systems, one always wants to know what the entropy is. Brains are always open systems, and the disposal of excess heat and entropy is so effective that the laws of thermodynamics put no limitation on normal brain function. Heat stroke is another matter.

Now we come to a variety of *modulations*, by the nervous system, of gradients imposed on it. One of the simplest and most straightforward techniques is to pass electric current across the tissues and measure the tissue resistance and reactance, in order to derive the *impedance* vector. If this is done properly in conjunction with anisotropicities, we need an impedance tensor. However, for our purposes, the noise level is so great that we are satisfied for the most part to assume a homeogeneous tissue resistivity. We do not even worry much about the reactance. In this connection, Ross Adey has just reminded us that one can also apply high-frequency ratio waves to the nervous system and study the effects of their modulation, effects we pick up with secondary electrodes.

Referring again to the magnetic field, there is also the technique of *nuclear magnetic resonance imaging*, which imposes a magnetic field on the brain, and then manipulates the field with radio-frequency currents. By using appropriate tags, one can study the concentrations and fluxes of a broad variety of substances in the brain. This is an area that has not been touched on at this conference but is quite likely to enrich our understanding extraordinarily in years to come. Neural activity, after all, is not carried only by flows of electric current manifested in electric and magnetic fields. It is also carried by concentrations of transmitter substances, and we can learn a great deal by measuring the dynamics of brain activity at that level and not merely in terms of synaptic and action potentials.

There are also a variety of delayed measures, deriving from the fact that the nervous system has to pay for its *metabolic* upkeep. We can measure the aftereffects of activity in terms of oxygen consumption and secondary effects on the partial pressure of oxygen in various part of the brain. This has been done since the 1950s, when Grey Walter measured them with platinum electrodes. It is currently done by imaging regional blood flow with any of half a dozen techniques, and also by the 2-deoxyglucose method, which in fact looks at the barn long after the horse has vanished, but has its uses.

Of all our many tools and techniques, the most important are the *models* we construct with which to interpret our observations. Let me recall a remark I recently quoted to Roy John, made by Lord Kelvin, I believe, at a meeting of the Royal Society in the 1880s when they still met as one group of biologists, physicists, and the rest. After a biologist had given a paper that reported the different number of scales there were on two species of fish, Kelvin stood up, said, "All science is measurement, but not all measurement is science," and sat down. Of course what he meant was that the significance and value of measurements depend directly on the degree to which they are related to a larger body of data. Such relating is done with models, as Dr. Orpheus said in the poem we heard at the opening of this session.

Again emphasizing *dynamics* as the theme of this conference, the classical language used for the description of dynamics in physics, chemistry, and other areas has been and still is *partial and integral differential equations* or their equivalent, in a number of variants. Similarly, when we approach brain dynamics in terms of actual observables, we need models expressed in differential equations to describe the *rates of change in time and space* of these varying quantities. We also need to specify the *time and distance scales*. Here we have a range of options. I think the best one is to opt for the particular scale that measures the psychological here-and-now of ourselves and our animals; that is to say, the time span of a second or of very few seconds, and the space of the immediate surround of an animal. In a way, this scale is brought into the nervous system by the receptor arrays, which characterize the dynamics of the nervous system in terms of seconds and millimeters.

We can in fact go to much smaller levels, as Ted Melnechuk alluded to in mentioning reported molecular connections between neurons, but this is getting into microstructure. It seems to me that we want to get at the dynamics of the brain at time scales that relate to the transactions that the brain has with the environment, which do not occur at the molecular level, except –.

Adey: Why do you say that?

Freeman: Because there is an inability, as I see it, for the larger context of the brain to identify the events occurring at single molecules, or even at parts of neurons or even single neurons, and to amplify these into the large-scale activity of the nervous system. A brain-behavior event must last in the order of 50–100 ms and cannot exist in an action potential, no more than an action potential lasting about 1 ms can exist in the few microseconds of a molecular event in the membrane.

Going on from considerations of scale, we also need to specify the *dimensionality* of the system under study: how many variables we want to observe, or think we are able to observe, or, in fact, must observe, in order to describe or account for the activities we are studying.

Models take numerous forms. I will classify them somewhat arbitrarily into *statistical, linear*, and *nonlinear*. Statistical models essentially are used for measurement, in order to define standard errors, degrees of confidence, and so forth. Statistical models are also used in classification, when, having obtained a collection of evoked potentials or segments of EEG or trains of units or of patterns of chemical reactivity, we want to classify them and apply statistics for that purpose – similarly with correlations. None of this statistical modeling has anything to do with a causal analysis; it is simply a statement of what goes with what, what is distinguishable from what, and how well, and how certain we are of the numbers that we have derived.

The entry into dynamics as I see it is best done in terms of the construction of differential equations, and the place to begin is with *linear* differential equations. Jim Wright and Bob Kydd gave a beautiful example of the manner in which such equations can be developed and used. In order to use them effectively, it appears to me that the first step is to demonstrate, in accordance with the superposition principle, the additivity and proportionality of responses in comparison to inputs. This is an old physiological criterion. Sherrington used it, in conjunction with what he called the "algebraic" function of the nervous system, and his method, paired shock testing, is still commonly used.

Once one has defined a linear domain, by use of superposition – and this is possible for a broad range of neural dynamic systems, including neural masses, single neurons, parts of neurons, and membranes – then it is necessary to derive and specify the characteristic exponents and frequencies. This is done essentially by the diagonalization of the matrix derived from measurement to get the eigenvectors and the eigenvalues. When you use the fast Fourier transform, this is implicitly the kind of information you are trying to extract from the data, though in this case it is a rather limited kind of information, having to do with what we call imaginary exponents. In using this, one all too commonly does not worry or even think about whether or not the the data that are being treated in this way come from a linear system or from a system that is performing in a linear range. Violations of the underlying assumption of linearity produce some strange and I think innately uninterpretable data.

The range across which linearization can be achieved, or in which a system can be put into a linear range of function, is sometimes excessively restricted, but one can, by systematic variation of parameters, establish a sequence of ranges within which linearization can be undertaken. Then, the use of *describing functions* can expand the domain in which linear analysis is feasible. This is such a powerful technique that by no means should it be neglected. It constitutes our vestibule to the real work, where we finally get down to what constitutes the serious approach to brain dynamics – the study of *nonlinear* systems.

In coming to full-blown nonlinear systems, whether subsets of neurons, single neurons, or assemblies of larger scale, up to and including whole sensory systems, whole motor systems, the limbic system, ultimately the whole brain (though God forbid we should attempt to try that in the near future), we have already to know what the connectivity is – how these nerve cells are connected together. We need to know their open-loop time and space constants, the ranges they communicate over, their membrane passive time constants, their dendritic cable delays. We need to know furthermore what the nature of the nonlinearity is – and I should say "nonlinearities," since they are multiple: are they static, are they time-varying, are they dynamic, as in the Hodgkin-Huxley equations? We need to know where they are located in the topology. This is the approach that is necessary for a serious and deliberate simulation of these complex networks. Clearly one has to start off with so-called simpler systems, which are not found merely in invertebrates; in fact, that phrase, "simpler systems," has been excessively devoted to invertebrates, which, if anything, are more complicated than vertebrates.

The goal of nonlinear analysis is a description of the expected properties of the neural system that we are observing. We describe these in terms of the *phase space*, which consists of a specification of the *attractors*, which is simply a statement of what the characteristic patterns of change are that these systems tend to follow; the *basins*, which are the initial conditions that allow the system to change toward one or another of the attractors; the *trajectories*, which are the sequences of changes – whenever you record an evoked potential, it is basically a display of a trajectory in nonlinear phase space; and finally the *bifurcations*, which are switchings from a tendency to follow one trajectory of change to a tendency to follow a different trajectory. The latter basically indicate that the nervous system has made some structural change; learning may facilitate the ability of the nervous system to do that.

Finally we have the need to *test* these models. A key feature here, which is often overlooked, is that models need not be born whole like Pallas Athena from the forehead of Zeus. We should rather recognize that we approach our goal in small steps, often wrong, but that with each experiment we do we learn a little more, we become a bit more enlightened, so that we can hope to optimize our understanding.

The first thing to optimize is not the model per se, it is the *measurement*. Alan Gevins has pointed out an excellent way to do this, which is first to test the measurements one makes against the observations one has made on behavior, in order to classify the behavior in respect to these measurements, and then to optimize the classification by adjusting the filters or whatever other data-reduction processes one has used. In other words, the most important question with the data is whether or not one has done the *reduction* right, for if it does not fit, most likely one either has an artifact or one did not analyze the data correctly to begin with. Only when one has explored that issue thoroughly can one go on to change the model and, one hopes, improve it.

The improving of one's model is an evolutionary process. However, in the phraseology of Stephen Jay Gould, this evolution is not continuous but "punctate", because what happens typically is that we settle on some model which seems to satisfy us for a period of a month or a year or a decade and we think that now we have got the truth, when in fact all we have got is a false energy minimum, a local region of stability that sooner or later will be upset by some physicist with a slide in his hand.

Melnechuk: Thank you, Dr. Freeman. Does anyone have a necessary comment? Dr. Adey.

Adey: Mr. Chairman, can Dr. Freeman say more about why he thinks the time course of molecular and membrane events is irrelevant to the organization of nervous function?

Melnechuk: Dr. Freeman?

Freeman: That is not quite what I said. What I was trying to say, and I did not say it correctly, is that the level at which we can find the information that relates to the transactions of the animal with the outside world is not to be found at the level of the time scales of single molecular reactions and single action potentials. I do not deny that photon capture is exceedingly important in the retina as a prelude to the process of identifying a picture as that of your grandmother. What I am saying is that the information available to the neurophysiologist is not going to be found by measuring any capture time of rods in the retina. It is to be found in some other process that extends over a period of something like 100 ms.

Adey: I could not agree more that molecular events do not directly yield an understanding of perception, any more than the Heisenberg equation can be directly expanded into an understanding of chemistry, and yet we know that the Heisenberg equation implicitly contains all of chemistry, as we know that the molecular level of brain dynamics is at least a necessary substrate, both for the perceptual level that Dr. Freeman has spent a lifetime studying in the rodent olfactory bulb and its associated cortex, and also for the level of human mentation.

If I may, I will epitomize findings made in our own and other laboratories in support of the idea that molecular processes do pertain in their time courses and even in other ways to cerebral mechanisms of cognitive processes, and that, counter to another doubt expressed by Dr. Freeman in his first talk in this session, wave mechanisms such as the EEG represent a fundamental aspect of cerebral information processing.

In brief, low-frequency EEG-like gradients impressed on neurons can (a) modulate behavioral states; (b) entrain EEG rhythms in monkeys, cats, and rabbits; (c) modulate neuronal firing patterns; (d) alter neuronal membrane potentials; (e) increase neuronal efflux of neurotransmitters and calcium; and (f) modulate the activity of intracellular enzymes that control messenger functions, metabolic energy production, and the synthesis of molecules involved in cell growth. EEG-like gradients have all these presumably related effects despite being six orders of magnitude smaller than the membrane potential gradient. This is inexplicable in terms of classical models based on equilibrium considerations. The windowed character of these responses with respect to frequency and intensity also argue their nonequilibrium and nonlinear nature.

In addition, low-frequency oscillations in the membrane potentials of cerebral neurons produce currents in the pericellular fluid that influence adjacent cells. Leakage of such oscillations plays a prime role in the genesis of the EEG in brain tissue. Intercellular communication based on such oscillations is a clear possibility.

Let me express two more thoughts regarding the connection of molecular time scales to those of cognition. First, I am mindful of Holger Hyden's remarkable lectures at the University of Birmingham in England in 1973 (Hyden 1973), in which he emphasized the importance of electrically sensitive macromolecular systems that would be distinct from synaptic, connectionist organization in the brain, and would be responsible for the rapid establishment of a gestalt over wide cortical domains as an essential step in the establishment of memory traces and in the processes of recall. He further emphasized that the only class of macromolecules with this capability are proteins, and the turnover in brain proteins may take months.

Secondly, the presence in the audience of Dr. Babloyantz reminds me that she works with Dr. Prigogine, who received a Nobel prize for work which, amongst other things, was related to dynamic phenomena like those of limit cycles, where very fast pacing at microsecond to nanosecond to picosecond levels determines much slower nonlinear processes, such as Zhabotinsky-Belousov reactions. These rapid phenomena are inherent in the way we believe that longer biomolecular cycles are determined.

For such reasons, modern cell biology in general and the nonlinear electrodynamics of biomolecular systems seem directly relevant to the problems of cerebral physiology and organization at various structural levels, and, beyond that, by implication and extrapolation, appear to some of us to have great relevance to the organization of behavior and mentation.

Melnechuk: You have just demonstrated why we had seven minds here – for we hope that, as the superposition of the seven colors of the spectrum yields white light, the already partially overlapping views presented here today will before long converge into an illuminating unity. In that regard, and with particular reference to your point about the coupling of small fast phenomena to large slow ones, I wish to mention the synthetic idea of Dr. Arnold J. Mandell, which he has since developed much further (e.g., Mandell 1983), as he once expressed it in the title of a book chapter, "Vertical Integration of Levels of Brain Function Through Parametric Symmetries Within Self-similar Stochastic Fields" (Mandell 1980).

References

Hyden H (1973) Changes in brain protein during learning. In: Ansell GB, Bradley PB (eds) Macromolecules and behavior. University Park Press, Baltimore, pp 3–26

Mandell AJ (1980) Vertical integration of levels of brain function through parametric symmetries within self-similar stochastic fields. In: Pinsker HM, Willis WD (eds) Information processing in the nervous system. Raven, New York, pp 177–197

Mandell AJ (1983) From chemical homology to topological temperature: a notion relating the structure and function of brain polypeptides. In: Başar E, Florh H, Haken H, Mandell AJ (eds) Synergetics of the brain. Springer, Berlin Heidelberg New York, pp 365–376

Epilogue

Brain Waves, Chaos, Learning, and Memory

E. Başar

Nine months went by after the Berlin conference. Then, at the beginning of May 1986 Walter Freeman visited my laboratories in Lübeck and gave a lecture on his recently obtained results on the nonlinear dynamics of brain potentials and strange attractors. At the beginning of June, I spent some time in La Jolla (San Diego) and had further fruitful discussions with Robert Galambos and Theodore Bullock about the results of the Berlin conference and especially the final workshop "How Brains May Work." Theodore Melnechuk had finished his correspondence with the seven speakers of the workshop and we discussed several issues that helped me to reconstruct several points that emerged on the last day of the conference. In Palo Alto I met Roy John and, during an interesting trip to San Francisco and a walk in Muir Woods, we discussed his conecept of the "hyperneuron" and the possibility of resonant propagation of fields in the brain.

Study of EEG Needs Large Windows

Having been together again with most of the panelists in Berlin, I am encouraged to write this epilogue, which in no way tries to summarize the conference. I think that this last task is the job of the conference participants and of the reader of this volume. Freeman had, anyway, given an excellent summary of relevant methodologies. My duty in writing these lines is to try to catalyze a large new effort to demonstrate the extent to which studies of the dynamics of brain field potentials, of the EEG and event-related potentials, may contribute to integrative neuroscience. The results of the conference showed us that this contribution could be very large and even explosive. But this development has a price: the research scientist should not use narrow windows. He should not analyze just the EEG or just power spectral analysis or just averaged evoked potentials, or emphasize only a particular pathway or group of neurons in a given brain structure. As Theodore Bullock would say, we have to use not only one tool but a battery of tools, we have to use several strategies.

Several examples of this kind of study are presented in the present volume. In order to give a few examples, I mention the work of Petsche who, years ago, started to explore the brain with an analysis of electrical activities in the Papez circuit and who now explores with his coworkers layers of the cat cortex on one hand and cognitive potentials of the human brain on the other hand. Wright and Kidd, in their studies on biochemical and physiological aspects, have introduced new theoretical ideas in relation to theoretical physics. Nunez injected highly important thoughts about neurophysics and its possible contributions to understanding brain function.

Springer Series in Brain Dynamics 1
Edited by Erol Başar

Already decades ago, the five scientist panelists, namely Adey, Bullock, Freeman, Galambos, and John, started their research activities by using not only spike analysis at the cellular and neuroanatomical level but also the interpretation of field potentials and the EEG as an indispensable window for understanding integrative functions of the CNS. The reader who will go through the papers and panel contributions of these scientists will find in their narrative explanations the key to the emerging new trend in brain research which presents excellent windows to supplement familiar connectivistic models. Those who know history can better predict the future.

The difficulties mentioned by Adey in the early years of cognitive EEG research have existed in every young science. Research on field potentials has survived despite the slogan "EEG Epiphenomenon." Its survival was based not only on the fruitful applications of research on event-related potentials in cognitive neuroscience and its tremendous application in clinical science but also on the fact that connectivistic models have some great shortcomings and limitations despite their postulated precision and reliability. Especially in cognitive neuroscience, field potential research seems to offer more physiological significance because it affords the possibility of performing behavioral experiments without application of anesthetic agents and of using the multichannel approach. On the other hand, the important new trend demonstrating that the EEG is not a *pure noise* but that it reflects a *chaotic behavior* stemming from deterministic signals should now cause much thinking on the part of those scientists who still link the EEG to the expression "epiphenomenon." It now seems probable that the EEG reflects one of the most basic properties of brain signalling.

All Science Is Measurement...

At this point I want to repeat Freeman's recall of a famous statement made by Kelvin around 1870: "All science is measurement, but not all measurement is science." Most of the speakers in the conference confirmed this view by discussing their data not just as measurements but as the basis for functional interpretations or for possible and plausible future applications (see for example the contributions by Heinze et al., Maurer, Picton, Saermark, Hari, and Hoke). In this epilogue, it is quite impossible to review all of these papers, which should be studied in detail. However, just to give an example, I refer to the papers by Heinze et al. and by Gevins to mention the vigor of critical thinking in the new methodologies. In the chapter "New Scopes at the Cellular Level," the contributions of Speckmann and Eckhorn present important new trends at the cellular level. Although the epilepsy model merits important consideration in brain research, we did not have an opportunity to discuss this issue in the workshop.

Memory and Dynamic Memory

There was a strong trend at the conference not to separate the sensory and cognitive processes of the brain as different functions. This trend was maintained during the conference through several contributions and their discussions. In the last session, this trend converged with another trend that related cognitive processes to such key concepts as those of dynamics, resonance, network, coherent fields, memory, and connectivistic (or nonconnectivistic) approaches. I did not ask the discussants to mention these key words repeatedly. I interpret this spontaneous behavior as a sign of an urgent scientific need to discuss these topics. Bergson (1939) tried to define memory as follows: "It is memory which makes the past and future real and therefore creates true duration and true time."

Karl Lashley spent 30 years of his industrious life striving to discover the nature of the "memory trace" in the animal brain, beginning with experimental investigations of the rat's brain and ending with the chimpanzee. He was hunting for the engram, the record; that is to say, "the structural impression that psychical experience leaves on protoplasm." He failed to find it. Lashley's experimental work in animals tried to identify the locus of memory in the cortex. He trained rats to run mazes and then removed various cortical regions. It was not possible to find a particular region corresponding to the ability to remember the way through a maze.

Evidence for the opposite point of view was developed in the late 1940s by the neurosurgeon Wilder Penfield who examined the reactions of epileptic patients whose brains were being operated on to remove epileptic foci. By inserting electrodes into various parts of their exposed brains, he electrically stimulated neurons or neuron assemblies. Penfield (1975) found that stimulation of given neural structures created specific images or sensations in the patient. These artificially elicited impressions were of events from some earlier time of life, such as a childhood birthday party. The set of locations which could trigger such specific events was extremely small. Penfield's results opposed Lashley's conclusions; they seemed to imply that local areas are responsible for specific memories. Did Lashley's interpretation point out the possibility of distributed memory in the brain, as described, for example, in one extent, by Roy John?

Because of such discrepancies in the study of the higher functions of the brain, several scientists have preferred a cell-by-cell analytical approach to brain function. The success of Hubel and Wiesel (1962) enhances the tendency of this type of analysis, which is certainly an extremly important tool. They showed that neurons in the lateral geniculate nucleus and visual cortex of cats were sensitive to the position, orientation, and motion of stimuli in the visual field. Because of the discovery of cells in the visual cortex that can be triggered by stimuli of ever-increasing complexity, some people have wondered if there exists in one's brain a "grandmother cell" that fires if and only if one's grandmother comes into view. Does such a "superhypercomplex cell" exist?

Now, I want to consider the issue of learning and memory at the cellular level. There have been significant advances in this area for more than 20 years. A recent article that reviewed several memory processes and described a molecular framework (Godet et al. 1986) led to three specific conclusions:

1. The same extracellular signal (neurotransmitter) that initiates short-term memory also initiates long-term memory
2. By contrast, there are genes and proteins necessary for the cellular mechanisms underlying long-term memory that are not required for the mechanisms of short-term memory
3. Whereas proteins and mRNAs critical for short-term memory are pre-existing and turn over slowly, certain proteins and mRBAs necessary for long-term memory must be either induced or, if constitutively expressed, only transiently accessible to modification

The emphasis of Galambos on mechanisms at the genetical level should be emphasized as a trend indicator in memory-related cognitive process. Remarks by Galambos illuminate necessary future trends. What can one make of all this?

Field Potentials and Memory

It would be an extremely difficult task for me to try to add a new synthesis about memory at the end of such a conference, which was not a conference specifically on memory models. However, I think that the reader of this book can develop an interesting insight into a new window on learning and memory by reading the articles and discussion contributions of Adey, Freeman, Galambos, and John. I strongly recommend that these articles be read and reread. I am sure that these readings would help to develop some new avenues in the direction of dynamic memories in the brain, which can now be described only globally and not yet in a uniform manner. The use of such expressions as superposition of fields, common modes, eigenvalues, resonances, wave-particle properties (see John in this volume), and solitons reminded me of the discussions of Einstein, Heisenberg, Bohr, and Schrödinger in the first decades of this century. The fruitful discussions by Adey and Freeman and the remark of Freeman about the contributions of John and Adey expressed attitudes similar to those expressed in discussions of physics at the beginning of the century.

As I was myself reading and rereading the chapters written by the workshop participants, Volkmar Weiss, an East German scientist with whom I have been corresponding for 1 year, sent me a recent article entitled "Memory as a Macroscopic Ordered State by Entrainment and Resonance in Energy Pathways." This highly interesting but very speculative article helped me to put together some of my own speculative ideas about the brain's use of resonances in information processing.[1] I will try to describe these thoughts in the following pages.

Resonances and Coherent States

Adey was the first investigator to recognize the importance of correlating the EEG with cognitive processes on a quantitative basis. His pioneering work consisted of

[1] Segments from the paper by Weiss, along with some thoughts on order, chaos and dissipative structures, are given in the appendices to this epilogue.

studies that sought to reveal a constancy of brain wave responses in relationship to particular task performances. Adey now indicates the need to develop concepts for the understanding of the nonlinear electrodynamics of transductive coupling. John assumes that a resonance is rapidly established between neural ensembles that are oscillatory in an common mode. In other words, John and Adey suggest that field potentials provide the causal mechanism for the establishment of coherences among large neural populations.

In my long-standing interpretations of sensory evoked and event-related potentials, I have considered the EEG as the key operator, which undergoes a quasi-structural change, going from a usually incoherent state to a coherent state. I tried to describe precisely how brain structures involved in signal processing evoked by a sensory stimulus would usually go to a lower level of entropy (Başar 1980, 1983).

For the time being, it is difficult and speculative to describe universal models of a dynamic type of memory. However, we can use a chain of steps towards such models. If we assume that fluctuating field potentials of the brain reflect properties of dissipative structures (neural oscillatory networks as dissipative structures), we are led to speculate about the existence of yet unknown simple communication mechanisms between structures. Discussions that arose in the workshop "How Brains May Work" mentioned some relevant experimental data.

I also want to mention again the expression "common modes" used by several participants in the symposium, namely John, the group of Wright and Kidd from New Zealand, and myself. The reader who is oriented in theoretical physics would immediately mention the Schrödinger equation and the eigenvalues that correspond to common modes. Fitting with this, the speculative remarks of Weiss (1986) link various models presented at the conference. The brain is not just a physical system, and it is also not just a biological system; the brain is a most complicated biophysical system, in which physical laws and neuronal structures are interwoven. Accordingly, one speculative step that can be taken is to mention that its resonating fields may reach very high electrical amplitudes in comparison to the usual fluctuations of field potentials. If in the brain we had to deal with a wave equation, say the Schrödinger equation, then we could also speculate as to the existence of tunneling effects, which would in turn create unimaginably high voltages and favor extracellular communication.

Such models are now very tentative, although some of the experimental data presented throughout the conference indicates that such phenomena are beyond possibility. Yet a clear theoretical explanation is missing in spite of data that should be taken into consideration very seriously.

If such mechanisms do exist, they could permit unimaginably speedy information transfer in the brain, and the related dynamic associations and/or dynamic memories would resonate, couple, and integrate all of the known types of memory. I know that here I go a bit too fast and too far. However, I feel that after participating in the brainstorming afternoon session of August 22, 1985, in Berlin, it is my duty to express what I think.

Can the brain use for communication what are probably the simplest physical mechanisms, such as dissipative structures? What do we know about molecular mechanisms? What did physicists know about gravitation or about far fields at the turn of the century? These questions do now establish a new trend in brain research.

I believe that the study of field potentials right now offers the only possibility of dealing with similar neural problems.

To be recommended strongly are the outstanding results of Freeman in which, during olfactory cognition, repeatable dynamic patterns in cortex are correlated to various percepts. Freeman's results, which indicate tuned oscillators with spectral resonances around 18–24 Hz and 40–70 Hz, should be discussed in conjunction with results of Galambos and coworkers showing interesting phenomenological responses in the 40-Hz range.

While visiting the laboratory of Petsche in October 1985, I had the opportunity to learn about the multielectrode approach developed by him and his coworkers. A visit by Ulla Mitzdorf and Reinhard Eckhorn in my laboratories during the winter of 1986 enabled me to hear more about their recent research. The more multielectrodes techniques develop, the more we will be able to link connectivistic models with global models. The multielectrode approach could also help us to learn more about ordered and chaotic states at the cellular level.

EEG Is Not Simple Noise

An important advance in studying the chaotic behavior of the EEG has been the description of the EEG dimension by means of the algorithm of Grassberger and Procaccia, as first used by Agnessa Babloyantz, and also described in this volume by Röschke and Başar for the intracranial structures of the cat brain. For the quantitative description of the erratic behavior observed in chaotic systems, it has become clear that simple spectral analysis of the measured signals does not provide an adequate tool for characterizing the origin of the observed chaos. Several investigators have tried to describe the entropy and information content of the brain in several other ways, often including analogies (see for example Başar 1980).

The measurement of other independent quantities like the Hausdorff-Besicovich or *fractal* dimension or the dynamical (Kolmogorov) entropy appears to be necessary for determining the "degree of chaos" which is presented in the experimental data (Mayer-Kress 1985).

Can We Measure Whether the Cortex Is More Complex than the Hippocampus?

The fractal dimension, roughly speaking, provides some measure of how many relevant degrees of freedom are involved in the dynamics of the system under consideration. Naturally, a system with many degrees of freedom is generally going to be more complex than one with only a few degrees of freedom. The higher the dimension, the better the approximation of "pure random noise." Since the EEG activity does show chaotic behavior, with a dimension of about 4–5, we have the fascinating possibility of attributing deterministic properties to the EEG. Moreover, the new algorithm enables us to state that, at least during the slow-wave sleep (SWS) stage, the cortex behaves more complexly than does the hippocampus or the reticular formation (Röschke and Başar, this volume). There seem to exist in the brain various strange attractors, even despite the great simultaneity of EEG waves during the hypersynchrony of SWS, as shown by the differentiated electrical behavior of various cranial structures. We expect higher-level information processing in the cortex. In fact, the cat cortex seems to be more chaotic than the hippocampus or reticular formation. At this point I want to mention Arnold Mandell's tremendous efforts ever the last half-dozen years to introduce nonlinear dynamics into biological science and especially into brain research, both at his colloquium at the University of California at San Diego and by means of several national and international conferences.

Magnetoencephalography

On the last day of the conference, we had a morning session on magnetoencephalography (MEG), with contributions from Deecke, Hari, Weinberg, Saermark, and Hoke, which showed the immense future possibilities of MEG work that still suffers from the facts that the measurement can usually be performed with only one channel and that the long measurements greatly reduce the time resolution of the evoked magnetic fields.

Session chairman Deecke emphasized that the EEG and MEG are, for the time being, complementary tools and that the MEG cannot entirely replace the EEG for short ranges of time. This session showed that we have to keep our eyes open for new technical developments in MEG research that could provide brain scientists with new abilities for highly extended results in field potential studies by providing much better resolution in topographical mapping.

The View of a Philosopher

Leinfellner emphasized the philosophical implications of the possibility of fast transneural communication, especially the possibilities of understanding the intracranial language. As a philosopher, he tried to explain a new model in which perceptions, memories, thoughts, and computations are encoded in specific patterns of brain waves. He drew his conclusions after much reading of the literature on field potentials and after visits to Adey's, John's and my laboratories. We spent an interesting day in Lübeck together last year. Although neither Adey, John, nor I could comment during the workshop about his presentation, I think that readers will appreciate his efforts to globally interpret new brain theories emerging from the study of field potentials.

Who Tells the True Story?

Bullock presented the first lecture in the Berlin conference and with his strong presence described in his contribution very important aspects of the analysis of neuronal networks. He has emphasized more and more in recent years that reductionist approaches to brain function will not suffice in understanding the brain. "If you want to understand how Budapest or New York work, it helps to study single individuals but it cannot suffice. You have to use some techniques that average individual differences and lump categories. The same is true if we want to understand how the human body works; single-cell studies are important but cannot suffice. The gap in mutual communication is regrettably wide between workers who study single neuronal units and those who study to compound field potentials of assemblies of neurons. This gap is unnecessary and counter-productive; neither approach can hope to tell the whole story. Even with both approaches working together and each extended in various ways, especially by multichannel recordings, we will be hard pressed for a long time to understand what is really going on in the brain." So stated Bullock (1983) in one of his recent studies in which he tried to emphasize the use not of simple methods but of strategies employing seven methods or ways of thinking (Bullock 1984).

At this time it is difficult to mention all of the important remarks made by Bullock. However, I find it very useful to mention here some of his recommendations for exploring the variety of brain potentials: "We should also have to show the dozen or more distinct kinds of potentials, including field potentials that integrate over some volume, the variety of differently spreading conductance changes, both increases and decreases, and the local differences in the regenerative properties of patches of membrane, plus the local proclivities to facilitation, fast or slow, or its opposite, to rebound or other aftereffects, to iterativeness or oscillation, to burst formation, autoinhibition and on and on. These examples are chosen at the unit level. However, we must recognize emergent properties of ensembles, such as synchronization, recruitment, kindling and the like".

Four years prior to the Berlin conference, Bullock wrote: "We are all the more in need of innovative schemes for dealing heuristically with a large number of variables of different dimension. *My message is simply this: let us keep our eyes open; there are discoveries to be made, principles to be induced, scientific revolutions to hatch"*.

I think that the Berlin conference brought the participants various ideas for developing new strategies for exploring the brain and perhaps for seeing the brain as an ensemble of cells with various of their own strategies.

The Beginning and the End

In organizing the conference, I hoped to catalyze a new advance by putting several leading scientists together to give lectures and develop extended discussions. I hoped to create with the conference a framework that would spark an explosive trend in brain research by combining several commensurable views. Several issues of the conference were made more explicit at the suggestion of Galambos, who advised me to be bold in my choice of themes and in motivating the senior lecturers and discussants. My collaboration on experiments with Bullock during the summer of 1984 in La Jolla gave me directly or indirectly new perspectives for handling an integrative approach to brain function and also for integrating several directions. I now reproduce a portion of my opening remarks at the conference:

The Prologuc

... One of the tasks facing the organizer of any conference is to select a conference theme, or as we say in Wagnerian language, a Leitmotiv. Theser are several leitmotives in this conference. However, by reading the titles and abstracts, it is easy to see that one of the central concerns in many of the contributions is with the integrated understanding of the various measures of brain information processing.

As you all know, the tempo of EEG-related research has increased tremendously since Hans Berger's pioneer contributions in 1929. In the early 1950's, Dawson's superposition techniques opened an era of evoked potential studies. The first averaging computer was developed at MIT in 1950. The early studies of Brazier and Barlow were followed from 1960 onwards by a veritable explosion of papers. In 1964 and 1965, further important discoveries emerged. In Bristol, Grey Walter and coworkers measured the CNV, in Germany Kornhuber and Deecke measured the readiness potentials, and in New York, Sutton, Braren, John, and Zubin measured the P300 wave. Ross Adey and coworkers pioneered establishing psychophysiological paradigms based on spectral analysis and the coherence function.

The early researchers took cognizance of ongoing EEG activity in their EP studies, but in the late 1960's and early 70's, a bifurcation developed between EEG studies and EP and ERP studies. This bifurcation can be attributed to the widespread marketing of evoked potential analyzers which treated the EEG activity as background noise that had to be eliminated by averaging.

I propose that amidst this gathering of most distinguished scientists working with a variety of EEG-derived measures, we should explore further, as one of the central themes, the notion that the variously named slow potential changes which are often studied in isolation, remain fundamentally manifestations of the ongoing EEG activity, which should be regarded as the groundwork from which, rather than the background against which, these phenomena arise.

Furthermore, I hope that emphasis will be given to the complementary information that is obtained from a combination of the more integrated measures of various slow wave potentials, and the spike activity measures from single and multiple unit microelectrode recordings.

The last day of the conference will give us the opportunity of learning more about the important and exciting new field of Magnetoencephalography.

What we have learned from the study of the human brain may not yet reflect the underlaying dynamics involved. In this regard, we may require information from nonhuman brains and integrate the lessons we learn. We may need to look through the window of Dr. Bullock's work on nonhuman brains to enrich our understanding of human brains.

Finally I would like to emphasize that imagination, courage, and theoretical speculation are also important in brain research. We should allow some time for visiting "castles in the air," even though it is expected that we will spend most of our time on solid ground. Few major discoveries have been made without some courageous and creative leap along the way, and in the last session, we have made provision for people to open for inspection their "castles in the air" and to discuss the architecture of these more ethereal dwellings.

I think that the happenings in the conference fitted the hoped-for framework.

Appendix 1

Excerpts from a translation of Dr. Volkmar Weiss's article (Weiss 1986).

Any cavity, here the brain, in which the waves are confined, has certain resonance frequencies (Planck 1900; Başar 1980). It is a general property of waves in a confined space that they exist only at definite frequencies which are integral multiples of π (Feynman et al. 1963). A sinusoidal motion in time is called a mode. According to Fourier's theory (1811, 1812) any information at all can be analyzed by assuming that it is the sum of the motions at all the different modes, combined with appropriate amplitudes and phases. If these modes are decoded, any information can always be analyzed as a superposition of n harmonic oscillators, with the frequencies $n\pi$ corresponding to the modes (Sinz 1976; Wright et al. 1985; Franaszczuk and Blinowska 1985).

The experimental data of the preceding papers are in principle in agreement with a number of models of phase transition and cooperative phenomena in brain advanced during the last years (e.g. Stuart et al. 1979; Başar 1983; Kaiser 1983; Del Giudice et al. 1985). These theories treat the brain as a mixed physical system in which the electrochemical part interacts with a macroscopic quantum-ordered state responsible for the creation and maintenance of memory. In any macroscopic quantum-ordered state, cooperative modes appear as Bose-quanta (bosons). They occur as dynamic products of the system and result from the action of the quantum exchange forces. Each of these exchange particles is a boson and is observed as a wave which establishes and maintains a highly cooperative effect among the basic elements of the system...

The ordered states are created by rearrangement of the symmetry attributes. For this to occur, a symmetry carrier within the brain dynamics itself must exist and has to be a quantum mode, otherwise stability is impossible. These symmetry-carrying agents are called "Goldstone bosons" (also called "collective modes") and undergo Bose-Einstein condensation into the resonant frequencies $n\pi$. In such a way information can be printed into the ground state as eigenvalues (Schrödinger 1926) of the Fourier transforms. A characteristic of the asymmetric ground state (long-term memory) is that the ground state is a macroscopic manifestation of quantum mechanics, analogous to the domain structure of a ferromagnet (and here realized by protein domains and their surroundings). Figuratively speaking, we may say that the brain consciously feels the presence of aligned vectors...

Besides the ground state, the brain has coherent states (Başar 1980), excited by external or internal stimuli. Attracting cycles start in the oscillators. The existence of these metastable excited states, which will quickly die off, can be associated with STM. The average phase of internal oscillations, expressed in values of the exciting oscillation, varies by π radian in the range of entrainment (Haken 1983). Whenever the system goes from the normal into this excited state, generalized long-range Goldstone bosons appear. At this moment of resonance, energy quanta are emitted having exactly the frequencies $n\pi$ as the frequencies which triggered them into leaving that state. And the waves feed back to the elements which originated their oscillations. Driven by stimulation, the nonlinear dissipative system brain develops new macroscopic orderings (Haken 1983).

Appendix 2

Chaos and Order

In Prigoginian terms, all systems contain subsystems, which are continually *fluctuating* (Prigogine and Stengers 1984). At times, a single fluctuation or a combination of them become so powerful, as a result of positive feedback, that it shatters the pre-existing organization. At this revolutionary moment – a singular moment or a bifurcation point – it is inherently impossible to determine in advance in which direction change will take: whether the system will disintegrate into "chaos" or leap to a now higher level of "order" or organization, which Prigogine called a "dissipative structure." (Such physical or chemical structures are termed "dissipative" because, compared with the simpler structures they replace, they require more energy to sustain them. Prigogine insists that order and organization can actually arise "spontaneously" out of disorder and chaos through a process of "self-organization."

Let us here repeat a *Gedankenexperiment* described by Prigogine and Stengers (1984): "Suppose we have two kinds of molecules, "red" and "blue". Because of the chaotic motion of the molecules we would expect that at a given moment we would have more red molecules, say, in the left part of a vessel. Then a bit later more blue molecules would appear, and so on. The vessel would appear to us as "violet", with occasional irregular flashes of red or blue. However, this is *not* what happens with, for example, a chemical clock; here the system is all blue, then it abruptly changes its color to red, then again to blue. Because all these changes occur at regular time intervals, we have a *"coherent process"*.

Such a degree of order stemming from the activity of billions of molecules seems incredible, and indeed, if chemical clocks had not been observed, no one would believe that such a process is possible. To change color all at once, molecules must have a way to *"communicate"*. The system has to act as a whole. *Dissipative structures introduce probably one of the simplest physical mechanisms for communication.*

References

Babloyantz A, Nicolis C, Salazar M (1985) Evidence of chaotic dynamics of brain activity during the sleep cycle. Phys Lett [A] 111:152–156

Başar E (1980) EEG-Brain dynamics. Relation between EEG and brain evoked potentials. Elsevier/North-Holland, Amsterdam

Başar E (1983) Toward a physical approach to integrative physiology. I. Brain dynamics and physical causality. Am J Physiol 245:R510–R533

Bergson H (1939) Matière et memoire. Essai sur la relation du corps à l'èsprit. Presses Universitaires de France, Paris

Bullock TH (1983) Electrical signs of activity in assemblies of neurons: compound field potentials as objects of study in their own right. Acta Morphol Hung 31(1–3):39–62

Bullock TH (1984) Comparative neuroscience holds promise for quiet revolutions. Science 225: 473–478

Del Guidice E, Doglia S, Milani M, Vitiello G (1985) A quantum field theoretical approach to the collective behavior of biological systems. Nucl Phys [B] 251:375–400

Feynman RP, Leighton RB, Sands M (1963) The Feynman lectures on physics. Addison-Wesley, Reading

Franaszczuk PJ, Blinowska KJ (1985) Linear model of brain electrical activity – EEG as a superposition of damped oscillatory modes. Biol Cybern 53:19–25

Godet P, Castellucci VF, Schacher S, Kandel ER (1986) The long and the short of long-term memory – a molecular framework. Nature 322:419–422

Haken H (1983) Advanced synergetics. Springer, Berlin Heidelberg New York

Hubel DH, Wiesel TN (1962) Receptive fields, binocular interaction and functional architecture in the cat's visual cortex. J Physiol (Lond) 160:106–154

Kaiser F (1983) Specific effects in externally driven self-sustained oscillating biophysical model systems. In: Fröhlich H, Kremer F (eds) Coherent excitations in biological systems. Springer, Berlin Heidelberg New York, pp 128–133

Mayer-Kress G (1985) Dimensions and entropies in chaotic systems. Springer, Berlin Heidelberg New York Tokyo

Penfield W (1975) The mystery of the mind. Princeton University Press, Princeton

Planck M (1900) Zur Theorie des Gesetzes der Energieverteilung im Normalspektrum. Verh Dtsch Phys Ges 2:237–245

Prigogine I, Stengers I (1984) Order out of chaos. Bantam, New York
Schrödinger E (1926) Quantisierung als Eigenwertproblem. Ann Phys 81:109–140
Sinz R (1976) Neuro- und psychophysiologische Aspekte des Gedächtnisses – ein Oszillator-Resonanzmodell extrasynaptischer Informationsspeicherung. Z Psychol 184:352–381
Stuart CIJM, Takahashi Y, Umezawa H (1979) Mixed system brain dynamics: neural memory as a macroscopic ordered state. Found Phys 9:301–327
Weiss V (1986) Memory as a macroscopic ordered state by entrainment and resonance in energy pathways. In: Weiss V et al (eds) Psychogenetik der Intelligenz. Modernes Lernen Borgmann, Dortmund
Wright JJ, Kydd RR, Lees GJ (1985) State-changes in the brain viewed as linear steady-states and non-linear transitions between steady-states. Biol Cybern 53:11–17